T0296492

LONDON MATHEMATICAL SOCIETY LECTURE NOTE SERIES

Managing Editor: Professor N.J. Hitchin, Mathematical Institute,
University of Oxford, 24–29 St Giles, Oxford OX1 3LB, United Kingdom

The titles below are available from booksellers, or, in case of difficulty, from Cambridge University Press.

London Mathematical Society Lecture Note Series. 274

The Mandelbrot Set, Theme and Variations

Edited by

Tan Lei
Université de Cergy-Pontoise

CAMBRIDGE
UNIVERSITY PRESS

CAMBRIDGE UNIVERSITY PRESS
Cambridge, New York, Melbourne, Madrid, Cape Town, Singapore,
São Paulo, Delhi, Dubai, Tokyo, Mexico City

Cambridge University Press
The Edinburgh Building, Cambridge CB2 8RU, UK

Published in the United States of America by
Cambridge University Press, New York

www.cambridge.org
Information on this title: www.cambridge.org/9780521774765

© Cambridge University Press 2000

First published 2000

A catalogue record for this publication is available from the British Library

ISBN 978-0-521-77476-5 Paperback

Contents

Contents

Introduction

Tan Lei

Complex dynamical systems is a fascinating field. It has both the visual appeal of endlessly varying and beautiful fractals, and the intellectual appeal of sophisticated mathematical developments. Its theoretical challenges have attracted mathematicians from around the world.

A central object of study in complex dynamics is the *Mandelbrot set M*, a fractal shape that classifies the dynamics of the quadratic polynomials $f_c(z) = z^2 + c$. The Mandelbrot has a remarkably simple definition:

$$M = \{c \in \mathbb{C} \ : \ f_c^n(z) \text{ remains bounded as } n \to \infty.\}.$$

Nevertheless M exhibits a rich geometric and combinatorial structure, with many intriguing details and many remaining mysteries. Although it is defined in terms of quadratic polynomials, the Mandelbrot set reappears in virtually every other family of rational maps, as can be observed in computer experiments.

The mathematical theory of the Mandelbrot set and related objects has undergone rapid development during the last two decades, sparked by the pioneering work of Douady and Hubbard. This volume provides a coherent perspective on the present state of the field.

Its articles range from the systematic exposition of current knowledge about the Mandelbrot set, to the latest research in complex dynamics. In addition to presenting new work, this collection documents for the first time important results hitherto unpublished or difficult to find in the literature.

1 A detailed description of the contents of this volume

The Preface, by J. Hubbard, gives an intriguing first-hand account of developments in the field during the period 1976-1982. It recounts the discovery of the Mandelbrot set and the introduction of basic tools, such as the Riemann mapping to the exterior of M, Hubbard trees, quasiconformal mappings, polynomial-like mappings, holomorphic motions, Thurston's theory, etc.

Part one: Universality of the Mandelbrot set

With the advent of computers, it has been easy to experimentally observe copies of Mandelbrot set in many families of complex analytic dynamical systems, but it was a mathematical challenge to explain this universality.

The first results in this direction were obtained by Douady-Hubbard using polynomial-like mappings. This volume contains three articles about the subject:

[McMullen] shows that small Mandelbrot sets are dense in the bifurcation locus for any holomorphic family of rational maps. As a consequence, one obtains general estimates for the dimension of the bifurcation locus, using Shishikura's result that the Hausdorff dimension of the boundary of M is 2.

[Douady-Buff-Devaney-Sentenac] studies the birth of infinitely many baby Mandelbrot sets in parameter space near a rational map with a parabolic point. This paper shows each baby Mandelbrot set sits in the heart of a nest of imploded cauliflowers.

[Haïssinsky] explains why small copies of the Mandelbrot set appear within M itself.

Part two: Quadratic Julia sets and the Mandelbrot set

One of the central open questions concerning the Mandelbrot set is the following: is M a locally connected set? A positive answer would imply that there is a complete topological description of M, and in particular that hyperbolic dynamical systems are dense in the quadratic family. Similarly, one can completely answer questions about the topology of $J(f_c)$ when the Julia set is known to be locally connected.

A major breakthrough, due to Yoccoz, states that $J(f_c)$ is locally connected and M is locally connected at c, for certain special values of $c \in M$. These are the values where f_c has no indifferent points and is not renormalizable. Yoccoz's original proof is not published, but this volume includes two articles, [Milnor] and [Roesch], treating its central points:

[Milnor] shows that the Julia set of f_c is locally connected, using Yoccoz puzzles;

[Roesch] continues from [Milnor] and shows that M is locally connected at such parameters c. The proof given is a new argument, due to Shishikura, using holomorphic motions.

Turn to the case where f_c does have an indifferent point, [Tan] proves that M is locally connected when f_c has a parabolic point of multiplier 1. This paper also surveys more general results on the local connectivity of M when c has an indifferent point.

The landing behavior of external rays is often a starting point in the study of local connectivity. In this volume, [Petersen-Ryd] gives a new and elementary proof that external rays to M with rational angle do land, and describes the dynamics of f_c for the endpoint c.

Dynamical systems on the interval or the circle are closely related to dynamics in one complex variable. This volume includes two papers on real dynamics:

[Luzzatto] presents an annotated account of Jakobson's theorem, stating that the set of $c \in \mathbb{R}$ such that $f_c(z)$ has chaotic dynamics has positive Lebesgue measure. More precisely, $c = -2$ is a point of Lebesgue density of the set of $c \in \mathbb{R}$ such that f_c has an absolutely continuous invariant measure;

[Petersen] presents the Herman-Swiatek's theorem, stating that an analytic circle homeomorphism with a critical point and irrational rotation number θ is quasi-symmetrically conjugate to a rigid rotation if and only if θ is of bounded type. Using this theorem, [Petersen] also establishes the new result that there exist quadratic Siegel polynomials with a rotation numbers of *unbounded type* whose Julia sets are locally connected.

The delicate theory of bifurcations of indifferent period points is studied in two related papers:

[Jellouli1] shows that small perturbations of quadratic Siegel polynomials with Diophantine rotation number are conjugate to their linear parts up to an arbitrarily small error term. This work bears on the study of the Lebesgue measure of nearby filled Julia sets;

[Jellouli2] discusses various important quantities and their relations appearing in the perturbation of a quadratic polynomial with a parabolic fixed point. These include the multiplier of the periodic orbit coming from the perturbation, its first and second derivatives with respect to the parameter, and the holomorphic indices.

Part three: Julia set of rational maps

A rational map f tends to be expanding on its Julia set $J(f)$, away from the orbits of recurrent critical points. This phenomenon, observed already by Fatou and Julia, was made precise in a strong form by Ricardo Mañé. In this volume:

[Shishikura-Tan] presents a new proof of Mañé's result, showing that any point $x \in J(f)$, which is neither a parabolic periodic point nor a limit point of a recurrent critical orbit, has a neighborhood which is contracted by the family $\{f^{-n}\}_{n \in \mathbb{N}}$.

[Yin] shows, as an application, that if all critical points in $J(f)$ are non-recurrent, then $J(f)$ is 'shallow' or 'porous'; consequently its Hausdorff dimension is less than two (as first shown by Urbanski).

The combinatorics of general rational maps, even of degree two, is still not well-understood. However many interesting rational maps can be constructed by *mating* pairs of polynomials of the same degree. If the polynomials are critically finite, then the existence of their mating can be reduced to a topological problem using Thurston's theory. In this volume:

[Shishikura1] provides a proof of a result, first developed by Mary Rees, showing that if a mating f exists combinatorially, then in fact f is topologically conjugate to a natural quotient of the dynamics of the two polynomials.

Part four: Foundational results

[Douady] gives a new proof of Ahlfors-Bers theorem on integrability of measurable complex structures. This theorem is of central importance for the surgery and deformation of holomorphic dynamical systems.

[Shishikura2] gives a thorough treatment of the theory of parabolic implosions (developed by Douady and Lavaurs from Ecalle-Voronin's theory). This powerful tool provides precise analytic information about the rational maps obtained by perturbing a parabolic cycle, and underlies Shishikura's proof that the boundary of the Mandelbrot set has Hausdorff dimension two. Other applications appear in [Douady-Buff-Devaney-Sentenac], [Jellouli2] and [Tan] in this volume. Refinements are included in the Appendix.

2 Techniques in complex dynamics

A wide range of modern methods of complex analysis and dynamics can be seen at work in the articles of this volume.

The most classical techniques, successfully used by Fatou and Julia, are the *Poincaré metric, Schwarz Lemma and Montel's theorem on normal families*. One can see these methods applied in [Petersen-Ryd], [Shishikura-Tan] and the appendix to [Haïssinsky].

The *Riemann mapping theorem, together with the Carathéodory theory*, is a central tool for understanding the topology of the Julia set and its complementary components. The rays coming from the Riemann representation often meet (land) at same points and cut the Julia set. They are preserved under iteration and transfer easily to the parameter spaces. Using external rays, one obtains a type of Markov partition called a *Yoccoz puzzle*. These methods appear in [Haïssinsky], [Milnor], [Roesch] and [Tan].

Quasiconformal deformations, the Beltrami equation, the Ahlfors-Bers theory and its refined version G. David's theorem. Starting with Sullivan's work, the integrability of *measurable* complex structures has found remarkable applications to complex dynamics. For example, it is essential to the Douady-Hubbard theory of *polynomial-like maps, Mandelbrot-like families*, which in turn forms the foundations of the theory of renormalization and the universality of the Mandelbrot set. These methods appear in [Douady], [McMullen], [Douady-Buff-Devaney-Sentenac] and [Haïssinsky].

Holomorphic motions. This tool for studying the variation of dynamics in families plays a crucial role in [Roesch]. See also [McMullen], [Douady-Buff-Devaney-Sentenac] and [Haïssinsky].

Grötzsch inequality about moduli of annuli. These conformal invariants play an essential role in work on local connectivity of Julia sets and the Mandelbrot set. See [Milnor] and [Roesch].

Yoccoz inequality. This application of the Grötzsch inequality is discussed in [Tan], Appendix D and in [Haïssinsky].

Parabolic implosion. This important modern theory is developed in detail in [Shishikura2], and is applied in [Tan], [Jellouli2] and [Douady-Buff-Devaney-Sentenac].

Expansion on subsets of the Julia set away from recurrent critical orbits. This is the content of Mañé's result and is the theme of [Shishikura-Tan]. See [Yin] for an example of application.

Jakobson's theorem. Collet-Eckmann maps. This is one of the key results in real dynamical systems having a complex flavor, and is exposed here in [Luzzatto].

Circle homeomorphisms, distortion of cross-ratios. They are important on their own right and have important applications in complex dynamics. Both features are exhibited in [Petersen].

Transferring results on the dynamical planes to the parameter space. This is the most common way to get information about the parameter space. See [McMullen], [Douady-Buff-Devaney-Sentenac], [Haïssinsky], [Roesch], [Tan], [Jellouli1] and [Luzzatto].

Thurston's theory, creating rational maps from combinatorial information. One example of this is mating of polynomials. And the fact that mating is an adequate way to describe topologically the dynamics is reported here in [Shishikura1].

Acknowledgements. This is above all a collective work. It is a privilege for me, the editor, to be able to thank all the contributors, who participated with enthusiasm, energy, rigour and dedication. Special thanks go to C. McMullen, who provided valuable help at every stage of the preparation, and A. Manning, R. Oudkerk and S. van Strien for their linguistic, technical and moral support.

Tan Lei, Department of Mathematics, University of Warwick, Coventry CV4 7AL, United Kingdom
e-mail: tanlei@maths.warwick.ac.uk

Preface

John Hubbard

Holomorphic dynamics is a subject with an ancient history: Fatou, Julia, Schroeder, Koenigs, Böttcher, Lattès, which then went into hibernation for about 60 years, and came back to explosive life in the 1980's.

This rebirth is in part due to the introduction of a new theoretical tool: Sullivan's use of quasi-conformal mappings allowed him to prove Fatou's no-wandering domains conjecture, thus solving the main problem Fatou had left open.

But it is also due to a genuinely new phenomenon: the use of computers as an experimental mathematical tool. Until the advent of the computer, the notion that there might be an "experimental component" to mathematics was completely alien. Several early computer experiments showed great promise: the Fermi-Pasta-Ulam experiment, the number-theoretic computations of Birch and Swinnerton-Dyer, and Lorenz's experiment in theoretical meteorology stand out. But the unwieldiness of mainframes prevented their widespread use.

The microcomputer and improved computer graphics changed that: now a mathematical field was behaving like a field of physics, with brisk interactions between experiment and theory.

I mention computer graphics because faster and cheaper computers alone would not have had the same impact; without pictures, the information pouring out of mathematical computations would have remained hidden in a flood of numbers, difficult if not impossible to interpret. For people who doubt this, I have a story to relate. Lars Ahlfors, then in his seventies, told me in 1984 that in his youth, his adviser Lindelöf had made him read the memoirs of Fatou and Julia, the prize essays from the *Académie des Sciences* in Paris. Ahlfors told me that they struck him at the time as "the pits of complex analysis": he only understood what Fatou and Julia had in mind when he saw the pictures Mandelbrot and I were producing. If Ahlfors, the creator of one of the main tools in the subject and the inspirer of Sullivan's no-wandering domains theorem, needed pictures to come to terms with the subject, what can one say of lesser mortals?

In this preface, I will mainly describe the events from 1976 to 1982, as I saw them.

The first pictures

For me at least, holomorphic dynamics started as an experiment. During the academic year 1976-77, I was teaching DEUG (first and second year calculus) at the University of Paris at Orsay. At the time it was clear that willy-nilly, applied mathematics would never be the same again after computers. So I tried to introduce some computing into the curriculum.

This was not so easy. For one thing, I was no computer whiz: at the time I was a complex analyst with a strong topological bent, and no knowledge of dynamical systems whatsoever. For another, the students had no access to computers, and I had to resort to programmable calculators. Casting around for a topic sufficiently simple to fit into the 100 program steps and eight memories of these primitive machines, but still sufficiently rich to interest the students, I chose Newton's method for solving equations (among several others).

This was fairly easy to program. But when a student asked me how to choose an initial guess, I couldn't answer. It took me some time to discover that no one knew, and even longer to understand that the question really meant: what do the basins of the roots look like?

As I discovered later, I was far from the first person to wonder about this. Cayley had asked about it explicitly in the 1880's, and Fatou and Julia had explored some cases around 1920. But now we could effectively answer the question: computers could draw the basins. And they did: the math department at Orsay owned a rather unpleasant computer called a *mini-6*, which spent much of the spring of 1977 making such computations, and printing the results on a character printer, with X's, 0's and 1's to designate points of different basins. Michel Fiollet wrote the programs, and I am extremely grateful to him, as I could never have mastered that machine myself.

In any case, Adrien Douady and I poured over these pictures, and eventually got a glimpse of how to understand some of them, more particularly Newton's method for $z^3 - 1$ and $z^3 - z$. At least, we understood their topology, and possibly the fact that we concentrated on the topology of the Julia sets has influenced the whole subject; other people looking at the same pictures might have focussed on other things, like Hausdorff dimension, or complex analytic features.

Also, I went for help to the IHES (*Institut des Hautes Etudes Scientifiques*), down the road but viewed by many at Orsay as alien, possibly hostile territory. There dynamical systems were a big topic: Dennis Sullivan and René Thom were in residence and Michael Shub, Sheldon Newhouse, and John Guckenheimer were visiting that spring.

I learned from Sullivan about Fatou and Julia, and especially about the fact that an attracting cycle must attract a critical point, and more generally that the behavior of the critical points dominates the whole dynamical picture. This suggested how to make parameter-space pictures, and the *mini-6* made many of these also. (These pictures are among the great-grand parents of the present volume, and whether the authors know it or not, they appear in the genealogical tree of most if not all the papers.) But having the pictures was no panacea: they looked chaotic to us, and we had no clear idea how to analyze them.

The year 1981-82: an ode to the cafés of Paris

At the end of 1977, I went to Cornell, where I thought and lectured about the results we had found. In particular, Mandelbrot saw the pictures that the *mini-6* had produced, and correctly calling them "rather poor quality," invited me to give a lecture at Yorktown Heights in 1978, saying that he had often thought about the "Fatou-Julia fractals," although he had never made pictures of them. But nothing much got proved until the next visit to France, for the academic year 1981–82.

By then, many things had subtly changed. Douady had understood that it was wiser to iterate polynomials before iterating Newton's method, as they are considerably simpler. His sister Véronique Gautheron had written programs to investigate the dynamics of polynomials. Computers had improved; Véronique used a machine, now long defunct, called a *Goupil* (later a small HP), but for me the arrival of the Apple II was decisive. Mandelbrot had access to the IBM computer facilities of Yorktown Heights; he had produced much better pictures of the Mandelbrot set than we had, and had published a paper about it. Feigenbaum had performed his numerical experiments, and the physicists were interested in the iteration of polynomials, more particularly renormalization theory.

Perhaps it was an illusion, but it seemed to me that holomorphic dynamics was in the air. Milnor and Thurston had long studied interval maps and were beginning to consider polynomials in the complex. In the Soviet Union (behind the iron curtain, very much in existence at the time) Lyubich and Eremenko, who were at the time just names we were vaguely aware of, were also starting to study holomorphic dynamics. In Brazil, Paolo Sad had produced a paper (hand-written) on the density of hyperbolic dynamics. The paper was wrong, and the result is still the main unproved question of the theory. (At the time I did not appreciate the importance of the result Sad had claimed, but I clearly recall coming back to our apartment on the Rue Pascal and hearing from my wife that Sullivan had telephoned all the way from Brazil with the message that Sad's paper "was coming apart at the seams.") But Sad's techniques led to one of the most important tools of the subject, the Mañe-Sad-Sullivan λ-lemma and holomorphic motions. In Japan, Ushiki had been an early advocate of computer graphics. His brilliant student Shishikura was beginning to take an interest in the field. In any case, the stage was set for a fertile year, and indeed 1981-82 was simply wonderful: I will describe three episodes, all of which occur in cafés, and all of which are somehow connected with computers.

• **The connectivity of the Mandelbrot set.** Mandelbrot had sent us a copy of his paper, in which he announced the appearance of islands off the mainland of the Mandelbrot set M. Incidentally, these islands were in fact not there in the published paper: apparently the printer had taken them for dirt on the originals and erased them. (At that time, a printer was a human being, not a machine.) Mandelbrot had penciled them in, more or less randomly, in the copy we had.

One afternoon, Douady and I had been looking at this picture, and wondering what happened to the image of the critical point by a high iterate of the polynomial $z^2 + c$ as c takes a walk around an island. This was difficult to imagine, and we had started to suspect that there should be filaments of M connecting the islands to the mainland. Overnight, Douady thought that such filaments could be detected as barriers: something had to happen along them, and found that is should be the arguments of the rays landing at 0.

In any case, Douady called the following morning, inviting me to join him at a café Le Dauphin on the Rue de Buci. He had realized that what we had discovered was that the Mandelbrot set was connected: over a croissant, he wrote the statement, $c \mapsto \phi_c(P_c(0))$ is an isomorphism $\mathbb{C} \setminus M \to \mathbb{C} \setminus \overline{\mathbb{D}}$.

Sullivan flew from the United States to hear Douady speak on our proof in the analytic dynamics seminar at Orsay that week. Sibony was also in the audience, as were Kahane and many others. The following week, Sibony announced he had an independent proof that the map proposed by Douady was a bijection.

Not long after, I made a list of all the quadratic polynomials whose critical point is periodic of period 5. Then I asked the computer (an Apple II, with a pen plotter attached) to draw all the Julia sets. Today this would be virtually instantaneous; at the time it took several hours. Then I looked really carefully at the drawings, trying to see what made them different from each other. After I marked the orbit of the critical point, in each case a tree was staring me in the face. A bit of reflection soon told me some necessary conditions a tree with marked points must satisfy in order to be a possible tree for a polynomial.

In a day or so I had drawn all the trees that could be drawn corresponding to the critical point being periodic of period 6. The fact that these did indeed correspond to the appropriate polynomials was strong evidence that the description was right. Not long afterwards Douady came up with the algorithm for external angles in terms of trees. The complex kneading sequence was born. These trees (now called Hubbard trees) together with external rays have now become a central tool of combinatorics and classifications.

• **Matings.** One night in the spring of 1982, I set the computer (the same Apple II, now equipped with a dot-matrix printer), to drawing Julia sets of rational functions of degree 2, running through a list of parameter values where the two critical points were periodic. Douady actively disapproved of this activity, thinking that we should focus on quadratic polynomials until they were better understood.

The next morning I had a pile of perhaps 40 such drawings (on those folded sheets with holes along the sides typical of dot-matrix printers), most of which looked like junk. But several evidently had some structure. I collected these, and met Douady at the local cafe (this time the café des Ursulines, near Ecole Normale, whose owner at the time was very welcoming to mathematicians). He looked at one of them, and after a while drew in two trees connecting the orbits of the critical point. One was the tree of the rabbit, the other the tree of the polynomial with the critical point periodic of period 4, with external angles 3/15 and 4/15. It was immediately clear that the picture really did represent this object.

This suggested many experiments to confirm the existence of matings, which we carried out; soon we came up with the mating conjecture: two quadratic polynomials can be mated unless they belong to conjugate limbs of the Mandelbrot set. But we had to wait for Thurston's theorem and the work of Silvio Levy, Mary Rees, Mitsu Shishikura, and Tan Lei, to see the mating conjecture proved for post-critically finite polynomials. As far as I know, there is still no reasonable mating conjecture in degree 3.

• **Polynomial-like mappings.** One of the great events of that year was Sullivan's proof of the non-existence of wandering domains for rational functions. Douady and I were both present at his first lecture on the subject, and immediately saw the power of his methods: invariant Beltrami forms and the Ahlfors-Bers theorem.

I had written my thesis (under Douady's direction) about Teichmüller spaces, and was quite familiar with these techniques, but had never thought of applying them to dynamics; Sullivan knew better. He had studied Ahlfors's work on Kleinian groups, and had realized that the same techniques could be used in holomorphic dynamics. In Sullivan's view, there is a dictionary relating Kleinian groups to holomorphic dynamics; the no-wandering domains theorem showed the power of this program. Trying to understand further parts of this dictionary has been an important motivating force for a lot of the research in the field, and more particularly Curt McMullen's.

In short order Douady and I used quasi-conformal mappings to show that the multiplier map gives a uniformization of the hyperbolic components of the interior of M, something we had conjectured but couldn't prove.

Soon thereafter, in some café in the north of Paris, Douady was ruminating about polynomial-like mappings. Of course, he didn't yet have a precise definition, but he had seen that if an analytic mapping mapped a disc $D \subset \mathbb{C}$ to $\mathbb{C}$ so that the boundary of D maps outside of D, winding around it several times, then many of the proofs about polynomials would still go through.

Thinking over what he had said, I saw that evening that if one could construct an appropriate invariant Beltrami form, then the polynomial-like mapping would be conjugate to a polynomial. This time I called Douady and asked him to meet me at the Café du Luxembourg, and presented my argument. My proof of the existence of the invariant Beltrami form was shaky, as I was requiring extra unnecessary conditions, but Adrien soon saw that if we got rid of these, the invariant Beltrami form was easy to construct. The straightening theorem was born.

The following year, back in Cornell, I was making parameter-space pictures of Newton's method for cubic polynomials. I saw Mandelbrot sets appearing on the screen, which wasn't really surprising, but was simply amazed to see a dyadic tree appearing around it. Looking carefully at this tree, I found that it reflected the digits of external angles of points in the Mandelbrot set written in base 2. Seeing that these angles were made by God, and not by man, was an extraordinary realization to me. I started my Harvard colloquium lecture two days later by saying I was changing its subject completely, because I had made an amazing discovery two days earlier. That colloquium was one of the best lectures I delivered in my life.

Quasi-conformal mappings have remained one of the central tools in holomorphic dynamics, eventually becoming a field in its own right called quasiconformal surgery, and providing an amazing flexibility in chopping up and reassembling the very rigid objects of the field. Applications are too numerous to list, but Shishikura's sharp bound on the number of nonrepelling cycles of a rational function, and his construction of Herman rings from Siegel discs, stand out as early monuments to the power of this method.

That spring, I went to the United States for a couple of weeks and gave a number of lectures about these results, at Columbia and Cornell, and the SUNY graduate school. I had never met John Milnor, though I had practically been raised on his books. But he came to the Columbia lecture, and about a week later, back in France, I received a letter from him pointing out that the proof of the connectivity of the Mandelbrot set also proved that the external argument of real polynomials in the boundary of M is a monotone function of the parameter, and that this settled the entropy conjecture of Metropolis, Stein and Stein from the 1950's.

Another important event of that year for Douady and me was a Séminaire Bourbaki lecture in which Malgrange presented work of Ecalle on indifferent fixed points. The theory of Ecalle cylinders and the parabolic implosion grew out of that lecture. These results, and many others from that year, such as the existence of Julia sets that are not locally connected, and the connection of polynomial-like mappings with renormalization, are clearly part of the genealogy of all the papers of this volume; I would call them the grandparents of the present volume.

The Orsay notes

In the fall of 1982, I returned to Cornell. The situation was then as follows: a huge amount had been discovered, and largely proved. In particular, we had proved that if the Mandelbrot set is locally connected, then hyperbolic quadratic polynomials are open and dense, and we had formulated the MLC conjecture (the Mandelbrot set is locally connected). We had also constructed the combinatorial model $\overline{M}$ of the Mandelbrot set and the canonical mapping $M \to \overline{M}$, and proved that if MLC holds, then this map is a homeomorphism.

We had complete proofs of the combinatorial description of Julia sets of strictly preperiodic polynomials and polynomials with attracting cycles, although there were gaps for polynomials with parabolic cycles, which led to gaps in understanding the landing of rays at roots of components.

But our proofs were handwritten, mainly understandable only to Douady and me; next to nothing had been published. The only publications I can think of from that year were two *Notes aux Comptes-Rendus*, one by Douady and me on the connectivity on the Mandelbrot set and trees, and one by Sullivan on the no-wandering domains theorem.

Getting this material organized and written was essentially Douady's work: he gave a course at ENS and Orsay the following year, delivering each week the same lecture in both places and then writing it, using the notes of of Pierrette Sentenac, Marguerite Flexor, Régine Douady and Letizia Herault. Tan Lei and Lavaurs were in the audience, and become the first generation of students in this field. On my side, Ben Wittner and Janet Head were also first generation students in the field at Cornell.

I was not present for most of the writing of the Orsay notes, and am always amazed at the extraordinary new inventions present in the notes. The *tour de valse* was in our hand-written notes (where it was called the *lemme du coup de fouet*; this name was considered offensive by Pierrette Sentenac), but the *arrivée au bon port* and the local connectivity of the Julia set of $z^2 + z$ are among Douady's inventions of that year. Others contributed to the notes, including Lavaurs, who provided some of the central ideas of the parabolic implosion and Tan Lei with her resemblance between the Mandelbrot set near a preperiodic point and the Julia set for that point.

Other publications followed: the Mañe-Sad-Sullivan paper in 1983, and the polynomial-like mappings paper by Douady and me, in 1985. Together, I would describe these and the Orsay notes as the parents of the papers of the present volume.

Of course, the present volume has another parent: Bill Thurston's topological characterization of rational functions. Sullivan had linked holomorphic dynamics with quasi-conformal mappings, and Thurston invented another great new technique by linking holomorphic dynamics to Teichmüller theory. For a postcritically finite polynomial, there are 3 ways of encoding the dynamics: my trees, the external arguments of the critical values, and the Thurston class. Only the last extends to rational functions, and with this theorem it became clear that *post-critically finite branched mappings* provide the right way to encode the combinatorics of rational mappings.

The work after this is no longer history, but current events, and I leave an introduction to the present volume to Tan Lei.

Department of Mathematics, Cornell University, Ithaca, NY 14853-7901, U.S.A. and Département de Mathématiques, Université de Aix-Marseille I, 13331 Marseille Cedex 3, France.
email: jhh8@cornell.edu and John.Hubbard@cmi.univ-mrs.fr

The Mandelbrot set is universal

Curtis T. McMullen[*]

Abstract

We show small Mandelbrot sets are dense in the bifurcation locus for any holomorphic family of rational maps.

1 Introduction

Fix an integer $d \geq 2$, and let $p_c(z) = z^d + c$. The *generalized Mandelbrot set* $M_d \subset \mathbb{C}$ is defined as the set of c such that the Julia set $J(p_c)$ is connected. Equivalently, $c \in M_d$ iff $p_c^n(0)$ does not tend to infinity as $n \to \infty$. The traditional Mandelbrot set is the quadratic version M_2.

A *holomorphic family of rational maps* over X is a holomorphic map

$$f : X \times \widehat{\mathbb{C}} \to \widehat{\mathbb{C}}$$

where X is a complex manifold and $\widehat{\mathbb{C}}$ is the Riemann sphere. For each $t \in X$ the family f specializes to a rational map $f_t : \widehat{\mathbb{C}} \to \widehat{\mathbb{C}}$, denoted $f_t(z)$. For convenience we will require that X is *connected* and that $\deg(f_t) \geq 2$ for all t.

The *bifurcation locus* $B(f) \subset X$ is defined equivalently as the set of t such that:

1. The number of attracting cycles of f_t is not locally constant;

2. The period of the attracting cycles of f_t is locally unbounded; or

3. The Julia set $J(f_t)$ does not move continuously (in the Hausdorff topology) over any neighborhood of t.

It is known that $B(f)$ is a closed, nowhere dense subset of X; its complement $X - B(f)$ is also called the *J-stable set* [MSS], [Mc2, §4.1].

As a prime example, $p_c(z) = z^d + c$ is a holomorphic family parameterized by $c \in \mathbb{C}$, and its bifurcation locus is ∂M_d. See Figure 1.

In this paper we show that *every* bifurcation set contains a copy of the boundary of the Mandelbrot set or its degree d generalization. The Mandelbrot sets M_d are thus *universal*; they are initial objects in the category of bifurcations, providing a lower bound on the complexity of $B(f)$ for all families f_t.

For simplicity we first treat the case $X = \Delta = \{t : |t| < 1\}$.

---

[*]Research partially supported by the NSF. 1991 Mathematics Subject Classification: Primary 58F23, Secondary 30D05.

1

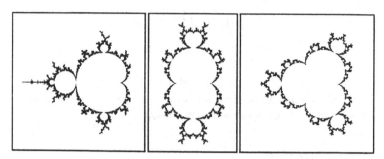

Figure 1. Mandelbrot sets of degrees 2, 3 and 4.

Theorem 1.1 *For any holomorphic family of rational maps over the unit disk, the bifurcation locus $B(f) \subset \Delta$ is either empty or contains the quasiconformal image of ∂M_d for some d.*

The proof (§4) shows that $B(f)$ contains copies of ∂M_d with arbitrarily small quasiconformal distortion, and controls the degrees d that arise. For example we can always find a copy of ∂M_d with $d \le 2^{2\deg(f_t)-2}$, and generically $B(f)$ contains a copy of ∂M_2 (see Corollary 4.4). Since the Theorem is local we have:

Corollary 1.2 *Small Mandelbrot sets are dense in $B(f)$.*

There is also a statement in the dynamical plane:

Theorem 1.3 *Let f be a holomorphic family of rational maps with bifurcations. Then there is a $d \ge 2$ such that for any $c \in M_d$ and $m > 0$, the family contains a polynomial-like map $f_t^n : U \to V$ hybrid conjugate to $z^d + c$ with $\mathrm{mod}(U - V) > m$.*

Corollary 1.4 *If f has bifurcations then for any $\epsilon > 0$ there exists a t such that $f_t(z)$ has a superattracting basin which is a $(1 + \epsilon)$-quasidisk.*

Proof. The family contains a polynomial-like map $f_t^n : U \to V$ hybrid conjugate to $p_0(z) = z^d$, a map whose superattracting basin is a round disk. Since $\mathrm{mod}(V - U)$ can be made arbitrarily large, the conjugacy can be made nearly conformal, and thus f_t has a superattracting basin which is a $(1 + \epsilon)$-quasidisk. ∎

For applications to Hausdorff dimension we recall:

Theorem 1.5 (Shishikura) *For any* $d \geq 2$, *the Hausdorff dimension of* ∂M_d *is two. Moreover* H. $\dim(J(p_c)) = 2$ *for a dense* G_δ *of* $c \in \partial M_d$.

This result is stated for $d = 2$ in [Shi2] and [Shi1] but the argument generalizes to $d \geq 2$. Quasiconformal maps preserve sets of full dimension [GV], so from Theorems 1.1 and 1.3 we obtain:

Corollary 1.6 *For any family of rational maps* f *over* Δ, *the bifurcation set* $B(f)$ *is empty or has Hausdorff dimension two.*

Corollary 1.7 *If* f *has bifurcations, then* H. $\dim(J(f_t)) = 2$ *for a dense set of* $t \in B(f)$.[1]

For higher-dimensional families one has (§5):

Corollary 1.8 *For any holomorphic family of rational maps over a complex manifold* X, *either* $B(f) = \emptyset$ *or* H. $\dim(B(f)) =$ H. $\dim(X) = 2 \dim_{\mathbb{C}} X$.

Similar results on Hausdorff dimension were obtained by Tan Lei, under a technical hypothesis on the family f [Tan].

A family of rational maps f is *algebraic* if its parameter space X is a quasi-projective variety (such as $\mathbb{C}^n$) and the coefficients of $f_t(z)$ are rational functions of t. For example, $p_c(z) = z^d + c$ is an algebraic family over $X = \mathbb{C}$. Such families almost always contain bifurcations [Mc1]:

Theorem 1.9 *For any algebraic family of rational maps, either*

1. *The family is trivial (* f_t *and* f_s *are conformally conjugate for all* $t, s \in X$ *); or*

2. *The family is affine (every* f_t *is critically finite and double covered by a torus endomorphism); or*

3. *The family has bifurcations (* $B(f) \neq \emptyset$ *).*

Corollary 1.10 *With rare exceptions, any algebraic family of rational maps exhibits small Mandelbrot sets in its parameter space.*

---

[1]This set of t can be improved to a dense G_δ using Shishikura's idea of hyperbolic dimension.

This Corollary was our original motivation for proving Theorem 1.1.

As another application, for $t \in \mathbb{C}^{d-1}$ let

$$f_t(z) = z^d + t_1 z^{d-2} + \cdots + t_{d-1}$$

and let

$$\mathcal{C}_d = \{t \ : \ J(f_t) \text{ is connected}\}$$

denote the *connectedness locus*. Then we have:

Corollary 1.11 (Tan Lei) *The boundary of the connectedness locus has full dimension; that is,* H. $\dim(\partial \mathcal{C}_d) = $ H. $\dim(\mathcal{C}_d) = 2d - 2$.

Proof. Consider the algebraic family $g_a(z) = z^d + az^{d-1}$, which for $a \neq 0$ has all but one critical point fixed under g_a. By Theorem 1.9, this family has bifurcations at some $a \in \mathbb{C}$. Then there is a neighborhood U of $(a, 0, \ldots, 0) \in \mathbb{C}^{d-2}$ such that for $t \in U$ all critical points of f_t save one lie in an attracting or superattracting basin. If $t \in B(f) \cap U$, then the remaining critical point has a bounded forward orbit under f_t, but under a small perturbation tends to infinity. It follows that $B(f) \cap U = \partial \mathcal{C}_d \cap U \neq \emptyset$, and thus $\dim(\partial \mathcal{C}_d) \geq \dim B(f) = 2d - 2$. ∎

Remark. Rees has shown that the bifurcation locus has positive measure in the space of all rational maps of degree d [Rees]; it would be interesting to known general conditions on a family f such that $B(f)$ has positive measure in the parameter space X.

Acknowledgements. I would like to thank Tan Lei for sharing her results which prompted the writing of this note. Special cases of Theorem 1.1 were developed independently by Douady and Hubbard [DH, pp.332-336] and Eckmann and Epstein [EE].

2   Families of rational maps

In this section we begin a more formal study of maps with bifurcations.

Definitions. A *local bifurcation* is a holomorphic family of rational maps $f_t(z)$ over the unit disk Δ, such that $0 \in B(f)$.

The following natural operations can be performed on f to construct new local bifurcations:

1. *Coordinate change:* replace f_t by $m_t \circ f_t \circ m_t^{-1}$, where $m : \Delta \times \widehat{\mathbb{C}} \to \widehat{\mathbb{C}}$ is a holomorphic family of Möbius transformations.

2. *Iteration:* replace $f_t(z)$ by $f_t^n(z)$ for a fixed $n \geq 1$.

3. *Base change:* replace $f_t(z)$ by $f_{\phi(t)}(z)$, where $\phi : \Delta \to \Delta$ is a nonconstant holomorphic map with $\phi(0) \in B(f)$.

The first two operations leave the bifurcation locus unchanged, while the last transforms $B(f)$ to $\phi^{-1}(B(f))$.

Marked critical points. We will also consider pairs (f, c) consisting of a local bifurcation and a *marked critical point*; this means $c : \Delta \to \widehat{\mathbb{C}}$ is holomorphic and $f_t'(c_t) = 0$. The operations above also apply to (f, c); a coordinate change replaces c_t with $m_t(c_t)$ and a base change replaces c_t with $c_{\phi(t)}$.

Misiurewicz points. A marked critical point c of f is *active* if its forward orbit

$$\langle f_t^n(c_t) \; : \; n = 1, 2, 3, \ldots \rangle$$

fails to form a normal family of functions of t on any neighborhood of $t = 0$ in Δ. A parameter t is a *Misiurewicz point* for (f, c) if the forward orbit of c_t under f_t lands on a repelling periodic cycle. If $t = 0$ is a Misiurewicz point, then either c is active or c_t is preperiodic for all t.

Proposition 2.1 *If c is an active critical point, then (f, c) has a sequence of distinct Misiurewicz points $t_n \to 0$.*

Proof. This is a traditional normal families argument. Choose any 3 distinct repelling periodic points $\{a_0, b_0, c_0\}$ for f_0, and let $\{a_t, b_t, c_t\}$ be holomorphic functions parameterizing the corresponding periodic points of f_t for t near zero. Since $\langle f_t^n(c_t) \rangle$ is not a normal family, it cannot avoid these three points, and any parameter t where $f_t^n(c_t)$ meets a_t, b_t or c_t is a Misiurewicz point. ∎

Ramification. Next we discuss the existence of univalent inverse branches for a single rational map $F(z)$. Let $d = \deg(F, z)$ denote the local degree of F at $z \in \widehat{\mathbb{C}}$; we have $d > 1$ iff z is a critical point of multiplicity $(d - 1)$. We say y is an *unramified preimage* of z if for some $n \geq 0$, $F^n(y) = z$ and $\deg(F^n, y) = 1$. We say z is *unramified* if it has infinitely many unramified preimages. In this case its unramified preimages accumulate on the full Julia set $J(F)$.

Proposition 2.2 *If z has 5 distinct unramified preimages then it has infinitely many.*

Proof. Let E be the set of all unramified preimages of z, and let C be the critical points of F. Then $F^{-1}(E) \subset E \cup C$, so if $|E|$ is finite then

$$d|E| = \sum_{z \in F^{-1}(E)} 1 + \mathrm{mult}(f', z) \leq |F^{-1}(E)| + 2d - 2 \leq |E| + 4d - 4$$

and therefore $|E| \leq 4$.   ∎

Corollary 2.3 *Let (f, c) be a local bifurcation with marked critical point. Then the set of t such that c_t is ramified for f_t is either discrete or the whole disk.*

Proof. By the previous Proposition, the ramified parameters are defined by a finite number of analytic equations in t.   ∎

Proposition 2.4 *After a suitable base change, any local bifurcation f can be provided with an active marked critical point c such that c_0 is unramified for f_0.*

Remark. It is possible that all the active critical points are ramified at $t = 0$. The base change in the Proposition will generally not preserve the central fiber f_0.

Proof. The set $C = \{(t, z) \in \Delta \times \widehat{\mathbb{C}} : f_t'(z) = 0\}$ is an analytic variety with a proper finite projection to Δ. By Puiseux series, after a base change of the form $\phi(t) = \epsilon t^n$ all the critical points of f can be marked by holomorphic functions $\{c_t^1, \dots, c_t^m\}$. Since $t = 0$ is in the bifurcation set, by [Mc2, Thm. 4.2], there is an i such that $\langle f_t^n(c_t^i) \rangle$ is not a normal family at $t = 0$. That is, c^i is an active critical point.

Next we show c^i can be chosen so that for generic t it is disjoint from the forward orbits of all other critical points. If not, there is a c^j and $n \geq 1$ such that $f_t^n(c_t^j) = c_t^i$ for all t. Then c^j is also active and we may replace c^i with c^j. If the replacement process were to cycle, then c^i would be a periodic critical point, which is impossible because it is active. Thus we eventually achieve a c^i which is generically disjoint from the forward orbits of the other critical points.

In particular, there is a t such that c_t^i is unramified for f_t. By Corollary 2.3, the set $R \subset \Delta$ of parameters where c_t^i is ramified is discrete. By Proposition 2.1, there are Misiurewicz points t_n for (f, c^i) with $t_n \to 0$. Choose n such that $t_n \notin R$, and make a base change moving t_n to zero; then c^i is active, and c_0^i is unramified for f_0.   ∎

Misiurewicz bifurcations. Let (f, c) be a local bifurcation with a marked critical point. We say (f, c) is a *Misiurewicz bifurcation* of degree d if

M1. $f_0(c_0)$ is a repelling fixed-point of f_0;

M2. c_0 is unramified for f_0;

M3. $f_t(c_t)$ is not a fixed-point of f_t, for some t; and

M4. $\deg(f_t, c_t) = d$ for all t sufficiently small.

Proposition 2.5 *For any local bifurcation (f, c) with c active and c_0 unramified, there is a base change and an $n > 0$ such that (f^n, c) is a Misiurewicz bifurcation.*

Remark. The delicate point is condition (M4). The danger is that for every Misiurewicz parameter t, the forward orbit of c_t might accidentally collide with another critical point before reaching the periodic cycle. We must avoid these collisions to make the degree of f_t^n at c_t locally constant.

Proof. There are Misiurewicz points $t_n \to 0$ for (f, c), and c_t is unramified for all t near 0, so after a base change and replacing f with f^n we can arrange that (f, c) satisfies conditions (M1), (M2) and (M3).

We can also arrange that $\deg(f_t, c_t) = d$ for all $t \neq 0$. However (M4) may fail because $\deg(f_t, c_t)$ may jump up at $t = 0$. This jump would occur if another critical of f_t coincides with c_t at $t = 0$.

To rule this out, we make a further perturbation of f_0. Let a_t locally parameterize the repelling fixed-point of f_t such that $f_0(c_0) = a_0$. Choose a neighborhood U of a_0 such that for t small, f_t is linearizable on U and U is disjoint from the critical points of f_t. (This is possible since $f_0'(a_0) \neq 0$.)

Let $b_t \in U - \{a_t\}$ be a parameterized repelling periodic point close to a_t. Then b_t has preimages in U accumulating on a_t. Choose s near 0 such that $f_s(c_s)$ hits one of these preimages (such an s exists by the argument principle and (M3)). For this special parameter, c_s first maps close to a_s, then remains in U until it finally lands on b_s. Since there are no critical points in U, we have $\deg(f_s^i, c) = d$ for all $i > 0$.

Making a base change moving s to $t = 0$, we find that (f^n, c) satisfies (M1-M4) for n a suitable multiple of the period of b_s. ∎

3 The Misiurewicz cascade

In this section we show that when a Misiurewicz point bifurcates, it produces a cascade of polynomial-like maps.

Definitions. A *polynomial-like map* $g : U \to V$ is a proper, holomorphic map between simply-connected regions with $\overline{U}$ compact and $\overline{U} \subset V \subset \mathbb{C}$ [DH]. Its *filled Julia set* is defined by

$$K(g) = \bigcap_1^\infty g^{-n}(V);$$

it is the set of points that never escape from U under forward iteration.

Any polynomial such as $p_c(z) = z^d + c$ can be restricted to a polynomial-like map $p_c : U \to V$ of degree d with the same filled Julia set. Moreover small analytic perturbations of $p_c : U \to V$ are also polynomial-like.

A degree d Misiurewicz bifurcation (f, c) gives rise to polynomial-like maps $f_t^n : B_0 \to B_n$, by the following mechanism. For small t, a small ball B_0 about the critical point c_t maps to a small ball B_1 close to, but not containing, the fixed-point of f_t. The iterates $B_i = f_t^i(B)$ then remain near the fixed-point for a long time, ultimately expanding by a large factor. Finally for suitable t, as B_i escapes from the fixed-point it maps back over B_0, resulting in a degree d map $f_t^n : B_0 \to B_n \supset B_0$. Since most of the images $\langle B_i \rangle$ lie in the region where f_t behaves linearly, the first-return map $f_t^n : B_0 \to B_n$ behaves like a polynomial of degree d.

This scenario leads to a cascade of families of polynomial-like maps, indexed by the return time n. Here is a precise statement.

Theorem 3.1 *Let (f, c) be a degree d Misiurewicz bifurcation, and fix $R > 0$. Then for all $n \gg 0$, there is a coordinate change depending on n such that $c_t = 0$ and*

$$f_t^n(z) = z^d + \xi + O(\epsilon_n)$$

whenever $|z|, |\xi| \le R$. Here $t = t_n(1 + \gamma_n\xi)$, t_n and γ_n are nonzero, and γ_n, t_n and ϵ_n tend to zero as $n \to \infty$.

The constants in $O(\cdot)$ above depend on f and R but not on n.

The proof yields more explicit information. Let $\lambda_0 = f_0'(f_0(c_0))$ be the multiplier of the fixed-point on which c_0 lands, and let r be the multiplicity of intersection of the graph of c_t and the graph of this fixed-point at $t = 0$. Then for $t = t_n$, the critical point c_t is periodic with period n, and we have:

$$t_n \sim C\lambda_0^{-n/r}, \tag{3.1}$$
$$\gamma_n = C'\lambda_0^{-n/(d-1)}, \quad \text{and} \tag{3.2}$$
$$\epsilon_n = n(|\lambda_0|^{-n/r} + |\lambda_0|^{-n/(d-1)}), \tag{3.3}$$

for certain constants C, C' depending on f. Due to the choice of roots, there are r possibilities for t_n and $(d-1)$ for γ_n; the Theorem is valid for all choices.

Finally for ξ fixed and $t = t_n(1 + \gamma_n \xi)$, the map f_t^n is polynomial-like near c_t for all $n \gg 0$, and in the *original* z-coordinate its filled Julia set satisfies

$$\operatorname{diam} K(f_t^n) \asymp |\lambda_0|^{-n/(d-1)}.$$

Notation. We adopt the usual conventions: $a_n = O(b_n)$, $a_n \asymp b_n$, $a_n \sim b_n$ and $n \gg 0$ mean $|a_n| < C|b_n|$, $(1/C)|b_n| < |a_n| < C|b_n|$, $a_n/b_n \to 1$ and $n \geq N$, where C and N are implicit constants.

Proof. We will make several constructions that work on a small neighborhood of $t = 0$. First, let a_t parameterize the repelling fixed-point of f_t such that $a_0 = f_0(c_0)$. Let $\lambda_t = f_t'(a_t)$ be its multiplier. There is a holomorphically varying coordinate chart $u = \phi_t(z)$ defined near $z = a_t$ such that

$$\phi_t \circ f_t \circ \phi_t^{-1}(u) = \lambda_t u \tag{3.4}$$

for u near 0. We call $u = \phi_t(z)$ the *linearizing coordinate*; note that $u = 0$ at a_t.

We next arrange that $u = 1$ is an unramified preimage of c_t. Since c_0 is unramified by (M2), its unramified preimages accumulate on a_0. Let b_0 be one such preimage, with $f_0^p(b_0) = c_0$ and b_0 in the domain of ϕ_0. Then b_0 prolongs to a holomorphic function b_t with $f_t^p(b_t) = c_t$. Replacing ϕ_t by $\phi_t(z)/\phi_t(b_t)$, we can assume $u = \phi_t(b_t) = 1$.

For small t, the composition $f_t^p \circ \phi_t^{-1}$ is univalent near $u = 1$. By applying a coordinate change $z \mapsto m_t(z)$, where m_t is a Möbius transformation depending on t, we can arrange that $c_t = 0$ and that

$$f_t^p \circ \phi_t^{-1}(u) = (u - 1) + O((u - 1)^2) \tag{3.5}$$

on $B(1, \epsilon)$.

Since $\deg(f_t, 0) = d$ for t near 0 by (M4), we have

$$\phi_t \circ f_t(z) = \sum A_i(t) z^i \tag{3.6}$$

$$= A_0(t) + A_d(0) z^d (1 + O(|z| + |t|)) \tag{3.7}$$

with $A_d(0) \neq 0$. Here $A_0(t) = f_t(0)$ is the u-coordinate of the critical value. By (M3), c_t is not pre-fixed for all t, so there is an $r > 0$ such that

$$A_0(t) = t^r B(t) \tag{3.8}$$

where $B(0) \neq 0$.

Next for $n \gg 0$ we choose t_n such that

$$f_t^{1+n+p}(c_t) = c_t \quad \text{when } t = t_n. \tag{3.9}$$

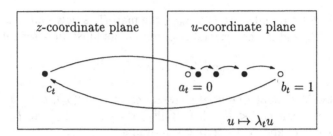

Figure 2. Visiting the repelling fixed-point

More precisely, for $t = t_n$ we will arrange that c_t maps first close to a_t, then lands after n iterates on b_t, and thus returns in p further iterates to c_t; see Figure 2. In the u-coordinate system, f_t is linear and $b_t = 1$, so the equation $f_t^{n+1}(c_t) = b_t$ becomes

$$\lambda_t^n A_0(t) = 1 \quad \text{when } t = t_n. \tag{3.10}$$

By the argument principle, for $n \gg 0$ this equation has a solution t_n close to any root of the approximation $\lambda_0^n t^r B(0) = 1$ obtained from (3.8). Moreover

$$t_n \sim B(0)^{-1} \lambda_0^{-n/r}$$

(verifying (3.1)), and t_n satisfies (3.9) because $f_t^p(b_t) = c_t$. (There are actually be r solutions for t_n for a given n; any one of the r solutions will do.)

We now turn to the estimate of $f_t^{1+n+p}(z)$ for (t, z) near $(t_n, 0)$. We will assume throughout that $t = t_n + s$ and that:

$$|z| \text{ and } |s/t_n| \text{ are } O(\Lambda^{-n/(d-1)}) \tag{3.11}$$

where $\Lambda = |\lambda_0| > 1$. (To see this is the correct scale at which to work, suppose $\operatorname{diam}(B) \asymp \operatorname{diam} f_t^{1+n+p}(B)$, where B is a ball centered at $z = 0$. Then $\operatorname{diam} f_t(B) \asymp (\operatorname{diam} B)^d$, and f_t^n is expanding by a factor of about Λ^n, while f_t^p is univalent, so we get $\operatorname{diam} B \asymp \Lambda^n(\operatorname{diam} B)^d$, or $\operatorname{diam} B \asymp \Lambda^{-n/(d-1)}$. Similarly $|f_t^{1+n+p}(0)| \asymp \Lambda^n(s/t_n)t_n^r \asymp (s/t_n) = O(\operatorname{diam} B)$ when s is as above.)

It is also convenient to set

$$\widetilde{\Lambda} = \min(\Lambda^{1/(d-1)}, \Lambda^{1/r}) > 1,$$

so that we may assert:

$$z \text{ and } t \text{ are } O(\widetilde{\Lambda}^{-n}). \tag{3.12}$$

By (3.11) the n iterates of $f_t(z)$ lie within the domain of linearization, so by (3.7) we have

$$\phi_t \circ f_t^{1+n}(z) = \lambda_t^n A_0(t) + \lambda_t^n A_0(d) z^d (1 + O(|z| + |t|)).$$

The first term is approximately 1. Indeed, $\lambda_t^n = \lambda_{t_n}^n (1 + O(ns))$, so by (3.8) we have

$$\begin{aligned}
\lambda_t^n A_0(t) &= \lambda_t^n (t_n + s)^r B(t_n + s) \\
&= \lambda_{t_n}^n (1 + O(ns)) \cdot t_n^r \left(1 + \frac{s}{t_n}\right)^r \cdot B(t_n)(1 + O(s)) \\
&= \lambda_{t_n}^n A_0(t_n) \left(1 + r\frac{s}{t_n} + O((s/t_n)^2) + O(ns)\right) \\
&= 1 + r\frac{s}{t_n} + O((s/t_n)^2) + O(ns)
\end{aligned}$$

by (3.10). Similarly, $\lambda_t^n = \lambda_0^n(1 + O(t))$, so

$$\begin{aligned}
&\phi_t \circ f_t^{1+n}(z) - 1 = \\
&\lambda_0^n A_0(d) z^d (1 + O(|z| + |nt|)) + r\frac{s}{t_n} + O((s/t_n)^2) + O(ns) = \\
&\lambda_0^n A_0(d) z^d + r\frac{s}{t_n} + O(n\Lambda^{-n/(d-1)}\widetilde{\Lambda}^{-n}),
\end{aligned}$$

using (3.11) and (3.12). The expression above as a whole is $O(\Lambda^{-n/(d-1)})$, so composing with the univalent map $f_t^p \circ \phi_t^{-1}$ introduces (by (3.5)) an additional error of size $O(\Lambda^{-2n/(d-1)})$, which is already accounted for in the $O(\cdot)$ above. Thus the expression above also represents $f_t^{1+n+p}(z)$.

Finally we make a linear change of coordinates of the form $z \mapsto \alpha_n z$, conjugating the expression above to

$$f_t^{1+n+p}(z) = \alpha_n^{1-d} \lambda_0^n A_0(d) z^d + \alpha_n r\frac{s}{t_n} + O(n\alpha_n \Lambda^{-n/(d-1)}\widetilde{\Lambda}^{-n}).$$

Setting $\alpha_n = (\lambda_0^n A_0(d))^{1/(d-1)}$ to normalize the coefficient of z^d, we have $|\alpha_n| \asymp \Lambda^{n/(d-1)}$ and thus:

$$\begin{aligned}
f_t^{1+n+p}(z) &= z^d + \alpha_n r\frac{s}{t_n} + O(n\widetilde{\Lambda}^{-n}) \\
&= z^d + \xi + O(\epsilon_n)
\end{aligned}$$

with $t = t_n(1 + \gamma_n\xi)$, γ_n and ϵ_n as in (3.2) and (3.3). Notice that if $|z|$ and $|\xi|$ are bounded by R in the expression above, then (3.11) is satisfied in our original coordinates. Reindexing n, we obtain the Theorem. ∎

4   Small Mandelbrot sets

We now show the Misiurewicz cascade leads to small Mandelbrot sets in parameter space. From this we deduce Theorems 1.1 and 1.3 of the Introduction.

Hybrid conjugacy. Let g_1, g_2 be polynomial-like maps of the same degree. A *hybrid conjugacy* is a quasiconformal map ϕ between neighborhoods of $K(g_1)$ and $K(g_2)$ such that $\phi \circ g_1 = g_2 \circ \phi$ and $\overline{\partial}\phi|K(g_1) = 0$. We say g_1 and g_2 are *hybrid equivalent* if such a conjugacy exists. By a basic result of Douady and Hubbard, every polynomial-like map g of degree d is hybrid equivalent to a polynomial of degree d, unique up to affine conjugacy if $K(g)$ is connected [DH, Theorem 1].

Theorem 4.1 *Let (f, c) be a degree d Misiurewicz bifurcation. Then the parameter space Δ contains quasiconformal copies $\mathcal{M}_d^n$ of the degree d Mandelbrot set M_d, converging to the origin, with $\partial \mathcal{M}_d^n$ contained in the bifurcation locus $B(f)$.*

More precisely, for all $n \gg 0$ there are homeomorphisms

$$\phi_n : M_d \to \mathcal{M}_d^n \subset \Delta$$

such that:

1. *f_t^n is hybrid equivalent to $z^d + \xi$ whenever $t = \phi_n(\xi)$;*

2. *$d(0, \mathcal{M}_d^n) \asymp |\lambda_0|^{-n/r}$;*

3. *$\mathrm{diam}(\mathcal{M}_d^n)/d(0, \mathcal{M}_d^n) \asymp |\lambda_0|^{-n/(d-1)}$;*

4. *ϕ_n extends to a quasiconformal map of the plane with dilatation bounded by $1 + O(\epsilon_n)$; and*

5. *$\psi_n^{-1} \circ \phi_n(\xi) = \xi + O(\epsilon_n)$, where $\psi_n(\xi) = t_n(1 + \gamma_n \xi)$.*

The notation is from (3.1) – (3.3).

We begin by recapitulating some ideas from [DH]. Let $\Delta(R) = \{z : |z| < R\}$, and let

$$g_\xi(z) = z^d + \xi + h(\xi, z)$$

be a holomorphic family of mappings defined for $(\xi, z) \in \Delta(R) \times \Delta(R)$, where $R > 10$ and $g_\xi'(0) = 0$. Let $\mathcal{M} \subset \Delta(R)$ be the set of ξ such that the forward orbit $g_\xi^n(0)$ remains in $\Delta(R)$ for all $n > 0$.

Lemma 4.2 *There is a $\delta > 0$ such that if $\sup |h(\xi, z)| = \epsilon < \delta$ then there is a homeomorphism*

$$\phi : M_d \to \mathcal{M}$$

such that for all $\xi \in M_d$, $g_{\phi(\xi)}$ is hybrid equivalent to $z^d + \xi$, $|\phi(\xi) - \xi| < O(\epsilon)$, and ϕ extends to a $1 + O(\epsilon)$-quasiconformal map of the plane.

Proof. Let $p_\xi(z) = z^d + \xi$. Since $R > 10$ we have $M_d \subset \Delta(R)$ and $K(p_\xi) \subset \Delta(R)$ for all $\xi \in M_d$; indeed these sets have capacity one, so their diameters are bounded by 4 [Ah]. In addition, for $\xi \in M_d$ the map $p_\xi : U \to \Delta(R)$ is polynomial-like, where $U = p_\xi^{-1}(\Delta(R))$. By continuity, when $\sup |h|$ is sufficiently small, $\mathcal{M}$ is compact and g_ξ is polynomial-like for all $\xi \in \mathcal{M}$.

By results of Douady and Hubbard, we can also choose δ small enough that $|h| < \delta$ implies there is a homeomorphism

$$\phi : M_d \to \mathcal{M}$$

such that $g_{\phi(\xi)}$ is hybrid equivalent to $z^d + \xi$ [DH, Prop. 21].

Now assume $|h| < \epsilon < \delta$. For $t \in \Delta$ let $\mathcal{M}_t$ denote the parameters where the critical point remains bounded for the family

$$g_{\xi, t} = z^d + \xi + t \frac{\delta}{\epsilon} h(\xi, z),$$

and define $\phi_t : M_d \to \mathcal{M}_t$ as above. Then ϕ_t is a family of injections, with $\phi_0(z) = z$, and $\phi_t(\xi)$ is a holomorphic function of t for every ξ. (For example this is clear at $\xi \in \partial M_d$ because Misiurewicz points are dense in ∂M_d; for the general case see [DH, Prop. 22].)

By a theorem of Słodkowski [Sl] (cf. [Dou], [BR]), $\phi_t(z)$ prolongs to a holomorphic motion of the entire plane, and its complex dilatation $\mu_t = \overline{\partial}\phi_t/\partial\phi_t$ gives a holomorphic map of the unit disk into the unit ball in $L^\infty(\widehat{\mathbb{C}})$. By the Schwarz lemma, $\|\mu_t\|_\infty \leq |t|$; since $\phi = \phi_{\epsilon/\delta}$, we obtain a quasiconformal extension of ϕ with dilatation $1 + O(\epsilon)$. The bound on $|\phi(\xi) - \xi|$ similarly results by applying the Schwarz Lemma to the map $\Delta \to \widehat{\mathbb{C}} - \{0, 1, \infty\}$ given by $t \mapsto \phi_t(\xi)$, once three points have been normalized to remain fixed during the motion. ∎

Proof of Theorem 4.1. Fix $R > 10$. For all $n \gg 0$, Theorem 3.1 provides a family of rational maps of the form

$$g_\xi(z) = f_t^n(z) = z^d + \xi + O(\epsilon_n)$$

defined for $(\xi, z) \in \Delta(R) \times \Delta(R)$, where $t = \psi_n(\xi) = t_n(1 + \gamma_n \xi)$. The preceding Lemma gives homeomorphisms $\widetilde{\phi}_n : M_d \to \widetilde{\mathcal{M}_d^n} \subset \Delta(R)$ for all $n \gg 0$. Setting $\phi_n = \psi_n \circ \widetilde{\phi}_n$, the Theorem results from the Lemma and the bounds (3.1) – (3.3). ∎

Example. The quadratic family $(f, c) = (z^2 + t - 2, 0)$ is a Misiurewicz bifurcation of degree $d = 2$, with $\lambda_0 = 2$ and $r = 1$. Thus M_2 contains small copies $\mathcal{M}_2^n$ of itself near $c = -2$, with $d(\mathcal{M}_2^n, -2) \asymp 4^{-n}$ and diam $\mathcal{M}_2^n \asymp 16^{-n}$.

Consequences. Assembling the preceding results, we may now prove the Theorems stated in the Introduction. Here is a more precise form of Theorem 1.1:

Theorem 4.3 *Let f be a holomorphic family of rational maps over the unit disk with bifurcations. Then there is a nonempty list of degrees*

$$D \subset \{2, 3, \ldots, 2^{2 \deg(f_t) - 2}\}$$

such that for any $\epsilon > 0$ and $d \in D$, $B(f)$ contains the image of ∂M_d under a $(1 + \epsilon)$-quasiconformal map.

If the critical points of f are marked $\{c_t^1, \ldots, c_t^m\}$ such that

$$(i \leq N) \iff c^i \text{ is active and } c_t^i \text{ is unramified for some } t,$$

then we may take

$$D = \{\inf_t \sup_k \deg(f_t^k, c_t^i) : i \leq N\}.$$

Proof. Let $B_0 = B(f)$. After a base change we can assume that f is a local bifurcation with critical points marked as above. By Proposition 2.4, there is at least one active, unramified critical point, so $N \geq 1$. For any $i \leq N$, we can make a base change so c_t^i is active and unramified; then by Proposition 2.5, a further base change makes (f^n, c^i) into a degree d Misiurewicz bifurcation.

Let $d_i = \inf_t \sup_k \deg(f_t^k, c_t^i)$. We claim $d = d_i \leq 2^{2 \deg(f_t) - 2}$. Indeed, $\deg(f_t^n, c_t^i)$ assumes its minimum outside a discrete set, and it is equal to d near $t = 0$, so $d_i \geq d$. On the other hand, c_0^i lands on a repelling periodic cycle, so $\deg(f_0^k, c_0^i) = d$ for all $k > n$, and therefore $d_i \leq d$. Finally d is largest if c^i hits all the other critical points of f before reaching the repelling cycle; in this case $d = (p_1 + 1)(p_2 + 1) \cdots (p_m + 1)$ for some partition $p_1 + p_2 + \cdots + p_m = 2d - 2$. The product is maximized by the partition $1 + 1 + \cdots + 1$, so $d \leq 2^{2 \deg(f_t) - 2}$.

By Theorem 4.1, the bifurcation locus $B(f)$ contains almost conformal copies $\partial \mathcal{M}_d^n$ of ∂M_d accumulating at $t = 0$, with diam$(\mathcal{M}_d^n) \ll d(0, \mathcal{M}_d^n)$. Letting $\phi : \Delta \to \Delta$ denote the composition of all the base-changes occurring so far, we have $B(f) = \phi^{-1}(B_0)$. Then ϕ is univalent and nearly linear on $\mathcal{M}_d^n$ for $n \gg 0$, so $\phi(\partial \mathcal{M}_d^n) \subset B_0$ is a $(1 + \epsilon)$-quasiconformal copy of ∂M_d. ∎

Let Rat$_d$ be the space of all rational maps of degree d; it is a Zariski-open subset of $\mathbb{P}^{2d+1}$. We now make precise the statement that a generic family contains a copy of the standard Mandelbrot set.

Corollary 4.4 *There is a countably union of proper subvarieties $R \subset \mathrm{Rat}_d$ such that for any local bifurcation, either $f_t \in R$ for all t, or $B(f)$ contains a copy of ∂M_2.*

Proof. On a finite branched cover X of Rat_d, the critical points of $f \in \mathrm{Rat}_d$ can be marked $\{c^1(f), \ldots, c^{2d-2}(f)\}$. Clearly $\deg(f^n, c_i(f)) = 2$ outside a proper subvariety $V_{n,i}$ of X. Let R be the union of the images of these varieties in Rat_d, and apply the preceding argument to see $D = \{2\}$ if some $f_t \notin R$. ∎

Proof of Theorem 1.3. The proof follows the same lines as that of Theorem 4.3; to get $\mathrm{mod}(V - U)$ large one takes R large in Theorem 3.1. Thus Theorem 1.3 also holds for all $d \in D$. ∎

5 Hausdorff dimension

In this section we prove Corollary 1.8: for any holomorphic family f of rational maps over a complex manifold X, we have $\mathrm{H.dim}(B(f)) = \mathrm{H.dim}(X)$ if $B(f) \neq \emptyset$.

   Recall that the *Hausdorff dimension* of a metric space X is the infimum of the set of $\delta \geq 0$ such that there exists coverings $X = \bigcup X_i$ with $\sum (\mathrm{diam}\, X_i)^\delta$ arbitrarily small.

Lemma 5.1 *Let Y be a metric space, X a subset of $Y \times [0,1]$. Then*

$$\mathrm{H.dim}(X) \geq 1 + \inf \mathrm{H.dim}(X_t)$$

where $X_t = \{y \,:\, (y,t) \in X\}$.

Here $Y \times [0,1]$ is given the product metric.

Proof. Fix δ with $\delta + 1 > \mathrm{H.dim}(X)$. For any n there is a covering $X \subset \bigcup B(y_i, r_i) \times I_i$ with $|I_i| = r_i$ and $\sum r_i^{\delta+1} < 4^{-n}$. Note that

$$X_t \subset \bigcup_{t \in I_i} B(y_i, r_i)$$

and

$$\int_0^1 \sum_{t \in I_i} r_i^\delta \, dt = \sum r_i^{\delta+1} < 4^{-n}.$$

Let E_n be the set of t where the integrand exceeds 2^{-n}; then $\sum m(E_n) < \sum 2^{-n} < \infty$. Thus almost every t belongs to at most finitely many E_n, so almost every X_t admits infinitely many coverings with $\sum r_i^\delta < 2^{-n} \to 0$. Therefore $\delta \geq \inf \mathrm{H.dim}(X_t)$, and the Theorem follows. ∎

16   C. McMullen

The Lemma above is related to the product formula

$$\text{H.}\dim(X \times Y) \geq \text{H.}\dim(X) + \text{H.}\dim(Y);$$

see [Fal, Ch. 5] and references therein.

Proof of Corollary 1.8. Suppose $B(f) \neq \emptyset$. Then there is a $t_0 \in B(f)$ and a locally parameterized periodic point $a(t)$ of period n such that $a(t)$ changes from attracting to repelling near t_0 [MSS], [Mc2, §4.1]. More formally this means the multiplier $\lambda(t) = (f^n)'(a(t))$ is not locally constant and $|\lambda(t_0)| = 1$.

Choosing local coordinates we can reduce to the case $X = \Delta^n$ and $t_0 = 0$. Let $\Delta_s = \Delta \times \{s\}$ for $s \in \Delta^{n-1}$. For coordinates in general position, $\lambda(t)$ is nonconstant on Δ_0. Shrinking the Δ^{n-1} factor, we can also assume $a(t)$ changes from attracting to repelling in the family $f|\Delta_s$ for all s. Then

$$B(f)_s = B(f) \cap \Delta_s \supset B(f|\Delta_s) \neq \emptyset$$

and $\text{H.}\dim B(f|\Delta_s) = 2$ by Corollary 1.6. Applying the Lemma above to $B(f) \subset \Delta \times \Delta^{n-1}$ we find

$$\text{H.}\dim(B(f)) \geq (2n-2) + \inf_s \text{H.}\dim B(f)_s = 2n = \text{H.}\dim(X).$$

∎

References

[Ah] L. Ahlfors. *Conformal Invariants: Topics in Geometric Function Theory.* McGraw-Hill Book Co., 1973.

[BR] L. Bers and H. L. Royden. Holomorphic families of injections. *Acta Math.* **157**(1986), 259–286.

[Dou] A. Douady. Prolongement de mouvements holomorphes (d'après Słodkowski et autres). In *Séminaire Bourbaki, 1993/94*, pages 7–20. Astérisque, volume 227, 1995.

[DH] A. Douady and J. Hubbard. On the dynamics of polynomial-like mappings. *Ann. Sci. Éc. Norm. Sup.* **18**(1985), 287–344.

[EE] J.-P. Eckmann and H. Epstein. Scaling of Mandelbrot sets generated by critical point preperiodicity. *Comm. Math. Phys.* **101**(1985), 283–289.

[Fal] K. J. Falconer. *The Geometry of Fractal Sets.* Cambridge University Press, 1986.

[GV] F. W. Gehring and J. Väisälä. Hausdorff dimension and quasiconformal mappings. *J. London Math. Soc.* **6**(1973), 504–512.

[MSS] R. Mañé, P. Sad, and D. Sullivan. On the dynamics of rational maps. *Ann. Sci. Éc. Norm. Sup.* **16**(1983), 193–217.

[Mc1] C. McMullen. Families of rational maps and iterative root-finding algorithms. *Annals of Math.* **125**(1987), 467–493.

[Mc2] C. McMullen. *Complex Dynamics and Renormalization*, volume 135 of *Annals of Math. Studies*. Princeton University Press, 1994.

[Rees] M. Rees. Positive measure sets of ergodic rational maps. *Ann. scient. Éc. Norm. Sup.* **19**(1986), 383–407.

[Shi1] M. Shishikura. The boundary of the Mandelbrot set has Hausdorff dimension two. In *Complex Analytic Methods in Dynamical Systems (Rio de Janeiro, 1992)*, pages 389–406. Astérisque, volume 222, 1994.

[Shi2] M. Shishikura. The Hausdorff dimension of the boundary of the Mandelbrot set and Julia sets. *Annals of Math.* **147**(1998), 225–267.

[Sl] Z. Słodkowski. Holomorphic motions and polynomial hulls. *Proc. Amer. Math. Soc.* **111**(1991), 347–355.

[Tan] Tan L. Hausdorff dimension of subsets of the parameter space for families of rational maps. *Nonlinearity*, **11**(1998), 233–246.

MATHEMATICS DEPARTMENT, HARVARD UNIVERSITY, 1 OXFORD ST, CAMBRIDGE, MA 02138-2901, U.S.A.
e-mail: ctm@abel.MATH.HARVARD.EDU

Baby Mandelbrot sets are born in cauliflowers

Adrien Douady

with the participation of

Xavier Buff, Robert L. Devaney and Pierrette Sentenac

Abstract

We show that at the neighborhood of a cusp point $c_0 \neq 1/4$ of the Mandelbrot set M, there is a sequence (M_n) of small quasi-conformal copies of M in M, tending to c_0. Moreover, for each n, the copy M_n is encaged in a set Γ_n which is homeomorphic to the Julia set J_ϵ of $z^2 + \frac{1}{4} + \epsilon$ for $\epsilon > 0$ small, in fact in a nested sequence of sets $(\Gamma_{n,m})_m$ which are homeomorphic to the preimage of J_ϵ by $z \mapsto z^{2^m}$, and accumulate to M_n. We prove that the distance from M_n to c_0 tends to 0 like $1/n^2$ and its diameter, with or without the decorations $\bigcup_m \Gamma_{n,m}$, tends to 0 like $1/n^3$.

1   The phenomena in the Mandelbrot set

When one looks closely at the Mandelbrot set M at the neighborhood of the cusp point c_0 of a primitive hyperbolic component, one can observe a sequence (M_n) of small copies of M in M, tending to c_0. Each copy M_n is encaged in a set Γ_n which resembles the Julia set J_ϵ of $z^2 + \frac{1}{4} + \epsilon$ for $\epsilon > 0$ small, in fact in a nested sequence of sets $(\Gamma_{n,m})_m$ which resemble the preimage of J_ϵ by $z \mapsto z^{2^m}$, and accumulate to M_n (Fig. 1).

The distance from M_n to c_0 tends to 0 like $1/n^2$ and its diameter, with or without the decorations $\bigcup_m \Gamma_{n,m}$, tends to 0 like $1/n^3$.

In this paper, we shall first give a model Υ for this figure, then we shall describe a more general situation in which Υ appears quasi-conformally in a set M_F defined in the parameter space. Then we have to prove on one hand that Υ does appear in M_F (section 10), and on the other that M_F appears in the Mandelbrot set in the neighborhood of c_0 (section 11).

A precise statement is given in Theorem 1, and in Theorem 2 for the more general situation. The proof of Theorem 2 will be given in sections 10. It relies on concepts and results concerning polynomial-like mappings which can be found in [DH Pol-like], with a complement by Lyubich [Ly], and on the description of parabolic implosion [La], [D/AMS], [Sh], [TL], [DSZ]. We shall also use Słodkowski's theorem on extension of holomorphic motions [Sł], [D/Sł]. Starting from these results, the main difficulty is in keeping track of notations.

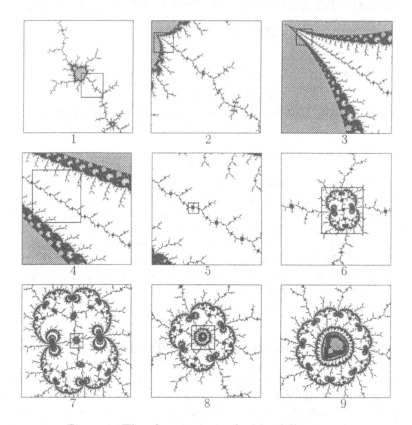

Figure 1: The phenomena in the Mandelbrot set.

But first of all, we have to specify what we mean by "to appear quasi-conformally" (section 2).

2 Conventions

We denote by T_s the translation by s, i.e. we set $T_s(z) = z + s$.

We denote by S_R^1 the circle of radius R in $\mathbf{C}$, and by $A_{R,R'}$ the annulus $\{z|\ R' < |z| < R\}$.

We denote by K_c the filled Julia set and by J_c the Julia set for $z^2 + c$, and by M the Mandelbrot set.

The Julia set $J_{\frac{1}{4}}$ is called the *cauliflower*; a Julia set $J_{\frac{1}{4}+\epsilon}$ with $\epsilon > 0$ small is called an *imploded cauliflower*. The Julia set $J_{\frac{1}{4}+\epsilon}$ (Fig. 2) does not tend to $J_{\frac{1}{4}}$ when $\epsilon \to 0$. It does not actually have a limit, but there is a family of compact sets $(J_{\frac{1}{4},\sigma})_{\sigma \in \mathbf{T}}$ (Julia-Lavaurs sets) such that, if a sequence $(J_{\frac{1}{4}+\epsilon_n})$

has a limit with $\epsilon_n > 0$ tending to 0, this limit is one of the sets $J_{\frac{1}{4},\sigma}$. The compact set $J_{\frac{1}{4},\sigma}$ depends continuously on σ for $\sigma \in \mathbf{T}$ [D/AMS].

Figure 2: A sequence of imploded cauliflowers.

Given a compact set K in $\mathbf{C}$, connected, full and containing more than one point, we denote by ϕ_K the isomorphism $\mathbf{C} \smallsetminus K \to \mathbf{C} \smallsetminus \overline{\mathbf{D}}$ such that $\lim(\phi_K(z)/z)$ is real > 0.

Given two compact sets X and Y in $\mathbf{C}$, we say that X *appears* quasi-conformally in Y if there is a quasi-conformal homeomorphism $\varphi : \mathbf{C} \to \mathbf{C}$ such that $\varphi(X) \subset Y$ and $\varphi(\partial X) \subset \partial Y$ (the last condition is there in order to exclude the case $\varphi(X) \subset \overset{\circ}{K}$, that one could always realize as soon as $\overset{\circ}{K} \neq \emptyset$).

3   The model

Consider an imploded cauliflower $J_{\frac{1}{4}+\epsilon}$ and choose $R > 1$ so that $J_{\frac{1}{4}+\epsilon}$ is contained in the annulus $A_{\sqrt{R},\frac{1}{\sqrt{R}}}$. Denote by $\Gamma_0(\epsilon)$ the set $J_{\frac{1}{4}+\epsilon}$ enlarged by the factor $R^{3/2}$, so that it is now contained in $A_{R^2,R}$. Denote by $\Gamma_m(\epsilon)$ the inverse image of $\Gamma_0(\epsilon)$ by the map $z \mapsto z^{2^m}$; these sets are contained in disjoint annuli and their union is the set of points whose orbit under $z \mapsto z^2$ has a (unique) point in $\Gamma_0(\epsilon)$.

We set (Fig. 3)
$$\Xi(\epsilon) = M \cup \phi_M^{-1}(\bigcup_m \Gamma_m(\epsilon)) \ .$$

Up to a quasi-conformal homeomorphism of $\mathbf{C}$, the set $\Xi(\epsilon)$ does not depend on the choice of R provided it satisfies the requirement. It is possible to choose R satisfying the requirement for all values of ϵ which are sufficiently small. Indeed the set $L = \bigcup_{\sigma \in \mathbf{T}} J_{\frac{1}{4},\sigma}$ is a compact set which does not contain 0, and R such that $L \subset A_{\sqrt{R},\frac{1}{\sqrt{R}}}$ will do.

We set
$$\Upsilon = \{0\} \cup \bigcup_{n \geq n_0} h_n(\Xi(\epsilon_n))$$

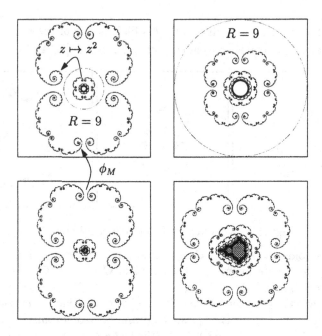

Figure 3: The model $\Xi(\epsilon)$.

where $\epsilon_n = \frac{\pi^2}{n^2}$, and $h_n(z) = \frac{\pi^2}{n^2} + \frac{C}{R^2 n^3}z$, the constant C being chosen small enough to ensure that the sets $h_n(\Xi(\epsilon_n))$ are contained in disjoint discs. Once again, up to a quasi homeomorphism of $\mathbf{C}$, the set Υ does not depend on the choice of n_0, R and C, provided they satisfy the requirements.

We shall prove

THEOREM 1.- *If c_0 is the root point of a primitive component of $\overset{\circ}{M}$, the set Υ appears quasi-conformally in M, with 0 (in Υ) corresponding to c_0 (in M).*

We shall first describe a more general situation in which the set Υ appears quasi-conformally in some set M_F defined in the parameter space. In section 11 we shall show by a renormalization argument that this situation occurs in that of Theorem 1.

4   A more general situation

Let Λ be an open set in $\mathbf{C}$ and $\lambda_0 \in \Lambda$, and let $\mathbf{f} = (f_\lambda : U'_\lambda \to U_\lambda)_{\lambda \in \Lambda}$ be an analytic family of quadratic-like maps, i.e. of polynomial-like maps of degree 2. We suppose that f_{λ_0} has a fixed point α_0 with derivative 1 ; this implies that f_{λ_0} is hybrid equivalent to $z^2 + \frac{1}{4}$, i.e. to $z + z^2$. We are interested only with values of λ which are close to λ_0.

Let $g_\lambda : V_\lambda \to U_\lambda$ be an analytic isomorphism, depending analytically on λ, with V_λ relatively compact in $U_\lambda \setminus \overline{U'_\lambda}$ (Fig.4). We suppose that the open sets U_λ, U'_λ, V_λ are Jordan domains with a C^1 boundary undergoing a holomorphic motion.

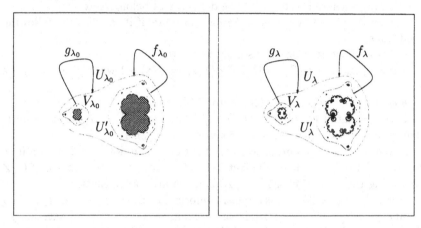

Figure 4: The more general situation.

For each $\lambda \in \Lambda$, the map f_λ has two fixed points α'_λ and α''_λ ; we may not be able to decide which is which, in a continuous way, but the function $\lambda \mapsto \delta(\lambda) = (\alpha'_\lambda - \alpha''_\lambda)^2$ is well defined and holomorphic on Λ . We suppose that the function δ vanishes at λ_0 with a non zero derivative.

We denote by ω_λ the critical point of f_λ, we set $a_n(\lambda) = f^n_\lambda(\omega_\lambda)$ and $b_\lambda = g^{-1}_\lambda(\omega_\lambda)$.

We denote by F_λ the map $U'_\lambda \cup V_\lambda \to U_\lambda$ which induces f_λ on U'_λ and g_λ on V_λ. We define $K(F_\lambda)$ as the set of points z such that $F^n_\lambda(z)$ is defined and belongs to $U'_\lambda \cup V_\lambda$ for all n. It is a full compact set in $\mathbf{C}$.

We denote by M_F the set of values of λ for which $\omega_\lambda \in K(F_\lambda)$. It is a closed set in Λ. Note that $\lambda \in M_F$ does not imply that $K(F_\lambda)$ is connected; in fact it is never connected, since $U'_\lambda \cap K(F_\lambda)$ and $V_\lambda \cap K(F_\lambda)$ are disjoint, not empty as they contain $K(f_\lambda)$ and $g^{-1}_\lambda(K(f_\lambda))$ respectively, and cover $K(F_\lambda)$.

Theorem 1 will be a consequence of the more general:

THEOREM 2.- *In this situation, the set Υ appears quasi-conformally in M_F.*

We can normalize the situation in the following way: we suppose $\lambda_0 = 0$, $\alpha_0 = 0$, and
$$f_\lambda(z) = z + z^2 + \lambda + O(z^3) \ .$$

Indeed, under the hypotheses made above, one can reduce $f_\lambda(z)$ to this form by a change of variables of the form $(\lambda, z) \mapsto (u(\lambda), A(\lambda).z + B(\lambda))$, with $u(\lambda_0) = 0$, $u'(\lambda_0) \neq 0$.

5   Fatou coordinates

Let $f : U' \to U$ be a holomorphic map and Ω a connected open set in $U \cup U'$. We denote by Ω/f the quotient of Ω by the equivalence relation identifying z with $f(z)$ whenever z and $f(z)$ are defined and belong to Ω.

A map $\phi : \Omega \to \mathbf{C}$ is called a *Fatou coordinate* if it satisfies the following conditions:

- ϕ is an analytic isomorphism of Ω onto its image;
- ϕ conjugates f to T_1 on Ω, i.e. $\phi(f(z)) = \phi(z) + 1$ when z and $f(z)$ are in Ω ;
- ϕ induces a bijection of Ω/f onto $\mathbf{C}/\mathbf{Z}$.

If ϕ is a Fatou coordinate on Ω, Ω/f has naturally a quotient analytic structure and this bijection is an analytic isomorphism.

Given Ω, a Fatou coordinate on Ω, if it exists, is unique up to an additive constant. This follows from the fact that a $\mathbf{C}$-analytic automorphism of $\mathbf{C}/\mathbf{Z}$ preserving the generator of $H^1(\mathbf{C}/\mathbf{Z})$ is necessarily a translation.

If $\Psi : U_1' \cup U_1 \to U_2' \cup U_2$ is a quasi-conformal conjugacy between $f_1 : U_1' \to U_1$ and $f_2 : U_2' \to U_2$, and Ω_1 is the domain of a Fatou coordinate for f_1, then $\Omega_2 = \Psi(\Omega_1)$ is the domain of a Fatou coordinate for f_2. Indeed Ω_2/f_2 is quasi-conformally homeomorphic to $\mathbf{C}/\mathbf{Z}$, thus analytically isomorphic to it.

In view of the straightening theorem for polynomial-like mappings and analytic families of polynomial-like mappings, many results that are known for the family of quadratic polynomials $P_\epsilon : z \mapsto z + z^2 + \epsilon$ (i.e. $z \mapsto z^2 + \frac{1}{4} + \epsilon$ up to a change of origin) can be transported to our situation.

For $P_0 : z \mapsto z + z^2$, one can define two Fatou coordinates $\Phi_{P_0}^+ : \Omega_{P_0}^+ \to \mathbf{C}$ and $\Phi_{P_0}^- : \Omega_{P_0}^- \to \mathbf{C}$ such that $\Omega_{P_0}^+$ (resp. $\Omega_{P_0}^-$) contains the disc $D_r^+ = \{z \mid |z - r| < r\}$ (resp. $D_r^- = \{z \mid |z + r| < r\}$) at right (resp. at left) of 0 for $r > 0$ small enough (in fact one can take $r = \frac{1}{4}$). A maximal domain for $\Phi_{P_0}^-$ is $B' \cup B'' \cup \,]\omega_{P_0}, 0[$, where B' and B'' are the two main squares of the parabolic chessboard, i.e. the connected components of $\overset{\circ}{K}(P_0) \smallsetminus \bigcup P_0^{-n}(\mathbf{R})$ which intersect D_r^- for $r > 0$ small, and $\omega_{P_0} = -\frac{1}{2}$ is the critical point of P_0. As a holomorphic function, $\Phi_{P_0}^-$ extends to all of $\overset{\circ}{K}(P_0)$, but is no longer injective there.

A maximal domain for $\Phi_{P_0}^+$ is $\mathbf{C} \smallsetminus \mathbf{R}_-$. Indeed there is a branch h of P_0^{-1} defined on $\mathbf{C} \smallsetminus \mathbf{R}_-$ and which maps it into itself; for any z in $\mathbf{C} \smallsetminus \mathbf{R}_-$ the point $h^n(z)$ tends to 0 tangentially to $\mathbf{R}_+$ when $n \to \infty$, so one can extend $\Phi_{P_0}^+$ from D_r^+ to $\mathbf{C} \smallsetminus \mathbf{R}_-$ by $\Phi_{P_0}^+(z) = \Phi_{P_0}^+(h^n(z)) + n$ for n big enough.

In our situation, the map f_0 is quasi-conformally conjugate to P_0. We can choose a Fatou coordinate ϕ_0^- whose domain Ω_0^- contains D_r^- (for $r > 0$ small enough) and the points $a_n = f_0^n(\omega_0)$ for $n \geq 1$.

PROPOSITION 1.- *There is a Fatou coordinate ϕ_0^+ for f_0 whose domain Ω_0^+*

contains $\overline{V}_0$.

Proof : Remind the way one constructs a quasi-conformal conjugacy between f_0 and P_0 in [DH Pol-like]. By the Caratheodory theorem, the map f_0 extends to a continuous map $\overline{U_0'} \to \overline{U_0}$ which is a local diffeomorphism except at ω_0. First we choose a $R > 1$, and C^1 diffeomorphisms $\eta : \partial U_0 \to S_R^1 = \{z \mid |z| = R\}$ and $\eta' : \partial U_0' \to S_{R'}^1$ where $R' = \sqrt{R}$, so that $\eta(f_0(z)) = (\eta'(z))^2$ for $z \in \partial U_0'$. Second we fill (η, η') into an arbitrary C^1 diffeomorphism ψ between $\overline{U}_0 \setminus U_0'$ and the closed annulus $\overline{A}_{R,R'}$. At that moment one can arrange so that $\psi^{-1}([-R, -R']) \cap \overline{V}_0 = \emptyset$. Later $]-\infty, -1[$ becomes the external ray of the cauliflower of argument $\frac{1}{2}$, i.e. itself. So $\overline{V}_0$ is contained in $\Psi^{-1}(\mathbf{C} \setminus \mathbf{R}_-)$ where Ψ is the quasi-conformal conjugacy between f_0 and P_0 constructed using ψ. ∎

We choose $r_0 > 0$ small and $\theta_0 \in]0, \pi[$, and we define the sector S as the set of values of λ such that $|\lambda| \le r_0$ and $|Arg(\lambda)| \le \theta_0$. We set $S^* = S \setminus \{0\}$. If r_0 has been chosen small enough we have $S \subset \Lambda$, and for $\lambda \in S$ there exist ([D/AMS],[DSZ]) Fatou coordinates $\phi_\lambda^- : \Omega_\lambda^- \to \mathbf{C}$ and $\phi_\lambda^+ : \Omega_\lambda^+ \to \mathbf{C}$ satisfying the following conditions:

- Ω_0^+ (resp. Ω_0^-) contains a disc D_r^+ (resp. D_r^-) ;
- $\overline{V}_0 \subset \Omega_0^+$;
- $a_n(0) = f_0^n(\omega_0) \in \Omega_0^-$ for $n \ge 1$;
- the fixed points α_λ', α_λ'' of f_λ belong to the boundaries $\partial \Omega_\lambda^+$, $\partial \Omega_\lambda^-$;
- the set $\Omega^\pm = \{(\lambda, z) \mid \lambda \in S \ z \in \Omega_\lambda^\pm\}$ is open in $S \times \mathbf{C}$;
- for $\lambda \in S^*$ the intersection $\Omega_\lambda^+ \cap \Omega_\lambda^-$ is the domain of a Fatou coordinate;
- (normalization) $\phi_\lambda^-(a_n(\lambda)) = n$, $\phi_\lambda^+(b_\lambda) = 0$.

Even though being the domain of a Fatou coordinate is a pretty restrictive condition for an open set, the conditions above do not determine the domains $\Omega_\lambda^\pm$. We can arrange so that $\Omega_\lambda^\pm$ is relatively compact in U_λ, bounded by a C^1 Jordan curve which depends continuously on λ. We could on the opposite arrange so that $\Omega_\lambda^+ = \Omega_\lambda^-$ for $\lambda \in S^*$.

Restricting S if necessary, we have $\overline{V}_\lambda \subset \Omega_\lambda^+$ for all $\lambda \in S$.

6 The phase

With the conditions above, the function $(\lambda, z) \mapsto \phi_\lambda^\pm(z)$ is continuous on $\Omega^\pm$, and $\mathbf{C}$-analytic on $\Omega^\pm \cap (S^* \times \mathbf{C})$ ([Sh],appendix).

For $\lambda \in S^*$, the functions ϕ_λ^+ and ϕ_λ^- both induce a Fatou coordinate on $\Omega_\lambda^+ \cap \Omega_\lambda^-$, so $\phi_\lambda^+ - \phi_\lambda^-$ is a constant with a value $\tau(\lambda) \in \mathbf{C}$. This defines a function $\tau : S^* \to \mathbf{C}$ called the *lifted phase*, which is holomorphic and tends to ∞ when λ tends to 0 ; we have [D/AMS], [Sh], [DSZ] :

$$\tau(\lambda) = \frac{-\pi}{\sqrt{\lambda}} + O(1) \quad .$$

Reducing slightly θ_0 and r_0 if necessary, the map τ induces a **C**-analytic isomorphism $\lambda \mapsto \tilde{\lambda}$ of S^* onto a closed set $\tilde{S}$ which contains a set of the form $\{Z \mid |Z| \geq R_0, \ |Arg(-Z)| \leq \theta_0'\}$.

We have $\tilde{\lambda} \sim \frac{-\pi}{\sqrt{\lambda}}$; $\lambda \sim \frac{\pi^2}{\tilde{\lambda}^2}$; $\frac{d\lambda}{d\tilde{\lambda}} \sim \frac{-2\pi^2}{\tilde{\lambda}^3}$ when λ tends to 0, or equivalently when $\tilde{\lambda} = \tau(\lambda)$ tends to ∞ in $\tilde{S}$.

Extending ϕ_0^- to a function on $\overset{\circ}{K}(f_0)$ by $\phi_0^-(z) = \phi_0^-(f_0^n(z)) - n$, we define the *Lavaurs map* L_σ for $\sigma \in \mathbf{C}$ ([La], [D/AMS], [DSZ]) by

$$L_\sigma = (\phi_0^+)^{-1} \circ T_\sigma \circ \phi_0^- \ ;$$

the domain of definition of L_σ being $(\phi_0^-)^{-1}(T_{-\sigma}(\phi_0^+(\Omega_0^+)))$. On each compact set of this domain, f_λ^n tends to L_σ uniformly when λ tends to 0 and n tends to ∞ in such a way that $\tau(\lambda) + n$ tends to σ.

7   A sequence of Mandelbrot-like families

We set $\tilde{V}_\lambda = \phi_\lambda^+(V_\lambda)$ for $\lambda \in S$, and $\tilde{W}_n = \{\tilde{\lambda} \in \tilde{S} \mid \tilde{\lambda} + n \in \tilde{V}_\lambda\}$ where $\lambda = \tau^{-1}(\tilde{\lambda})$. We set $W_n = \tau^{-1}(\tilde{W}_n)$. The sets W_n are disjoint. Indeed, for $\lambda \in W_n$ we have $a_n(\lambda) \in V_\lambda$. More precisely W_n is the set of values of λ such that $a_i(\lambda) \in \Omega_\lambda^- \cup \Omega_\lambda^+$ for $1 \leq i \leq n$ and $a_n(\lambda) \in V_\lambda$.

We choose *centers*, i. e. base points in all these sets: ω_λ for U_λ, b_λ for V_λ, $0 = \phi_\lambda^+(b_\lambda)$ for $\tilde{V}_\lambda$, $\tilde{\lambda}_n = -n$ for $\tilde{W}_n$, and $\lambda_n = \tau^{-1}(\tilde{\lambda}_n)$ for W_n, so that $f_{\lambda_n}^n(\omega_{\lambda_n}) = b_{\lambda_n}$.

When n tends to infinity, the open set $T_n(\tilde{W}_n)$ tends to $\tilde{V}_0$, with convergence of the boundaries. In particular $diam(\tilde{W}_n)$ tends to $\delta_0 = diam(\tilde{V}_0)$. It follows that

$$\lambda_n \sim \frac{\pi^2}{n^2} \quad \text{and} \quad diam(W_n) \sim \frac{2\pi^2 \delta_0}{n^3} \ .$$

For $\lambda \in W_n$, we denote by V_λ' the connected component of $f_\lambda^{-n}(V_\lambda)$ which contains ω_λ, and we set $G_\lambda = g_\lambda \circ f_\lambda^n : V_\lambda' \to U_\lambda$ (note that n is determined by λ since the sets W_n are disjoint).

For $\sigma \in T_n(\tilde{W}_n)$, we set $\tilde{G}_{\sigma,n} = G_\lambda$ with $\lambda = \tau^{-1}(\sigma - n)$. For $\sigma \in \tilde{V}_0$, we denote by $V_{0,\sigma}'$ the connected component of $L_\sigma^{-1}(V_0)$ which contains ω_0, and define $\tilde{G}_\sigma : V_{0,\sigma}' \to U_0$ by $\tilde{G}_\sigma = g_0 \circ L_\sigma$.

PROPOSITION 2.- *a) Take n big enough. Then for $\lambda \in W_n$, the map $G_\lambda : V_\lambda' \to U_\lambda$ is quadratic-like, and the family $(G_\lambda)_{\lambda \in W_n}$ is Mandelbrot-like.*

b) For $\sigma \in \tilde{V}_0$, the map $\tilde{G}_\sigma : V_{0,\sigma}' \to U_0$ is quadratic-like, and the family $(\tilde{G}_\sigma)_{\sigma \in \tilde{V}_0}$ is Mandelbrot-like.

c) When n tends to infinity, the family $(\tilde{G}_{\sigma,n})_{\sigma \in T_n(\tilde{W}_n)}$ tends to $(\tilde{G}_\sigma)_{\sigma \in \tilde{V}_0}$, uniformly on each compact in its domain of definition in $T_n(\tilde{W}_n) \times \mathbf{C}$.

Proof : a,1) : For $i \leq n$, denote by $V'_{\lambda,i}$ the component of $f_\lambda^{-1}(V_\lambda)$ which contains a_i. If $1 \leq i \leq n-1$, the map $f_\lambda : V'_{\lambda,i} \to V'_{\lambda,i+1}$ is holomorphic and proper with no critical point, thus is an isomorphism. The map $f_\lambda : V'_\lambda \to V'_{\lambda,1}$ is a ramified covering of degree 2. It follows that $f_\lambda^n : V'_\lambda \to V_\lambda$ is a ramified covering of degree 2, and $G_\lambda = g_\lambda \circ f_\lambda^n : V'_\lambda \to U_\lambda$ is quadratic-like.

a,2) : For n big enough, when λ ranges on a curve which follows closely the boundary ∂W_n, the point $\tilde{\lambda}$ follows $\partial \tilde{W}_n$, the point $\tilde{a}_n(\lambda) = \tilde{\lambda} + n$ ranges on a curve close to $\partial \tilde{V}_0$, thus $a_n(\lambda)$ on a curve close to ∂V_0 , so finally $G_\lambda(\omega_\lambda) = g_\lambda(a_n(\lambda))$ ranges on a curve in $U_0 \setminus \overline{U'_0}$, which makes 1 turn around 0.

b,1) : Define $V'_{0,\sigma,1}$ as the connected component of $L_{\sigma-1}^{-1}(V_0)$ which contains $a_1(0) = f_0(\omega_0)$. Then $L_{\sigma-1} : V'_{0,\sigma,1} \to V_0$ is an isomorphism, while $f_0 : V'_{0,\sigma} \to V'_{0,\sigma,1}$ is a ramified covering of degree 2. So $\tilde{G}_\sigma = g_0 \circ L_\sigma : V'_{0,\sigma} \to U_0$ is quadratic-like.

b,2) : When σ follows $\partial \tilde{V}_0$, the point $L_\sigma(\omega_0) = (\phi_0^+)^{-1}(\sigma)$ follows ∂V_0, and $\tilde{G}_\sigma(\omega_0)$ ranges on a curve in $U_0 \setminus \overline{U'_0}$, making 1 turn around 0.

c) : This follows from the fact that f_λ^n, where $\lambda = \tau^{-1}(\sigma - n)$, tends to L_σ when n tends to ∞. ∎

8 About Mandelbrot-like families

In this section, we consider a Mandelbrot-like family $\mathbf{h} = (h_\lambda)_{\lambda \in W}$, provided with a horizontally analytic tubing Θ ([DH Pol-like]). More precisely we assume:

- W is a Jordan domain in $\mathbf{C}$, with C^1 boundary.
- For each $\lambda \in W$, the map Θ_λ is a quasi-conformal embedding of $\overline{A}_{R^2,R}$ in $\mathbf{C}$; for each $Z \in \overline{A}_{R^2,R}$ the point $\Theta_\lambda(Z)$ depends holomorphically on λ.
  We set $C_\lambda = \Theta_\lambda(S^1_{R^2})$ and $C'_\lambda = \Theta_\lambda(S^1_R)$; we denote by U_λ (resp U'_λ) the Jordan domain bounded by C_λ (resp. C'_λ).
- h_λ is a quadratic-like mapping $U'_\lambda \to U_\lambda$; the map $\mathbf{h} : (\lambda, z) \mapsto (\lambda, h_\lambda(z))$ is $\mathbf{C}$-analytic and proper $\mathcal{U}' \to \mathcal{U}$, where $\mathcal{U}$ is the open set $\{(\lambda, z) \mid \lambda \in W$ and $z \in U_\lambda\}$ and $\mathcal{U}'$ is defined similarly ; we denote by ω_λ the critical point of h_λ.
- We have $\Theta_\lambda(Z^2) = h_\lambda(\Theta_\lambda(Z))$ for $Z \in S^1_R$.
- The map $\mathbf{h}$ extends continuously to a map $\overline{\mathcal{U}'} \to \overline{\mathcal{U}}$, and $\Theta : (\lambda, Z) \mapsto (\lambda, \Theta_\lambda(Z))$ extends continuously to a map $\overline{W} \times \overline{A}_{R^2,R} \to \overline{\mathcal{U}}$, such that Θ_λ is injective on $A_{R^2,R}$ for $\lambda \in \partial W$.
  The map $\lambda \mapsto \omega_\lambda$ extends continuously to $\overline{W}$.
- For $\lambda \in \partial W$, we have $h_\lambda(\omega_\lambda) \in C_\lambda$.
- When λ ranges over ∂W making 1 turn, the vector $h_\lambda(\omega_\lambda) - \omega_\lambda$ makes 1 turn around 0.

We denote by $M_{\mathbf{h}}$ the connectedness locus of the family $\mathbf{h}$:

$$M_{\mathbf{h}} = \{\lambda \in W \mid K(h_\lambda) \text{ connected}\} = \{\lambda \mid \omega_\lambda \in K(h_\lambda)\} \quad .$$

It is proved in [DH Pol-like] that $M_{\mathbf{h}}$ is homeomorphic to the Mandelbrot set M. More precisely, for any $\lambda \in M_{\mathbf{h}}$, there is a unique $c = \chi(\lambda) \in M$ such that h_λ is hybrid-equivalent to $z \mapsto z^2 + c$, and the map χ so defined is a homeomorphism of $M_{\mathbf{h}}$ onto M.

Using the tubing Θ, we can extend χ to a homeomorphism χ_Θ of W onto $W_M = \{c \in \mathbf{C} \mid \mathcal{G}_M(c) < 2Log(R)\}$, where $\mathcal{G}_M$ is the potential function defined by M.

The map χ_Θ is defined in the following way:

We first define $\mathcal{G}_\Theta$ on $\mathcal{U}$. For $\lambda \in W$ and $z \in U_\lambda \smallsetminus K(h_\lambda)$, there is a k such that $z_k = h_\lambda^k(z) \in \overline{U}_\lambda \smallsetminus U'_\lambda = \Theta_\lambda(\overline{A}_{R^2,R})$, then we set $\mathcal{G}_\lambda(z) = \frac{1}{2^k}Log|\Theta_\lambda^{-1}(z_k)|$. If there are two values for k, they give the same value for $\mathcal{G}_\lambda(z)$. We set $\mathcal{G}_\lambda(z) = 0$ for $z \in K(h_\lambda)$.

We denote by H_λ the set of points z such that $\mathcal{G}_\lambda(z) \leq \mathcal{G}_\lambda(\omega_\lambda)$: if $\lambda \in M_{\mathbf{h}}$, we have $H_\lambda = K(h_\lambda)$, and if $\lambda \in W \smallsetminus M_{\mathbf{h}}$ the set H_λ is limited by an 8-shaped curve.

We define $\phi_\lambda : U_\lambda - H_\lambda \to A_{R^2,\rho_\lambda}$, where $Log(\rho_\lambda) = \mathcal{G}_\lambda(\omega_\lambda)$, by the conditions :

- $\phi_\lambda = \Theta_\lambda^{-1}$ on $U_\lambda \smallsetminus U'_\lambda$;
- $\phi_\lambda(h_\lambda(z)) = (\phi_\lambda(z))^2$ for $z \in U'_\lambda \smallsetminus H_\lambda$;
- ϕ_λ is continuous on $U_\lambda \smallsetminus H_\lambda$.

For $\lambda \in W \smallsetminus M_{\mathbf{h}}$, we have $h_\lambda(\omega_\lambda) \in U_\lambda \smallsetminus H_\lambda$, and $\chi_\Theta(\lambda)$ is defined by

$$\chi_\Theta(\lambda) = \phi_M^{-1}(\phi_\lambda(h_\lambda(\omega_\lambda))) \quad .$$

For $\lambda \in M_{\mathbf{h}}$, we set $\chi_\Theta(\lambda) = \chi(\lambda)$.

It is proved in [DH Pol-like, IV-4] that the map χ_Θ defined this way is a homeomorphism of W onto an open set in $\mathbf{C}$, which is easily seen to be W_M under the hypotheses made here, and that χ_Θ is quasi-conformal on $W' \smallsetminus M_{\mathbf{h}}$ for any W' relatively compact in W (Prop. 20).

Lyubich has proved ([Ly]) that there exists a quasi-conformal mapping $\tilde{\chi}$ of a neighborhood W^* of $M_{\mathbf{h}}$ in W onto a neighborhood W_M^* of M in $\mathbf{C}$ extending χ. It then follows from Rickmann's lemma ([DH Pol-like I-5, Lemma 2],[Ri]) that χ_Θ is quasi-conformal on W' for any W' relatively compact in W, as conjectured in [DH Pol-like].

Let now Γ be an arbitrary compact set in $A_{R^2,R}$, and denote by Γ_m the preimage of Γ by $Z \mapsto Z^{2^m}$.

PROPOSITION 3.- *We have*

$$\chi_\Theta^{-1}(\phi_M^{-1}(\Gamma_m)) = \{\lambda \in W \mid h_\lambda^{m+1}(\omega_\lambda) \in \Theta_\lambda(\Gamma)\} \quad .$$

Proof : Setting $c = \chi_\Theta(\lambda)$ and $Q_c(z) = z^2 + c$, the following conditions are equivalent: $\phi_M(c) = \phi_c(c) \in \Gamma_m$; $\phi_c(Q_c^{m+1}(0)) \in \Gamma$; $h_\lambda^{m+1}(\omega_\lambda) \in \Theta_\lambda(\Gamma)$. ∎

COROLLARY.- *We have*

$$M_h \cup \{\lambda \mid (\exists k) \ h_\lambda^k(\omega_\lambda) \in \Theta_\lambda(\Gamma)\} = \chi_\Theta^{-1}\left(M \cup \bigcup_m \phi_M^{-1}(\Gamma_m)\right).$$

9 Appearance of $\Xi(\epsilon_n)$

Let us come back now to the situation of sections 4 to 7. We shall apply the above Corollary to the family $\mathbf{G}_n = (G_\lambda : V'_\lambda \to U_\lambda)_{\lambda \in W_n}$, or equivalently to the family $(\tilde{G}_{\sigma,n})_{\sigma \in \tilde{V}_0}$.

We fix $R > 1$. For each n, we shall choose a tubing $\Theta_n = (\Theta_\lambda)_{\lambda \in W_n}$ for $\mathbf{G}_n$, with $\Theta_\lambda : \overline{A}_{R^2,R} \to \overline{U}_\lambda \setminus V'_\lambda$ quasi-conformal.

We first take a tubing Θ_f for $\mathbf{f}$ defined on $\Lambda' \times \overline{A}_{R^2,R_f}$ with Λ' a neighborhood of 0 in Λ. Then the straightening theorem provides a homeomorphism $\psi_\lambda : J(f_\lambda) \to J_{\epsilon(\lambda)}$ induced by a map which is K_f-quasi-conformal with a K_f independent of λ for λ in a neighborhood Λ'' of 0. In particular for the center λ_n of W_n defined in section 7, we get $\psi_{\lambda_n} : J(f_{\lambda_n}) \to J_{\epsilon_n}$ with some ϵ_n. This ϵ_n is determined by $\tau_P(\epsilon_n) = -n$, where τ_P is the lifted phase for the family of polynomials (P_ϵ), taking $\psi_\lambda(b_\lambda)$ as a base point on the right side. It follows that $-\frac{\pi}{\sqrt{\epsilon_n}} \sim -n$ and $\epsilon_n \sim \frac{\pi^2}{n^2}$.

We denote by $\Gamma_0(\epsilon_n)$ the set J_{ϵ_n} enlarged by the factor $R^{\frac{3}{2}}$ as in section 3, and by $\Gamma_m(\epsilon_n)$ the preimage of $\Gamma_0(\epsilon_n)$ by $Z \mapsto Z^{2^m}$, agreeing with the notations of section 3 and of Prop. 3.

We can then choose for each n big enough a homeomorphism $\Theta_n^0 : \overline{A}_{R^2,R} \to \overline{U}_{\lambda_n} \setminus V'_{\lambda_n}$ so that

- Θ_n^0 is K-quasi-conformal with a K independent of n ;
- $\Theta_n^0(Z^2) = f_{\lambda_n}(\Theta_n^0(Z))$ for $|Z| = R$;
- $\Theta_n^0(Z) = \psi_{\lambda_n}^{-1}(R^{-\frac{3}{2}}.Z)$ for $Z \in \Gamma_0(\epsilon_n)$.

When λ ranges in W_n, the set $J(f_\lambda)$ remains a Cantor set contained in $U'_\lambda \setminus \overline{V'_\lambda}$ and undergoes a holomorphic motion. The sets ∂U_λ and $\partial V'_\lambda$ also undergo a holomorphic motion. By Słodkowski's theorem, one can find a holomorphic motion (ι_λ) of $\overline{U}_{\lambda_n} \setminus V'_{\lambda_n}$ which induces the above mentioned ones on ∂U_{λ_n}, ∂V_{λ_n} and $J(f_{\lambda_n})$. Setting $\Theta_\lambda = \iota_\lambda \circ \Theta_n^0$, we obtain a horizontally analytic tubing Θ_n for $\mathbf{G}_n$ such that $\Theta_\lambda(\Gamma_0(\epsilon_n)) = J(f_\lambda)$ for each λ in W_n.

PROPOSITION 4.- *The set $\Xi(\epsilon_n)$ appears quasi-conformally in $M_F \cap W_n$.*

Proof : By [DH Pol-like] completed by [Ly], the map χ_{Θ_n} is a quasi-conformal homeomorphism of W_n onto W_M. Taking $\Gamma = \Gamma_0(\epsilon_n)$, the set $M \cup \bigcup_m \phi_M^{-1}(\Gamma_m)$ is precisely the one denoted by $\Xi(\epsilon_n)$ in section 2, so the Corollary of Prop. 3 tells that the image of $\Xi(\epsilon_n)$ by $\chi_{\Theta_n}^{-1}$ is the set

$$X_n = M_{\mathbf{G}_n} \cup \{\lambda \in W_n \mid (\exists k)\ G_\lambda^k(\omega_\lambda) \in J(f_\lambda)\}\quad .$$

It is clear that $X_n \subset M_F \cap W_n$.

The points $\xi \in \Xi(\epsilon_n)$ which are either Misiurewicz points in M or parameter values for which $R^{-\frac{3}{2}}.(\phi_M(\xi))^{2^m}$ is a (repelling) periodic point in J_{ϵ_n} form a dense set in the boundary $\partial \Xi(\epsilon_n)$. For such a point ξ, the point $\lambda = \chi_{\Theta_n}^{-1}(\xi)$ is a parameter value such that, for some k, the point $G_\lambda^k(\omega_\lambda)$ is a repelling point either under G_λ or under f_λ ; so $\xi \in \partial M_F$. Therefore $\chi_{\Theta_n}^{-1}(\partial \Xi(\epsilon_n)) \subset \partial M_F$. ∎

10　Proof of Theorem 2

We have seen that $\chi_n = \chi_{\Theta_n} : W_n \to W_M$ is a quasi-conformal map which induces a homeomorphism $X_n \to \Xi(\epsilon_n)$ with $\epsilon_n \sim \frac{\pi^2}{n^2}$.

Choose $C > 0$ such that $C.\overline{W}_M$ is contained in the open disc $D_{\frac{1}{2}}$, and define $\tilde{\chi}_n : \tilde{W}_n \to T_{-n}(C.W_M)$ by $\tilde{\chi}_n = \tilde{h}_n \circ \chi_n \circ \tau^{-1}$, where $\tilde{h}_n(Z) = C.Z - n$, and $\tilde{\chi} : \bigcup_{n \geq n_0} \tilde{W}_n \to \bigcup_{n \geq n_0} T_{-n}(C.W_M)$ by $\tilde{\chi}_{|\tilde{W}_n} = \tilde{\chi}_n$.

Set $\tau_0(\epsilon) = -\frac{\pi}{\sqrt{\epsilon}}$ for $\epsilon \in S^*$, and $\eta = \tau_0 \circ \tau^{-1} : \tilde{S} \to \tilde{S}_0 = \tau_0(S^*) = \{Z \mid |Z| \geq \frac{\pi}{\sqrt{r_0}}\ |Arg(-Z)| \leq \frac{\theta_0}{2}\}$. Remind that $\eta(Z)/Z$ tends to 1 when Z tends to ∞ in $\tilde{S}$; reducing θ_0 if necessary we get that $\eta'(Z)$ also tends to 1.

LEMMA.- *The map $\tilde{\chi}$ can be extended to a quasi-conformal homeomorphism $\tilde{\psi} : \tilde{S} \to \tilde{S}_0$ which induces η on $\partial \tilde{S}$.*

Proof : The map $T_n \circ \tilde{\chi}_n \circ T_{-n} : T_n(\tilde{W}_n) \to C.W_M$ has a limit $\tilde{\chi}_\infty : \tilde{V}_0 \to C.W_n$ when n tends to infinity, obtained by applying the straightening theorem for Mandelbrot-like families to the family $(\tilde{G}_\sigma)_{\sigma \in \tilde{V}_0}$, and multiplying by C.

The map $\tilde{\chi}_\infty$ extends continuously to closures, inducing on the boundaries the C^1-diffeomorphism

$$C.\phi_M^{-1} \circ \Theta_0^{-1} \circ g_0 \circ \tau^{-1} : \partial \tilde{V}_0 \to \partial V_0 \to \partial U_0 \to S_{R^2}^1 \to \partial(C.W_M)\quad .$$

We denote by γ_∞ the inverse diffeomorphism $\partial(C.W_M) \to \partial\tilde{V}_0$.

For each n, the map $T_n \circ \chi_n^{-1} \circ T_{-n}$ also extends continuously to closures, inducing on the boundaries a diffeomorphism $\gamma_n : \partial(C.W_M) \to \partial(T_n(\tilde{W}_n))$, and γ_n tends to γ_∞ in the C^1 topology on $\partial(C.W_M)$ when $n \to \infty$.

Let $\mathcal{L}^0$ be the vertical line defined by $Re(Z) = -\frac{1}{2}$, which separates $C.\overline{W}_M$ from $T_{-1}(C.\overline{W}_M)$, and let $\mathcal{B}^0$ be the vertical strip bounded by $\mathcal{L}^0$ and $T_1(\mathcal{L}^0)$, so that $\mathcal{B}^0 \supset C.\overline{W}_M$.

Let $\mathcal{L}$ be a C^1 curve which coincides with $\mathcal{L}^0$ outside of a compact set, and separates the closures of $\tilde{V}_0$ and $T_{-1}(\tilde{V}_0)$, and let $\mathcal{B}$ be the domain limited by $\mathcal{L}$ and $T_1(\mathcal{L})$, so that $\mathcal{B}$ contains the closure of $\tilde{V}_0$. Let β be a diffeomorphism $\mathcal{L}^0 \to \mathcal{L}$ which coincides with the identity outside a compact set.

One can find a diffeomorphism $v_\infty : \overline{\mathcal{B}}_0 \smallsetminus C.W_M \to \overline{\mathcal{B}} \smallsetminus \tilde{V}_0$ which coincides with γ_∞ on $\partial(C.W_M)$, with β and $T_1 \circ \beta \circ T_{-1}$ on $\partial\mathcal{B}_0$, and with the identity outside a compact set. This map v_∞ is quasi-conformal.

There is a n_0 such that, for $n \geq n_0$, the closure of $T_n(\tilde{W}_n)$ is contained in $\mathcal{B}$, and one can find a C^1-diffeomorphism $v_n : \overline{\mathcal{B}}_0 \smallsetminus C.W_M \to \overline{\mathcal{B}} \smallsetminus T_n(\tilde{W}_n)$ which coincides with γ_n on $\partial(C.W_M)$, with β and $T_1 \circ \beta \circ T_{-1}$ on $\partial\mathcal{B}_0$, and with the identity outside a compact set independent of n. These maps v_n can be chosen so that they are K-quasi-conformal with a K independent of n.

Then the maps $(T_{-n} \circ v_n \circ T_n)_{n \geq n_0}$ and $(\tilde{\chi}_n^{-1})_{n \geq n_0}$ match up to a quasi-conformal homeomorphism $v : \mathcal{H}^0 \to \mathcal{H}$, where $\mathcal{H}^0$ and $\mathcal{H}$ are the regions on the left of $T_{-n_0+1}(\mathcal{L}^0)$ and $T_{-n_0+1}(\mathcal{L})$ respectively. This map coincides with the identity outside a horizontal strip.

One can easily extend v to a quasi-conformal homeomorphism $\hat{v} : \mathbf{C} \to \mathbf{C}$ inducing $\tilde{\chi}_n^{-1}$ also for $n < n_0$. We can then modify $\hat{v}$ on a neighborhood of $\partial\tilde{S}_0$, keeping it quasi-conformal, so that it agrees with η^{-1} on $\partial\tilde{S}_0$. Then the inverse map $\tilde{\psi}$ satisfies the required properties:

- $\tilde{\psi}$ induces a quasi-conformal homeomorphism $\tilde{S} \to \tilde{S}_0$;
- $\tilde{\psi}_{|\tilde{W}_n} = \tilde{\chi}_n$ for each n ;
- $\tilde{\psi}$ coincides with η on $\partial\tilde{S}$. ∎

Now $\tau_0 \circ \tilde{\psi} \circ \tau$ is a quasi-conformal homeomorphism of S^* onto itself which coincides with the identity on the boundary, we extend it to a quasi-conformal homeomorphism $\psi : \mathbf{C} \to \mathbf{C}$ by taking the identity on $\mathbf{C} \smallsetminus S^*$.

The map ψ^{-1} induces a homeomorphism of

$$\Upsilon' = \{0\} \cup \tau^{-1}\left(\bigcup_{n \geq n_0} T_{-n}(\Xi(\epsilon_n))\right)$$

onto $\{0\} \cup \bigcup_{n \geq n_0} X_n$, so Υ' appears quasi-conformally in M_F.

The set Υ defined in section 3 differs from Υ' by two slight modifications:

- the sequence (ϵ_n) which is equivalent to $\frac{\pi^2}{n^2}$ is replaced by $\frac{\pi^2}{n^2}$ exactly ;
- the map τ_0^{-1} is replaced on $T_{-n}(W_M)$ by its affine tangent map at $-n$.

These do not affect the quasi-conformal type, in other words there is a quasi-conformal homeomorphism of $\mathbf{C}$ onto itself which maps Υ onto Υ'.

This finishes the proof of Theorem 2.

11   Proof of Theorem 1

Let c_0 be the cusp point of a primitive hyperbolic component of the interior of the Mandelbrot set M, and denote by Q_λ the map $z \mapsto z^2 + c_0 + \lambda$. Let Δ be the component of $\overset{\circ}{K}_{c_0}$ which contains 0, and k be the period of Δ, so that Q_0^k induces a proper map $\Delta \to \Delta$ of degree 2. We suppose $c_0 \neq \frac{1}{4}$. There is a parabolic point $\alpha_0 \in \partial\Delta$ of period k, with $(Q_0^k)'(\alpha_0) = 1$.

One can find a sequence $(b_n)_{n \in \mathbf{Z}}$ such that :

(BO1) : $b_{n+1} = Q_0(b_n)$ for all n ;

(BO2) : $b_N = 0$ for some N ;

(BO3) : $b_n \notin \Delta$ for $n < N$;

(BO4) : $b_{-mk} \to \alpha_0$ when $m \to \infty$.

Indeed, there is a Fatou coordinate ϕ_{Q_0} for Q_0^k defined on a domain $\Omega_{Q_0}^+$ with

- $\alpha_0 \in \partial\Omega_{Q_0}^+$;
- $D(\alpha_0, r) \smallsetminus \Delta \subset \Omega_{Q_0}^+$ for some $r > 0$;
- $\phi_{Q_0}^+(\Omega_{Q_0}^+) = \mathbf{C}^- = \{Z \mid Re(Z) < 0\}$.

Let b' be the preimage of 0 under Q_0 which is not in $Q_0^{k-1}(\Delta)$. Since $\alpha_0 \in J(Q_0)$, there is a $b \in D(\alpha_0, r)$ and an N such that $Q_0^{N-1}(b) = b'$, thus $Q_0^N(b) = 0$. Then $b \in \Omega_{Q_0}^+$, and we can define $b_{-mk} \in \Omega_{Q_0}^+$ by $\phi_{Q_0}^+(b_{-mk}) = \phi_{Q_0}^+(b) - m$. We define b_n by $b_n = Q_0^\nu(b_{-mk})$ if $n = -mk + \nu$ with $\nu \geq 0$ (this is independent of the way we write n as $-mk + \nu$). Then $(b_n)_{n \in \mathbf{Z}}$ satisfies (BO1) to (BO4).

Note that, for a sequence obtained by this construction, we have $b_{-mk} \in \Omega_{Q_0}^+$ for $m \geq 0$. But more generally, for an arbitrary sequence $(b_n)_{n \in \mathbf{Z}}$ satisfying (BO1) to (BO4) and an arbitrary Ω^+ which is the domain of a Fatou coordinate with range containing $\mathbf{C}^-$, there is a m_0 such that $b_{-mk} \in \Omega^+$ for $m \geq m_0$.

Remark : We shall actually prove a result which is stronger that the one stated in Theorem 1. We shall prove that, for **any** sequence (b_n) satisfying (BO1) to (BO4), there is a ν and a map $\mathbf{C} \to \mathbf{C}$ which makes Υ appear quasi-conformally in M, and which for each m maps the center $h_m(0)$ of the m-th copy of M in Υ to the point $b_{-mk+\nu}$.

In this situation, we have (see Fig. 6):

LEMMA.- *One can find Jordan domains U_0, U_0', V_0 with C^1 boundary such that :*

(JD1) : $\overline{\Delta} \subset U_0' \subset \overline{U_0'} \subset U_0$;

(JD2) : Q_0^k induces a proper map $f_0 : U_0' \to U_0$ of degree 2 ;

(JD3) : $\overline{V_0} \subset U_0 \smallsetminus \overline{U_0'}$;

(JD4) : $b_\ell \in V_0$ for some $\ell = -mk < N$, and $Q_0^{N-\ell}$ induces an isomorphism $g_0 : V_0 \to U_0$.

Proof : One can find Jordan domains U and U' with C^1 boundary such that $\overline{\Delta} \subset U'$ and that Q_0^k induces a quadratic-like mapping $U' \to U$, and one can choose U contained in an arbitrary small neighborhood of $\overline{\Delta}$. The sets $Q_0^i(\overline{\Delta})$ for $0 \leq i < k$ are disjoint (due to the fact that c_0 is the root of a *primitive* hyperbolic component), and one can choose U so that the sets $Q_0^i(\overline{U})$ are disjoint.

For $n < N$, denote by V_n the connected component of $Q_0^{-(N-n)}(U)$ which contains b_n. Then the sets $\overline{V_n}$ are disjoint, for each n the set V_n is a Jordan domain and Q_0^{N-n} induces a homeomorphism $\overline{V_n} \to \overline{U}$. For the Poincaré metric of $\mathbf{C} \smallsetminus \bigcup_{0 \leq i < k} Q_0^i(\overline{\Delta})$, the diameter of $\overline{V_n}$ is bounded, since there is a branch of Q_0^{-1} defined on the universal covering of this set and mapping it into itself ; so the euclidean diameter of $\overline{V}_{-mk}$ tends to 0 when $m \to \infty$. There is an $m_1 \geq m_0$ such that $\overline{V}_{-mk} \subset U$ for $m \geq m_1$.

Now the domains U, U' and V_{-mk} satisfy (JD1), (JD2) and (JD4) for any $m \geq m_1$.

SUBLEMMA.- *Let $P \subset \mathbf{C}^+ = \{Z \mid Re(Z) > 0\}$ be a topological closed disc such that $P \cap 2^m.P = \emptyset$ for $m \neq 0$. Then, given $s_0 > 0$, one can find a C^1-curve C such that :*

(AP1) : $C \subset \Sigma = \{Z \mid 0 < Re(Z) \leq s_0\}$;

(AP2) : $T_{2\pi i}(C) = C$;

(AP3) : $\frac{1}{2}.C \cap C = \emptyset$;

(AP4) : $(\forall m \in \mathbf{Z})\ C \cap 2^m.P = \emptyset$.

Proof of sublemma : Take $\mu > \sup_{x+yi \in P} \frac{|y|}{x}$. Let J be a C^1-arc joining $s_0 - \mu s_0 i$ to $s_0 + \mu s_0 i$ in $\{Z = x + yi \mid |y| < \mu x\}$ and avoiding $\bigcup_{m \in \mathbf{Z}} 2^m.P$. Set $J_m = 2^{-m}.J$. Take $s_1 \leq s_0$ such that $2\mu s_1 \leq \frac{\pi}{2}$. Consider the segments $J_m' = [2^{-m}(s_0 + \mu s_0 i)\ \ s_1 + 2\mu s_1 i]$, $J'' = [s_1 + 2\mu s_1 i\ \ s_1 + (2\pi - 2\mu s_1)i]$, $J_m''' = [s_1 + (2\pi - 2\mu s_1)i\ \ 2\pi i + 2^{-m}(s_0 - \mu s_0 i)]$. Then, for m big enough, J_m is contained in $\{Z \mid 0 < Re(Z) < \inf(2s_1, s_0)\}$, and

$$\check{C} = \bigcup_{k \in \mathbf{Z}} T_{2k\pi i}(J_m \cup J_m' \cup J'' \cup J_m''')$$

is a piecewise C^1-curve satisfying (AP1) to (AP4). One can smooth it and obtain a C^1-curve C still satisfying (AP1) to (AP4). ∎

End of proof of the lemma : Proceeding like in section 8 and taking logarithm, one can find a C^1-diffeomorphism h of the universal covering of

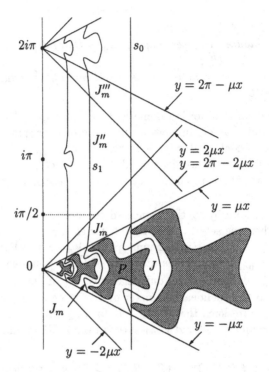

Figure 5: The proof of the sublemma.

$\overline{U} \smallsetminus K(f_0)$ onto a vertical strip $\Sigma = \{Z \mid 0 < Re(Z) \leq s_0\}$ which conjugates f to $Z \mapsto 2.Z$ and the deck transformation to $T_{2\pi i}$. Take $P = h(\check{V})$ with $\check{V}$ a lift of $\overline{V}_{-m_1 k}$. Note that P is a topological closed disc, with $P \cap 2^m.P = \emptyset$ for $m \neq 0$.

By the sublemma, one can find a C^1-curve C in Σ, invariant under $T_{2\pi i}$, avoiding $\bigcup 2^{-m}.P$, and such that $\frac{1}{2}.C$ lies between $i.\mathbf{R}$ and C. There is a unique m_2 such that the set $2^{-m_2}.P$ lies in the strip between C and $\frac{1}{2}.C$. Set $\ell = -(m_1 + m_2)k$. The projections of $h^{-1}(C)$ and $h^{-1}(\frac{1}{2}.C)$ are C^1 Jordan curves which bound Jordan domains U_0 and U_0' satisfying (JD1) and (JD2), and taking $V_0 = V_\ell \cap Q_0^{-(N-\ell)}(U_0)$ the conditions (JD3) and (JD4) also are satisfied.   ∎

We can now finish the proof of Theorem 1. Take for instance $U_\lambda = U_0$ for every $\lambda \in \mathbf{C}$. For λ close to 0, let U_λ' (resp. V_λ) be the Jordan domain bounded by the component of $Q_\lambda^{-k}(\partial U_\lambda)$ (resp. $Q_\lambda^{-\ell}(\partial U_\lambda)$) close to $\partial U_0'$ (resp. to ∂V_0) (see Fig. 6).

There is a simply connected neighborhood Λ of 0 such that $\partial U_\lambda'$ and ∂V_λ undergo a holomorphic motion when λ ranges in Λ, and that the inclusions

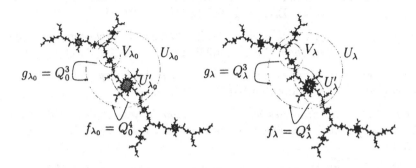

Figure 6: The mappings F_{λ_0} and F_λ.

$\overline{U'_\lambda} \subset U_\lambda$, $\overline{V_\lambda} \subset U_\lambda \smallsetminus \overline{U'_\lambda}$ hold for $\lambda \in \Lambda$. Then Q^k_λ and Q^ℓ_λ induce maps $f_\lambda : U'_\lambda \to U_\lambda$ and $g_\lambda : V_\lambda \to U_\lambda$, and we are in the situation described in section 4. The condition $\delta'(0) \neq 0$ follows from [DH/O] Exposé XII (ABP), Cor. 2 du Th. 1, or [TL].

By Theorem 2, the set Υ appears quasi-conformally in M_F. But M_F appears in M by $\lambda \mapsto c_0 + \lambda$. Indeed, if $\lambda \in M_F$, the critical point $\omega_\lambda = 0$ has an orbit under F_λ which stays in U_λ, so $Q^n_\lambda(0)$ does not tend to ∞ and $c_0 + \lambda \in M$. If $\lambda \in \partial M_F$, the sequence of functions $(\lambda' \mapsto F^n_{\lambda'}(\omega_{\lambda'}))_n$ is not normal at λ, the same holds for $(Q^n_{\lambda'}(0))$ and $c_0 + \lambda \in \partial M$.

This finishes the proof of Theorem 1.

Final remarks : To say that Υ appears quasi-conformally in M is a rather weak statement as we don't give a bound for the dilation ratio. Computer pictures seem to indicate that, if the radius R is chosen in an appropriate way, this dilation ratio is not too big. But we don't have an estimate.

The statement of quasi-conformal appearance does not imply the estimates stated in section 1 on the diameter of the sets M_n and their distance to c_0. These could be derived from the computations made. We leave that to the courageous reader.

The drawings are done by X. Buff using algorithms and programs by Christian Henriksen.

References.

[D/AMS] A. Douady, *Does a Julia set depend continuously on the Polynomial ?*, Proceedings of Symposia in Applied Math., **49**, 91-139, (1994).

[DH Pol-like] A. DOUADY & J. HUBBARD *On the dynamics of polynomial-like mappings*, Ann. Sci. ENS Paris **vol** 18 (1985), 287–343.

[DSZ] A. DOUADY, P. SENTENAC & M. ZINSMEISTER, *Implosion parabolique et dimension de Hausdorff*, C.R.A.S. t.325, Série I, 765–772, (1997).

[D/Sł] A. DOUADY, *Prolongement de mouvements holomorphes [d'après Słodkowski et autres*, Séminaire Bourbaki 775, (1993).

[La] P. LAVAURS, *Systèmes dynamiques holomorphes, Explosion de points périodiques paraboliques*, Thèse de doctorat, Université Paris-Sud in Orsay, (1989).

[Ly] M. LYUBICH, *Feigenbaum-Coullet-Tresser Universality and Milnor's Hairiness Conjecture*, Preprint IHES/M/96/61, to appear in Annals of Math.

[Sh] M. SHISHIKURA, *Bifurcation of parabolic fixed points*, in this volume.

[Sł] Z. SŁODKOWSKI, *Extensions of holomorphic motions*, Prépublication IHES/M/92/96, (1993).

[TL] TAN LEI, *Similarity between the Mandelbrot set and Julia sets*, Comm. Math. Phys., **134**, 587-617, (1990).

Adrien Douady, Département de Mathématique, Université de Paris-Sud, Bât. 425, 91405 Orsay, France.
    e-mail : Adrien.Douady@ens.fr

Xavier Buff, Universite Paul Sabatier, Laboratoire Emile Picard, UFR MIG, 118 route de Narbonne, 31062 Toulouse Cedex, France.
    e-mail : buff@topo.math.u-psud.fr

Robert Devaney, Department of Mathematics, Boston University, Boston MA 02215, U.S.A.
    e-mail : bob@math.bu.edu

Pierrette Sentenac, Département de Mathématique, Université de Paris-Sud, Bât. 425, 91405 Orsay, France.
    e-mail : sentenac@topo.math.u-psud.fr

Modulation dans l'ensemble de Mandelbrot

Peter Haïssinsky

July 20, 1999

Abstract

We prove A. Douady's and J.H. Hubbard's Tuning Theorem from the sketch given by A. Douady. This result explains why the Mandelbrot set contains "little copies" of itself, and where they can be found. We also give a topological model of the tuned Julia sets, and some further corollaries.

Le but de cet article est de montrer le Théorème de Modulation de A. Douady et J.H. Hubbard, ainsi que de corollaires immédiats. Ce résultat explique la présence de petites copies de l'ensemble de Mandelbrot dans lui-même. Nous donnons aussi un modèle topologique de l'ensemble de Julia d'un polynôme modulé, ainsi que quelques corollaires.

Nous commençons cette introduction par des rappels afin de situer le contexte et pouvoir ensuite énoncer les résultats principaux.

Rappels

On se restreint à la famille quadratique, bien que de nombreux résultats restent vrais pour des polynômes de degré plus élevé. La particularité des polynômes est l'existence à l'infini d'un point super-attractif, de bassin connexe, et dont le bord est leur ensemble de Julia.

On considère la famille de polynômes

$$Q_c : \begin{array}{c} \mathbb{C} \to \mathbb{C} \\ z \mapsto z^2 + c, \end{array} \quad c \in \mathbb{C}.$$

On note $\left| \begin{array}{l} K_c = \{z \in \mathbb{C}, \ Q_c^n(z) \not\to \infty\} \text{ l'ensemble de Julia rempli, et} \\ J_c = \partial K_c \text{ l'ensemble de Julia de } Q_c \, ; \end{array} \right.$

Ce sont des compacts totalement invariants, et K_c est de plus plein. Nous allons énumérer les principaux résultats dont on aura besoin. Les démonstrations se trouvent dans [9], sauf mention différente.

Proposition 0.1. *(I, Chap.3, Prop.1)*

De deux choses l'une : $\left\{ \begin{array}{l} \text{soit } 0 \in K_c \text{ et } K_c \text{ est connexe,} \\ \text{soit } K_c \text{ est un Cantor.} \end{array} \right.$

Pour tout c dans $\mathbb{C}$, on peut définir la coordonnée de Böttcher ϕ_c au voisinage de l'infini qui conjugue Q_c à $z \mapsto z^2$, ϕ_c tangente à l'identité à l'infini. Si K_c est connexe, alors $\phi_c : \mathbb{C} \setminus K_c \xrightarrow{\sim} \mathbb{C} \setminus \overline{\mathbb{D}}$ est une représentation conforme de $\mathbb{C} \setminus K_c$.

On définit le potentiel de Q_c par $G_c = \log|\phi_c|$. On prolonge G_c à $\mathbb{C}$ comme suit :

- pour $z \in \mathbb{C} \setminus K_c$, on utilise la relation $G_c(Q_c(z)) = 2G_c(z)$,
- pour $z \in K_c$, on prolonge par continuité en posant $G_c(z) = 0$.

On définit les *équipotentielles* par $G_c^{-1}(\eta)$, $\eta > 0$; si $\eta > G_c(0)$ alors c'est un cercle topologique. Le *rayon externe* $\mathcal{R}_c(\theta)$, où θ est compté en nombre de tours, est $\phi_c^{-1}(\eta e^{2i\pi\theta})$, pour η assez grand. On le prolonge tant qu'on ne "bute" pas sur un point précritique ω de Q_c, *i.e.* $\mathrm{Arg}\phi_c(\omega) = \theta$. Pour tout z de $\mathcal{R}_c(\theta)$, on dit que z a *pour argument* θ. Si $\lim_{\eta \to 0} \phi_c^{-1}(\eta e^{2i\pi\theta})$ existe alors $\mathcal{R}_c(\theta)$ *aboutit*.

Exemples.

Figure 1. *Ensembles de Julia quadratiques.*

Si K_c est connexe et localement connexe, on peut définir *le lacet de Carathéodory* par l'application $\gamma_c : \mathbb{T} \to \partial K_c$, fonction continue et surjective, qui à θ associe le point où $\mathcal{R}_c(\theta)$ aboutit.

Proposition 0.2. *(I, Chap.8, 2, Prop.2)*
$\forall(c, \theta) \in \mathbb{C} \times \mathbb{Q}/\mathbb{Z}$, $\mathcal{R}_c(\theta)$ *aboutit ou bute sur un point précritique de* Q_c.

Soit $M = \{c \in \mathbb{C},\ tel\ que\ K_c\ soit\ connexe\} = \{c \in \mathbb{C}, 0 \in K_c\}$ *l'ensemble de Mandelbrot*. Cet ensemble est compact connexe et plein (I, Chap.8, Théorème 1).

L'application

$$\phi : \quad \mathbb{C} \times \text{voisinage de l'infini} \quad \to \mathbb{C}$$
$$(c, z) \qquad\qquad\qquad \mapsto \phi_c(z)$$

est holomorphe par rapport à (c, z), partout où ϕ_c est définie.
Cette application induit l'uniformisante

$$\Phi_M : \ \mathbb{C} \setminus M \to \mathbb{C} \setminus \overline{\mathbb{D}} \,,$$
$$c \mapsto \phi_c(c)$$

de potentiel $G_M = \log |\Phi_M|$.
On définit les équipotentielles et les rayons externes de M de la même façon.
Pour $k \geq 1$ fixé, $\{c \in \mathbb{C} \ tq \ \exists \ k\text{-cycle attractif pour } Q_c\} \subset \overset{\circ}{M}$, et chaque composante est appelée *composante hyperbolique* de M de période k.
Remarque. $\forall c \in \mathbb{C}$, Q_c admet au plus un cycle non répulsif (car $\deg Q_c = 2$).

Pour toute composante hyperbolique W, on a la représentation conforme $\rho_W : W \to \mathbb{D}$, qui à c associe le multiplicateur du cycle attractif de Q_c (II, Chap.19, Th.1). L'application ρ_W se prolonge continûment au bord ; on note $\rho_W^{-1}(0)$ le *centre* de W et $\rho_W^{-1}(1)$ sa *racine*.
Définitions.
$\mathcal{D}_0 = \{$centres des composantes hyperboliques$\} = \{c \in \mathbb{C}, \ 0 \ soit \ périodique\}$;
$\mathcal{D}_1 = \{$racines des composantes hyperboliques$\} = \{points \ paraboliques\}$;
$\mathcal{D}_2 = \{$points de Misiurewicz$\} = \{c \in \mathbb{C}, \ 0 \ soit \ strictement \ prépériodique\}$.

Proposition 0.3. *(II, Chap.13, Th.1 et [19]) $\forall \theta \in \mathbb{Q}/\mathbb{Z}$, $\mathcal{R}_M(\theta)$ aboutit à $c \in \partial M$;*
- $c \in \mathcal{D}_1$ si θ est à dénominateur impair,
- $c \in \mathcal{D}_2$ sinon.

Théorème 0.4. (Relations entre arguments dynamiques et arguments externes)
- (I, Chap.8, I, Cor.2) : Pour $c \notin M$, $Arg_c c = Arg_M c$.
- (II, Chap.13, Th.1, Complément, Chap.18, Prop.1) : Pour $c \in \mathcal{D}_0$, il existe $\mathcal{R}_M(\theta_-)$ et $\mathcal{R}_M(\theta_+)$ aboutissant à la racine de la composante de c. Ils aboutissent dans le plan dynamique à la racine de la composante de $\overset{\circ}{K}_c$ contenant c, en étant adjacents à cette composante.
- (II, Chap.13, Th.1, Complément et [21]) : Pour $c \in \mathcal{D}_1$, il existe $\mathcal{R}_M(\theta_-)$ et $\mathcal{R}_M(\theta_+)$ aboutissant à c. Ils aboutissent dans le plan dynamique à la racine de la composante de $\overset{\circ}{K}_c$ contenant c, qui est un point parabolique, en étant adjacents à cette composante.
- (I, Chap.8, III, Th.2) : Pour $c \in \mathcal{D}_2$, il existe $\theta_1, \ldots, \theta_\nu$ tels que leurs rayons externes aboutissent à c dans M. Ils y aboutissent aussi dans le plan dynamique, et ce sont les seuls.

Je suggère aussi au lecteur de se reporter à l'approche de J. Milnor ([17]).

Applications à allure polynomiale, mouvements holomorphes et figuration rigide. Nous décrivons ici les outils analytiques dont nous aurons besoin.

Définitions.
▷ Une *application à allure polynomiale* de degré $d \geq 2$ est la donnée d'un triplet (U', U, g) où U' et U sont des disques topologiques tels que $U' \subset\subset U$ et tels qu'on ait $g : U' \to U$ holomorphe propre de degré d.
▷ $K(g) = \cap g^{-n}(U')$ est *l'ensemble de Julia rempli*;
▷ $J(g) = \partial K_g$ est *l'ensemble de Julia*.
Définition. Deux applications à allure polynomiale (U', U, f) et (V', V, g) sont *hybridement équivalentes* s'il existe une application quasiconforme $\varphi : U \to V$ qui conjugue f à g et telle que $\bar{\partial}\varphi = 0$ *pp.* sur $K(f)$. On note $f \sim_{hyb} g$.

Pour la définition et les propriétés des applications quasiconformes on peut se reporter à [1].

Indépendamment des propositions suivantes, on peut montrer que de nombreuses propriétés des ensembles de Julia (remplis) de polynômes se transposent à ce cadre. Ils sont d'ailleurs étroitement liés par le théorème suivant :

Théorème 0.5. *(Théorème de Redressement, [10] ; I, Th.1)*
Soit (U', U, g), une application à allure polynomiale de degré $d \geq 2$. Il existe un polynôme P de degré d tel que $P \sim_{hyb} g$.
De plus, si $K(g)$ est connexe, alors P est unique à conjugaison affine près. En particulier, si g est quadratique avec un ensemble de Julia connexe, alors il existe un unique $c \in M$ tel que $Q_c \sim_{hyb} g$.

Définition (famille analytique). Soit Λ une variété complexe et soit $g = (g_\lambda : U'_\lambda \to U_\lambda)_{\lambda \in \Lambda}$ une famille d'applications à allure polynomiale de degré $d \geq 2$. On pose $\mathcal{U} = \{(\lambda, z), z \in U_\lambda\}$, $\mathcal{U}' = \{(\lambda, z), z \in U'_\lambda\}$ et on définit $g(\lambda, z) = g_\lambda(z)$. Alors g est une *famille analytique* si les conditions suivantes sont satisfaites :
(i) $\mathcal{U}'$ et $\mathcal{U}$ sont homéomorphes à $\Lambda \times \mathbb{D}$ au-dessus de Λ,
(ii) la projection de $\overline{\mathcal{U}'}$ dans Λ est propre, où la fermeture de $\mathcal{U}'$ est prise dans $\mathcal{U}$,
(iii) l'application $g : \mathcal{U}' \to \mathcal{U}$ est $\mathbb{C}$-analytique et propre.
Remarque. Si le bord des domaines $U_\lambda, \lambda \in \Lambda$, sont paramétrés par un mouvement holomorphe (voir ci-dessous) alors les conditions (i) et (ii) ci-dessus sont satisfaites.

Théorème 0.6. *(Dépendance par rapport aux paramètres, [10] ; II, Th.2, Prop.14 ; IV, Prop.21)*
Soit $(g_\lambda)_{\lambda \in \Lambda}$ une famille analytique d'applications à allure quadratique.
(i) Il existe $\chi : \Lambda \to \mathbb{C}$ continue telle que $\forall \lambda \in \Lambda$, $g_\lambda \sim_{hyb} Q_{\chi(\lambda)}$.
(ii) On suppose que Λ est homéomorphe à $\mathbb{D}$. On définit $M_g = \{\lambda \in \Lambda, K(g_\lambda)$ connexe$\}$.
Si $\chi : M_g \to \chi(M_g)(\subset M)$ est propre et $\chi(M_g)$ est connexe alors χ est

un revêtement ramifié. De plus, si $M_g \subset\subset \Lambda$, alors $\chi : M_g \to M$ est un revêtement ramifié de degré d fini que l'on peut calculer en comptant le nombre de tours de $g_\lambda(\omega_\lambda) - \omega_\lambda$ (ω_λ : point critique de g_λ) quand on parcourt une fois un lacet entourant M_g.

Définitions. (a) Sous les hypothèses de (ii), si M_g est compact dans Λ et le degré de χ est $d = 1$, alors χ est un homéomorphisme et on dit que la famille est *mandelbrotesque*.

(b) Si $\chi : M_g \to M \setminus \{1/4\}$ est un homéomorphisme alors g est dite *semi-mandelbrotesque*.

Complément. M. Lyubich a démontré que χ est quasiconforme lorsque la famille est mandelbrotesque ([15]), *i.e.* il existe une extension quasiconforme au plan de χ.

Rappelons qu'un mouvement holomorphe d'un ensemble $X \subset \overline{\mathbb{C}}$ au-dessus d'une variété complexe Λ est une application $\iota : \Lambda \times X \to \overline{\mathbb{C}}$ qui vérifie les 3 conditions suivantes :

    (i) il existe $\lambda_0 \in \Lambda$ tel que pour tout $x \in X$, $\iota(\lambda_0, x) = x$;

    (ii) à $x \in X$ fixé, $\lambda \mapsto \iota(\lambda, x)$ est holomorphe ;

    (iii) à $\lambda \in \Lambda$ fixé, $x \mapsto \iota(\lambda, x)$ est injective.

Le résultat le plus fort sur les mouvements holomorphes est un théorème de Z. Słodkowski ([20]) :

Théorème 0.7. *Un mouvement holomorphe au-dessus du disque unité se prolonge en un mouvement holomorphe de toute la sphère. De plus, à λ fixé, l'application $z \mapsto \iota(\lambda, z)$ est quasiconforme.*

Enfin, nous généralisons la notion d'application à allure polynomiale :

Définition. Soit f une fraction rationnelle et soit P un polynôme ayant son ensemble de Julia rempli $K(P)$ connexe. On dit que P figure de manière *rigide* dans f s'il existe un voisinage U de $K(P)$, une application continue injective $\varphi : U \to \mathbb{C}$ tels que :

(a) pour tout $z \in K(P)$, $\varphi \circ P(z) = f \circ \varphi(z)$;

(b) l'application $\varphi \in W_{loc}^{1,p}$ pour tout $p < 2$, *i.e.* φ admet des dérivées partielles au sens des distributions L^p intégrables ;

(c) $\overline{\partial}\varphi = 0$ pp. sur $K(P)$.

Pour plus de précisions sur les figurations rigides, on peut se reporter à [11].

Résultats principaux

La première partie de cet article consiste à démontrer le théorème ci-dessous selon l'esquisse de [6].

Dans la suite, W désigne une composante hyperbolique de centre c_0 et de racine c_1. On dit que W est *primitive* si la fleur du point parabolique de c_1 n'est formée que d'un pétale attractif.

Théorème 1. de Modulation (A. Douady & J.H. Hubbard)

(a) Si W est une composante primitive de M de période k, on peut trouver un voisinage Λ de $\overline{W}$, et deux familles d'ouverts $(U'_c, U_c)_{c \in \Lambda}$ tels que $(g_c = Q_c^k|_{U'_c} : U'_c \to U_c)_{c \in \Lambda}$ soit une famille mandelbrotesque.

(b) Si W n'est pas primitive, on peut seulement construire Λ, les U'_c et les U_c de façon que Λ soit un voisinage de $\overline{W} \setminus \{c_1\}$, et que χ induise une famille semi-mandelbrotesque. De plus, $Q_{1/4}$ figure de manière rigide dans Q_{c_1}.

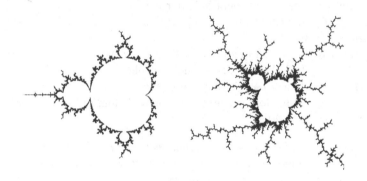

Figure 2. *M et un zoom.*

On note $M_{c_0} = \chi_{c_0}^{-1}(M)$ obtenu par ce théorème.

Complément (M. Lyubich, [15]). *Si W est primitive, la petite copie de M est quasiconforme à M.*

La démonstration consiste à construire des ouverts U'_c et U_c afin que les Q_c^k soient des applications à allure quadratique. Ensuite, on applique la théorie des applications à allure polynomiale qui est bien adaptée si W est primitive. Dans le cas non primitif, on n'obtient pas $M_g \subset\subset \Lambda$: une étude plus détaillée est nécessaire.

Dans la seconde partie, nous étudions l'opération de modulation. À $c_0 \in \mathcal{D}_0$ fixé, on note

$$c_0 \perp \cdot : \quad M \to M_{c_0}$$
$$c \mapsto \chi_{c_0}^{-1}(c).$$

Nous montrerons le théorème suivant (voir §6 pour un énoncé précis):

Théorème 2. *L'opération de modulation agit sur les ensembles de Julia remplis comme suit : $K_{c_0 \perp c}$ est homéomorphe à K_{c_0} dans lequel on a remplacé les composantes connexes de l'intérieur par des "copies" de K_c.*

En appendice, nous étudions la dépendance du point d'aboutissement d'un rayon en fonction du paramètre c.

Notes. Cet article est tiré de [11]. Les représentations d'ensembles de Julia, ainsi que de l'ensemble de Mandelbrot ont été faites à l'aide d'un logiciel de C.T. McMullen.

Remerciements. Je tiens à remercier Tan Lei pour son encouragement et pour tout le temps qu'elle a accordé à cet article. Ses nombreux commentaires ont grandement clarifié cette exposition, et je lui dois aussi des simplifications de démonstration.

Partie I : Théorème de modulation

1   Première construction

Soit W une composante de période k, de centre c_0 et de racine c_1. On suppose que $c_0 \neq 0$ et $c_1 \neq 1/4$. Dans ces conditions, il existe deux rayons externes $\mathcal{R}_M(\theta_1)$ et $\mathcal{R}_M(\theta_2)$, $\theta_1 < \theta_2$, qui aboutissent à c_1 (Figure 2). D'après le Théorème 0.4, $\mathcal{R}_{c_0}(\theta_1)$ et $\mathcal{R}_{c_0}(\theta_2)$ aboutissent à y_1, qui est k-périodique, et qui est la racine de la composante connexe U_1 de $\overset{\circ}{K}_{c_0}$ contenant c_0 (Figure 1). Soit $\eta > 0$. On appelle A l'ouvert délimité par $\mathcal{R}_{c_0}(\theta_1) \cup \{y_1\} \cup \mathcal{R}_{c_0}(\theta_2) \cup (G_c = \eta)$ et contenant U_1. On note A' la composante connexe de $Q_{c_0}^{-k}(A)$ contenant U_1, qui est bien définie car U_1 est k-périodique.

Proposition 1.1. *A' est délimité par quatre rayons externes d'angle θ_1, θ_2, θ_1' et θ_2' et par l'équipotentielle de niveau $\eta' = \eta/2^k$. De plus, $A' \subset A$ et $Q_{c_0}^k : A' \to A$ est propre de degré deux, et sa valeur critique est c_0.*

DÉMONSTRATION. On pose $U_i = Q_{c_0}^{i-1}(U_1)$; tant que $1 \leq i \leq k-1$, $Q_{c_0}|_{U_i} : U_i \to U_{i+1}$ est conjugué à l'identité sur $\mathbb{D}$, car U_i ne contient pas de point critique. En revanche, $Q_{c_0}|_{U_k = U_0} : U_0 \to U_1$ est conjugué à $z \mapsto z^2$ dans $\mathbb{D}$, qui se prolonge au bord. Le point y_1 a donc deux préimages, y_1' d'argument interne $1/2$ et lui-même. Comme $Q_{c_0}^k$ est univalente au voisinage de y_1', il existe $\theta_1' > \theta_2'$ tels que leurs rayons aboutissent en y_1' et tels que
$$\begin{cases} 2^k \theta_1' = \theta_1 \ (mod\ 1), \\ 2^k \theta_2' = \theta_2 \ (mod\ 1). \end{cases}$$

On définit A_1' selon l'énoncé. On a :
$$\left. \begin{array}{l} Q_{c_0}^k(\partial A_1') = \partial A \\ U_1 \subset Q_{c_0}^k(A_1') \\ Q_{c_0}^k(A_1') \subset A \end{array} \right\} \implies A' = A_1' \text{ et } Q_{c_0}^k : A' \to A \text{ est propre car c'est un}$$
polynôme.

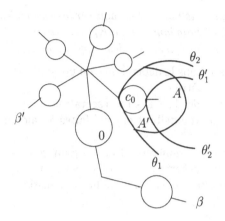

Figure 1: K_{c_0}.

Comme A est connexe, $Q_{c_0}^k|_{A'}$ a un degré qui est le même que $Q_{c_0}^k|_{U_1}$, *i.e.* 2.

De plus, le point c_0 n'a qu'un seul antécédent par $Q_{c_0}^{-k}|_A$: c'est donc la valeur critique. ∎

Proposition 1.2. *Il existe un ouvert Λ_1, voisinage de $\overline{W} \setminus \{c_1\}$, tel que pour tout $c \in \Lambda_1$, $\mathcal{R}_c(\theta_s)$ (resp. $\mathcal{R}_c(\theta'_s)$), $s = 1, 2$, aboutissent à un même point $y_1(c)$ k-périodique (resp. $y'_1(c)$, préimage de $y_1(c)$ par Q_c^k), ce qui permet de construire $Q_c^k : A'(c) \to A(c)$ propre de degré deux. De plus, pour tout $c \in \Lambda_1$, le point $y_1(c)$ est répulsif et c est la valeur critique de Q_c^k.*

DÉMONSTRATION. On considère l'ouvert Λ_1 délimité par les rayons $\mathcal{R}_M(\theta_1)$ et $\mathcal{R}_M(\theta_2)$, et l'équipotentielle η contenant W (voir Figure 2). Par le Théorème A.3 (voir Appendice), il existe une fonction holomorphe $c \mapsto y_1(c)$ telle que, pour tout $c \in \Lambda_1$, on a $y_1(c) = \gamma_c(\theta_1) = \gamma_c(\theta_2)$ et $y_1(c)$ est répulsif.

Les rayons d'angle θ'_1 et θ'_2 aboutissent pour tout $c \in \Lambda_1$ car leurs orbites n'entrent plus jamais dans $]\theta_1, \theta_2[$ (*cf* Appendice, Cor. A.5 et Prop. A.6). D'après la Proposition A.1, il existe aussi une fonction holomorphe $c \mapsto y'_1(c)$ telle que, pour tout $c \in \Lambda_1$, on a $y'_1(c) = \gamma_c(\theta'_1) = \gamma_c(\theta'_2)$.

Soit $A(c)$, $c \in \Lambda_1$, l'ouvert délimité par $\mathcal{R}_c(\theta_1) \cup \{y_1(c)\} \cup \mathcal{R}_c(\theta_2) \cup \{G_c = \eta\}$ et contenant c. On note $A'(c)$ la composante connexe de $Q_c^{-k}(A(c))$ ayant $y_1(c)$ et $y'_1(c)$ dans sa fermeture. Soit $A'_1(c)$ l'ouvert défini comme A_1 dans K_{c_0} (voir Figure 3). $Q_c^k(\partial A'_1(c)) = \partial A(c)$, donc $A'_1(c) = A'(c)$.

Pour tout $c \in \Lambda_1$, $Q_c^k : A'(c) \to A(c)$ est holomorphe propre, donc admet un degré ($A(c)$ connexe) que l'on définit par $d(c) = \frac{1}{2i\pi} \int_{\partial A'(c)} \frac{(Q_c^k)'(z)}{Q_c^k(z) - c} dz$ qui

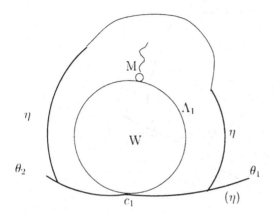

Figure 2: Λ_1.

est holomorphe et à valeur entière, donc constante sur Λ_1 connexe: $d(c) = d(c_0) = 2$. On montre de même que c est la valeur critique. ∎

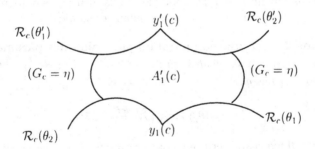

Figure 3: $A_1'(c)$.

Jusqu'ici, on a défini une famille $(A'(c) \xrightarrow{Q_c^k} A(c))_{c \in \Lambda_1}$ holomorphe propre de degré deux. Ses éléments ne sont pas à allure polynomiale car $A'(c)$ n'est pas relativement compact dans $A(c)$. Il est donc nécessaire de changer un peu leurs définitions. Pour cela, nous allons distinguer le cas primitif du cas non primitif.

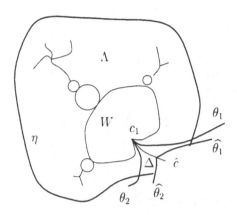

Figure 4: Λ.

2 W primitive

D'après le Théorème 0.4, $\mathcal{R}_{c_1}(\theta_1)$ et $\mathcal{R}_{c_1}(\theta_2)$ aboutissent au même point $y_1(c_1)$ parabolique de vraie période k ; on construit de la même façon $A'(c_1)$ et $A(c_1)$. Comme c_1 est la racine de W primitive, $y_1(c_1)$ n'a qu'un seul pétale répulsif dont l'axe est contenu dans $\mathbb{C} \setminus A'(c_1)$. Nous allons utiliser les résultats de [21] obtenus par implosion parabolique. Nous utiliserons d'abord un énoncé sur le plan des paramètres, puis son pendant dynamique.

Théorème 2.1. (Implosion parabolique - plan des paramètres)
Avec les notations précédentes, pour tout $\varepsilon > 0$, il existe $\widehat{\theta}_1 \in]\theta_1 - \varepsilon, \theta_1[$ et $\widehat{\theta}_2 \in]\theta_2, \theta_2 + \varepsilon[$ tels que

$$\gamma_M(\widehat{\theta}_1) = \gamma_M(\widehat{\theta}_2) \overset{def}{=} \hat{c} \in \mathcal{D}_2 \, .$$

Soit $\varepsilon > 0$ fixé assez petit que l'on déterminera ultérieurement. On définit Λ l'ouvert borné du plan des paramètres bordé par les rayons $\mathcal{R}_M(\widehat{\theta}_1)$ et $\mathcal{R}_M(\widehat{\theta}_2)$, par $\hat{c}$, ainsi que par l'équipotentielle $(G_M = \eta)$.

D'après le Théorème A.3, pour tout $c \in \Lambda$, les rayons dynamiques $\mathcal{R}_c(\widehat{\theta}_1)$ et $\mathcal{R}_c(\widehat{\theta}_2)$ aboutissent à un même point $y_2(c)$, qui est prépériodique répulsif et qui dépend $\mathbb{C}$-analytiquement de c.

Pour $c \in \Lambda$, on peut alors définir l'ouvert U_c contenant $A(c)$ et bordé par les rayons $\mathcal{R}_c(\widehat{\theta}_1)$ et $\mathcal{R}_c(\widehat{\theta}_2)$, par $y_2(c)$, ainsi que l'équipotentielle $(G_c = \eta)$; et on note aussi U'_c la composante connexe de $Q_c^{-k}(U_c)$ contenant $A'(c)$.

Si ε est assez petit, alors $y_2(c_1)$ appartient au pétale répulsif de $y_1(c_1)$, et son image réciproque par la branche de Q_c^{-k} qui envoie $y_1(c_1)$ sur $y'_1(c_1)$ envoie

$y_2(c_1)$ à l'intérieur de $A(c_1)$. On choisit ε ainsi. Ceci implique notamment que $(g_c = Q_c^k|_{U_c'} : U_c' \to U_c)_{c \in \Lambda}$ est une famille d'applications à allure quadratique.

Notons $\mathcal{U} = \{(c,z), \ c \in \Lambda, \ z \in U_c\}$, $\mathcal{U}' = \{(c,z), \ c \in \Lambda, \ z \in U_c'\}$ et $g : \mathcal{U}' \to \mathcal{U}, \quad (c,z) \mapsto (c, g_c(z))$.

Comme les domaines U_c bougent holomorphiquement, on en déduit que les conditions (i) et (ii) de la définition d'une famille analytique sont vérifiées. De plus, g est propre car $Q_c^k(z)$ est polynomiale.

D'après le Théorème 0.6, on peut définir $\chi : \Lambda \to \mathbb{C}$ telle que $g_c \sim_{hyb} Q_{\chi(c)}$; de plus, χ est continue sur Λ et analytique sur $\overset{\circ}{M_{c_0}}$ où $M_{c_0} = \{c \in \Lambda, K(g_c)$ connexe$\}$.

Montrons que M_{c_0} est un compact de Λ :

▷ M_{c_0} est fermé dans Λ car $M_{c_0} = \cap_{n \geq 0}\{c \in \Lambda, \ g^n(c,c) \in \overline{\mathcal{U}'}\}$.

▷ Afin de montrer que $M_{c_0} \subset\subset \Lambda$, on va montrer que pour tout c assez proche de $\partial\Lambda$, $K(g_c)$ est un Cantor.

Pour ceci, on s'appuie sur un résultat dynamique de [21] :

Théorème 2.2. (Implosion parabolique - plan dynamique)
Soit $\Delta = \Lambda - \Lambda_1$; pour tout $c \in \Delta$, il existe $n \geq 0$ tel que $Q_c^{nk}(c)$ soit séparé de c par $\mathcal{R}_c(\hat\theta_1) \cup \{y_2(c)\} \cup \mathcal{R}_c(\hat\theta_2)$.

On en déduit que pour tout $c \in \Delta$, $K(g_c)$ est totalement discontinu. Quant au cas $c \notin \Delta$ proche de $\partial\Lambda$, ceci découle du fait que $\partial\Lambda \subset (\mathbb{C} - M) \cup \{\hat{c}\}$ et $K(g_c) \subset K_c$.

Donc, $M_{c_0} \subset\subset \Lambda$.

Remarque. $\chi(c_1) = 1/4$.

En effet, soit $c \in W$; il existe $\tilde\varphi_c$ tel que $g_c \sim_{hyb} Q_{\chi(c)}$ et il existe $z_a(c) \in U_c'$ fixe attractif tel que $g_c'(z_a(c)) = \rho_W(c)$. Soit U un voisinage relativement compact dans le bassin immédiat de $z_a(c)$: $\tilde\varphi_c : U \to \tilde\varphi_c(U)$ est biholomorphe. Donc, $\tilde\varphi_c(z_a(c))$ est un point fixe de $Q_{\chi(c)}$ et $Q_{\chi(c)}'(\tilde\varphi_c(z_a(c))) = g_c'(z_a(c))$.

Par suite, on a
$$\chi : \begin{aligned} W &\to W_0 \\ c &\mapsto \rho_{W_0}^{-1} \circ \rho_W(c), \end{aligned}$$

et comme χ est continue, $\chi(c_1) = 1/4$.

Par construction, $c \in U_c$ pour tout $c \in \Lambda$: il s'agit donc de la valeur critique de g_c (tout point d'image c est précritique pour Q_c, et g_c est quadratique). Notons $\omega_c = g_c^{-1}(c)$ le point critique de g_c. D'après le Théorème 0.6, $\chi : M_{c_0} \to M$ a un degré d égal au nombre de tours de $g_c(\omega_c) - \omega_c = c - \omega_c$ quand c parcourt une fois un lacet qui tourne (une fois) autour de M_{c_0}.

Puisque g_c est quadratique, on peut appliquer le théorème des fonctions implicites pour montrer que $c \mapsto \omega_c$ est holomorphe (point critique simple). Posons $h(c) = c - \omega_c$. Cette fonction est donc holomorphe et ne s'annule pas au voisinage de $\partial\Lambda$ (car M_{c_0} est compact dans Λ). Pour montrer que

$d = 1$, nous pouvons donc utiliser le principe de l'argument pour montrer que d est le nombre des racines de $h(c) = 0$ compté avec multiplicité. Nous allons procéder en deux étapes : d'abord montrer que la seule solution dans Λ est c_0 et ensuite qu'il s'agit d'une racine simple.

Soit c' une racine ; ceci signifie que l'application $g_{c'}$ fixe le point critique. En comptant ses points fixes, on en déduit que l'un est c' et l'autre est le point répulsif au bord de son bassin : il coïncide avec $\gamma_{c'}(\theta_1)$ et $\gamma_{c'}(\theta_2)$. Or ces conditions caractérisent c_0, donc $c' = c_0$. Le fait que c_0 est une racine simple a été montré par Gleason (voir [9], II, Chap. 19, Lemme 2).

Par suite, $d = 1$ et la famille est mandelbrotesque.

3 W non primitive

Nous allons changer la définition des $A'(c)$ et $A(c)$ en deux temps.

On fixe $\varepsilon_0 > 0$ et on définit dans le plan des paramètres :
- $L_M(\theta_1) := \{c \in \Lambda_2,\ arg_M(c) = \theta_1 + \varepsilon_0 G_M^2(c)\}$,
- $L_M(\theta_2) := \{c \in \Lambda_2,\ arg_M(c) = \theta_2 - \varepsilon_0 G_M^2(c)\}$,
- $\Lambda \subset \Lambda_1$, l'ouvert contenant W et délimité par $L_M(\theta_1)$ et $L_M(\theta_2)$.

Lemme 3.1. *Soit $0 < \varepsilon < \varepsilon_0$; pour tout c dans Λ,*

$$\begin{cases} L(\theta_1) = \{z,\ arg_c(z) = \theta_1 + \varepsilon G_c^2(z),\ 0 < G_c(z) < \eta\} \\ L(\theta_2) = \{z,\ arg_c(z) = \theta_2 - \varepsilon G_c^2(z),\ 0 < G_c(z) < \eta\} \end{cases} \quad (\subset A(c))$$

sont bien définies jusqu'en $y_1(c)$.

DÉMONSTRATION. Si $c \in \Lambda \cap M$, il n'y a pas de problème.
Soit donc $c \in \Lambda \setminus M$; $L(\theta_1)$ est mal définie si et seulement si il existe une préimage ω de c dont l'argument t est dans $]\theta_1, \theta_2 + \varepsilon\eta^2[$ et le potentiel s est plus grand que $\sqrt{\frac{t-\theta_1}{\varepsilon}}$.

Dans le plan dynamique de Q_{c_0}, il revient au même de considérer $z_0 = \varrho_{c_0}^{-1} \circ \Phi_M(c)$ et d'étudier l'existence d'une préimage $w_0 = \phi_{c_0}^{-1}(e^{s+2i\pi t})$ de z_0 par Q_{c_0} telle que $0 < s < \eta'$ et $\theta_1 < t < \theta_1 + \varepsilon s^2$.

Soit $\Omega' = \{z = \phi_{c_0}^{-1}(e^{s+2i\pi t}),\ 0 < s < \eta'$ et $\theta_1 < t < \theta_1 + \varepsilon s^2\}$; il suffit de montrer que l'orbite d'un point de Ω' ne tombe jamais dans le domaine de définition de z_0, à savoir $B \overset{def}{=} \{\phi_{c_0}^{-1}(e^{s+2i\pi t}),\ 0 < s < \eta_1$ et $\theta_1 + \varepsilon_0 s^2 < t < \theta_2 - \varepsilon s^2\}$.

On a $\Omega' \cap B = \emptyset$. L'application de premier retour est donnée par $Q_{c_0}^k$ car $\Omega' \subset A'(c)$. On note $\Omega = Q_{c_0}^k(\Omega') = \{z = \phi_{c_0}^{-1}(e^{s+2i\pi t}),\ 0 < s < \eta$ et $\theta_1 < t < \theta_1 + \varepsilon s^2/2^k\}$; étant donné que $\Omega \cap A'(c) \subset \Omega'$ et $\Omega \cap B = \emptyset$, $L(\theta_1)$ est bien définie. Il en est de même pour $L(\theta_2)$. ∎

On a choisi ε de telle sorte que $L(\theta_s)$ ne contienne pas la valeur critique. Etant dans $A(c)$, $L(\theta_s)$ admet deux préimages lisses par Q_c^k, l'une $L'(\theta_s)$ aboutissant à $y_1(c)$, l'autre $L(\theta'_s)$ à $y'_1(c)$.

Soit $z \in L'(\theta_s)$: $2^k Arg_c z = \theta_s \pm \varepsilon 2^{2k} G_c^2(z)$.

▷ $Arg_c z > \theta_1 + \varepsilon G_c^2(z)$ et $L'(\theta_1)$ est au-dessus de $L(\theta_1)$.

▷ $Arg_c z < \theta_2 - \varepsilon G_c^2(z)$ et $L'(\theta_2)$ est au-dessous de $L(\theta_2)$.

On note $A_2(c)$ l'ouvert contenant $y'_1(c)$ limité par $L(\theta_s)$, $s = 1,2$ et $\{G_c = \eta\}$, et $A'_2(c) = Q_c^{-k}(A_1(c) \cap A'(c))$ limité par $L(\theta'_s)$, $L'(\theta_s)$, $s = 1,2$ et $\{G_c = \eta'\}$ (voir Figure 5).

On a $A'_2(c) \subset A_2(c)$. Maintenant, $\forall c \in \Lambda$, $\partial A'_2(c)$ et $\partial A_2(c)$ n'ont que $y_1(c)$ en commun et $Q_c^k : A'_2(c) \to A_2(c)$ est holomorphe propre de degré deux.

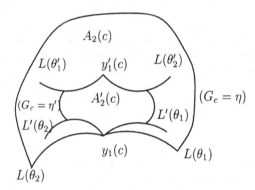

Figure 5: $A'_2(c) \subset A_2(c)$.

Second temps

Le point $y_1(c)$ étant répulsif, on peut rajouter un petit disque linéarisable de manière à construire une application à allure polynomiale g_c. Cependant, il nous faut choisir ces disques plus soigneusement pour avoir une famille analytique.

Lemme 3.2. *Pour tout $c \in \Lambda$, il existe un secteur invariant $\Delta_1(c)$ issu de $y_1(c)$, disjoint de $A(c)$, et une application continue $\iota : \Lambda \times \Delta_1(c_0) \to \mathbb{C}$ telle que, pour tout $c \in \Lambda$, $\iota(c, \Delta_1(c_0)) = \Delta_1(c)$.*

DÉMONSTRATION. Notons $\rho(c) = (Q_c^k)'(y_1(c))$. Pour tout $c \in \Lambda$, nous considérons la coordonnée linéarisante de $y_1(c)$ pour Q_c^k, que l'on prolonge au plan : on a donc une application holomorphe $\kappa_c : \mathbb{C} \to \mathbb{C}$, univalente au voisinage de 0, telle que $\kappa_c(0) = y_1(c)$ et $\kappa_c(\rho(c) \cdot z) = Q_c^k \circ \kappa_c(z)$ pour tout $z \in \mathbb{C}$. Nous normalisons κ_c par $\kappa'_c(0) = 1$. Par suite, κ_c dépend $\mathbb{C}$-analytiquement de c.

Notons Γ_c la composante connexe de $\kappa_c^{-1}(\mathcal{R}_c(\theta_1) \cup \mathcal{R}_c(\theta_2))$ contenant 0 et qui est invariante par la multiplication par $\rho(c)$. On transporte dans ce plan les restrictions aux rayons des applications $\phi_c^{-1} \circ \phi_{c_0}$ afin d'avoir un mouvement holomorphe $j : \Lambda \times \Gamma_{c_0} \to \overline{\mathbb{C}}$. Chaque Γ_c est un quasi-disque.

Nous voudrions prolonger j en un mouvement holomorphe global tout en gardant la conjugaison avec les multiplications par $\rho(c)$. Ceci semble être un problème difficile si on ne veut pas rétrécir Λ. En revanche, en utilisant l'extension barycentrique de A. Douady et C. J. Earle (voir [8]), on peut trouver un prolongement $j : \Lambda \times \overline{\mathbb{C}} \to \overline{\mathbb{C}}$ continue, qui est quasiconforme à c fixé, et qui satisfait $\rho(c) \cdot j(c,z) = j(c, \rho(c_0) \cdot z)$ pour tout (c, z).

On considère maintenant $S(c_0)$ la composante connexe de $\mathbb{C} - \Gamma_{c_0}$ ne contenant pas au voisinage de l'origine $\kappa_{c_0}^{-1}(A(c_0))$. Il existe $\lambda > 1$ et une application conforme $h : S(c_0) \to \Sigma$, où Σ est un secteur droit, tels que $h(\rho(c_0) \cdot z) = \lambda h(z)$. On note alors $\delta(r) = h^{-1}(|z| = r) \subset S_{c_0}$ pour $r > 0$.

Comme les domaines linéarisables de $y_1(c)$ bougent continûment, il existe une application continue $c \mapsto r(c) > 0$ telle que κ_c soit univalente sur le secteur $Sec(c)$ borné, issu de l'origine, et bordé par $\Gamma(c) \cup j(c, \delta(r(c)))$. Notons $\Delta_1(c) = \kappa_c(Sec(c))$. On pose

$$\iota(c, z) = \kappa_c \circ j(c, (h \circ \kappa_{c_0})^{-1}(z \cdot r(c)/r(c_0))).$$

■

Nous prolongeons ι par $\phi_c^{-1} \circ \phi_{c_0}$ sur $(A(c_0) \setminus A_2(c_0))$. On a bien $\Delta_1'(c) \subset\subset \Delta_1(c)$. Ceci permet donc de définir des domaines U_c' et U_c qui varient continûment, et d'obtenir une famille analytique d'applications à allure quadratique dans Λ (*cf* le cas primitif). Par le Théorème 0.6, nous obtenons M_{c_0}, lieu de connexité de la famille, et une application $\chi : M_{c_0} \to M$ continue.

Lemme 3.3. *Pour* $c \in M_{c_0}$, *on a* $\lim_{c \to c_1} \chi(c) = 1/4$.

Remarque. Par la construction de Λ, l'adhérence dans $\mathbb{C}$ de M_{c_0} est c_1.
DÉMONSTRATION. On commence par montrer que $\chi_{|W} = \rho_{W_0}^{-1} \circ \rho_W$. Il s'ensuit que le lemme est vrai sur W, et même sur $\overline{W}$. On considère une suite (c_n) de M_{c_0} de limite c_1. Par un corollaire des inégalités de Yoccoz (voir *e.g.* [14]), les $\chi(c_n)$ appartiennent à des membres de diamètres tendant vers zéro, *i.e.* ils ont même limites que les racines de ces membres, à savoir $1/4$. ■

Par suite, χ est propre et son extension $\chi : M_{c_0} \cup \{c_1\} \to M$ est un revêtement ramifié de degré fini (Théorème 0.6). D'après la remarque ci-dessus, on peut considérer un lacet basé à la racine c_1 qui fait une fois le tour de M_{c_0} dans $\Lambda \cup \{c_1\}$. En adaptant la démonstration du Théorème 0.6 dans le cas $M_{c_0} \subset\subset \Lambda$, on montre que le degré de χ est égal au nombre de tours de $(c - \omega_c)$ quand c parcourt une fois ce lacet. Comme dans le cas primitif, la fonction $c \mapsto c - \omega_c$ est holomorphe et c_0 est la seule racine comptée avec multiplicité. Par conséquent, χ est un homéomorphisme. On en déduit que $(g_c)_{c\in\Lambda}$ est une famille semi-mandelbrotesque.

Bien qu'on ne puisse pas définir d'application à allure polynomiale à la racine c_1 (comme dans le cas primitif), on a cependant la

Proposition 3.4. *Le polynôme $Q_{1/4}$ figure de manière rigide dans Q_{c_1}.*

DÉMONSTRATION. D'après [12], les composantes connexes de $\overset{\circ}{K}_{c_1}$ sont des quasidisques. On note U_1 celle qui contient c_1.

Soit $\varphi : U_1 \to \mathbb{D}$ une application conforme telle que $\left\{ \begin{array}{l} \varphi(y_1(c_1)) = 1, \\ \varphi(\omega_{c_1}) = 0. \end{array} \right.$

Cette application conjugue $Q_{c_1}^k$ à

$$B : z \mapsto \frac{z^2 + 3}{1 + 3z^2}.$$

Comme U_1 est un quasidisque, φ se prolonge quasiconformément à $\overline{\mathbb{C}}$, et il ne reste plus qu'à montrer que $Q_{1/4}$ figure dans B.

Pour cela, nous procédons comme dans la démonstration de la Prop. 2.1 de [12].

Notons S_0 un secteur invariant issu de $1/2$ à l'extérieur du chou-fleur, que l'on découpe en domaines fondamentaux D_n, et soit T_0 un secteur issu de 1 correspondant à S_0. On note S_n l'image réciproque de S_0 issu de $Q_{1/4}^{-n}(1/2)$. Sur une réunion de quadrilatères, on construit une application quasiconforme ψ par *pull-back*. Sur les secteurs restant, on prolonge par des arcs de cercles.

Pour montrer que ψ est ACL (voir [1]), il suffit de considérer une suite d'applications (ψ_n) de $W^{1,\nu}$ uniformément bornée qui converge uniformément vers ψ. Cette suite est en fait donnée par la construction de ψ. Pour montrer qu'elle est uniformément bornée, il suffit donc de calculer la norme L^ν de $\partial_z\psi$, à l'extérieur du chou-fleur.

La restriction de ψ à un domaine D_n a une eccentricité

$$K_\psi \asymp \left| \log \frac{\sqrt{n+1}}{\sqrt{n}} \right|^{-1} \asymp n.$$

Pour conclure, montrons :

Lemme 3.5. *On a $\psi \in W^{1,\nu}$ pour tout $\nu \in]1,2[$.*

DÉMONSTRATION. On note $P = Q_{1/4}$.

$$\|\psi\|_{L^\nu}^\nu \asymp \int (K_\psi Jac\,\psi)^{\nu/2} = \sum_n \int_{S_n} (K_\psi Jac\,\psi)^{\nu/2} + C.$$

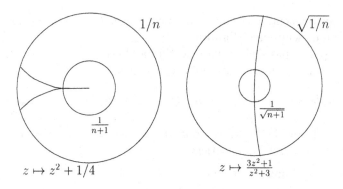

Figure 6: Estimation des distances.

Or

$$\int_{S_n} (K_\psi Jac\,\psi)^{\nu/2} = \int_{S_0} (K_\psi(P^{-n})Jac\,\psi(P^{-n}))^{\nu/2} |(P^{-n})'|^2$$
$$= \int_{S_0} (K_\psi Jac\,\psi)^{\nu/2} |(B^{-n})'|^\nu |(P^{-n})'|^{2-\nu}$$

car $Jac\,\psi(P^{-n})|(P^{-n})'|^2 = Jac\,\psi|(B^{-n})'|^2$.

Notons $\alpha \in]1,2[$ la dimension de Hausdorff de J_P et soit $p \in]1, \frac{\alpha}{\alpha+\nu-2}[$. Par le théorème de Koebe,

$$\|\psi\|_{L^\nu}^\nu \le C \int_{S_0} (K_\psi Jac\,\psi)^{\nu/2} \sum |(B^{-n})'(1)|^\nu \cdot |(P^{-n})'(1/2)|^{2-\nu},$$

et par l'inégalité de Hölder, cette somme est bornée par

$$\left(\sum |(B^{-n})'(1)|^{p\nu}\right)^{1/p} \left(\sum |(P^{-n})'(1/2)|^{\frac{2-\nu}{1-1/p}}\right)^{1-1/p}.$$

D'après [18], comme $p\nu > 1$ et $\frac{2-\nu}{1-1/p} > \alpha$, les deux sommes sont finies. De plus,

$$\int_{S_0} (K_\psi Jac\,\psi)^{\nu/2} \le C \sum_n n^{\nu/2} (\text{Aire } D_n)^{1-\nu/2} \left(\int_{D_n} Jac\,\psi\right)^{\nu/2}.$$

On trouve

$$\int_{S_0} (K_\psi Jac\,\psi)^{\nu/2} \le C \sum_n (1/n)^{4-(3/2)\nu} < \infty.$$

Par suite, l'application est dans $W^{1,\nu}$ et $Q_{1/4}$ figure de manière rigide dans B.　　　　　　　　　　　　　　　　　　　　　　　　　　　　■

Remarque. Réciproquement, $\psi^{-1} \in W^{1,2}$ car $\|\psi\|_{L^2}^2 \asymp \sum n \int_{\psi(D_n)} Jac\,\psi^{-1} \asymp \sum 1/n^3 < \infty$.

Partie II : Opération de modulation

Soient $c_0 \in \mathcal{D}_0$ de période $k \geq 1$, et W sa composante hyperbolique. On considère $\chi_{c_0} : M_{c_0} \to M$. Si W n'est pas primitive, on prolonge χ_{c_0} par continuité en posant $\chi_{c_0}(c_1) = 1/4$.

On définit alors

$$c_0 \perp . : \quad M \to M_{c_0}$$
$$c \mapsto c_0 \perp c = \chi^{-1}(c).$$

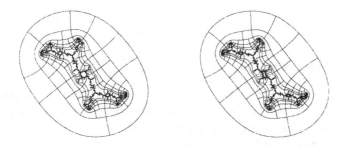

Figure 3. *Deux ensembles de Julia de la même copie.*

Dans cette partie, nous allons analyser l'opération de modulation afin d'obtenir des liens de caractère topologique ou/et combinatoire entre les différents ensembles de Julia. Le théorème principal est le suivant (voir §6 pour un énoncé précis) :

Théorème 2 *L'opération de modulation agit sur les ensembles de Julia remplis comme suit : $K_{c_0 \perp c}$ est homéomorphe à K_{c_0} dans lequel on a remplacé les composantes connexes de l'intérieur par des "copies" de K_c.*

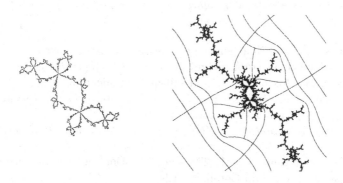

Figure 4. *Modulateur et zoom d'un polynôme modulé.*

Nous commencerons tout d'abord par étudier la combinatoire pour en extraire des invariants de M_{c_0}. Avant de démontrer le théorème, nous construirons un modèle abstrait de modulation de compacts. Ensuite, nous définirons un homéomorphisme entre ce modèle et l'ensemble de Julia du polynôme modulé. Une fois le théorème démontré, on énoncera quelques corollaires immédiats.

Le but est de comprendre la combinatoire de l'ensemble de Julia modulé à partir de celles de l'ensemble de départ et de c_0. A cette fin, nous allons étudier des *puzzles* associés à la copie M_{c_0} isolant les *petits* ensembles de Julia.

4   Puzzle associé à une copie M_{c_0}

La copie M_{c_0} de M diffère essentiellement de M en ce que les rayons dyadiques qui aboutissaient de manière unique à un point de M, ne sont plus les seuls qui aboutissent dans la copie. Plus précisément :

Définition. Soit $\hat{\mathcal{D}}_2 = \{c \in M_{c_0}, \exists n \geq 1, Q_c^n(c) = y_1(c)\}$.

Cet ensemble comprend exactement les paramètres de M_{c_0} qui représentent les points d'aboutissement des rayons dyadiques dans M.

Commençons par une proposition préliminaire qui va nous permettre de définir les puzzles :

Proposition 4.1. *Il existe une suite décroissante d'ouverts connexes $\Lambda_n, n \geq 1$, tels que :*
(i) $\Lambda_\infty \cap M = M_{c_0} \setminus (\hat{\mathcal{D}}_2 \cup \{c_1\})$, où $\Lambda_\infty = \cap \Lambda_n$;
(ii) $\forall z_0 \in Q_{c_0}^{-n}(y_1)$, il existe une fonction holomorphe $z : \Lambda_n \to \mathbb{C}$ holomorphe telle que $Q_c^n(z(c)) = y_1(c)$ avec $z(c_0) = z_0$; de plus, si $t \in \mathbb{Q}/\mathbb{Z}$ est l'angle d'un rayon qui aboutit à z_0, alors le rayon correspondant pour Q_c aboutit à $z(c)$.

DÉMONSTRATION. Nous allons procéder par récurrence sur n.
▷ D'abord, on considère Λ_0 délimité par $\mathcal{R}_M(\theta_1) \cup \{c_1\} \cup \mathcal{R}_M(\theta_2)$ et contenant c_0. D'après la Proposition 1.2, on sait que l'application de Λ_0, $y_1 : c \mapsto y_1(c)$ est bien définie et vérifie la propriété (ii). On note A_1 la composante connexe de $Q_{c_0}^{-k}(A)$ (*cf* Prop. 1.1) qui contient c_0.
Assertion 0-A. *Soit $t \in \gamma_{c_0}^{-1}(Q_{c_0}^{-i}(y_1)), 0 \leq i \leq k$; alors $\mathcal{R}_c(t)$ aboutit pour tout $c \in \Lambda_0$.*
En effet, pour tout $m > 1$, $2^m t \notin]\theta_1, \theta_2[$. Donc les rayons correspondant aboutissent en vertu du Corollaire A.5.

Par la Proposition A.1, on en déduit la propriété (ii) pour ces points.

Assertion 0-B. *Les rayons* $\mathcal{R}_M(\theta_1')$ *et* $\mathcal{R}_M(\theta_2')$ *aboutissent à un même point* $c_2 \in \hat{\mathcal{D}}_2$.

D'après la Proposition 0.3, $\mathcal{R}_M(\theta_1')$ aboutit à $c_2 \in \mathcal{D}_2$, car θ_1' est strictement prépériodique. Dans K_{c_2}, $\mathcal{R}_{c_2}(\theta_1')$ aboutit à c_2 (voir Théorème 0.4). Or, $c_2 \in \Lambda_0$, donc $\mathcal{R}_{c_2}(\theta_1')$ et $\mathcal{R}_{c_2}(\theta_2')$ aboutissent à $y_1'(c_2) = c_2$: on en déduit que $\mathcal{R}_{c_2}(\theta_2')$ aboutit aussi à c_2.

Ceci nous permet de définir $\Lambda_1 \subset \Lambda_0$ délimité par $\mathcal{R}_M(\theta_1') \cup \{c_2\} \cup \mathcal{R}_M(\theta_2')$.

▷ On suppose que l'on a un ouvert Λ_n bordé par les rayons préimages de θ_1 et θ_2 par un itéré au plus kn-ième. Ces rayons aboutissent à des points de $\hat{\mathcal{D}}_2$. On considère aussi A_n la composante connexe de $Q_{c_0}^{-nk}(A)$ qui contient c_0.

Assertion n-A. *Soit* $t \in \gamma_{c_0}^{-1}(Q_{c_0}^{-i}(y_1)), 0 \le i \le (n+1)k$; *alors* $\mathcal{R}_c(t)$ *aboutit pour tout* $c \in \Lambda_n$.

En effet, pour tout $m > 1$, $2^m t$ n'appartient pas aux intervalles dont les rayons sont dans Λ_n par construction. Donc les rayons correspondant aboutissent en vertu du Corollaire A.5.

Par la Proposition A.1, on en déduit la propriété (ii) pour ces points.

Assertion n-B. *Les rayons qui aboutissent à un point* $z_{n+1} \in (Q_{c_0}^{-k(n+1)}(y_1) \cap A_n)$ *aboutissent dans le plan des paramètres à un même point* $c_{n+1} \in \hat{\mathcal{D}}_2$.

La démonstration est la même que 0-B.

On peut donc définir $\Lambda_{n+1} \subset \Lambda_n$ l'ouvert bordé par ces rayons et contenant c_0.

Si $c \in \Lambda_\infty \cap M$, cela signifie d'une part que son orbite par Q_c reste bornée, et d'autre part que $c \in Q_c^{nk}(A(c)) = g_c^{-n}(A(c))$ pour tout $n \ge 1$, et donc $c \in K(g_c)$: ce qui implique $c \in M_{c_0}$. ∎

Définitions. En suivant l'idée des *puzzles* de Branner-Hubbard et Yoccoz (voir [2], [14]), on considère la suite de partitions définies comme suit, pour $c \in \Lambda_\infty$:

- on fixe $\eta > 0$ et on pose $\Gamma_0 = \cup_{0 \le n \le k}(\mathcal{R}_c(2^n \theta_1) \cup \{Q_c^n(y_1)\} \cup \mathcal{R}_c(2^n \theta_2)) \cup (G_c = \eta)$, dont la partition définit le *puzzle de niveau 0*, noté $\mathcal{P}_0$: la fermeture des composantes connexes bornées du complémentaire sont les *pièces* de $\mathcal{P}_0$. On définit par récurrence $\Gamma_n = Q_c^{-1}(\Gamma_{n-1})$, qui induit le puzzle $\mathcal{P}_n$ de niveau n. Q_c envoie une pièce de $\mathcal{P}_n$ sur une pièce de $\mathcal{P}_{n-1}$.

- un *bout* $x = (X_0 \supset X_1 \supset ...)$ est une suite décroissante de pièces telle que $\forall n \ge 0$, $X_n \in \mathcal{P}_n$. Q_c agit sur les bouts. On note $\mathcal{E}_\mathcal{P}$ l'ensemble des bouts. On munit $\mathcal{E}_\mathcal{P}$ de la distance ultramétrique d_u comme suit : si $x = (X_0 \supset X_1 \supset ...)$ et $y = (Y_0 \supset Y_1 \supset ...)$, on note $\nu = \sup\{n \in \mathbb{N} \cup \{\infty\}, X_n = Y_n\}$ et $d_u(x, y) = 1/2^\nu$.

- on définit enfin *l'impression* du bout x par $I(x) = \cap X_n$, compact connexe. On munit l'ensemble des impressions de la métrique de Hausdorff d_H sur les compacts non vides de $\mathbb{C}$.

À tout $z \in K_c$ qui n'est pas dans la grande orbite de $y_1(c)$, on associe le
bout $N(z) = (X_0 \supset X_1 \supset \ldots)$ tel que $\forall n \geq 0$, $z \in X_n$. Si z est une préimage
de y_1, on définit $N(z)$ comme l'ensemble des bouts dont l'impression contient
z. N est donc une application multivaluée surjective. Elle est continue en ce
sens que si z_k tend vers z alors tout point d'accumulation de $N(z_k)$ appartient
à $N(z)$. De plus, on a $z \in I \circ N(z)$.

La Proposition 4.1 entraîne le résultat suivant :
- $\forall n \geq 0$,

$$\iota_n \; : \; \Lambda_n \times \cup_{0 \leq p \leq n} \Gamma_p(c_0) \to \mathbb{C}$$
$$(c, z) \mapsto \phi_c^{-1} \circ \phi_{c_0}(z)$$

est un mouvement holomorphe de point base c_0 tel que $\iota_n(\Gamma_p(c_0)) = \Gamma_p(c)$.
- ces mouvements holomorphes induisent un homéomorphisme $\iota_c : \mathcal{E}_\mathcal{P}(c_0) \to$
$\mathcal{E}_\mathcal{P}(c)$ (c'est en fait une isométrie) conjugant Q_{c_0} à Q_c.

Une adaptation de la démonstration du Théorème 5.7 b) de [14] conduit
à (voir aussi [16]) :

Proposition 4.2. *Soit $c \in M_{c_0} \setminus \hat{\mathcal{D}}_2$, alors pour tout bout x de Q_c, ou bien
$I(x)$ est un point, ou bien $I(x)$ est homéomorphe à $K_{\chi_{c_0}(c)}$.*

Remarque. Si $c \in \hat{\mathcal{D}}_2$, alors l'impression de tout bout est un point. D'une
part, les rayons aboutissant aux points précritiques vont morceler les petites
copies du Julia ; d'autre part, comme le polynôme est sous-hyperbolique, il
existe une métrique telle que Q_c soit fortement dilatante, et par conséquent
les pièces ont un diamètre qui décroît exponentiellement ($\mathring{K}_c = \emptyset$).

Corollaire 4.3. *Pour tout c dans M_{c_0}, l'ensemble de Julia J_c est localement
connexe en tout point qui n'appartient pas à une petite copie.*

On vérifie aussi que l'application $I : (\mathcal{E}_\mathcal{P}, d_u) \to (Comp(\mathbb{C}), d_H)$ est semi-
continue supérieurement et continue aux bouts dont l'impression est un sin-
gleton. Elle est de plus injective.

Corollaire 4.4. *Si $c \in M_{c_0} \setminus \hat{\mathcal{D}}_2$, on note $S_c = \{z \in K_c$, tel que $\{z\}$ soit
l'impression d'un bout$\}$; alors il existe un homéomorphisme de S_{c_0} sur S_c qui
conjugue Q_{c_0} et Q_c.*

5   Modulation de compacts du plan

Soient $K \subset \mathbb{C}$ un compact connexe, localement connexe, plein, d'intérieur non
vide et $L \subset \mathbb{C}$ compact connexe, plein. On se propose de définir K_L comme
étant K auquel on a remplacé les composantes connexes de son intérieur par
des copies de L (Figure 7).

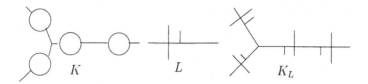

Figure 7: Modulation de compacts.

5.1 Introduction

D'après [9, 7], si K est compact, connexe, localement connexe et plein, $\overset{\circ}{K}$ a au plus un nombre dénombrable de composantes connexes $\{U_i\}_{i \in \mathbb{N}} = \pi_0(\overset{\circ}{K})$; chacune est un domaine de Jordan, et l'intersection des fermetures de deux d'entre elles contient au plus un point. Pour chaque U_i, on choisit un *centre* x_i. On considère aussi des uniformisantes $\varphi_i : \mathbb{D} \to U_i$ telles que $\varphi(0) = x_i$ et $\varphi'(0) > 0$. On appelle *rayon interne* de U_i tout segment $\varphi(re^{2i\pi t})$, où $t \in \mathbb{R}/\mathbb{Z}$ est fixé et r parcourt $[0, 1]$. Ceci nous permet de définir des *arcs réglementaires*:

▷ On définit sur K la relation d'équivalence fermée $\sim$ engendrée par:

$$x \sim y \text{ si } x = y \text{ ou s'il existe } U \in \overset{\circ}{K} \text{ tel que } x, y \in \overline{U}.$$

Comme K est localement connexe, pour tout $\varepsilon > 0$, il n'y a qu'un nombre fini de classes de diamètre supérieur à ε. On en déduit que son espace quotient $\hat{K}$ est compact et métrisable (voir [3], Chap. 1). On note $g : K \to \hat{K}$ la projection, qui est continue et propre. Les fibres sont ou bien des points, ou bien des réunions fermées et connexes de fermetures de composantes de $\overset{\circ}{K}$; en particulier, les fibres sont connexes.

Remarque. Il se peut que $\hat{K}$ soit un singleton, comme dans le cas du "lapin" de A. Douady.

▷ Pour tout couple (z, w) de K, il existe un unique chemin de $\hat{K}$ qui relie leurs projections. On définit alors l'arc réglementaire $]z, w[_K$ comme le relevé de cet arc qui joint z à w, homéomorphe à $]0, 1[$, qui traverse l'intérieur de K de manière minimale le long des rayons internes.

▷ Pour chaque $U \in \pi_0(\overset{\circ}{K})$, on définit $\pi_U : K \to \overline{U}$ continue surjective par:
- si $x \in \overline{U}$, $\pi_U(x) = x$,
- sinon, il existe un unique $y \in \overline{U}$ tq $\begin{cases} y \in [x, x_U]_K \\ ]x, y[_K \cap U = \emptyset \end{cases}$ et on pose $\pi_U(x) = y$.

π_U est localement constante sur $K \setminus \overline{U}$.

On définit l'application continue

$$\varphi : \quad K \to \hat{K} \times \Pi\overline{U}_i$$
$$x \to (g(x), (\pi_i(x))).$$

φ est injective : soient $x \neq y$, et donc $]x, y[_K \neq \emptyset$;

- si $]x, y[_K \cap \overset{\circ}{K} = \emptyset$ alors $g(x) \neq g(y)$;
- si $]x, y[_K \cap \pi_0(\overset{\circ}{K}) \neq \emptyset$, alors il existe U telle que $]x, y[_K \cap U \neq \emptyset$, donc $\pi_U(x) \neq \pi_U(y)$.

Par suite, on peut plonger K dans $\hat{K} \times \Pi\overline{U}_i$ à l'aide de φ :
$\varphi(K) = \tilde{K} = \{(\xi, \xi_i) \in \hat{K} \times \Pi\overline{U}_i, \xi_i = \pi_i(g^{-1}(\xi))$ si $g(U_i) \neq \xi$, et sinon $\exists k$ tq $g(U_k) = \xi$ et $\xi_i = \pi_i(x_k)$ pour $i \neq k\}$.

Remarque. Si $x \in \overline{U} \cap \overline{V}$, alors on peut aussi bien choisir x_U que x_V comme centre.

Ce modèle de K fournit une façon de définir K_L dans $\hat{K} \times \Pi L_i$ tel que les L_i soient des copies de L.

5.2 Lemme et définition

Pour chaque composante U de l'intérieur de K, on considère $\gamma_U : \mathbb{T} \to \partial U$ l'extension de φ_i au bord, qui est un homéomorphisme.

Soient $L \subset \mathbb{C}$ un compact connexe plein, et γ_L son lacet de Carathéodory, bien défini pp. (d'après P. Fatou).

Pour chaque composante U, on suppose que l'application

$$\psi_U : \quad \pi_U(K \setminus \overline{U}) \to \partial L$$
$$z \mapsto \gamma_L \circ \gamma_U^{-1}(z)$$

est bien définie.

Lemme 5.1. et Définition. *On pose* $K_L = \{(\xi, \xi_i) \in \hat{K} \times \Pi L_i, \xi_i = \psi_i \pi_i(g^{-1}(\xi))$ *si* $g(U_i) \neq \xi$, *et sinon* $\exists k$ *tq* $g(U_k) = \xi$ *et* $\xi_i = \psi_i \pi_i(x_k)$ *pour* $i \neq k\}$.
Alors K_L *est compact et on l'appelle "K modulé par L".*

DÉMONSTRATION. $\hat{K} \times \Pi L_i$ étant un espace métrique compact, il suffit de vérifier que K_L est fermé par les suites : soit $\tilde{\xi}^n = (\xi^n, (\xi_i^n)_{i \geq 0})$, $n \geq 0$, une suite de K_L convergeant vers $\tilde{\xi}^\infty = (\xi^\infty, (\xi_i^\infty)_{i \geq 0})$.

On définit $\Omega := \cup_{i \geq 0}\{g(x_i)\}$, qui correspond aux classes d'équivalence ayant plus d'un élément.

À chaque $n \geq 0$, on associe un point $y_n \in K$ comme suit : ou bien $\xi^n \notin \Omega$, et on pose $y_n = g^{-1}(\xi^n)$; ou bien $\xi^n \in \Omega$, alors il existe k_n tel que $g(x_{k_n}) = \xi^n$ et pour tout $i \neq k_n$, $\xi_i^n = \psi_i \pi_i(x_{k_n})$, et dans ce cas on pose $y_n = x_{k_n}$.

On extrait une sous-suite (y_{n_k}) convergeant vers $y_\infty \in K$. En particulier, $\zeta'' = g(y_n) \to g(y_\infty) = \xi^\infty$.

On considère les deux cas suivants :

premier cas : $g(y_\infty) \notin \Omega$. Comme g est continue, la suite y_n est convergente. Pour chaque i, il existe un voisinage connexe de y_∞ disjoint de $\overline{U}_i$. Par suite, pour n assez grand, $\pi_i(y_n) = \pi_i(y_\infty)$ et $\xi_i^\infty = \psi_i \pi_i(y_\infty)$.

Donc $\tilde{\xi}^\infty \in K_L$.

second cas : $g(y_\infty) \in \Omega$.

On note $\Omega(y_\infty)$ l'ensemble des U_i qui contiennent y_∞ dans leur adhérence. On choisit $U \in \Omega(y_\infty)$ vérifiant la propriété suivante si elle est satisfaite : il existe une sous-suite (y_{n_j}) telle que $U \cap]y_\infty, y_\infty[\neq \emptyset$ et $y_{n_j} \longrightarrow y_\infty$.

Si $U_i \neq U$, alors $\pi_i(y_\infty) = \pi_i(x_U)$ car $y_\infty \in \overline{U}$ et $g(y_\infty) = g(x_U)$.

Soit $U_i \notin \Omega(y_\infty)$; pour $k \gg 1$, y_{n_k} appartient a un voisinage connexe de y_∞ disjoint de $\overline{U}_i$ et donc $\pi_i(y_{n_k}) = \pi_i(y_\infty)$ et on en déduit que $\xi_i^\infty = \psi_i \pi_i(y_\infty) = \psi_i \pi_i(x_U)$.

Soit $U_i \in \Omega(y_\infty) \setminus \{U\}$; pour $j \gg 1$, on a aussi $\pi_i(y_{n_j}) = \pi_i(y_\infty)$ et donc $\xi_i^\infty = \psi_i \pi_i(x_U)$.

On en déduit que K_L est compact. ∎

Remarque. K_L est bien le modèle que l'on cherchait en ce sens qu'on peut définir les injections suivantes :

$$\varpi : \quad K \setminus (\cup_{i \geq 0} \overline{U}_i) \quad \to \quad K_L$$
$$x \qquad\qquad \mapsto \quad (g(x), \psi_i \pi_i(x))$$

(voir Figure 7)

$$\varpi_i : \quad L \quad \to \quad K_L$$
$$x \quad \mapsto \quad g(U_i), \left| \begin{array}{l} \psi_j \pi_j(x_i)), \text{ pour } j \neq i \\ x \in L_i \end{array} \right.$$

et $K_L = (\cup \varpi_i(L)) \cup \varpi(K \setminus (\cup_{i \geq 0} \overline{U}_i))$.

6   Théorème 2

Enoncé. *Soit $c_0 \in \mathcal{D}_0 \setminus \{0\}$ de période k et soit $c \in M$; on note $\tilde{c} = c_0 \perp c$. K_{c_0}, K_c, $K_{\tilde{c}}$ les ensembles de Julia remplis de c_0, c, $\tilde{c}$ respectivement. Alors $K_{\tilde{c}}$ est homéomorphe à $K_{c_0 K_c}$.*

Nous allons définir un homéomorphisme en utilisant notre travail sur les bouts. Nous remarquons que, pour tout $x \in \mathcal{E}_{c_0}$, g est constante sur $I(x)$; ceci nous permet de définir $\mathcal{G} : \mathcal{E}_{c_0} \to \hat{K}_{c_0}$ par $\mathcal{G}(x) = g(I(x))$. Cette application est continue car $d_H(I(x), X_n) \to 0$.

On définit alors $\tilde{g} : K_{\tilde{c}} \to \hat{K}_{c_0}$ par $\tilde{g}(z) = \mathcal{G} \circ \iota \circ N(z)$. Cette application est bien définie car pour tout $x \in N(z)$, $z \in I(x)$ et donc $\mathcal{G}(\iota(I(x))) = g(\iota(z))$. De plus, elle est surjective et continue d'après les propriétés de continuité de $\mathcal{G}$, ι et N.

DÉMONSTRATION DU THÉORÈME 2. On numérote les copies de K_c dans $K_{\tilde{c}}$ de manière cohérente avec les U_i.

On définit $\tilde{\varphi} : K_{\tilde{c}} \to K_{c_0 K_c}$ par :
- si $x \notin (\cup K_c(i))$, $\xi = \tilde{g}(x)$, et $\xi_j = \psi_j \pi_j(g^{-1}\tilde{g}(x))$;
- si $x \in K_c(i)$, $\xi = \tilde{g}(x)$, $\xi_i = x$, et pour $j \neq i$, $\xi_j = \psi_j \pi_j(x_i)$.

Montrons que $\tilde{\varphi}$ est un homéomorphisme.

▷ Continuité : il s'agit d'une adaptation de la démonstration du Lemme 5.1. On considère une suite (x_n) de $K_{\tilde{c}}$ convergente. On lui associe une suite (y_n) de K_{c_0} en s'appuyant sur la Proposition 4.1, puis on procède au même type de discussion, selon la position d'un point d'accumulation de (y_n).

▷ Surjectivité : soit $\tilde{\xi} = (\xi, \xi_i) \in K_{c_0 K_c}$;
· si $\xi \notin \Omega$, $\tilde{g}$ est injective et $\tilde{g}^{-1}(\xi)$ répond à la question.
· si $\xi = g(U_j)$, $\exists k$ tel que $g(U_k) = \xi$ et $\xi_i = \psi_i \pi_i(x_k), i \neq k$; d'où $\tilde{\varphi}(\xi_k) = \tilde{\xi}$ avec $\xi_k \in K_c(k)$.

▷ Injectivité : soient x, y tels que $\tilde{\varphi}(x) = \tilde{\varphi}(y) = (\xi, \xi_i)$;
· si $\xi \notin \Omega$ alors $x, y \in K_c \setminus (\cup K_c(i))$ où $\tilde{g}$ est injective, d'où $x = y$.
· sinon, il existe k tel que $g(x_k) = \xi$ et pour tout $i \neq k$, $\xi_i = \psi_i \pi_i(x_k)$. Du coup, x et y appartiennent à un même $K_c(k)$. Par conséquent, $x = \tilde{\varphi}(x)_k = \xi_k = \tilde{\varphi}(y)_k = y$. ∎

7 Corollaires

Dans ce paragraphe, nous déduisons des corollaires du Théorème 2.

Corollaire 7.1. *Soient $c_0 \in \mathcal{D}_0$, $c \in \mathcal{D}_0 \cup \mathcal{D}_2$; on note $\tilde{c} = c_0 \perp c$, et H_{c_0}, H_c, $H_{\tilde{c}}$ les arbres de Hubbard de c_0, c, $\tilde{c}$ respectivement. $H_{\tilde{c}}$ s'obtient en remplaçant les composantes connexes de $\overset{\circ}{H}_{c_0}$ par H_c en recollant par β, β'. Les points marqués sont alors déterminés comme suit :*
- $x_0 = 0$;
- *si $x_i \in H_c(j)$ est défini,* $\left| \begin{array}{l} j \neq 0, \ x_{i+1} = x_i \ dans \ H_c(j+1), \\ j = 0, \ x_{i+1} = Q_c(x_i) \ dans \ H_c(1). \end{array} \right.$

Corollaire 7.2. *Soient $c_0 \in \mathcal{D}_0$, $c \in \mathcal{D}_2$; on note $\tilde{c} = c_0 \perp c$; c a des arguments externes $t_1, \ldots, t_\nu$ dans M. Les arguments externes de $\tilde{c}$ dans M sont les arguments dans K_{c_0} des points de ∂U_1 d'arguments internes $t_1, \ldots, t_\nu$.*

DÉMONSTRATION. Soit $\mathcal{R}_M(t)$ aboutissant à $\tilde{c}$, alors $\mathcal{R}_{\tilde{c}}(t)$ aboutit aussi à $\tilde{c}$ ($\tilde{c} \in \mathcal{D}_2$).

Or. $t \in \gamma_{\tilde{c}}^{-1}(\tilde{c}) = \gamma_{c_0}^{-1}(\gamma_{U_1}(\gamma_c^{-1}(c)))$ par le Théorème 2. On conclut du fait que

$$\left| \begin{array}{l} \gamma_c^{-1}(c) = \gamma_M^{-1}(c), \\ \gamma_{\tilde{c}}^{-1}(\tilde{c}) = \gamma_M^{-1}(\tilde{c}). \end{array} \right.$$

∎

Corollaire 7.3. *Soient* $c_0 \in \mathcal{D}_0 \setminus \{0\}$, $c \in \mathcal{D}_1 \setminus \{1/4\}$; *on note* $\tilde{c} = c_0 \perp c$; *c a deux arguments externes* t_1 *et* t_2 *dans* M. *Les arguments externes de* $\tilde{c}$ *dans* M *sont les arguments dans* K_{c_0} *des points de* ∂U_1 *d'arguments internes* t_1 *et* t_2.

DÉMONSTRATION. On suppose que $\mathcal{R}_M(t)$ aboutit à $\tilde{c}$, alors $\mathcal{R}_{\tilde{c}}(t)$ aboutit à $y_1(\tilde{c})$.
Or. $t \in \gamma_{\tilde{c}}^{-1}(y_1(\tilde{c})) = \gamma_{c_0}^{-1}(\gamma_{U_1}(\gamma_c^{-1}(y_1(c))))$. On conclut par le Théorème 0.4.
∎

Définition. Soient $\left| \begin{array}{l} \theta_1 < \theta_2 \in \mathbb{Q}/\mathbb{Z} \\ t \in \mathbb{T} \end{array} \right.$ qu'on décompose en base deux :

$$\theta_i = 0.\overline{u_1^i...u_k^i}, \ i = 1, 2 \ et$$
$$t = 0.t_1...t_n... ;$$

alors $(\theta_1, \theta_2) \perp t := u_1^{t_1}...u_k^{t_1} u_1^{t_2}...u_k^{t_2} u_1^{t_3}...$

Théorème 7.4. *Soient* $c_0 \in \mathcal{D}_0$, $c \in \mathcal{D}_1 \cup \mathcal{D}_2$; *on note* $\tilde{c} = c_0 \perp c$; *soit* t *un argument externe de* c *dans* M, *alors il lui correspond un argument externe* $\tilde{t}$ *de* $\tilde{c}$ *dans* M, *à savoir* $\tilde{t} = (\theta_1, \theta_2) \perp t$.

Lemme 7.5. *Soit* $z \in \partial U_1$ *d'argument interne* t ; $\mathcal{R}_{c_0}(\theta_1), \mathcal{R}_{c_0}(\theta_2)$ *adjacents à* U_1 *en* y_1. *alors* $Arg_{c_0} z = (\theta_1, \theta_2) \perp t$.

La preuve du lemme est incluse dans la démonstration de la Proposition 1, Section 7 de [5]. Le théorème est alors un corollaire immédiat de ce qui précède.

Remarque. Étant donné qu'en base deux, les nombres dyadiques ont deux écritures, on obtient deux valeurs de $\tilde{t}$. Ceci est dû au fait que M s'étend au-delà de M_{c_0} en $\tilde{c}$ (présence de filament(s)).

A   Aboutissement de rayons rationnels

Dans cet appendice, nous montrons que les rayons rationnels aboutissent à des points qui dépendent holomorphiquement du paramètre dans de "grands" domaines du plan.

Proposition A.1. _Soient_ $\theta \in \mathbb{Q}/\mathbb{Z}$ _et_ $\Lambda \subset \mathbb{C}$ _un ouvert connexe tels que, pour tout_ $c \in \Lambda$, $\mathcal{R}_c(\theta)$ _aboutit; alors_ $c \mapsto \gamma_c(\theta)$ _est holomorphe et_ $\gamma_c(\theta)$ _est prépériodique répulsif._

S'il existe $\theta' \in \mathbb{Q}/\mathbb{Z}$ _tel que_ $\mathcal{R}_c(\theta')$ _aboutit pour tout_ $c \in \Lambda$ _et, pour un paramètre_ $\hat{c} \in \Lambda$, _on a_ $\gamma_{\hat{c}}(\theta) = \gamma_{\hat{c}}(\theta')$, _alors on a égalité sur tout_ Λ.

DÉMONSTRATION. On considère les applications $h_n(c) = \phi_c^{-1}(e^{1/n+2i\pi\theta})$. Cette suite (h_n) converge simplement vers $\gamma_c(\theta)$. Il suffit donc de montrer que cette suite est uniformément convergente sur tout compact.

Soit $\eta > 0$; pour $c \in \Lambda \cap \{G_M < \eta\}$, $\phi_c^{-1} : \{|z| > e^\eta\} \to \mathbb{C}$ est conforme et tangente à l'identité à l'infini, donc $\phi_c^{-1}(\{|z| > e^\eta\}) \supset \{|z| > 4.e^\eta\}$ par le théorème du quart de Koebe. Par conséquent, pour $n \gg 1$, $|h_n(c)| \le 4.e^\eta$. D'après le théorème de Montel, (h_n) est une famille normale, donc sa limite, $\gamma_c(\theta)$, est une fonction holomorphe de c.

Comme θ est rationnel, il existe $\ell \ge 0$ et $k \ge 1$ (minimaux) tels que $Q_c^{\ell+k}(\gamma_c(\theta)) = Q_c^\ell(\gamma_c(\theta))$. On considère l'application suivante:

$$\rho : \Lambda \to \mathbb{C}$$
$$c \mapsto (Q_c^k)'(Q_c^\ell(\gamma_c(\theta)))$$

qui est holomorphe. De plus, $Q_c^\ell(\gamma_c(\theta))$ appartient à l'ensemble de Julia J_c, donc $|\rho(c)| \ge 1$. D'après le Lemme A.2 ci-dessous, ρ est ouverte, et on en déduit que $|\rho(c)| > 1$.

Par hypothèse, on a $\gamma_{\hat{c}}(\theta) = \gamma_{\hat{c}}(\theta')$, et ces points sont solutions de l'équation $F(c,z) := Q_c^{\ell+k}(z) - Q_c^\ell(z) = 0$. Comme

$$\frac{\partial F}{\partial z}|_{(\hat{c},\gamma_{\hat{c}}(\theta))} = (Q_{\hat{c}}^{\ell+k})'(\gamma_{\hat{c}}(\theta)) - (Q_{\hat{c}}^\ell)'(\gamma_{\hat{c}}(\theta))$$
$$= (Q_{\hat{c}}^\ell)'(\gamma_{\hat{c}}(\theta))) \cdot ((Q_{\hat{c}}^k)'(Q_{\hat{c}}^\ell(\gamma_{\hat{c}}(\theta)))) - 1) \ne 0$$

(car $Q_{\hat{c}}^\ell(\gamma_{\hat{c}}(\theta))$ est répulsif), ces points coïncident aussi sur un voisinage de $\hat{c}$ par l'unicité du théorème des fonctions implicites; donc sur tout Λ par prolongement analytique. ∎

Lemme A.2. _Soit_ $X = \{(c,z) \in \mathbb{C}^2, Q_c^k(z) = z\}$; _l'application_ $\rho : X \to \mathbb{C}$ _définie par_ $\rho(c,z) = (Q_c^k)'(z)$ _est ouverte._

DÉMONSTRATION. X est une courbe algébrique de $\mathbb{C}^2$, donc ses composantes sont non bornées. De plus, l'application $\sigma : X \to X$, définie par $\sigma(c,z) = (c, Q_c(z))$ est périodique et propre.

Or $\rho(c,z) = 2^k \Pi_{0 \le j < k} \sigma^j(c,z)$. Donc, quand $(c,z) \in X \to \infty$, $\rho(c,z)$ aussi et cette application n'est pas constante, donc ouverte. ∎

Une application de la Proposition A.1 est le théorème suivant.

Théorème A.3. _Soient_ $0 < \theta_1 < \theta_2 < ... < \theta_\nu < 1 \in \mathbb{Q}$; _on suppose que leur rayons_ $\mathcal{R}_M(\theta_i), i = 1, ..., \nu$, _aboutissent à un même point_ $\hat{c}$ _et que ce sont_ *les seuls. Soit Λ la composante connexe de $\mathbb{C} \setminus (\mathcal{R}_M(\theta_1) \cup \mathcal{R}_M(\theta_\nu) \cup \{\hat{c}\})$ qui ne contient pas l'origine. Alors, pour tout $c \in \Lambda$, les rayons dynamiques $\mathcal{R}_c(\theta_i)$ aboutissent tous au même point $\zeta(c)$ et l'application $c \mapsto \zeta(c)$ est holomorphe.*

Dans sa démonstration, nous utiliserons la Proposition 0.2 sous la forme :

Corollaire A.4. *Soit $\theta \in \mathbb{Q}/\mathbb{Z}$ et soit $c \notin M$ d'argument t ; si l'orbite positive de θ ne contient pas t, alors $\mathcal{R}_c(\theta)$ aboutit.*

Nous aurons aussi besoin de la proposition suivante :

Proposition A.5. **([9], I, Chap.4, Prop.4)** *Soit $c_0 \in \mathcal{D}_0 \cup \mathcal{D}_2$. Ce point est une extrémité de son arbre de Hubbard.*

DÉMONSTRATION DU THÉORÈME A.3. Si $c \in M$, comme K_c est connexe, la grande orbite de c est dans K_c et tout rayon rationnel aboutit (Proposition 0.2). Supposons donc $c \notin M$ et notons $t \in]\theta_1, \theta_\nu[$ son argument externe. D'après la Proposition 0.3, ou bien $\hat{c} \in \mathcal{D}_2$, ou bien $\hat{c} \in \mathcal{D}_1$. Distinguons ces deux cas.

Cas parabolique $(\hat{c} \in \mathcal{D}_1)$. On peut extraire une démonstration de ce cas des travaux de J. Milnor (sections 2 à 5 de [17]), où il utilise aussi la Prop. A.1 (voir Théorème 3.1 de [17] ou Théorème D.8 de [21]). Nous préférons présenter une approche différente. D'après le Théorème 1 de [19], les rayons $\mathcal{R}_{\hat{c}}(\theta_i), i = 1, \ldots, \nu$, aboutissent au point parabolique $\hat{\alpha}$ qui attire $\hat{c}$.

Soient $0 < \theta^- < \theta^+ < 1$ les arguments des rayons qui aboutissent à $\hat{\alpha}$ en étant adjacents à la composante de Fatou qui contient $\hat{c}$. D'après le Théorème D.2 de [21], les rayons $\mathcal{R}_M(\theta^\pm)$ aboutissent à $\hat{c}$, et leurs arguments font donc partie des θ_i.

On peut définir $\Lambda_\pm$ la composante connexe de $\mathbb{C} \setminus (\mathcal{R}_M(\theta^-) \cup \{\hat{c}\} \cup \mathcal{R}_M(\theta^+))$ qui ne contient pas l'origine. Montrons le Théorème A.3 avec $\Lambda_\pm$ à la place de Λ et les $\theta^\pm$ à la place des θ_i. D'après le Théorème principal de [13], I (voir aussi [9], II, Chap. 18, Prop. 1, [17] Cor. 4.3 ou [21], Th. D.3.b)), $\hat{c}$ est la racine d'une composante de centre $c_0 \in \mathcal{D}_0 \cap \Lambda$. Pour ces paramètres, les ensembles de Julia sont homéomorphes et les rayons $\mathcal{R}_{c_0}(\theta^-)$ et $\mathcal{R}_{c_0}(\theta^+)$ séparent l'origine de c_0. Comme la valeur critique est une extrémité de son arbre de Hubbard (Prop. A.5), les orbites des $\theta^\pm$ par doublement de l'angle n'entrent jamais dans l'intervalle $]\theta^-, \theta^+[$. Le Corollaire A.4 montre donc que ces rayons aboutissent pour tout $c \in \Lambda$. De surcroît, comme les rayons aboutissent à un même point pour le centre c_0, on conclut par la Proposition A.1 que les rayons aboutissent à un même point pour tout $c \in \Lambda$ avec une dépendance holomorphe. En travaillant avec $\Lambda_\pm$, on peut montrer qu'il n'y a pas d'autres rayons que les $\mathcal{R}_M(\theta^\pm)$ qui aboutissent à $\hat{c}$ (Cor. D.6 de [21] ou Cor. 5.4 de [17]), et donc que $\nu = 2$ et $\Lambda = \Lambda_\pm$. Ceci termine la démonstration dans le cas parabolique.

Cas de Misiurewicz $(\hat{c} \in \mathcal{D}_2)$. D'après le Théorème 1 de [19], les rayons aboutissent dans le plan dynamique de $\hat{c}$ à $\hat{c}$. Ce point $\hat{c}$ étant strictement prépériodique et l'extrémité de son arbre de Hubbard (Prop. A.5), on en déduit que les orbites positives des θ_i n'atteignent jamais t. Par suite, le Corollaire A.4 s'applique et les rayons $\mathcal{R}_c(\theta_i)$, $i = 1, ... \nu$, aboutissent pour tout $c \in \Lambda$. Tous ces rayons sont prépériodiques et aboutissent en un point pré-répulsif. Par le Corollaire A.4 et l'unicité du théorème des fonctions implicites, ils aboutissent aussi dans un voisinage connexe $\hat{\Delta}$ de $\hat{c}$. Par suite, on a $\gamma_c(\theta_1) = ... = \gamma_c(\theta_\nu)$ pour tout $c \in \Lambda \cup \hat{\Delta}$ (connexe) par la Proposition A.1. ∎

Remarque. Dans le cas parabolique, on montre même un peu plus : tous les rayons dynamiques qui aboutissent aux points paraboliques de $\hat{c}$ continuent à aboutir à un même point dans tout Λ.

References

[1] L.V. Ahlfors, *Lectures on quasiconformal mappings*, Van Nostrand, 1966.

[2] B. Branner & J.H. Hubbard, *The iteration of cubic polynomial, part II : Patterns and parapatterns*, Acta Math., vol. 169, pp. 229-325, 1992.

[3] R. Daverman, *Decompositions of manifolds*, Pure and applied mathematics, 124, Academic Press, Inc., Orlando, Fla., 1986.

[4] G. David. *Solutions de l'équation de Beltrami avec* $\|\mu\|_\infty = 1$, Ann. Acad. Sci. Fenn. Ser. A 1 Math., 13, pp. 25-70, 1988.

[5] A. Douady, *Algorithms for computing angles in the Mandelbrot set*, in Chaotic Dynamics and Fractals, Atlanta 1985, Notes Rep. Math. Sci. Engrg., vol. 2, pp. 155-168, 1986.

[6] A. Douady. *Applications de la chirurgie holomorphe*, Proc. Int. Congr. Math., Berkeley, pp. 724-738, 1986.

[7] A. Douady, *Descriptions of compact sets of* $\mathbb{C}$, Topological methods in modern mathematics (Stony Brook, 1991), pp. 429-465, 1993.

[8] A. Douady & C.J. Earle, *Conformally natural extension of homeomorphisms of the circle*, Acta Math., vol. 157, pp. 23-48, 1986.

[9] A. Douady & J.H. Hubbard, *Etude dynamique des polynômes complexes I & II*, Pub. math. d'Orsay 84-02 et 85-05, 1984/85.

[10] A. Douady & J.H. Hubbard, *On the dynamics of polynomial-like mappings*, Ann. Sci. ENS Paris, vol. 18, pp. 287-343, 1985.

[11] P. Haïssinsky, *Applications de la chirurgie holomorphe, notamment aux points paraboliques*, Thèse de l'Université de Paris-Sud, 1998.

[12] P. Haïssinsky, *Parabolic points and μ-conformal maps*, prépublication de l'ENS de Lyon, 1998.

[13] P. Haïssinsky, *Déformation J-équivalente de polynômes géométriquement finis - Pincements de polynômes*, prépublication de l'ENS de Lyon, 1998.

[14] J.H. Hubbard, *Local connectivity of Julia sets and bifurcation loci: three theorems of J.C. Yoccoz*, Topological methods in modern mathematics (Stony Brook, 1991), pp. 467-511, 1993.

[15] M. Lyubich, *Feigenbaum-Coullet-Tresser universality and Milnor's hairiness conjecture*, preprint IHES, 1996, à paraître.

[16] J. Milnor, *Local connectivity of Julia sets: expository lectures*, dans ce volume.

[17] J. Milnor, *Periodic orbits, external rays and the Mandelbrot set: an expository account*, preprint Stony Brook, Astérisque, à paraître.

[18] C.T. McMullen, *Hausdorff dimension and conformal dynamics I & II*, preprint, 1997.

[19] C. Petersen & G. Ryd, *Convergence of rational rays in parameter spaces*, dans ce volume.

[20] Z. Słodkowski, *Holomorphic motions and polynomial hulls*, Proc. AMS, vol. 111, pp. 347-355, 1991.

[21] Tan L., *Local properties of the Mandelbrot set at parabolic points*, dans ce volume.

Peter Haïssinsky, Department of Mathematics, University of Jyväskylä, PL 35, Fin-40351 Jyväskylä, Finland.
e-mail: phaissin@math.jyu.fi

Local connectivity of Julia sets:
expository lectures

John Milnor

Abstract

An exposition of results of Yoccoz, Branner, Hubbard and Douady
concerning polynomial Julia sets. The contents are as follows:

Introduction

The following notes provide an introduction to work of Branner, Hubbard,
Douady, and Yoccoz on the geometry of polynomial Julia sets. They are an
expanded version of lectures given in Stony Brook in Spring 1992.

Section 1 describes unpublished work by J.-C. Yoccoz on local connectivity
of quadratic Julia sets. (Compare [Hu3].) It presents only the "easy" part of
his theory, in the sense that it considers only non-renormalizable polynomials,
and makes no effort to describe the more difficult arguments which are needed
to deal with local connectivity in parameter space. It is based on second hand
sources (Hubbard [Hu1] together with lectures by Branner and Douady), and
uses the language of Branner and Hubbard. The presentation is quite different
from that of Yoccoz. (Compare Problem 1-e at the end of §1.)

Section 2 describes the analogous arguments used by Branner and Hubbard [BH2] to study higher degree polynomials for which all but one of the
critical orbits escape to infinity. In this case, the associated Julia set J is
never locally connected. The basic problem is rather to decide when J is
totally disconnected. This Branner-Hubbard work came before Yoccoz, and
its technical details are not as difficult. However, in these notes their work is
presented simply as another application of the same geometric ideas.

Section 3 complements the Yoccoz results by describing a family of examples, due to Douady and Hubbard, showing that an infinitely renormalizable
quadratic polynomial may have non-locally-connected Julia set. An Appendix
describes needed tools from complex analysis, including the Grötzsch inequality.

We will assume that the reader is familiar with the basic properties of Julia sets and the Mandelbrot set. (For general background, see for example [Be], [Br2], [CG], [D1], [D2], [EL], [L1], or [St]. A brief survey of the Mandelbrot set is given in §3.) In particular, we will make use of *external rays* for a polynomial Julia set $J(f) \subset \mathbb{C}$. (Compare [DH1], [DH2], [D5], [GM], [M2], [M3].)

1. Local Connectivity of Quadratic Julia Sets

This section will prove the following.

> **Theorem 1.1 (Yoccoz).** *If* $f_c(z) = z^2 + c$ *is a quadratic polynomial such that:*
>
> (1) *the Julia set* $J(f_c)$ *is connected,*
>
> (2) *both fixed points are repelling, and*
>
> (3) f_c *is not simply renormalizable,*
>
>   *then* $J(f_c)$ *is locally connected.*

(For the definition of renormalizability, see 1.4, as well as §3.) In terms of the familiar parameter space picture for the family of quadratic maps $f_c(z) = z^2 + c$, condition (1) says that the parameter value c belongs to the Mandelbrot set M, while (2) says that c does not belong to the closure of the central region bounded by the cardioid, and (3) says that c does not belong to any one of the many small copies of M which are scattered densely around the boundary of M. (Compare Figure 1.)

Remarks: The proof gives much more, since it effectively describes the Julia set by a new kind of symbolic dynamics. With a little more work, the Yoccoz method can also deal with the finitely renormalizable case. Since the case of a map with attracting or parabolic fixed point had been understood much earlier [DH2], we see that conditions (2) and (3) can be actually replaced by the following weaker pair of conditions:

   (2') f has no irrationally indifferent periodic points (Cremer or Siegel points), and

   (3') f is not *infinitely* renormalizable.

These modified conditions (1), (2') and (3') are all essential. In fact:

   Condition (1): For $c \notin M$ the Julia set $J(f_c)$ is a Cantor set, which is certainly not locally connected.

   Condition (2'): Sullivan and Douady showed that a polynomial Julia set with a Cremer point is never locally connected. (Compare [Su], [D1],

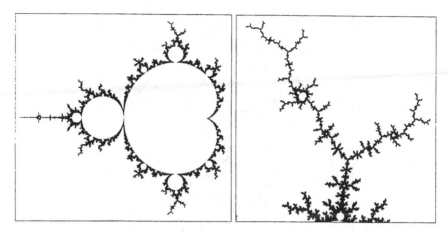

Figure 1. Boundary of the Mandelbrot set M, and a detail of the one-third limb showing several small copies of M. (Compare §3.)

[M2]. For a more explicit description of non local connectivity, see [Sø1-2].) Similarly, Herman has constructed quadratic polynomials with a Siegel disk having no critical point on its boundary. The corresponding Julia set cannot be locally connected. (See [He,§17.1], [D1,II,5], [D4].)

Condition (3'): In §3, following unpublished work by Douady and Hubbard, we will describe infinitely renormalizable polynomials for which J is not locally connected.

Thus the sharpened version of Theorem 1.1 comes fairly close to deciding exactly which quadratic polynomial Julia sets are locally connected. Yoccoz has also proved a corresponding result in parameter space: For c in the Mandelbrot set M, if the polynomial $f_c(z) = z^2 + c$ is not infinitely renormalizable, then M is locally connected at c. For proofs of this more difficult result, see [Hu3], [K].

The Yoccoz puzzle. The proof of Theorem 1.1 begins as follows. If a connected quadratic Julia set has two repelling fixed points, then one fixed point (to be called β) is the landing point of the zero external ray, and the other (called α) is the landing point of a cycle of q external rays, where $q \geq 2$. (Compare [M2], [Pe1].) Let $0 = c_0 \mapsto c_1 \mapsto \cdots$ be the critical orbit. The *Yoccoz puzzle* of *depth zero* consists of q non-overlapping closed topological disks $P_0(c_0), P_0(c_1), \ldots, P_0(c_{q-1})$, called "*puzzle pieces*", which are obtained by cutting the region $G \leq 1$ along the q external rays landing at α. Here G is the Green's function (or canonical potential function) which vanishes only on the filled Julia set, and satisfies $G(f(z)) = 2G(z)$; while $P_0(c_i)$ denotes the unique puzzle piece of depth zero which contains

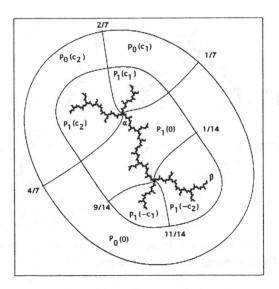

*Figure 2. Julia set for $f(z) = z^2 + i$, showing the Yoccoz puzzles
of depth zero and one. Here $q = 3$. The 1/7, 2/7 and 4/7 rays
land at the fixed point α.*

the post-critical point $c_i = f^{\circ i}(0)$.

Inductive Construction: If $P_d^{(1)}, \ldots, P_d^{(m)}$ are the puzzle pieces of
depth d, then the connected components of the sets $f^{-1}(P_d^{(i)})$ are the puzzle
pieces $P_{d+1}^{(j)}$ of depth $d + 1$. As an example, there are always $2q - 1$
pieces of depth one. These consist of q pieces $P_1(c_0), P_1(c_1), \ldots, P_1(c_{q-1})$
which touch at the fixed point α, together with $q - 1$ additional pieces
$P_1(-c_1), \ldots, P_1(-c_{q-1})$ which touch at the pre-image $-\alpha$.

Since the α-fixed point is used to construct this puzzle, the points in its
grand orbit will have to be dealt with by a special argument. (See 1.9.) It
will be convenient to introduce the abbreviated notation

$$J_0 = \bigcup_{n \geq 0} f^{-n}(\alpha)$$

for the set of all points in the Julia set which belong to this excluded grand
orbit. For any $z \in J \smallsetminus J_0$, it is easy to see that there is a unique sequence
of puzzle pieces

$$P_0(z) \supset P_1(z) \supset P_2(z) \supset \cdots$$

which contain z.

Main Problem. Let $z \in J \smallsetminus J_0$ be any point which belongs to the Julia set $J(f)$ but not to the countable set $\bigcup f^{-n}(\alpha) \subset J(f)$, and let $P_d(z)$ be the unique puzzle piece of depth d which contains this point z. *Does the intersection $\bigcap_d P_d(z)$ of these puzzle pieces consist of the single point z?*

The associated annuli. Following Branner and Hubbard, consider the puzzle pieces $P_d(z) \supset P_{d+1}(z)$ of two consecutive depths containing some given $z \in J \smallsetminus J_0$. With luck, the smaller puzzle piece $P_{d+1}(z)$ will be contained in the interior of $P_d(z)$. In this case the difference set

$$A_d(z) = \text{Interior}\big(P_d(z)\big) \smallsetminus P_{d+1}(z)$$

will be an annulus, whose modulus $\mathbf{mod}\, A_d(z)$ is a well defined positive real number. (Compare the Appendix.) However, we must also allow for the possibility that the boundary of $P_d(z)$ intersects P_{d+1}. In that case, we describe $A_d(z)$ as a "*degenerate annulus*", and define $\mathbf{mod}\, A_d(z)$ to be zero.[1] As an example, in Figure 2 the annulus $A_0(-c_1)$ has positive modulus, but the annulus $A_0(0)$ around the critical point is degenerate.

With these definitions we can reformulate the Main Problem as follows:

Modified Main Problem (Branner and Hubbard). *Given a point $z \in J \smallsetminus J_0$, is the sum*

$$\sum_d \mathbf{mod}\, A_d(z)$$

infinite? If so, using the Grötzsch inequality, it is not difficult to prove that the intersection $\bigcap P_d(z)$ must consist of the single point z. (See A.7 in the Appendix.)

The sequence of puzzle pieces

$$P_0(0) \supset P_1(0) \supset P_2(0) \supset \cdots$$

containing the critical point will play a special role. For this sequence to be well defined, we need the following.

Standing Hypothesis 1.2. *We assume that $0 \notin J_0$, so that the orbit of the critical point does not eventually land on the fixed point α.*

This is no real restriction. For in the contrary case where $0 \in J_0$ so that the critical orbit does end at α, we are in the post-critically finite case, and local connectivity can be established by more classical methods. (Compare [DH2] or [M2].)

---

[1]It will be convenient to refer to $A_d(z)$ loosely as an "*annulus*" even when it may be degenerate, and hence consist of one or more simply connected regions. In fact it might be clearer to work with the closed ring $\widehat{A}_d(z) = P_d(z) \smallsetminus \text{Interior}\big(P_{d+1}(z)\big)$ which has interior $A_d(z)$, and which has the homotopy type of a circle even in the degenerate case.

We can compare the sequence $\{P_d(z)\}$ for an arbitrary point $z \in J \smallsetminus J_0$ with the sequence $\{P_d(0)\}$ as follows. *Define the semi-critical depth* $S(z)$ *to be the largest integer* $d \geq 0$ *for which* $P_d(z) = P_d(0)$, *taking* $S(z)$ *to be* $+\infty$ *if* $P_d(z) = P_d(0)$ *for all* d, *and setting* $S(z) = -1$ *if* $P_d(z) \neq P_d(0)$ *for all* d. Thus $S(z)$ is large whenever z is close to the critical point.

To each point $z_0 \in J \smallsetminus J_0$ we can associate the orbit $z_0 \mapsto z_1 \mapsto z_2 \mapsto \cdots$, and the sequence of numbers $S(z_0), S(z_1), S(z_2), \ldots$. (The sequence associated with the critical orbit $0 = c_0 \mapsto c_1 \mapsto \cdots$ will be of particular interest.) Following Branner and Hubbard, this sequence can be displayed graphically by introducing the associated *tableau*. (Compare Figures 4, 6, 9.)

Definition. The *tableau* associated with an orbit $z_0 \mapsto z_1 \mapsto z_2 \mapsto \cdots$ in J_0 is an array with one column associated with each z_i and one row associated with each depth in the Yoccoz puzzle. We will draw a solid vertical line at depth d in the j-th column whenever $d < S(z_i)$, and a double vertical line when $d = S(z_i)$. Such single or double vertical line segments correspond to puzzle pieces $P_d(z_i)$ which coincide with the critical puzzle piece $P_d(0)$, and a long chain of such vertical segments indicates an orbit point z_i which is very close to the critical point. Diagonal arrows of length m, pointing north-east, correspond to iterates $f^{\circ m}$, mapping some puzzle piece $P_d(z_i)$ by a branched covering map onto $P_{d-m}(z_{i+m})$.

Given any orbit $z_0 \mapsto z_1 \mapsto z_2 \mapsto \cdots$ in J_0, note that f carries any puzzle piece $P_d(z_i)$ of depth $d > 0$ onto the puzzle piece $P_{d-1}(z_{i+1})$. This map $P_d(z_i) \to P_{d-1}(z_{i+1})$ is either a conformal isomorphism or a two-fold branched covering according as $P_d(z_i)$ does or does not contain the critical point. Now consider the restriction of f to the annulus

$$A_d(z_i) = \text{Interior}\big(P_d(z_i)\big) \smallsetminus P_{d+1}(z_i).$$

We can distinguish three cases, as follows.

Critical Case. If $d < S(z_i)$ so that $0 \in P_{d+1}(z_i)$ (indicated by a vertical line segment in the tableau), then $A_d(z_i)$ coincides with $A_d(0)$, and will be called a *critical annulus*. In the non-degenerate case, note that $A_d(z_i)$ surrounds the critical point 0, and that f maps the annulus $A_d(z_i)$ onto $A_{d-1}(z_{i+1})$ by a 2-fold covering map. It follows easily that

$$\text{mod}\, A_d(z_i) = \tfrac{1}{2}\,\text{mod}\, A_{d-1}(z_{i+1}).$$

Off-Critical Case. If $d > S(z_i)$ so that $0 \notin P_d(z_i)$ (indicated by a blank in the tableau), then f maps $A_d(z_i)$ biholomorphically onto $A_{d-1}(z_{i+1})$ and the moduli are equal. In this case, we will say that the annulus $A_d(z_i)$ is *off-critical*.

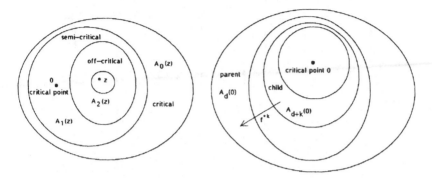

Figure 3. Critical, semi-critical and off-critical annuli (left); and a "child" (right).

Semi-Critical Case. Finally, if $d = S(z_i)$ so that $0 \in A_d(z_i)$ (corresponding to a double line in the tableau), we will say that $A_d(z_i)$ is *semi-critical*. Here the topology is more complicated, since f maps the annulus $A_d(z_i)$ onto the entire puzzle piece $P_{d-1}(z_{i+1})$. However, it is easy to check that $A_d(z_i)$ is non-degenerate if and only if the annulus $A_{d-1}(z_{i+1})$ is non-degenerate. For an inequality between moduli, see Problem 1-b.

As examples, in the schematic picture Figure 3 (left), the annulus $A_0(z)$ is critical, while $A_1(z)$ is semi-critical, and $A_2(z)$ is off-critical. In Figure 2, the annuli $A_0(-c_1)$ and $A_0(-c_2)$ are both semi-critical.

Definition. The critical annulus $A_{d+k}(0)$ in the Yoccoz puzzle will be called a *child* of the critical annulus $A_d(0)$ if and only if $A_{d+k}(0)$ is an unramified two-fold covering of $A_d(0)$ under the map $f^{\circ k}$. (See Figure 3 (right), and Figure 4. Our terminology is based on [Hu1], but with several modifications.)

Note that the modulus of such a child is always exactly half the modulus of the parent. In a tableau, starting with any critical annulus, represented by a single vertical segment, we can locate the parent (necessarily unique if it exists), by drawing a line to the north-east and extending until it hits another vertical segment. Our strategy for solving the Modified Main Problem can now be summarized as follows:

> *Find a critical annulus of positive modulus, and prove that it has so many descendents, children and grandchildren and so on, that the modulus sum is infinite.*

Branner and Hubbard summarize the tableau properties which are needed to carry out this argument in three rules, as follows.

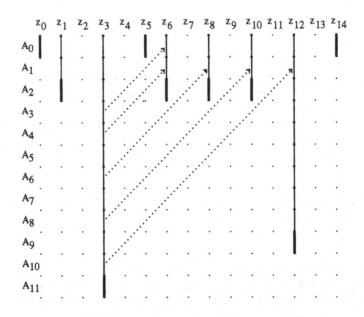

Figure 4. An example: the tableau associated with the orbit of the point $z_0 = 1$ for the map $f(z) = z^2 - 1.6$. Here we see by following the top diagonal arrow that the critical annulus $A_3(z_3) = A_3(0)$ is an unramified two-fold covering of the critical annulus $A_0(z_6) = A_0(0)$. In particular, the critical annulus at depth 3 is a child of the critical annulus at depth 0. Similarly, the critical annuli at depths $4, 6, 8, 10$ are children of the critical annulus of depth 1.

First tableau rule: *Every column of a tableau is either all critical, or all off-critical, or has exactly one semi-critical depth* $S(z_i)$, *and is critical above and off-critical below.*

Now let us compare the tableau of the critical orbit

$$0 = c_0 \mapsto c_1 \mapsto c_2 \mapsto \cdots$$

with the tableau of some given orbit $z_0 \mapsto z_1 \mapsto \cdots$ in J_0. (The case $z_0 = c_0$ is not excluded.) If the tableau of $\{z_j\}$ is critical or semi-critical at depth d in column m, draw a line "north-east" from this critical or semi-critical annulus, as indicated by the dotted line on the right, and draw a corresponding line north-east from depth d of column zero in the critical tableau.

Second tableau rule: *Everything strictly above the diagonal line on the left must be copied above the diagonal line on the right.*

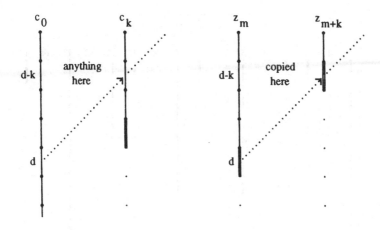

Figure 5. Illustration for the second and third tableau rules, with the critical tableau on the left, and the tableau for $z_0 \mapsto z_1 \mapsto \cdots$ on the right.

The proofs of these two rules are easily supplied. $\square$

Now suppose that the critical annulus of depth d is a child of the critical annulus of depth $d - k$, as indicated in the figure, and suppose that the tableau of $\{z_j\}$ is semi-critical at depth d of column m.

Third tableau rule: *Following the diagonal arrow from this semi-critical annulus of depth d in the tableau of $\{z_j\}$, we must reach a semi-critical annulus at depth $d - k$, as illustrated.*

Proof. According to the hypothesis, $f^{\circ k}$ maps $A_d(0)$ onto $A_{d-k}(0)$, where the point z_m is an element of this annulus $A_d(0)$. Therefore $f^{\circ k}(z_m) = z_{m+k}$ must belong to $A_{d-k}(0)$. $\square$

Definition. We will say that a critical annulus $A_d(0)$ is *excellent* if it contains no post-critical points, or equivalently if there is no semi-critical annulus in the d-th row of the critical tableau.

In Figure 6, note that the critical annuli at depths $1, 3, 4, 6, 7, 8, 9, \ldots$ are excellent. Each one has exactly two children, which are also excellent. On the other hand, $0, 2, 5, 10, 18$ (Fibonacci numbers minus three) are not excellent. Each of these has only one child; however this child is excellent.

Definition: We will say that the critical tableau is *recurrent* if the numbers $S(c_k)$ with $k > 0$ are unbounded, so that there are columns which go arbitrarily far down. It is *periodic* if $S(c_k) = \infty$ for some $k > 0$, so that the

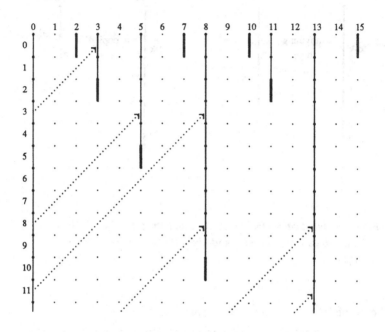

Figure 6. The "Fibonacci" critical tableau. (Compare [BH2].) Here the closest recurrences of the critical orbit come after a Fibonacci number of iterations:

$$1,\ 2,\ 3,\ 5,\ 8,\ 13,\ 21,\ 34,\ 55,\ \ldots.$$

If u_n is the n-th Fibonacci number, then the semi-critical annulus for column number u_n occurs at depth $u_{n+1} - 3$. The diagonal dotted lines illustrate the genealogies

$$0 \leftarrow 3 \leftarrow 8 \begin{smallmatrix}\leftarrow & \cdots \\ \leftarrow & \cdots\end{smallmatrix}$$
$$\nwarrow$$
$$11 \begin{smallmatrix}\leftarrow & \cdots \\ \leftarrow & \cdots\end{smallmatrix},$$

with arrows from child to parent. Since $A_3(0)$ is the only child of $A_0(0)$, it follows from 1.3 that every descendent of $A_3(0)$ has at least two children.

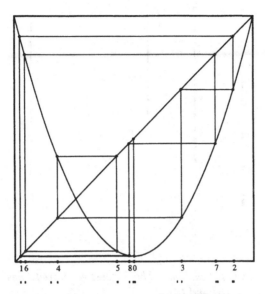

Figure 7. Graph of the unimodal map $x \mapsto x^2 - 1.8705286\cdots$ *which realizes the Fibonacci tableau. (Compare [ML].) The first eight points on the critical orbit are marked. The critical orbit closure is a rather thin Cantor set, which is plotted underneath the graph. (Compare Problem 1-f.)*

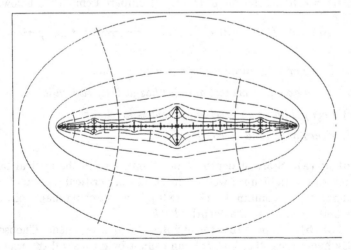

Figure 8. The Yoccoz puzzle at depths zero through five for this quadratic Fibonacci map, drawn to the same scale. Note that the critical pieces are the biggest ones at any specified depth.

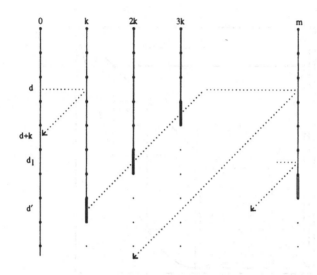

Figure 9. Finding Children. The successive dotted lines illustrate Lemma 1.3 (a), (b) and (d).

corresponding column goes all the way down. In this case, it follows easily that $P_d(c_i) = P_d(c_{i+k})$ for all d and i. (Compare Lemma 1.5 below.)

Lemma 1.3. *If the critical tableau is recurrent but not periodic, then:*

(a) *Every critical annulus has at least one child.*

(b) *Every excellent critical annulus has at least two children.*

(c) *Every child of an excellent parent is excellent.*

(d) *Every only child is excellent.*

Proof of (a). Start at depth d of column zero in the critical tableau, and march to the right until we first meet another critical annulus (= solid vertical line), say at column k. (Figure 9.) Now marching diagonally southwest, we find the first child at depth $d + k$.

Proof of (b). Suppose that the annulus $A_d(0)$ is excellent. Choose k as in (a). By hypothesis, the k-th column cannot be critical all the way down. Hence it must be semi-critical at some depth $d' > d$. Starting at column k and depth d', proceed diagonally right (north-east). By the tableau rules, column $2k$ must be semi-critical at depth $d' - k$. Similarly column $3k$ must be semi-critical at depth $d' - 2k$, and so on, until we again reach depth

d. Furthermore, as we follow this diagonal, we do not meet any other critical or semi-critical annuli. In particular, at the point where we reach depth d, there cannot be a critical or semi-critical annulus. (The hypothesis that d is excellent comes in at this point. Compare Figure 9.) Now let us again march to the right at depth d until we reach a critical annulus, say at column m. Again turning $135°$ and proceeding diagonally south-west, we cannot meet any critical or semi-critical annulus until we are back at column 0. In this way we prove that the critical annulus of depth $m + d$ is also a child of d.

Proof of (c). This is clear, since if $f\big(A_{d+k}(0)\big) = A_d(0)$ where $A_{d+k}(0)$ contains a post-critical point, then $A_d(0)$ does also.

Proof of (d), by contradiction. Consider a child d' which is not excellent, and let $d_1 = n' - k$ be the parent. Then for some $k' \geq k$ the k'-th column has semi-critical annulus at depth d'. (The case $k' = k$ is illustrated in Figure 9.) Following the diagonal up from column k' and depth d', we must meet a semi-critical annulus at depth d_1 by the third tableau rule. (Thus the parent is not excellent.) Now proceed to the right at depth d_1 until we meet a critical annulus, say at column m. Then it follows as above that $d_1 + m$ is a second child; hence d' was not an only child. $\square$

Definition 1.4. (Compare [DH3], [Mc2].) A quadratic polynomial f is *renormalizable* if there exists a closed topological disk Δ_0 around the critical point, and an integer $p \geq 2$ (called the *renormalization period*), with the following four properties:

(1) Δ_0 should be centrally symmetric about the critical point so that f carries Δ_0 onto the image $\Delta_1 = f(\Delta_0)$ by a ramified 2-fold covering map.

(2) f should induce conformal isomorphisms

$$\Delta_1 \xrightarrow{\cong} \Delta_2 \xrightarrow{\cong} \cdots \xrightarrow{\cong} \Delta_p \,,$$

where $\Delta_i = f^{\circ i}(\Delta_0)$. In particular, the critical point should be disjoint from $\Delta_1 \cup \cdots \cup \Delta_{p-1}$.

(3) However, the disk Δ_p should contain Δ_0 in its interior.

(4) Finally, the entire orbit of the critical point under the map $f^{\circ p}$ should be contained in the original disk Δ_0.

(Compare the discussion of "polynomial-like mappings" in §2, and the discussion of "tuning" in §3.) If these conditions are satisfied, then the *small filled Julia set* $K_p = K(f^{\circ p}|_{\Delta_0})$ can be defined as the compact connected set consisting of points whose orbits remain in Δ_0 under all iterations of $f^{\circ p}$. This is a proper subset of the filled Julia set $K(f)$ of the original map

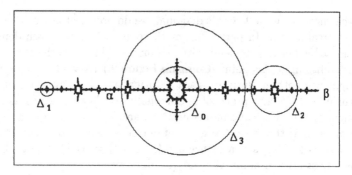

Figure 10. An example of simple renormalization: the Julia set for
$f(z) = z^2 - 1.75$*, with renormalization period* $p = 3$*. (The pa-*
rameter value $c = -1.75$ *is the root point of a small copy of the*
Mandelbrot set.) The filled Julia set $K(f^{\circ 3}|_{\Delta_0})$ *is a topological disk*
bounded by a "cauliflower".

f. Following McMullen [Mc2], the map f is called *simply renormalizable* of period p if also:

(5) Each intersection $f^{\circ i}(K_p) \cap f^{\circ j}(K_p)$ with $0 \le i < j < p$ is either vacuous, or consists of a single point which does not disconnect either of these two sets.

We will be interested only in the simply renormalizable case. See Figure 10 for a simply renormalizable example, and Figure 11 for a map which is renormalizable but not simply renormalizable. (Compare [RS].)

We can now state two basic lemmas.

Lemma 1.5. *If the critical tableau associated with* f *is periodic, then* f *is simply renormalizable. More precisely, if* $P_d(c_p) = P_d(0)$ *for all depths* d*, with* p *minimal, and therefore* $P_d(c_{i+p}) = P_d(c_i)$ *for all* i *and* d *by the Second Tableau Rule, then* f *is simply renormalizable of period* p*.*

Lemma 1.6. *If the critical orbit lies completely within the union*

$$P_1(c_0) \cup P_1(c_1) \cup \cdots \cup P_1(c_{q-1})$$

of those puzzle pieces of depth one which touch the fixed point α*, then* f *is simply renormalizable of period* q*. (See Figure 2.)*

The following construction will be needed to prove Lemmas 1.5 and 1.6. Recall that each of the puzzle pieces $P_0(c_i)$ of depth zero consists of points in

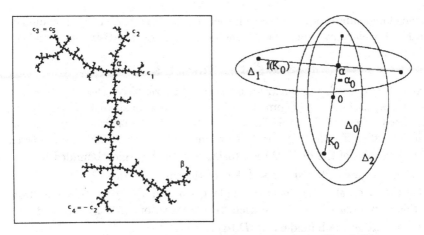

*Figure 11. An example of crossed renormalization: The Julia set
for $f(z) = z^2 + .41964338 + .60629073\,i$ is shown on the left,
with a schematic diagram of its renormalization on the right. The
associated small Julia set $K(f^{\circ 2}|_{\Delta_0})$ is a simple arc from c_2 to c_4.
Our discussion will always exclude such examples.*

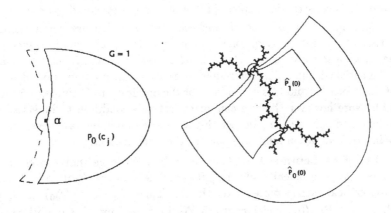

*Figure 12 Left: A Puzzle Piece and Thickened Piece of Level 0.
Right: Thickened Pieces of Level 0 and 1 for $z \mapsto z^2 + i$.*

some closed subset of the filled Julia set $K(f)$, together with points outside of $K(f)$ which have potential G and external angle t satisfying inequalities of the form

$$0 < G \leq 1, \qquad t_i \leq t \leq t_i'.$$

(Here $0 \leq i < q$.) We construct a *thickened puzzle piece* $\widehat{P}_0(c_i) \supset P_0(c_i)$ in two steps, as follows. (Figure 12.) First choose a small disk $D_\epsilon(\alpha)$ about the fixed point α. Second, choose $\eta > 0$ so small that every external ray whose angle differs from t_i or t_i' by at most η intersects this disk. Now let $\widehat{P}_0(c_i)$ consist of the disk $D_\epsilon(\alpha)$, together with the region bounded by:

(1) the segment $t_i - \eta \leq t \leq t_i' + \eta$ of the equipotential $G = 1$,

(2) the external ray segments of angle $t_i - \eta$ and $t_i' + \eta$ which extend from this equipotential $G = 1$ to their first intersection with $D_\epsilon(\alpha)$, and

(3) an arc of the boundary of $D_\epsilon(\alpha)$.

Evidently this thickened puzzle piece $\widehat{P}_0(c_i)$ contains the original $P_0(c_i)$. We now construct thickened puzzle pieces of greater depth by the usual inductive procedure: If $\widehat{P}_d^{(j)}$ is a thickened puzzle piece of depth d, then each component of $f^{-1}(\widehat{P}_d^{(j)})$ is a thickened puzzle piece of depth $d + 1$.

The virtue of these thickened pieces is the following statement, which is easily proved by induction: *If a puzzle piece* $P_d^{(j)}$ *contains* $P_{d+1}^{(k)}$, *then the corresponding thickened piece* $\widehat{P}_d^{(j)}$ *contains* $\widehat{P}_{d+1}^{(k)}$ *in its interior.* In other words, this construction replaces all of our annuli by non-degenerate annuli.

We will only make use of thickened pieces which are small enough to satisfy the following additional restriction: *If* $\widehat{P}_d(z)$ *contains the critical point, then the interior of* $P_d(z)$ *must already contain this critical point.* If 1.2 is satisfied, then this requirement is easily satisfied for any bounded values of the depth d, and this will suffice for the applications (Lemmas 1.5 through 1.8). Note however that we cannot expect this condition to be satisfied for unbounded values of d. In fact this would be impossible whenever the fixed point α is an accumulation point for the critical orbit.

Proof of Lemma 1.5. Choose $d \geq p$ so large that the critical annulus $A_d(0)$ is a child of $A_{d-p}(0)$, and let $\Delta_0 = \widehat{P}_d(0)$ be the thickened critical puzzle piece of depth d. Then $\Delta_p = f^{\circ p}(\Delta_0)$ is equal to $\widehat{P}_{d-p}(c_p) = \widehat{P}_{d-p}(0)$, which contains Δ_0 in its interior. The hypothesis guarantees that the successive points c_p, c_{2p}, c_{3p}, ... all belong to Δ_0. Since p is minimal, the various puzzle pieces $P_{d-i}(c_i)$ with $0 \leq i < p$ cannot overlap, so any intersection point must belong to the boundary of these puzzle pieces, and therefore cannot disconnect the small filled Julia set. Therefore f is simply renormalizable. $\square$

Proof of Lemma 1.6. (suggested by M. Lyubich). Let $\Delta_0 = \widehat{P}_1(0)$. Then it is easy to check that the successive images $\Delta_i = f^{\circ i}(\Delta_0)$ are disjoint from the critical point for $0 < i < q$, and that Δ_q contains Δ_0 in its interior. (Compare Figure 2.) We will prove inductively that $c_{qi} \in P_1(0) \subset \Delta_0$ for every i. It certainly follows from this inductive hypothesis that $c_{qi+1} \in P_0(c_1) \cap K(f) \subset P_1(c_1)$, and similarly that $c_{qi+j} \in P_0(c_j) \cap K(f) \subset P_1(c_j)$ for $0 < j < q$. In particular, c_{qi+q-1} must belong to $P_1(c_{q-1})$, hence $c_{qi+q} \in f\big(P_1(c_{q-1})\big) = P_0(0)$. By the hypothesis of 1.6, the point c_{qi+q} is known to lie in one of the puzzle pieces $P_1(c_h)$ which lie around α. Evidently it can only lie in $P_1(0)$, as required. $\square$

Corollary 1.7. *If f is not simply renormalizable, then there exists a critical annulus of positive modulus.*

Proof. According to 1.6, the critical orbit must visit one of the puzzle pieces $P_1(-c_1)$, ..., $P_1(-c_{q-1})$ which surround the pre fixed point $-\alpha$. Suppose for example that $c_d \in P_1(-c_i)$. It is easy to check that the corresponding annulus $A_0(c_d) = A_0(-c_i)$ has positive modulus. (Compare Figure 2.) Pulling this annulus back inductively along the critical orbit, it follows that $A_d(0)$ also has positive modulus. $\square$

As another another application of thickened puzzle pieces, we can study orbits which stay away from the critical point.

Lemma 1.8. *Let $z_0 \mapsto z_1 \mapsto \cdots$ be an orbit in $J \smallsetminus J_0$ which is disjoint from some critical puzzle piece $P_N(0)$. Then the intersection $\bigcap P_d(z_0)$ of the puzzle pieces containing z_0 reduces to the single point z_0.*

Proof. Let $U_i = \text{Interior}\big(\widehat{P}_{N-1}^{(i)}\big)$, where the various puzzle pieces of depth $N-1$ have been numbered so that the critical value c_1 belongs to $P_{N-1}^{(0)}$. Each puzzle piece of depth $\geq N$ is contained in some unique $P_{N-1}^{(i)}$, and we give it the Poincaré metric $\text{dist}_i(x,y)$ associated with the corresponding U_i. For $i > 0$, note that there are exactly two branches of f^{-1} on U_i, call them g_1 and g_2, and that each of these maps g_κ on U_i is univalent and carries U_i onto a proper subset of some U_j. Thus each g_κ is Poincaré distance decreasing. Since every puzzle piece of depth N is a compact subset of the associated U_i (we need the thickening at this point), it follows that

$$\text{dist}_j\big(g_\kappa(x), g_\kappa(y)\big) \leq \lambda \, \text{dist}_i(x,y)$$

throughout any such puzzle piece, provided that $i > 0$, where $\lambda < 1$ is a uniform constant. Let δ be the maximum of the Poincaré diameters of the puzzle pieces of depth N. Then for any $h > 0$, since none of the puzzle pieces

$$P_{N+h}(z_0) \to P_{N+h-1}(z_1) \to \cdots \to P_{N+1}(z_{h-1}) \to P_N(z_h)$$

(except possibly the first) can be contained in the critical value puzzle piece $P_{N-1}^{(0)}$, it follows that the Poincaré diameter of $P_{N+h}(z_0)$ is at most $\lambda^h \delta$, which tends to zero as $h \to \infty$. $\square$

In order to deal with the possibility that $f^{\circ d}(z_0) = \alpha$, so that the puzzle piece $P_d(z_0)$ is not uniquely defined, we will need the following.

Definition. For any point z in the Julia set let $P_d^*(z)$ be the union of the puzzle pieces of depth d containing z. (In most cases, $P_d^*(z)$ is equal to the unique depth d puzzle piece $P_d(z)$ which contains z. However, if $f^{\circ d}(z) = \alpha$ then $P_d^*(z)$ is a union of q distinct puzzle pieces.)

We can now state and prove the principal result of this section.

Theorem 1.9. *Suppose as usual that f is quadratic, with connected Julia set, with both fixed points repelling, and not simply renormalizable. Suppose further that the critical orbit is disjoint from the fixed point α. Then for any $z \in J(f)$ we have*
$$\bigcap_d P_d^*(z) = \{z\}.$$

Proof. Since we assume that f is not simply renormalizable, we know from Corollary 1.7 that there exists some critical annulus $A_m(0)$ which has positive modulus. We will first prove that $\bigcap P_d(0) = \{0\}$, then prove that $\bigcap P_d(z) = \{z\}$ for any $z \in J(f)$ which is not an iterated pre-image of α, and finally prove the corresponding result when $f^{\circ n}(z) = \alpha$.

Critically Recurrent Case. Suppose that the critical orbit is recurrent, so that 1.3 applies. First consider the puzzle pieces $P_d(0)$ around the critical point. If the non-degenerate annulus $A_m(0)$ has at least 2^k descendents in the k-th generation for each k, then each of these contributes exactly $\mathbf{mod}\, A_m(0)/2^k$ to the sum $\sum_d \mathbf{mod}\, A_d(0)$. Hence this sum is infinite, as required. On the other hand, if there are fewer descendents in some generation, then one of them must be an only child, hence excellent by Lemma 1.3d. Using Lemma 1.3b and 1.3c, we again see that $\sum_d \mathbf{mod}\, A_d(0)$ is infinite. Therefore in either case the intersection $\bigcap P_d(0)$ reduces to the single point 0.

Now consider a point $z_0 \neq 0$ of the Julia set. We assume that the orbit $z_0 \mapsto z_1 \mapsto \cdots$ is disjoint from α, so that the puzzle pieces $P_d(z_0)$ are well defined. If the orbit of z_0 does not accumulate at zero, then we have $\bigcap P_d(z_0) = \{z_0\}$ by Lemma 1.8. Suppose, on the other hand, that the origin is an accumulation point of $\{z_n\}$. In other words, suppose that the tableau of the point z_0 has critical annuli reaching down to all depths. For each depth d, let us start at column zero (corresponding to the point z_0 itself) and advance to the right until we first hit a critical annulus at column n, then proceed diagonally back down until we reach column zero at depth $n + d$.

It follows from this construction that the annulus $A_{n+d}(z_0)$ is conformally isomorphic to $A_d(0)$. Furthermore, distinct values of d must correspond to distinct values of $n + d$. Thus the sum $\sum \mathbf{mod}\, A_d(z_0)$ is also infinite, hence $\bigcap P_d(z_0) = \{z_0\}$, as required.

Critically Non-recurrent Case. Now suppose that the critical orbit is not recurrent. Then the critical value $f(0)$ satisfies the hypothesis of Lemma 1.8. Hence $\bigcap P_d\big(f(0)\big) = \{f(0)\}$, and it follows easily that $\bigcap P_d(0) = \{0\}$. Next consider a point $z_0 \neq 0$. Again we may assume that the orbit $\{z_n\}$ accumulates at zero, since otherwise the conclusion would follow from Lemma 1.8. Again 1.7 tells us that there exists one depth m such that $A_m(0)$ has positive modulus. The corresponding depth m for the tableau of z_0 must have infinitely many columns k which are critical. For each of these, let us proceed diagonally back down in the tableau of z_0, ignoring whatever critical or semi-critical annuli we may meet, until we reach column zero at depth $m + k$. Each time we meet a critical or semi-critical annulus, we lose up to half of the modulus. (Problem 1-b below.) However, the hypothesis that the critical orbit is non-recurrent guarantees that such losses will only occur a bounded number of times. Thus $\sum \mathbf{mod}\, A_d(z_0)$ has infinitely many summands which are bounded away from zero. Hence this sum is infinite, and we have proved that $\bigcap P_d(z_0) = \{z_0\}$ in this case also.

Iterated Pre-images of α. If some forward image $z_n = f^{\circ n}(z_0)$ is equal to the fixed point α, then the above arguments do not make sense as stated. In this case, there are q distinct puzzle pieces $P_n^{(i)}$ of depth n which have z_0 as common boundary point. Each of these is contained in a unique sequence of nested puzzle pieces $P_n^{(i)} \supset P_{n+1}^{(i)} \supset \cdots$ which have z_0 as common boundary point. **Assertion:** *For each one of these q nested sequences the intersection $\bigcap_d P_d^{(i)}$ reduces to the single point z_0.* In fact, the proof of Lemma 1.8 applies equally well to this situation. Evidently the statement that $\bigcap_d P_d^*(z_0) = \{z_0\}$ follows immediately.

Thus we have proved Theorem 1.9, $\bigcap_d P_d^*(z) = \{z\}$, in all cases. Theorem 1.1, as stated at the beginning of this section, is a straightforward consequence. (Compare Problem 1-a below.) □

For further developments on the puzzle technique, see for example [F], [L2], [L3], [L4], [Pe3].

Here are some problems for the reader.

Problem 1-a. Local Connectivity. Prove that the intersection of $J(f)$ with each puzzle piece is connected. Conclude that $J(f)$ is locally connected at z whenever $\bigcap P_d^*(z) = \{z\}$.

Problem 1-b. Semi-Critical Annuli. Show that a semi-critical annulus $A_d(z_i)$ of depth $d > 0$ is non-degenerate if and only if $A_{d-1}(z_{i+1})$ is non-

degenerate. In the non-degenerate case, show that $A_d(z_i)$ is the union of
  (1) a ramified two-fold covering of $A_{d-1}(z_{i+1})$, and
  (2) a conformal copy of $P_d(z_{i+1})$.
Splitting $A_{d-1}(z_{i+1})$ by a carefully chosen simple closed curve through the
critical value, use the Grötzsch inequality to prove that $\mathbf{mod}\, A_d(z_i) >$
$\frac{1}{2}\,\mathbf{mod}\, A_{d-1}(z_{i+1})$.

Problem 1-c. Non-degenerate Annuli. Show that an annulus $A_0(z)$
of depth zero is non-degenerate if and only if it is semi-critical, and show that
$A_d(z_m)$ is non-degenerate if and only if $A_0(z_{d+m})$ is non-degenerate.

Problem 1-d. Illustrating the Tableau Rules. Suppose that we try
to modify the Fibonacci tableau of Figure 6 by changing just one column.
For example, suppose that we place the semi-critical annulus for column 5
at depth 3 or 4 or at depth ≥ 6, instead of at depth 5, without changing
columns 0 through 4. Show that the tableau rules would then force column
8 to end already at the semi-critical depth 0 or 1 or 2 respectively.

Problem 1-e. The Yoccoz τ-function. For each depth $d > 0$,
consider the sequence of puzzle pieces $f^{\circ i}\big(P_d(0)\big) = P_{d-i}(c_i)$ with $0 < i \leq d$.
If one of these pieces is equal to the critical puzzle piece $P_{d-i}(0)$, then choose
the smallest such i and set $\tau(d) = d - i$. Otherwise, if there is no such
i, set $\tau(d) = -1$. Thus $-1 \leq \tau(d) < d$ for all $d > 0$. Show that
$\tau(d+1) \leq \tau(d) + 1$ in all cases, and show that the annulus $A_d(0)$ is a
child of $A_n(0)$ if and only if $n = \tau(d)$ with $\tau(d+1) = \tau(d) + 1 > 0$.

Problem 1-f. Persistent Recurrence. Show that the critical tableau is
recurrent if and only if the function $\tau(d)$ is unbounded, and that it is periodic
if and only if $\tau(d+1) = \tau(d)+1$ for large d. By definition, the critical orbit is
persistently recurrent if the tableau is non-periodic with $\lim_{d\to\infty} \tau(d) = +\infty$.
Show that the Fibonacci tableau is persistently recurrent. In the persistently
recurrent case, show that the critical orbit is bounded away from the fixed
point β. (Otherwise, for any depth d we could choose a post-critical point
$c_n \in P_d(\beta)$, then choose the smallest k with $0 \in P_{d+k}(c_{n-k})$, and conclude
that $\tau(d+k) = 0$.) Show that the critical orbit is bounded away from every
iterated pre-image of β, and hence that its closure is nowhere dense in the
Julia set. Similarly show that the critical orbit is bounded away from the
fixed point α; and use this fact to show that the critical orbit closure is a
Cantor set. More generally, show that the critical orbit is bounded away from
any periodic or pre-periodic point. (For further information, see [L2].)

Problem 1-g. The Critical Orbit is Generically Dense. It is convenient to say that a property of certain points in a compact set is *generically*
true if it is true throughout a countable intersection of dense relatively open
subsets. *For a generic point $c \in \partial M$ in the boundary of the Mandelbrot set,
show that the critical orbit for the map f_c is everywhere dense in the Julia*

set. (Proof outline, suggested by McMullen and by J. Kwapisz: Let $M(p, \epsilon)$ be the set of $c \in \partial M$ such that the critical orbit for f_c comes within ϵ of every periodic orbit of period p. Using Montel's Theorem, show that this set is dense for $p \geq 2$. It is clearly open. Therefore a generic c belongs to the intersection of the $M(p, \epsilon)$, which implies that the critical orbit is dense in J.) As one corollary, it follows that a generic f_c is not persistently recurrent, and also there are no excellent annuli in the generic case. If f_c is renormalizable of period k, show that the union of the small Julia sets $f_c^{\circ i}(J_k)$ is a proper subset of the Julia set $J(f_c)$. Conclude that f_c is not renormalizable in the generic case.

2. Polynomials For Which All But One of the Critical Orbits Escape

This section will describe easier applications of ideas from §1. Its object is to describe the Branner-Hubbard theory of polynomials for which only one critical point has bounded orbit. However, we begin with the simpler case where *no* critical point has bounded orbit. The following is well known.

Theorem 2.1. *Let f be a polynomial of degree $n \geq 2$ for which all critical orbits escape to infinity. Then the filled Julia set K (in this case equal to the Julia set $J = \partial K$) is a Cantor set of measure zero[2]. Furthermore, the dynamical system $(J, f|_J)$ is homeomorphic to the one-sided shift on n symbols[3].*

Proof. Let $\omega_1, \ldots, \omega_{n-1}$ be the (not necessarily distinct) critical points of f, and let $G : \mathbb{C} \to \mathbb{R}_+$ be the canonical potential function ($=$ Green's function), which satisfies $G(f(z)) = n\, G(z)$ and vanishes precisely on the filled Julia set K. Thus $G(\omega_i) > 0$ by hypothesis. The critical points of G in $\mathbb{C} \smallsetminus K$ are the points ω_i and also all of their preimages under iterates of f. Hence the critical values of G (other than zero) are the numbers of the form $G(\omega_i)/n^k$ with $k \geq 0$.

The *puzzle* of f is constructed as follows. Choose a number γ, not of the form $G(\omega_i)/n^k$, so that $0 < \gamma < \mathrm{Min}\{G(f(\omega_i))\}$. Then the region $X_0 = G^{-1}(0, \gamma]$ contains no critical values of f, and is bounded by smooth curves. Similarly, each locus

$$X_d = f^{-d}(X_0) = G^{-1}[0, \gamma/n^d]$$

---

[2]Shishikura has shown that the Hausdorff dimension of this Cantor set can be arbitrarily close to two.

[3]Both Przytycki [Pr] and Makienko [Ma] have proved the sharper result that any rational Julia set which is totally disconnected and contains no critical point must be isomorphic to a one-sided shift.

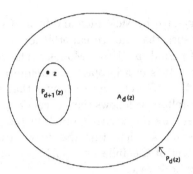

Figure 13. Nested puzzle pieces, and the annulus $A_d(z)$.

is bounded by smooth curves. Note that the complementary region $\mathbb{C} \smallsetminus X_d$ cannot have any bounded component, since the harmonic function G cannot have a local maximum. It follows that the locus X_d is the disjoint union of a finite number of closed topological disks. By definition, each of these closed disks will be called a *puzzle piece* P_d of depth d. *Since these puzzle pieces contain no critical value of f, it follows easily that f maps each P_d of depth $d > 0$ by a conformal isomorphism onto some puzzle piece $f(P_d)$ of depth $d - 1$.*

If $P_d \supset P_{d+1}$ are nested puzzle pieces of depths d and $d + 1$, then the set

$$A_d = \text{Interior}(P_d) \smallsetminus P_{d+1}$$

is a well defined annulus of strictly positive modulus. (Figure 13.) We will call such an A_d an *annulus of depth* d in the puzzle. Evidently f^{od} maps each annulus of depth d by a conformal isomorphism onto an annulus of depth zero. *Hence the moduli of all annuli constructed in this way are uniformly bounded away from zero.*

For any point z in the filled Julia set we can form the nested sequence of puzzle pieces $P_0(z) \supset P_1(z) \supset \cdots$, all containing z. Since the moduli of the annuli $\text{Interior}\, P_d(z) \smallsetminus P_{d+1}(z)$ are bounded away from zero, it follows that the intersection $\bigcap P_d(z)$ reduces to the single point z. Since each boundary circle ∂P_d is disjoint from K, this implies that K is totally disconnected.

The proof that J has measure zero will be based on the *McMullen inequality*

$$\text{area}(P_{d+1}) \leq \frac{\text{area}(P_d)}{1 + 4\pi \, \text{mod}\,(A_d)},$$

with $A_d = \text{Interior}(P_d) \smallsetminus P_{d+1}$ as above. (Compare the Appendix.) However, to apply this inequality in a useful way, we will need to construct annuli somewhat differently. Choose a number $\epsilon > 0$ which is small enough so that

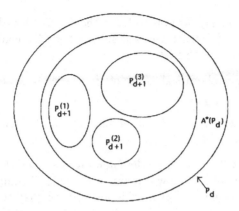

Figure 14. The "thin annulus" $A^*(P_d) \subset A_d \subset P_d$.

there are no critical values of G within the closed interval $[\gamma - \epsilon, \gamma]$. Then each connected component of $G^{-1}[(\gamma - \epsilon)/n^d, \gamma/n^d]$ is an annulus, and there is one such "thin annulus"

$$A^*(P_d) \;=\; P_d \cap G^{-1}\!\left[\frac{\gamma - \epsilon}{n^d}, \frac{\gamma}{n^d}\right]$$

within each connected component P_d of $G^{-1}[0, \gamma/n^d]$. The construction is such that:

(a) these annuli $A^*(P_d)$ have modulus uniformly bounded away from zero, say $\mathbf{mod}\, A^*(P_d) \geq \epsilon > 0$, and

(b) every puzzle piece P_{d+1} which is contained in P_d is actually contained in the smaller disk $P_d \smallsetminus A^*(P_d)$ (the bounded component of $\mathbb{C} \smallsetminus A^*(P_d)$).

Thus the McMullen inequality takes the following sharper form. For each fixed puzzle piece P_d,

$$\sum_{P_{d+1} \subset P_d} \mathbf{area}(P_{d+1}) \;\leq\; \frac{\mathbf{area}(P_d)}{1 + 4\pi\,\mathbf{mod}\,\big(A^*(P_d)\big)} \;\leq\; \frac{\mathbf{area}(P_d)}{1 + 4\pi\,\epsilon},$$

to be summed over those puzzle pieces of depth $d+1$ which are contained in the given P_d. It now follows inductively that the sum of the areas of *all* puzzle pieces of depth d satisfies

$$\sum \mathbf{area}(P_d) \;\leq\; \sum \mathbf{area}(P_0)/(1 + 4\pi\,\epsilon)^d.$$

Since this tends to zero as $d \to \infty$, and since $J \subset \bigcup P_d$, it follows that J has area zero.

To prove that $(J, f|_J)$ is isomorphic to the one-sided d-shift, we proceed as follows. We will construct a closed subset $\Gamma \subset \mathbb{C}$, consisting of paths leading out to infinity, such that:

(1) Γ contains all critical values of f,

(2) $f(\Gamma) \subset \Gamma$, and

(3) the complement $U = \mathbb{C} \smallsetminus \Gamma$ is simply-connected, and contains the Julia set.

The construction will be based on the following. By a *loxodrome* in $\mathbb{C} \smallsetminus K$ will be meant a smooth curve which cuts the family of equipotential curves at some constant angle. If we fix this angle, note that there is a unique loxodrome through any point of $\mathbb{C} \smallsetminus K$ which is not a critical point of G. The required Γ will be a locally finite union of loxodromes, all with the same fixed angle, each starting a some postcritical point of f and extending out to infinity. If we choose the angle badly, then such a loxodrome may run into some critical point of G. However, for all but countably many choices of angle this problem will not occur. There remains the possibility that some postcritical point z is itself a precritical point of f, and hence a critical point of G. In this case, there will be finitely many distinct loxodromes of given angle starting at z, and we can simply choose any one of them, making compatible choices along the orbit of z. The required properties (1), (2), (3) are now easily verified.

Since $f(\Gamma) \subset \Gamma$, it follows that $f^{-1}(U) \subset U$. Furthermore, since U is simply connected, and contains no critical values of f, it follows that every one of the d branches of f^{-1} near a point of U extends uniquely to a holomorphic map $g_i : U \to U$. The images

$$g_1(U), \ldots, g_n(U) \subset U$$

are disjoint open sets covering the Julia set. We will show that the intersections

$$J_i = J \cap g_i(U)$$

are disjoint compact sets which form the required *Bernoulli partition* $J = J_1 \cup \cdots \cup J_n$ of the Julia set. That is:

> *For each sequence of integers* $i_0, i_1, \ldots$ *between 1 and n there exists one and only one orbit* $z_0 \mapsto z_1 \mapsto z_2 \mapsto \cdots$ *in the Julia set with* $z_k \in J_{i_k}$ *for every k.*

In fact z_0 can be described as the intersection of the nested sequence of sets $g_{i_0} \circ g_{i_1} \circ \cdots \circ g_{i_k}(J)$. To prove this statement, we use the Poincaré metric on U. Since each g_i restricted to the compact set $J \subset U$ shrinks Poincaré distances by a factor bounded away from one, it follows that each such intersection $\bigcap_k g_{i_0} \circ g_{i_1} \circ \cdots \circ g_{i_k}(J)$ consists of a single point. This proves 2.1. $\square$

Maps with exactly one bounded critical orbit

This will be an exposition of results due to Branner and Hubbard [BH3]. We now suppose that exactly one of the $n - 1$ critical points of f, counted with multiplicity, has bounded orbit, while the orbits of the remaining $n - 2$ critical points escape to infinity. (Thus we exclude examples such as $z \mapsto z^3 + i$, for which a double critical point has bounded orbit; however, a double critical point escaping to infinity is fine.) Furthermore, we assume that $n \geq 3$, so that at least one critical orbit does escape. Then, according to Fatou and Julia, the Julia set is disconnected, with uncountably many connected components. Let

$$c_0 \mapsto c_1 \mapsto c_2 \mapsto \cdots$$

be the unique bounded critical orbit.

As in the proof of Theorem 2.1, choose a number $\gamma > 0$ which is not a critical value of G, and so that the region $G^{-1}(0, \gamma]$ contains no critical value of f. Again we define the puzzle pieces of depth d to be the connected components P_d of the locus

$$\bigcup P_d = G^{-1}\left[0, \gamma/n^d\right] .$$

Thus each point $z \in K$ determines a nested sequence $P_0(z) \supset P_1(z) \supset \cdots$, and the central problem is to decide whether or not $\bigcap_d P_d(z) = \{z\}$. Again we look at the intermediate annuli

$$A_d(z) = \text{Interior } P_d(z) \smallsetminus P_{d+1}(z) .$$

As in the Yoccoz proof, such an annulus is said to be semi-critical, critical, or off-critical according as the critical point c_0 belongs to the annulus itself, or to the bounded or the unbounded component of its complement. (For this purpose, we ignore the other critical points, whose orbits escape to infinity.) This Branner-Hubbard puzzle is easier to deal with than the Yoccoz puzzle for three reasons:

(a) All of the annuli $A_d(z)$ are non-degenerate, with strictly positive modulus.

(b) The various puzzle pieces of depth d are pairwise disjoint.

(c) For each $z \in K$, the intersection $\bigcap_d P_d(z)$ of the puzzle pieces containing z has an immediate topological description: It is equal to the connected component of the filled Julia set K which contains the given point z. For this intersection is clearly connected, and no larger subset can be connected since each boundary $\partial P_d(z)$ is disjoint from K.

The *tableau* of an orbit $z_0 \mapsto z_1 \mapsto \cdots$ is defined just as in §1, and can be described as a record of exactly which of the annuli $A_d(z_i)$ are critical or

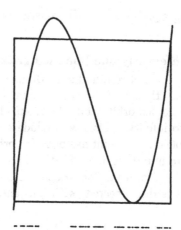

Figure 15. An example: Graph of the function $f(x) = x(x - c_0)^2/(1 - c_0)^2$ on the unit interval $I = [0,1]$, with critical points $c_0 = .76$ and $c_0/3$. There is just one bounded critical orbit $c_0 \mapsto 0 \mapsto 0 \mapsto \cdots$. Since the iterated pre-images of any point of J are dense in J, and since there are three distinct branches of f^{-1} mapping I into itself, it follows that the Julia set J is completely contained in the real interval I. This Julia set is plotted underneath the graph. Since c_0 and 0 evidently belong to distinct connected components of J, it follows from Theorem 2.2 that J is totally disconnected.

semi-critical or off-critical. First suppose that the tableau of the critical orbit is not periodic. We continue to assume that the degree is $n \geq 3$ and that exactly one of the $n - 1$ critical points has bounded orbit.

Theorem 2.2. *If the critical tableau is not periodic, or equivalently if the post-critical points $c_1, c_2, \ldots$ are all disjoint from the critical component $\bigcap_d P_d(c_0)$, then for every point z_0 of the filled Julia set K the sum $\sum_d \operatorname{mod} A_d(z_0)$ is infinite, hence $\bigcap_d P_d(z_0) = \{z_0\}$. It follows that $J = K$ is a totally disconnected set of area zero.*

Thus J is again homeomorphic to a Cantor set. However, in this case $(J, f|_J)$ is not isomorphic to a shift, or even a sub-shift. For there are critical points in J, hence f is not locally one-to-one on J.

The proof of this theorem is quite similar to the proof of the Yoccoz theorem. However there are simplifications, leading to the sharper result which is stated: $\sum \operatorname{mod} A_d(z) = \infty$ for *all* $z \in K$. In particular, 1.8 is replaced by the following statement, which is true whether or not the critical tableau is periodic. Consider an orbit $z_0 \mapsto z_1 \mapsto \cdots$.

Lemma 2.3. *Suppose that the points z_1, z_2, ... are all disjoint from some neighborhood $P_N(c_0)$ of the critical point c_0. Then the annuli $A_d(z_0)$ have modulus uniformly bounded away from zero, hence $\sum_d \operatorname{mod} A_d(z_0) = \infty$.*

Proof. Each annulus $A_d(z_i)$ of depth $d > 0$ has modulus at least half of the modulus of $A_{d-1}(z_{i+1})$. In fact, if $i > 0$ and $d > N$ then these two annuli are conformally isomorphic. Thus

$$\operatorname{mod} A_d(z_0) \geq \operatorname{mod} A_0(z_d)/2^{N+1} \,,$$

where the right side is bounded away from zero since there are only finitely many annuli of depth zero. □

Proof of 2.2. If the critical tableau is recurrent, then the proof proceeds exactly as is the Yoccoz argument: Every critical annulus either has at least 2^n descendents in the n-th generation for every n, or else has a descendent with this property. Since all annuli are non-degenerate, it follows that $\sum \operatorname{mod} A_d(c_0) = \infty$. On the other hand, if the critical tableau is non-recurrent, then it follows from 2.3 that $\sum \operatorname{mod} A_d(c_0) = \infty$. In particular, it follows that the collection of puzzle pieces $\{P_d(c_0)\}$ forms a fundamental system of neighborhoods of the critical point c_0. The corresponding statements for any iterated pre-image of c_0 follow immediately.

Now consider a point $z_0 \in K$, with orbit $z_0 \mapsto z_1 \mapsto \cdots$ which never meets the critical point c_0. If this orbit does not accumulate at c_0, then the statement that $\sum_d \operatorname{mod} A_d(z_0) = \infty$ follows from 2.3. On the other hand, if this orbit does accumulate at c_0, then since $\sum \operatorname{mod} A_d(c_0) = \infty$, an easy tableau argument shows that $\sum_d \operatorname{mod} A_d(z_0) = \infty$ also. Thus $\bigcap_d P_d(z) = \{z\}$ in all cases.

Since each boundary $\partial P_d(z)$ is disjoint from the filled Julia set K, it follows that K is totally disconnected, and hence that $J = K$. To prove that this set has measure zero, we proceed as in the proof of Theorem 2.1. Choose $\epsilon > 0$ so that the interval $[\gamma - \epsilon, \gamma]$ contains no critical values of the function $G : \mathbb{C} \to \mathbb{R}_+$. Then each puzzle piece $P_d(z)$ contains a unique component $A^*(P_d(z))$ of the set $G^{-1}[(\gamma - \epsilon)/n^d, \gamma/n^d]$. These annuli $A^*(P_d(z)) \subset A_d(z)$ are also non-degenerate, and the proof above shows equally well that $\sum_d \operatorname{mod} A^*(P_d(z)) = \infty$. Hence, just as in the proof of 2.1, for each fixed puzzle piece P_d, the total area of the puzzle pieces P_{d+1} of depth $d + 1$ which are contained in P_d satisfies

$$\sum_{P_{d+1} \subset P_d} \operatorname{area} P_{d+1} \leq \frac{\operatorname{area} P_d}{1 + 4\pi \operatorname{mod} A^*(P_d)} \,.$$

Define the ratio $\mu(P_d)$ by the formula

$$\mathbf{area}\, P_d \;=\; \mu(P_d) \sum_{P_{d+1} \subset P_d} \mathbf{area}\, P_{d+1}$$

so that this McMullen inequality takes the form $1 + 4\pi \, \mathbf{mod}\, A^*(P_d) \leq \mu(P_d)$. Then, substituting this formula inductively, we can write

$$\sum_{P_0} \mathbf{area}\, P_0 \;=\; \sum_{P_0 \supset P_1} \mu(P_0)\, \mathbf{area}(P_1) \;=\; \cdots$$

$$=\; \sum_{P_0 \supset \cdots \supset P_d} \mu(P_0) \cdots \mu(P_{d-1})\, \mathbf{area}(P_d) \;,$$

where the left hand expression is to be summed over all puzzle pieces of depth zero, the next over all pairs $P_0 \supset P_1$, and so on. (If $P_0 \supset \cdots \supset P_d$, note that $P_0, \ldots, P_{d-1}$ are uniquely determined by P_d.) Let η_d be the minimum value of the product $\mu(P_0) \cdots \mu(P_{d-1})$ as $P_0, \ldots, P_{d-1}$ varies over all sequences of nested puzzle pieces $P_0 \supset P_1 \supset \cdots \supset P_{d-1}$. Then we see from this last equality that

$$\sum \mathbf{area}(P_0) \;\geq\; \eta_d \sum \mathbf{area}(P_d) \;,$$

to be summed over all puzzle pieces of depth zero or d respectively. Thus, if we can prove that $\eta_d \to \infty$ as $d \to \infty$, then it will follow that

$$\mathbf{area}(J) \;\leq\; \sum \mathbf{area}(P_d) \;\leq\; \sum \mathbf{area}(P_0)/\eta_d \;\to\; 0 \;,$$

hence $\mathbf{area}(J) = 0$ as required.

Clearly $1 < \eta_1 < \eta_2 < \cdots$. If these numbers tended to a finite limit $L < \infty$, then for each d we could find puzzle pieces $P_0(d) \supset P_1(d) \supset \cdots \supset P_{d-1}(d)$ so that $\mu(P_0(d)) \cdots \mu(P_{d-1}(d)) \leq L$. Hence we could choose a puzzle piece P_0 which occurs infinity often as $P_0(d)$, then choose $P_1 \subset P_0$ which occurs infinitely often as $P_1(d)$, and so on. In this way, we could find a sequence

$$P_0 \supset P_1 \supset P_2 \supset \cdots$$

with $\mu(P_0) \cdots \mu(P_{d-1}) \leq L < \infty$ for every d. Since $1 + 4\pi \, \mathbf{mod}\, \big(A^*(P_i)\big) \leq \mu(P_i)$, this would imply that

$$1 + 4\pi \sum \mathbf{mod}\, \big(A^*(P_i)\big) \;\leq\; L < \infty \;,$$

contradicting our statement that $\sum \mathbf{mod}\, A_d(z) = \infty$ for all $z \in K$. This completes the proof of 2.2. $\square$

On the other hand, if the critical tableau is periodic, then we will prove the following.

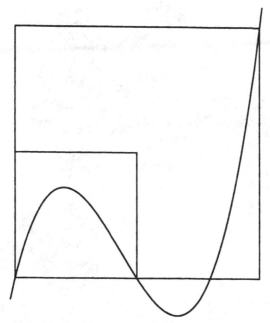

Figure 16. Example for Theorem 2.4: Graph of the map $f(x) = x(2x - 1)(5x - 4)$ on the unit interval. In this case the connected interval $[0, \frac{1}{2}]$ is contained in the filled Julia set K. The orbit of the critical point $\frac{2}{3}$ escapes to $-\infty$.

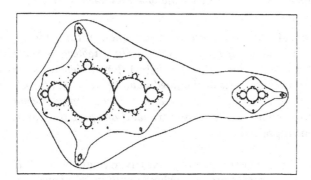

Figure 17. Julia set for this map, drawn to the same scale, showing the puzzle pieces of depth zero and one. Each non-trivial component of J is homeomorphic to a quadratic Julia set (the "basilica").

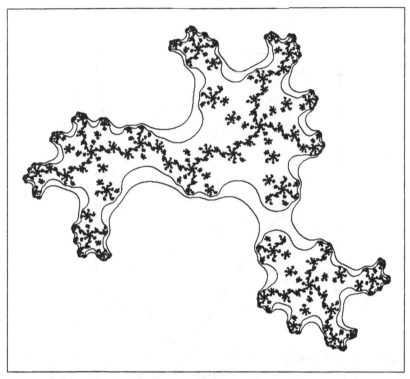

Figure 18. Julia set for $z \mapsto z^3 + a z^2 + 1$, with $a = -1.10692 + .63601\,i$, showing the puzzle pieces of depth zero and one. Each non-trivial component of J is homeomorphic to the Julia set for the quadratic map $z \mapsto z^2 + i$.

Theorem 2.4. *Still assuming that just one critical orbit $c_0 \mapsto c_1 \mapsto \cdots$ is bounded, if the associated critical tableau is periodic of period $p \geq 1$, so that $P_d(c_0) = P_d(c_p)$ for all depths d, then the connected component of the filled Julia set $K = K(f)$ which contains c_0 is non-trivial, that is, consists of more than one point. In fact, a component of K is non-trivial if and only if it contains some iterated pre-image of c_0.*

Thus there there are countably many non-trivial components of K. These countably many components are everywhere dense in K, since the iterated pre-images of any point of the Julia set are dense in the Julia set. Since a disconnected Julia set necessarily has uncountably many components, it follows that there are uncountably many single points components. (The Julia set of a rational function may have uncountably many non-trivial components. Compare [Mc1]. However, in the polynomial case no such example is known.)

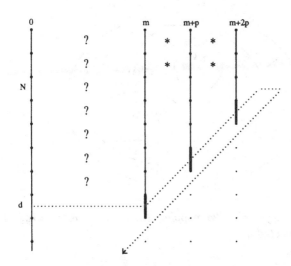

Figure 19. Tableau for an orbit $z_0 \mapsto z_1 \mapsto \cdots$ which intersects every critical puzzle piece. Here the stars stand for entries copied from the periodic critical tableau.

Proof of Theorem 2.4. If $P_d(c_0) = P_d(c_p)$ for all d, then it follows from the tableau rules that $P_d(c_i) = P_d(c_{p+i})$ for all i and d. Hence the entire orbit

$$c_0 \mapsto c_p \mapsto c_{2p} \mapsto \cdots$$

of c_0 under $f^{\circ p}$ is contained in the critical component $\bigcap_d P_d(c_0) \subset K(f)$. This intersection certainly has more than one point. For either it contains $c_0 \neq c_p$, or else c_0 is a superattracting point, in which case some entire neighborhood of c_0 belongs to $\bigcap_d P_d(c_0)$.

It follows easily that every pre-critical point in $K(f)$ also belongs to a non-trivial connected component. For if the orbit $z_0 \mapsto z_1 \mapsto \cdots$ intersects the critical component $\bigcap_d P_d(c_0)$, then we can choose the smallest $\ell \geq 0$ for which z_ℓ belongs to this critical component. It follows easily that $f^{\circ \ell}$ maps the component $\bigcap_d P_d(z_0)$ containing z_0 homeomorphically onto this critical component.

Now consider an orbit $z_0 \mapsto z_1 \mapsto \cdots$ in $K(f)$ which is disjoint from this critical component. This means that the tableau of this orbit has no columns which are completely critical. The proof is now divided into two cases:

Case 1. Suppose that there exists a fixed puzzle piece $P_d(c_0)$ which is disjoint from this orbit $\{z_n\}$. Then according to Lemma 2.3 we have $\sum_d \mathrm{mod}\, A_d(z_0) = \infty$, hence $\bigcap_d P_d(z_0) = \{z_0\}$.

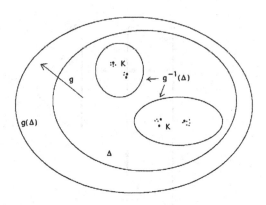

Figure 20. A polynomial-like mapping (g, Δ) *with* $K = K(g, \Delta)$
totally disconnected.

Case 2. If the orbit $\{z_n\}$ intersects *every* critical puzzle piece, then we use a tableau argument as follows. Choose a depth N so that the periodic critical tableau has no semi-critical annuli at depths $\geq N$. By hypothesis, there are infinitely many pairs (d, m), with $d \geq N$, so that the d-th row of the tableau for z_0 is semi-critical in column m and off-critical in earlier columns. (Compare Figure 19.) Using the tableau rules, we can then compute the tableau in column $m+i$ and depth $d-i$ for $0 < i < d-N$. In fact the entries in column $m+jp$ and depth $d-jp$ are semi-critical, and the others are off-critical. It now follows that the annulus $A_{d+m+1}(z_0)$ is conformally isomorphic to an annulus of depth N. Hence its modulus is bounded away from zero. It follows that $\sum_\ell \mathbf{mod}\, A_\ell(z_0) = \infty$, which completes the proof of Theorem 2.4. $\square$

We can understand this proof better by introducing the following concepts, which are due to Douady and Hubbard [DH3].

Definition. By a *polynomial-like map* is meant a pair (g, Δ) where $\Delta \subset \mathbb{C}$ is a closed topological disk and g is a continuous mapping, holomorphic on the interior of Δ, which carries Δ onto a closed topological disk $g(\Delta)$ which contains Δ in its interior, such that g maps boundary points of Δ to boundary points of $g(\Delta)$.

The *degree* $\delta \geq 1$ of such a polynomial-like mapping is a well defined topological invariant. Note that almost every point of $g(\Delta)$ has precisely δ pre-images in Δ. The *filled Julia set* $K(g, \Delta)$ is defined to be the compact set consisting of all $z_0 \in \Delta$ such that the entire orbit $z_0 \mapsto z_1 \mapsto \cdots$ of z_0 under g is defined and is contained in Δ.

Lemma 2.5. *Such a polynomial-like map of degree δ has $\delta - 1$ critical points, counted with multiplicity, in the interior of Δ. The filled Julia set $K(g, \Delta)$ is connected if and only if it contains all of these $\delta - 1$ critical points.*

Proof. Consider the nested sequence of compact sets

$$g(\Delta) \supset \Delta \supset g^{-1}(\Delta) \supset g^{-2}(\Delta) \supset \cdots$$

with intersection $K(g, \Delta)$. First suppose that the boundary $\partial\Delta$ contains no post-critical points, that is points $g^{\circ k}(\omega_i)$ with $k > 0$ where ω_i is a critical point of g. Then clearly each $g^{-k}(\Delta)$ is a compact set bounded by one or more closed curves. In fact, each $g^{-k}(\Delta)$ is either a closed topological disk or a finite union of closed topological disks, each of which maps onto the entire disk Δ under $g^{\circ k}$. To see this, note that for each component B of $g^{-k}(\Delta)$ the image $g^{\circ k}(B)$ is compact, and that $g^{\circ k}$ maps the boundary ∂B into $\partial\Delta$ and maps the interior of B onto an open subset of the interior of Δ. Since the interior of Δ is connected, this implies that $g^{\circ k}(B) = \Delta$. If some such component B were not simply-connected, then some component B' of $\mathbb{C} \smallsetminus g^{-k}(\Delta)$ would be bounded. A similar argument would then show that $g^{\circ k}$ must map B' onto the complementary disk $\widehat{\mathbb{C}} \smallsetminus \text{Interior}(\Delta)$, which is impossible.

Applying the Riemann-Hurwitz formula to the ramified covering $\Delta \to g(\Delta)$, we see that the number of critical points of g in the interior of Δ, counted with multiplicity, is equal to $n\chi(g(\Delta)) - \chi(\Delta) = n \cdot 1 - 1$. (Here χ is the Euler characteristic.) Similarly, applying this formula to $g^{-1}(\Delta) \to \Delta$, we see that the number of critical points in $g^{-1}(\Delta)$ is equal to $n\chi(g^{-1}(\Delta)) - \chi(\Delta)$. Thus all of the $n - 1$ critical points are contained in $g^{-1}(\Delta)$ if and only if $\chi(g^{-1}(\Delta)) = 1$, so that $g^{-1}(\Delta)$ consists of a single topological disk. Similarly, it follows by induction that all of the critical points are contained in $g^{-k}(\Delta)$ if and only if $g^{-k}(\Delta)$ is a topological disk. If every $g^{-k}(\Delta)$ is a disk, then it follows that the set $K = \bigcap g^{-k}(\Delta)$ is connected. On the other hand, if some $g^{-k}(\Delta)$ consists of two or more disks, then each one of these disks must contain a point of K, which is therefore disconnected.

To complete the proof, we must allow for the possibility that $\partial\Delta$ may contain some post-critical point of g. Clearly there can be at most $d - 1$ post-critical points in the annular region $g(\Delta) \smallsetminus \Delta$. Hence we can choose a disk $\Delta_1 \subset g(\Delta)$ whose boundary avoids these post-critical points. If Δ_1 contains Δ in its interior, and also contains all critical values $g(\omega_i)$ in its interior, then it is easy to check that the pair $(g, g^{-1}(\Delta_1))$ is a polynomial-like map of the same degree, and with the same filled Julia set, but with no post-critical points in the disk boundary. The proof then proceeds as above.
$\square$

Remark. Douady and Hubbard prove much sharper statements: If (g, Δ) is polynomial-like of degree $\delta \geq 2$, with $K(g, \Delta)$ connected, then there exists a polynomial map ψ of degree δ so that ψ on some neighborhood of $K(\psi)$ is quasi-conformally conjugate to g on a neighborhood of $K(g, \Delta)$. Furthermore, this quasi-conformal conjugacy can be chosen so as to satisfy the Cauchy-Riemann equations (in an appropriate sense) on the compact set $K(\psi)$. The polynomial map ψ is then uniquely determined up to affine conjugacy. In the case $\delta = 2$, one has the further statement that ψ depends continuously on (g, Δ).

Now let us return to the situation of Theorem 2.4.

Lemma 2.6. *If the critical tableau is periodic of period* $p \geq 1$, *then for any critical puzzle piece* $P_d(c_0)$ *with d sufficiently large, the pair* $\left(f^{\circ p}, P_d(c_0)\right)$ *is polynomial-like of degree two. Furthermore, the critical orbit*

$$c_0 \mapsto c_p \mapsto c_{2p} \mapsto \cdots$$

under $f^{\circ p}$ *is completely contained in* $P_d(c_0)$, *so that the filled Julia set* $K\left(f^{\circ p}, P_r(c_0)\right)$ *is connected. In fact* $K\left(f^{\circ p}, P_d(c_0)\right)$ *is equal to the intersection of the critical puzzle pieces* $\bigcap_k P_k(c_0)$, *and hence is precisely equal to the connected component of* $K(f)$ *which contains* c_0.

This is proved by a straightforward tableau argument. Details will be left to the reader. □

In this way, Branner and Hubbard show that each non-trivial component of $K(f)$ is homeomorphic to an appropriate quadratic Julia set. As examples, Figures 17 and 18 illustrate the case $p = d = 1$.

3. An Infinitely Renormalizable Non Locally Connected Julia Set

This section describes an unpublished example of Douady and Hubbard. It begins with an outline, with few proofs, of results from Douady and Hubbard [DH1], [DH2], [DH3], [D3]. (See also [M1], [D5], [Sch], [M3].)

Hyperbolic components in the Mandelbrot set

Let $f_c(z) = z^2 + c$. By definition, the *Mandelbrot set* M is the compact set consisting of all parameter values $c \in \mathbb{C}$ such that

the Julia set $J(f_c)$ is connected $\Longleftrightarrow$ 0 has bounded orbit.

However, it is often convenient to identify M with the corresponding set of polynomials f_c . A map $f_c \in M$ is hyperbolic (on its Julia set) if and only if it has a necessarily unique attracting periodic orbit. The hyperbolic maps in M form an open subset of the plane, and each connected component H in this open set is called a *hyperbolic component* in M . Let p be the period of the attracting orbit, and let $\lambda_p = \lambda_p(f_c) \in \mathbb{D}$ be its multiplier. The basic facts about hyperbolic components in M are as follows:

(1) *Any two maps in the same hyperbolic component $H \subset M$ have attracting orbits of the same period p . We call $p = p_H$ the* period *of H .*

(2) *Each hyperbolic component H is conformally isomorphic to the open unit disk $\mathbb{D}$ under the correspondence $f_c \mapsto \lambda_p(f_c)$. In fact this correspondence extends uniquely to a homeomorphism between the closure $\overline{H}$ and the closed unit disk $\overline{\mathbb{D}}$.*

In particular, each H has a unique *center point* c_H which maps to $\lambda_p(c_H) = 0$, and each boundary ∂H contains a unique *root point* $r_H \in \partial H$ which maps to $\lambda_p(r_H) = 1$. If the map f_c has a superattracting periodic orbit, then evidently c is the center point for one and only one hyperbolic component.

(3) *Similarly, if f_r has a parabolic periodic orbit, then r is the root point for one and only one hyperbolic component H . If the period of this orbit is p and the multiplier is $\lambda_p = \exp(2\pi i m/p')$ then the period of H is pp' .*

By definition, the *principal hyperbolic component* $H_\heartsuit$ is the set of f_c having an attracting fixed point. If the multiplier is $\lambda_1 \in \mathbb{D}$ then a brief computation shows that $c = \lambda_1(2 - \lambda_1)/4$, hence

$$\lambda_1(c) = 1 - \sqrt{1 - 4c} ,$$

taking that branch of the square root which lies in the right half plane. The boundary $\partial H_\heartsuit$ is the *cardioid*, consisting of all points which have the form $c = e^{i\theta}(2 - e^{i\theta})/4$, so that $\lambda_1(c) = e^{i\theta}$.

Satellites. Given any hyperbolic component H of period $p \geq 1$ and any root of unity $e^{2\pi i m/p'} \neq 1$, it follows from (2) that there is a unique point $r \in \partial H$ with $\lambda_p(r) = e^{2\pi i m/p'}$. According to (3), this r is the root point of a new hyperbolic component H' of period pp' . We say that H' is a *satellite*, which is *attached* to H at *internal angle* m/p' . In the special case where H is the principal hyperbolic component $H_\heartsuit$, we will use the notation $H' = H(m/p')$ for this satellite, and in general we will use the notation $H' = H * H(m/p')$. (Compare the discussion of "tuning" below.)

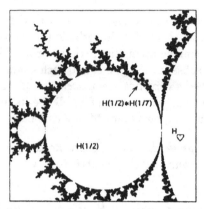

Figure 21. The period two component $H(1/2)$ in the Mandelbrot set, with an arrow pointing to its satellite $H(1/2) * H(1/7)$.

As an example, taking $m/p' = 1/2$, the point $r = -3/4$ is the root point of the period two component $H(1/2)$, consisting of all c with $|c + 1| < 1/4$.

External rays. We consider not only external rays for the Julia set in the z-plane (= *dynamic plane*), but also external rays for the Mandelbrot set in the c-plane (= *parameter plane*). Let $H \subset M$ be a hyperbolic component of period $p > 1$. Then exactly two external rays in $\mathbb{C} \smallsetminus M$ land at the root point r_H . Let $0 < a < b < 1$ be their angles. These angles have period exactly p under doubling, and hence can be expressed as fractions of the form $n/(2^p - 1)$. As an example, for the satellite component $H(1/p)$ we have $a = 1/(2^p - 1)$ and $b = 2/(2^p - 1)$.

For any $c \in H \cup \{r_H\}$ the corresponding external rays $R_a(c)$ and $R_b(c)$ in $\mathbb{C} \smallsetminus J(f_c)$ land at a periodic point which can be described as the "root point" of the Fatou component containing c . See Figure 22 for a schematic picture of the way these external rays are arranged in the parameter plane and in the dynamic plane. In particular

$$0 < a/2 < b/2 \ \leq \ a < b \ \leq \ 1 + a/2 < (1+b)/2 < 1 \,,$$

where the $a/2$-ray and the $b/2$-ray always land at distinct points: exactly one of these two rays is periodic.

Renormalization and Tuning. Recall that a map $f_c \in M$ is *simply renormalizable* if there is an integer $p \geq 2$ called the renormalization period, and a closed topological disk Δ in the z-plane, centrally symmetric about the origin, so that:

(1) $f^{\circ p-1}$ maps the image $f(\Delta)$ by a conformal isomorphism onto a disk $f^{\circ p}(\Delta)$ which contains Δ in its interior,

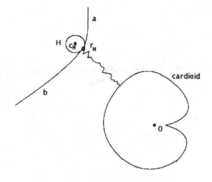

Figure 22. Schematic picture of external rays for a hyperbolic component $H \subset M$ in the c-plane,

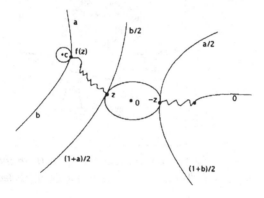

and a corresponding picture in the dynamic plane.

(2) the entire orbit of 0 under $f^{\circ p}$ is contained in Δ, and

(3) the images $f^{\circ i}(K_p)$ and $f^{\circ j}(K_p)$ do not disconnect each other for $0 \le i < j < p$. Here K_p denotes the small filled Julia set consisting of points of Δ whose orbits under $f^{\circ p}$ remain in Δ forever.

The following facts are surely true (compare [M1], [Mc2], [D5], [M3]), although I am not aware of any published proof.

The set of all $f_c \in M$ which are simply renormalizable of period p consists of a finite number of small copies of M (sometimes with the root point deleted). Each of these small copies contains a unique hyperbolic component of period p. Conversely, each

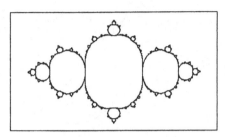

Figure 23. Julia set for the root point of $H(1/2)$,

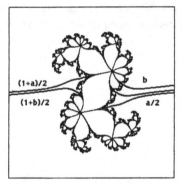

for the root point of $H(1/7)$ *(here the middle ray* $a = b/2$ *has not been labeled),*

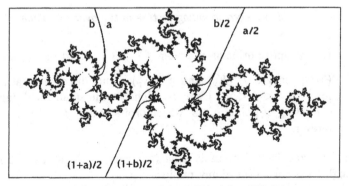

and for the root point of $H(1/2) * H(1/7)$.

hyperbolic component H of period p ≥ 2 determines a small copy of M which can be described as the image of an associated mapping c ↦ H ∗ c , which embeds M homeomorphically onto a proper subset H ∗ M ⊂ M . The elements of this form (with the possible exception of the root point r_H), are precisely the elements of M which are simply renormalizable of period p .

This embedding c ↦ H∗c maps each hyperbolic component H′ of period p′ ≥ 1 conformally onto a hyperbolic component H∗H′ of period pp′ in such a way that the multiplier $\lambda_{pp'}(H∗c)$ is equal to $\lambda_{p'}(c)$. In particular, this embedding maps center points to center points and root points to root points. It carries the principal hyperbolic component $H_\heartsuit$ onto H itself, and maps the satellite $H(m/p')$ of $H_\heartsuit$ onto the satellite $H ∗ H(m/p')$ of H . This ∗-product operation between hyperbolic components is associative, and has the principal hyperbolic component $H_\heartsuit$ as two-sided identity element. Thus it makes the collection of hyperbolic components into a monoid (=associative semigroup with identity) which operates as a semigroup of embeddings of M into itself. This monoid is free non-commutative.

Intuitively, the Julia set for $H ∗ c'$ is obtained from the Julia set of a map f_c belonging to the hyperbolic component H by replacing each bounded component of $\mathbb{C} \smallsetminus J(f_c)$ by a copy of $J(f_{c'})$ and defining the dynamics appropriately so that only the copy which is pasted in place of the critical component is mapped non-homeomorphically. The result is described as f_c *tuned by* $f_{c'}$.

In terms of external angles, let $a < b$ be the two external angles for the root point of H . These angles have periodic binary expansions of the form $.a_1 \cdots a_p a_1 \cdots a_p \cdots$ and $.b_1 \cdots b_p b_1 \cdots b_p \cdots$, both with period exactly equal to p . **Assertion:** If the point $c' \in \partial M$ is the landing point of a ray with angle $t \in \mathbb{R}/\mathbb{Z}$, then the image $H ∗ c'$ is the landing point of a ray whose binary expansion can be obtained by inserting the p-tuple $a_1 \cdots a_p$ in place of each zero in the binary expansion of t and the p-tuple $b_1 \cdots b_p$ in place of each one. (See [D3].)

The example

Consider a sequence of integers $1 < p_1 < p_2 < \cdots$.

Theorem 3.1. *If this sequence diverges to infinity sufficiently rapidly, then the sequence of subsets $H(1/p_1) ∗ H(1/p_2) ∗ \cdots ∗ H(1/p_k)∗M$ intersects in a single point $\omega \in M$ with the property that the Julia set $J(f_\omega)$ is not locally connected.*

The proof begins as follows. It will be convenient to use the abbreviation $r(k)$ for the root point $r_{H(1/p_k)}$ on the cardioid. Start with any $p_1 > 1$ and consider the embedding $c \mapsto H(1/p_1) * c$ from M into itself, which carries the period 1 root point $r_\heartsuit = 1/4$ to the root point $r(1) = r_{H(1/p_1)}$. Since this embedding is continuous, we can place $H(1/p_1) * c$ as close as we like to $r(1)$ by choosing c close to $1/4$. In particular, if p_2 is sufficiently large, then the entire $1/p_2$-limb will be close to $1/4$ by the Yoccoz inequality, hence every point $c \in H(1/p_1) * H(1/p_2) * M$ will certainly be close to $r(1)$. Under such a small perturbation, note that the parabolic fixed point of $f_{r(1)}$ splits up into a repelling fixed point for f_c, together with a nearby period p_1 orbit. Thus, by choosing p_2 large, we can place this entire period p_1 orbit into an ϵ-neighborhood of the parabolic fixed point of $f_{r(1)}$.

Similarly, choosing p_3 even larger, for any $c \in H(p_1) * H(p_2) * H(p_3) * M$ we can guarantee that the parabolic period p_1 orbit for $H(1/p_1) * r(2)$ is replaced by a nearby period $p_1 p_2$ orbit. In particular, we can guarantee that this new orbit lies within the $\epsilon + \epsilon/2$ neighborhood of the original parabolic fixed point. Continuing inductively, we can guarantee that all of the new orbits which are constructed lie within the 2ϵ neighborhood of the original fixed point. Furthermore, we can easily guarantee that these successive copies of M have diameter shrinking to zero.

Thus, if ω is the unique point in the intersection, we know that the Julia set $J(f_\omega)$ contains orbits of period 1, p_1, $p_1 p_2$, $\ldots$, all lying within the 2ϵ neighborhood of the original parabolic fixed point. These various periodic points can be described as the landing points of external rays of angles $a_k < b_k$ where $0 < a_1 < a_2 < \cdots < b_2 < b_1$. Furthermore, it is easy to check that the difference $b_k - a_k$ tends rapidly to zero, so that $\lim_{k \to \infty} a_k = \lim_{k \to \infty} b_k$. On the other hand, as noted in the preceding section, the external rays of angle $a_k/2$ and $b_k/2$ land at different points, which are negatives of each other. Each such landing point is either close to the original parabolic fixed point, or close to its negative. Thus we have a sequence of angles $a_k/2$ and $b_k/2$ tending to a common limit, yet the landing points of the corresponding external rays for $J(f_\omega)$ do not tend to a common limit. According to Caratheodory, this implies that $J(f_\omega)$ is not locally connected. $\square$

More generally, if H_1, H_2, $\ldots$ is any sequence of hyperbolic components which are not centered along the real axis, then it is again true that $\lim_{k \to \infty} a_k = \lim_{k \to \infty} b_k$. Let $\omega \in M$ be any element of the intersection $\bigcap_k H_1 * H_2 * \cdots * H_k * M$. Whenever the landing points in $J(f_\omega)$ of the rays of angle $a_k/2$ fail to converge to the critical point, it follows that $J(f_\omega)$ is not locally connected. In particular, this is true whenever the sequence of components H_k converges sufficiently rapidly to the root point $r_\heartsuit = 1/4$.

This seems to be the only known obstruction to local connectivity in the infinitely renormalizable case. Thus whenever the landing points of the $a_k/2$ rays do converge to the critical point, we can ask whether $J(f_\omega)$ is in fact locally connected.

For further papers related to local connectivity of Julia sets, see for example [Ki], [Pe2], [PM], [Sø1], [Sø2], [Z].

Appendix: Length-Area-Modulus Inequalities

The most basic length-area inequality is the following. Let $I^2 \subset \mathbb{C}$ be the open unit square consisting of all $z = x + iy$ with $0 < x < 1$ and $0 < y < 1$. By a *conformal metric* on I^2 we mean a metric of the form

$$ds = \rho(z)|dz|$$

where $z \mapsto \rho(z) > 0$ is any strictly positive continuous real valued function on the open square. In terms of such a metric, the *length* of a smooth curve $\gamma : (a, b) \to I^2$ is defined to be the integral

$$L_\rho(\gamma) = \int_a^b \rho\big(\gamma(t)\big)|d\gamma(t)| \,,$$

and the *area* of a region $U \subset I^2$ is defined to be

$$\mathbf{area}_\rho(U) = \int\int_U \rho(x + iy)^2 dx\, dy \,.$$

In the special case of the Euclidean metric $ds = |dz|$, with $\rho(z)$ identically equal to 1, the subscript ρ will be omitted.

Theorem A.1. *If* $\mathbf{area}_\rho(I^2)$ *(the integral over the entire square) is finite, then for Lebesgue almost every* $y \in (0, 1)$ *the length* $L_\rho(\gamma_y)$ *of the horizontal line* $\gamma_y : t \mapsto (t, y)$ *at height* y *is finite. Furthermore, there exists* y *so that*

$$L_\rho(\gamma_y)^2 \leq \mathbf{area}_\rho(I^2) \,. \tag{1}$$

In fact, the set consisting of all $y \in (0, 1)$ *for which this inequality is satisfied has positive Lebesgue measure.*

Remark 1. Evidently this inequality is best posible. For in the case of the Euclidean metric $ds = |dz|$ we have

$$L(\gamma_y)^2 = \mathbf{area}(I^2) = 1 \,.$$

Remark 2. It is essential here that we use a square, rather than a rectangle. If we consider instead a rectangle R with base Δx and height Δy, then the corresponding inequality would be

$$L_\rho(\gamma_y)^2 \le \frac{\Delta x}{\Delta y} \mathbf{area}_\rho(R) \qquad (2)$$

for a set of y with positive measure.

Proof of A.1. We use the Schwarz inequality

$$\left(\int_a^b f(x)g(x)\, dx \right)^2 \le \left(\int_a^b f(x)^2\, dx \right) \cdot \left(\int_a^b g(x)^2\, dx \right) ,$$

which says (after taking a square root) that the inner product of any two vectors in the Euclidean vector space of square integrable real functions on an interval is less than or equal to the product of their norms. We may as well consider the more general case of a rectangle $R = (0, \Delta x) \times (0, \Delta y)$. Taking $f(x) \equiv 1$ and $g(x) = \rho(x, y)$ for some fixed y, we obtain

$$\left(\int_0^{\Delta x} \rho(x, y)\, dx \right)^2 \le \Delta x \int_0^{\Delta x} \rho(x, y)^2\, dx ,$$

or in other words

$$L_\rho(\gamma_y)^2 \le \Delta x \int_0^{\Delta x} \rho(x, y)^2 dx ,$$

for each constant height y. Integrating this inequality over the interval $0 < y < \Delta y$ and then dividing by Δy, we get

$$\frac{1}{\Delta y} \int_0^{\Delta y} L_\rho(\gamma_y)^2 dy \le \frac{\Delta x}{\Delta y} \mathbf{area}_\rho(A) . \qquad (3)$$

In other words, the *average* over all y in the interval $(0, \Delta y)$ of $L_\rho(\gamma_y)^2$ is less than or equal to $\frac{\Delta x}{\Delta y}$ $\mathbf{area}_\rho(A)$. Further details of the proof are straightforward. $\square$

Now let us form a cylinder C of circumference Δx and height Δy by gluing the left and right edges of our rectangle together. More precisely, let C by the quotient space which is obtained from the infinitely wide strip $0 < y < \Delta y$ in the z-plane by identifying each point $z = x + iy$ with its translate $z + \Delta x$. Define the *modulus* $\mathbf{mod}(C)$ of such a cylinder to be the ratio $\Delta y / \Delta x$ of height to circumference. By the *winding number* of a closed curve γ in C we mean the integer

$$w = \frac{1}{\Delta x} \oint_\gamma dx .$$

Theorem A.2 (Length-Area Inequality for Cylinders). *For any conformal metric $\rho(z)|dz|$ on the cylinder C there exists some simple closed curve γ with winding number $+1$ whose* *length* $L_\rho(\gamma) = \oint_\gamma \rho(z)|dz|$ *satisfies the inequality*

$$L_\rho(\gamma)^2 \leq \mathbf{area}_\rho(A)/\mathbf{mod}(A) .\qquad (4)$$

Furthermore, this result is best possible: If we use the Euclidean metric $|dz|$ *then*

$$L(\gamma)^2 \geq \mathbf{area}(A)/\mathbf{mod}(A)\qquad (5)$$

for every such curve γ .

Proof. Just as in the proof of A.1, we find a horizontal curve γ_y with

$$L_\rho(\gamma_y)^2 \leq \frac{\Delta x}{\Delta y}\mathbf{area}_\rho(\mathcal{C}) = \frac{\mathbf{area}_\rho(\mathcal{C})}{\mathbf{mod}(\mathcal{C})} .$$

On the other hand in the Euclidean case, for any closed curve γ of winding number one we have

$$L(\gamma) = \oint_\gamma |dz| \geq \oint_\gamma dx = \Delta x ,$$

hence $L(\gamma)^2 \geq (\Delta x)^2 = \mathbf{area}(\mathcal{C})/\mathbf{mod}(\mathcal{C})$. $\square$

Definitions. A Riemann surface A is said to be an *annulus* if it is conformally isomorphic to some cylinder. An embedded annulus $A \subset \mathcal{C}$ is said to be *essentially embedded* if it contains a curve which has winding number one around $\mathcal{C}$.

Here is an important consequence of Theorem A.2.

Corollary A.3 (An Area-Modulus Inequality). *Let* $A \subset \mathcal{C}$ *be an essentially embedded annulus in the cylinder* $\mathcal{C}$, *and suppose that* A *is conformally isomorphic to a cylinder* $\mathcal{C}'$. *Then*

$$\mathbf{mod}(\mathcal{C}') \leq \frac{\mathbf{area}(A)}{\mathbf{area}(\mathcal{C})} \mathbf{mod}(\mathcal{C}) .\qquad (6)$$

In particular:

$$\mathbf{mod}(\mathcal{C}') \leq \mathbf{mod}(\mathcal{C}) .\qquad (7)$$

Proof. Let $\zeta \mapsto z$ be the embedding of $\mathcal{C}'$ onto $A \subset \mathcal{C}$. The Euclidean metric $|dz|$ on $\mathcal{C}$, restricted to A , pulls back to some conformal metric $\rho(\zeta)|d\zeta|$ on $\mathcal{C}'$, where $\rho(\zeta) = |dz/d\zeta|$. According to A.2, there exists a curve γ' with winding number 1 about $\mathcal{C}'$ whose length satisfies

$$L_\rho(\gamma')^2 \leq \mathbf{area}_\rho(\mathcal{C}')/\mathbf{mod}(\mathcal{C}') .$$

This length coincides with the Euclidean length $L(\gamma)$ of the corresponding curve γ in $A \subset \mathcal{C}$, and $\mathbf{area}_\rho(\mathcal{C}')$ is equal to the Euclidean area $\mathbf{area}(A)$, so we can write this inequality as

$$L(\gamma)^2 \leq \mathbf{area}(A)/\mathbf{mod}(\mathcal{C}') .$$

But according to (5) we have

$$\mathbf{area}(\mathcal{C})/\mathbf{mod}(\mathcal{C}) \; \leq \; L(\gamma)^2 \; .$$

Combining these two inequalities, we obtain

$$\mathbf{area}(\mathcal{C})/\mathbf{mod}(\mathcal{C}) \; \leq \; \mathbf{area}(A)/\mathbf{mod}(\mathcal{C}') \; ,$$

which is equivalent to the required inequality (6). □

Corollary A.4. *The modulus of a cylinder is a well defined conformal invariant.*

Proof. If $\mathcal{C}'$ is conformally isomorphic to $\mathcal{C}$ then (7) asserts that $\mathbf{mod}(\mathcal{C}') \leq \mathbf{mod}(\mathcal{C})$, and similarly $\mathbf{mod}(\mathcal{C}) \leq \mathbf{mod}(\mathcal{C}')$. □

It follows that the *modulus* of an annulus A can be defined as the modulus of any conformally isomorphic cylinder. Furthermore, if A is essentially embedded in some other annulus A', then $\mathbf{mod}(A) \leq \mathbf{mod}(A')$.

Corollary A.5 (Grötzsch Inequality). *Suppose that $A' \subset A$ and $A'' \subset A$ are two disjoint annuli, each essentially embedded in A. Then*

$$\mathbf{mod}(A') + \mathbf{mod}(A'') \; \leq \; \mathbf{mod}(A) \; .$$

Proof. We may assume that A is a cylinder $\mathcal{C}$. According to (6) we have

$$\mathbf{mod}(A') \; \leq \; \frac{\mathbf{area}(A')}{\mathbf{area}(\mathcal{C})} \, \mathbf{mod}(\mathcal{C}) \; , \qquad \mathbf{mod}(A'') \; \leq \; \frac{\mathbf{area}(A'')}{\mathbf{area}(\mathcal{C})} \, \mathbf{mod}(\mathcal{C}) \; .$$

where all areas are Euclidean. Using the inequality

$$\mathbf{area}(A') + \mathbf{area}(A'') \leq \mathbf{area}(\mathcal{C}) \; ,$$

the conclusion follows. □

Up to this point, we have only considered cylinders or annuli of finite modulus. If we take an arbitrary Riemann surface with free cyclic fundamental group, then it is always conformally isomorphic to some cylinder, providing that we allow also the possibility of a one-sided infinite or two-sided infinite cylinder (that is, either the upper half-plane or the full complex plane modulo the identification $z \equiv z + 1$). By definition, a Riemann surface conformally isomorphic to such an infinite cylinder will be called an *annulus of infinite modulus*.

Now consider the following situation. Let $U \subset \mathbb{C}$ be a bounded simply connected open set, and let $K \subset U$ be a compact subset such that the difference $A = U \smallsetminus K$ is an annulus (which may have finite or infinite modulus).

Corollary A.6. *Suppose that $K \subset U$ as described above. Then K reduces to a single point if and only if the annulus $A = U \smallsetminus K$ has infinite modulus. Furthermore, the diameter of K is bounded by the inequality*

$$4 \operatorname{diam}(K)^2 \leq \frac{\operatorname{area}(A)}{\operatorname{mod}(A)} \leq \frac{\operatorname{area}(U)}{\operatorname{mod}(A)} . \tag{8}$$

Proof. According to A.2, there exists a curve with winding number one about A whose length satisfies $L^2 \leq \operatorname{area}(A)/\operatorname{mod}(A)$. Since K is enclosed within this curve, it follows easily that $\operatorname{diam}(K) \leq L/2$, and the inequality (8) follows. Conversely, if K is a single point then using (7) we see easily that $\operatorname{mod}(A) = \infty$. $\square$

Corollary A.7. (Branner-Hubbard). *Let $K_1 \supset K_2 \supset K_3 \supset \cdots$ be compact subsets of $\mathbb{C}$ with each K_{n+1} contained in the interior of K_n. Suppose further that each interior K_n^o is simply connected, and that each difference $A_n = K_n^o \smallsetminus K_{n+1}$ is an annulus. If $\sum_1^\infty \operatorname{mod}(A_n)$ is infinite, then the intersection $\bigcap K_n$ reduces to a single point.*

Proof. It follows inductively from the Grötzsch inequality that the modulus $\operatorname{mod}(K_1^o \smallsetminus K_n)$ tends to infinity as $n \to \infty$. Hence by (7) and A.7 the intersection of the K_n is a point. $\square$

Corollary A.8 (McMullen Inequality). *Again suppose that $K \subset U \subset \mathbb{C}$, and that $A = U \smallsetminus K$ is an annulus. Then*

$$\operatorname{area}(K) \leq \operatorname{area}(U)/e^{4\pi \operatorname{mod}(A)} . \tag{9}$$

Proof. (Compare [BH, II].) We will first prove the weaker inequality

$$\operatorname{area}(K) \leq \frac{\operatorname{area}(U)}{1 + 4\pi \operatorname{mod}(A)} . \tag{10}$$

The *isoperimetric inequality* asserts that the area enclosed by a plane curve of length L is at most $L^2/(4\pi)$, with equality if and only if the curve is a round circle. (See for example [CR].) Combining this with the proof of A.6, we see that

$$\operatorname{area}(K) \leq \frac{L^2}{4\pi} \leq \frac{\operatorname{area}(A)}{4\pi \operatorname{mod}(A)} .$$

Writing this inequality as $4\pi \operatorname{mod}(A) \leq \operatorname{area}(A)/\operatorname{area}(K)$ and adding $+1$ to both sides we obtain the completely equivalent inequality $1 + 4\pi \operatorname{mod}(A) \leq \operatorname{area}(U)/\operatorname{area}(K)$. This in turn is equivalent to (10).

To sharpen this inequality, we proceed as follows. Cut the annulus A up into n concentric annuli A_i, each of modulus equal to $\mathbf{mod}(A)/n$. Let K_i be the bounded component of the complement of A_i, and assume that these annuli are nested so that $A_i \cup K_i = K_{i+1}^o$ with $K_1 = K$. Let $K_{n+1}^o = A \cup K = U$. Then

$$\mathbf{area}(K_{i+1})/\mathbf{area}(K_i) \geq 1 + 4\pi\,\mathbf{mod}(A)/n$$

by (10), hence

$$\mathbf{area}(U)/\mathbf{area}(K) \geq (1 + 4\pi\,\mathbf{mod}(A)/n)^n,$$

where the right hand side converges to $e^{4\pi\,\mathbf{mod}(A)}$ as $n \to \infty$. $\square$

As a final illustration of modulus-area inequalities, consider a *flat torus* $\mathbf{T} = \mathbb{C}/\Lambda$. Here $\Lambda \subset \mathbb{C}$ is to be a 2-dimensional *lattice*, that is an additive subgroup of the complex numbers, spanned by two elements λ_1 and λ_2 where $\lambda_1/\lambda_2 \notin \mathbb{R}$. Let $A \subset \mathbf{T}$ be an embedded annulus.

By the *"winding number"* of A in $\mathbf{T}$ we will mean the lattice element $w \in \Lambda$ which is constructed as follows. Under the universal covering map $\mathbb{C} \to \mathbf{T}$, the central curve of A lifts to a curve segment which joins some point $z_0 \in \mathbb{C}$ to a translate $z_0 + w$ by the required lattice element. We say that $A \subset \mathbf{T}$ is an *essentially embedded annulus* if $w \neq 0$.

Corollary A.9 (Bers Inequality). *If the annulus A is embedded in the flat torus $\mathbf{T} = \mathbb{C}/\Lambda$ with winding number $w \in \Lambda$, then*

$$\mathbf{mod}(A) \leq \frac{\mathbf{area}(T)}{|w|^2}. \tag{11}$$

Roughly speaking, if A winds many times around the torus, so that $|w|$ is large, then A must be very skinny. A slightly sharper version of this inequality is given in Problem A-3 below.

Proof. Choose a cylinder C' which is conformally isomorphic to A. The Euclidean metric $|dz|$ on $A \subset \mathbf{T}$ corresponds to some metric $\rho(\zeta)|d\zeta|$ on C', with

$$\mathbf{area}_\rho(C') = \mathbf{area}(A).$$

By A.2 we can choose a curve γ' of winding number one on C', or a corresponding curve γ on $A \subset \mathbf{T}$, with

$$L(\gamma)^2 = L_\rho(\gamma')^2 \leq \frac{\mathbf{area}_\rho(C')}{\mathbf{mod}(C')} = \frac{\mathbf{area}(A)}{\mathbf{mod}(A)} \leq \frac{\mathbf{area}(T)}{\mathbf{mod}(A)}.$$

Now if we lift γ to the universal covering space $\mathbb{C}$ then it will join some point z_0 to $z_0 + w$. Hence its Euclidean length $L(\gamma)$ must satisfy $L(\gamma) \geq |w|$. Thus

$$|w|^2 \leq \frac{\mathbf{area}(T)}{\mathbf{mod}(A)},$$

which is equivalent to the required inequality (11). □

Concluding Problems:

Problem A-1. In the situation of Theorem A.1, show that more than half of the horizontal curves γ_y have length $L_\rho(\gamma_y) \leq \sqrt{2\,\mathbf{area}_\rho(I^2)}$. (Here "more than half" is to be interpreted in the sense of Lebesgue measure.)

Problem A-2. Show that the converse to A.7 is false: A nest $K_1 \supset K_2 \supset K_3 \supset \cdots$ of compact subsets of $\mathbb{C}$ may intersect in a point even if $\sum_1^\infty \mathbf{mod}(K_n^o \setminus K_{n+1})$ is finite. (For example, do this by showing that a closed disk $\overline{\mathbb{D}}'$ of radius $1/2$ can be embedded in the open unit disk $\mathbb{D}$ so that the complementary annulus $A = \mathbb{D} \setminus \overline{\mathbb{D}}'$ has modulus arbitrarily close to zero.)

Problem A-3 (Sharper Bers Inequality). If the flat torus $\mathbf{T} = \mathbb{C}/\Lambda$ contains several disjoint annuli A_i, all with the same "winding number" $w \in \Lambda$, show that

$$\sum \mathbf{mod}(A_i) \leq \mathbf{area}(\mathbf{T})/|w|^2 .$$

If two essentially embedded annuli are disjoint, show that they necessarily have the same winding number.

References

[A] L. Ahlfors, "Conformal Invariants", McGraw-Hill 1973.

[Be] A. Beardon, "Iteration of Rational Functions", Grad. Texts Math. **132**, Springer 1991.

[Bl] P. Blanchard, Disconnected Julia sets, pp. 181-201 of "Chaotic Dynamics and Fractals", ed. Barnsley and Demko, Academic Press 1986.

[BDK] P. Blanchard, R. Devaney and L. Keen, The dynamics of complex polynomials and automorphisms of the shift, Inv. Math **104** (1991) 545-580.

[Br1] B. Branner, The parameter space for complex cubic polynomials, pp. 169-179 of "Chaotic Dynamics and Fractals", ed. Barnsley and Demko, Academic Press 1986.

[Br2] B. Branner, The Mandelbrot set, pp. 75-105 of "Chaos and Fractals", edit. Devaney and Keen, Proc. Symp. Applied Math. **39**, Amer. Math. Soc. 1989.

[BH1] B. Branner and J. H. Hubbard, The iteration of cubic polynomials, Part I: the global topology of parameter space, Acta Math. **160** (1988), 143-206;

[BH2] B. Branner and J. H. Hubbard, The iteration of cubic polynomials, Part II: patterns and parapatterns, Acta Math. **169** (1992), 229-325.

[CG] L. Carleson and T. Gamelin, "Complex Dynamics", Springer 1993.

[CR] R. Courant and H. Robbins, "What is Mathematics?", Oxford U. Press 1941.

[D1] A. Douady, Systèmes dynamiques holomorphes, Séminar Bourbaki, 35e année 1982-83, n° 599; Astérisque **105-106** (1983) 39-63.

[D2] A. Douady, Julia sets and the Mandelbrot set, pp. 161-173 of "The Beauty of Fractals", edit. Peitgen and Richter, Springer 1986.

[D3] A. Douady, Algorithms for computing angles in the Mandelbrot set, pp. 155-168 of "Chaotic Dynamics and Fractals", ed. Barnsley & Demko, Acad. Press 1986.

[D4] A. Douady, Disques de Siegel et anneaux de Hermann, Séminar Bourbaki, 39e année 1986-87, n° 677; Astérisque **152-153** (1987-88) 151-172.

[D5] A. Douady, Chirurgie sur les applications holomorphes, Proc. Int. Cong. Math. Berkeley, A.M.S. (1987) 724-738.

[DH1] A. Douady and J. H. Hubbard, Itération des polynômes quadratiques complexes, C. R. Acad. Sci. Paris **294** (1982) 123-126.

[DH2] A. Douady and J. H. Hubbard, Étude dynamique des polynômes complexes I & II, Publ. Math. Orsay (1984-85).

[DH3] A. Douady and J. H. Hubbard, On the dynamics of polynomial-like mappings, Ann. Sci. Ec. Norm. Sup. (Paris) **18** (1985), 287-343.

[EL] A. Eremenko and M. Lyubich, The dynamics of analytic transformations, Leningr. Math. J. **1** (1990) 563-634.

[F] D. Faught, Local connectivity in a family of cubic polynomials, thesis, Cornell 1992.

[GM] L. Goldberg and J. Milnor, Fixed point portraits of polynomial maps II, Ann. Sci. Éc. Norm. Sup. **26** (1993), 51–98.

[He] M. Herman, Recent results and some open questions on Siegel's linearization theorem of germs of complex analytic diffeomorphisms of C^n near a fixed point, pp. 138-198 of Proc 8th Int. Cong. Math. Phys., World Sci. 1986.

[Hu1] J. H. Hubbard, Puzzles and quadratic tableaux (according to Yoccoz), preprint 1990.

[Hu2] J. H. Hubbard, Parapuzzles and local connectivity in the parameter plane (according to Yoccoz), preprint 1990.

[Hu3] J. H. Hubbard, Local connectivity of Julia sets and bifurcation loci: three theorems of J.-C. Yoccoz, pp. 467-511 of "Topological Methods in Modern Mathematics", ed. Goldberg and Phillips, Publish or Perish 1993.

[Ka] J. Kahn, Holomorphic Removability of Julia Sets, Stony Brook I.M.S. Preprint 1998#11.

[Ki] J. Kiwi, Non-accessible Critical Points of Cremer Polynomials, Stony

Brook I.M.S. Preprint 1995#2.

[L1] M. Lyubich, The dynamics of rational transforms: the topological picture, Russian Math. Surveys **41:4** (1986), 43-117.

[L2] M. Lyubich, Milnor's attractors, persistent recurrence and renormalization, pp. 513-541 of "Topological Methods in Modern Mathematics", ed. Goldberg and Phillips, Publish or Perish 1993.

[L3] M. Lyubich, Renormalization ideas in conformal dynamics. Current developments in mathematics, 1995 (Cambridge, MA), 155-190, Internat. Press, Cambridge, MA, 1994.

[L4] M. Lyubich, Dynamics of quadratic polynomials. I, II. Acta Math. **178** (1997) 185-247, 247-297.

[LM] M. Lyubich and J. Milnor, The Fibonacci unimodal map, J. Amer. Math. Soc. **6** (1993), 425-457.

[Ma] P. Makienko, Totally disconnected Julia sets, MSRI preprint 042-95.

[Mc1] C. McMullen, Automorphisms of rational maps, pp. 31-60 of "Holomorphic Functions and Moduli", edit. Drasin et al., Springer 1988.

[Mc2] C. McMullen, "Complex Dynamics and Renormalization," Ann. Math. Studies **135**, Princeton U. Press, 1994

[Mc3] C. McMullen, "Renormalization and 3-manifolds which fiber over the circle", Ann. Math. Studies **142**, Princeton U. Press, 1996.

[M1] J. Milnor, Self-similarity and hairiness in the Mandelbrot set, pp. 211-257 of "Computers in Geometry and Topology", edit. Tangora, Lect. Notes Pure Appl. Math. **114**, Dekker 1989

[M2] J. Milnor, "Dynamics in One Complex Variable, Introductory Lectures", Vieweg 1999, to appear.

[M3] J. Milnor, Periodic orbits, external rays and the Mandelbrot set: an expository account, Stony Brook I.M.S. Preprint 1999#3; Astérisque to appear.

[Pe1] C. L. Petersen, On the Pommerenke-Levin-Yoccoz inequality, Ergodic Theory Dynamical Systems **13** (1993) 785-806.

[Pe2] C. L. Petersen, Local connectivity of some Julia sets containing a circle with an irrational rotation. Acta Math. 177 (1996), 163-224.

[Pe3] C. L. Petersen, Puzzles and Siegel disks, "Progress in Holomorphic Dynamics", pp. 50-85, Pitman Res. Notes Math. Ser., 387, Longman, Harlow, 1998.

[PM] R. Perez-Marco, Fixed points and circle maps, Acta Math. **179** (1997), 243-294.

[Pr] F. Przytycki, Iterations of rational functions: which hyperbolic components contain polynomials? Fund. Math. **149** (1996), 95-118.

[RS] J. Riedl and D. Schleicher, On Crossed Renormalization of Quadratic Polynomials, "Problems on Complex Dynamical Systems" (Proceedings of conference on holomorphic dynamics), Kyoto 1997.

[Sch] D. Schleicher, Rational parameter rays of the Mandelbrot set, Stony Brook I.M.S. Preprint 1997#13.

[Sh] M. Shishikura, The Hausdorff Dimension of the Boundary of the Mandelbrot Set and Julia Sets, Ann. of Math. **147** (1998), 225-267.

[Sø1] D. E. K. Sørensen, Accumulation theorems for quadratic polynomials, Ergodic Theory Dynam. Systems **16** (1996), 555–590.

[Sø2] D. E. K. Sørensen, Describing quadratic Cremer point polynomials by parabolic perturbations. Ergodic Theory Dynam. Systems **18** (1998), 739–758.

[St] N. Steinmetz, "Rational Iteration: Complex Analytic Dynamical Systems", de Gruyter 1993.

[Su] D. Sullivan, Conformal dynamical systems, pp. 725-752 of "Geometric Dynamics", edit. Palis, Lecture Notes Math. **1007**, Springer 1983.

[Z] S. Zakeri, Biaccessibility in Quadratic Julia Sets II: The Siegel and Cremer Cases, Stony Brook I.M.S. Preprint 1998#1.

Institute for Mathematical Sciences,
State University of New York at
Stony Brook NY 11794-3660, U.S.A.
e-mail: jack@math.sunysb.edu

Holomorphic motions and puzzles (following M. Shishikura)

Pascale Rœsch*

Abstract

We prove that the Mandelbrot set is locally connected at Yoccoz parameters, following an idea of Shishikura.

The aim of this article is to present a short proof, due to M. Shishikura (unpublished), of the following theorem first obtained by J.-C. Yoccoz:

Main Theorem (Yoccoz). *The Mandelbrot set M is locally connected at parameters c for which $f_c(z) = z^2 + c$ is a non-renormalizable map whose fixed points are repelling and whose critical point is recurrent.*

For parameters c satisfying the above conditions, another theorem of Yoccoz (see the paper by J. Milnor in this volume) asserts that the Julia set J_c of f_c is locally connected, and in particular exhibits a fundamental system of neighborhoods $P_n(c)$ of c—in the dynamical plane of f_c—whose closures have connected intersections with J_c. We derive the Main Theorem from this dynamical result in the following way. Given a prescribed parameter $c_0 \in M$, we use the Douady-Hubbard conformal representation of $\mathbb{C} \smallsetminus M$ to construct neighborhoods $\mathcal{P}_n$ of c_0—in the parameter plane—whose closures have connected intersections with M. Then, to show that $\{c_0\} = \bigcap_n \mathcal{P}_n$, we consider the annuli $\mathcal{A}_n = \mathcal{P}_n \smallsetminus \overline{\mathcal{P}_{n+1}}$ and prove that the sum of their moduli is infinite. To get this estimate, we compare the modulus of $\mathcal{A}_n$ with the modulus of the "corresponding" annulus $A_n(c_0) = P_n(c_0) \smallsetminus \overline{P_{n+1}(c_0)}$. In M. Shishikura's proof, this key comparison relies on quasiconformality properties of holomorphic motions and the observation that, for $c \in \mathcal{P}_n$, the dynamical piece $P_n(c)$ moves holomorphically.

Section 1 briefly recalls the construction of Yoccoz puzzles and some pertinent results which will be needed later (for proofs, the reader is referred to Milnor's article [M2]). Section 2 deals with parallel constructions in the parameter plane; it provides a para-puzzle "corresponding" to the dynamical Yoccoz puzzle (see also [B, H] for similar constructions). Section 3 completes the proof of the Main Theorem, and in particular explains how to compare the modulus of a parameter annulus with the modulus of a critical annulus in the dynamical plane.

---

*Research partially supported by the ESF.

I am very grateful to M. Shishikura for explaining this proof that I adapted in [R] to study local connectivity in the parameter plane of cubic Newton maps.

1 Puzzles in the dynamical planes

1.1 External rays

We consider the family of quadratic polynomials $f_c(z) = z^2 + c$ parameterized by $c \in \mathbb{C}$. By definition, the filled Julia set K_c of f_c is the complement of the open set

$$B_c(\infty) = \left\{ z \in \mathbb{C} \mid f_c^n(z) \xrightarrow[n \to \infty]{} \infty \right\}$$

which is the basin of infinity. The potential function

$$G_c \colon \mathbb{C} \longrightarrow \mathbb{R}_+, \quad z \longmapsto \lim_{n \to \infty} 2^{-n} \log^+ \left| f_c^n(z) \right|,$$

(where $\log^+ x = \sup\{0, \log x\}$) is a proper harmonic function whose zero set is K_c and whose critical values are bounded by $G_c(0)$. Hence, the open set $U_c = \{ z \mid G_c(z) > G_c(0) \}$ is a punctured disc and the classical Böttcher Theorem provides a unique conformal representation

$$\phi_c \colon U_c \longrightarrow \{ z \mid |z| > r_c \}, \quad \text{where} \quad \log r_c = G_c(0),$$

which conjugates f_c to $z \mapsto z^2$ and satisfies $G_c = \log|\phi_c|$.

The level curve $\{ z \in \mathbb{C} \mid G_c(z) = v \}$, $v > 0$, is called the *equipotential* of level v and is denoted by $E_c(v)$. The *external ray* of angle $t \in \mathbb{R}/\mathbb{Z}$, denoted by $R_c(t)$, is the gradient line of G_c stemming from infinity with the angle t (measured via the Böttcher coordinate ϕ_c). When K_c is connected, *i.e.*, when c belongs to the Mandelbrot set, ϕ_c is a conformal representation from $B_c(\infty)$ to $\mathbb{C} \smallsetminus \overline{\mathbb{D}}$ and external rays are just the inverse images by ϕ_c of the rays in $\mathbb{C} \smallsetminus \overline{\mathbb{D}}$.

An external ray is an analytic curve that either hits a critical point of G_c or accumulates on the Julia set J_c. Moreover, if t is rational and $R_c(t)$ hits no critical point, then $R_c(t)$ lands at (*i.e.* converges to) some point of J_c that is a repelling or parabolic (eventually) periodic point [DH1]. Conversely, for $c \in M$, each repelling periodic point is the landing point of at least one (and at most finitely many) external ray(s) (see [M1]).

1.2 Wakes of the Mandelbrot set

We now briefly describe the regions in the parameter plane where the fixed points of f_c are both repelling and one of them is the landing point of a prescribed cycle of external rays. These conditions define parameters for which the Yoccoz puzzle can be constructed.

Theorem 1.1 [DH1]. *The map*

$$\Phi_M : \begin{cases} \mathbb{C} \smallsetminus M \longrightarrow \mathbb{C} \smallsetminus \overline{\mathbb{D}} \\ \quad c \longmapsto \Phi_M(c) = \phi_c(c) \end{cases}$$

is a conformal representation—where ϕ_c is the Böttcher coordinate on $U_c \ni c$.

Thus, the conformal homeomorphisms ϕ_c and ϕ_M have the same image, $\mathbb{C} \smallsetminus \overline{\mathbb{D}}$, and they map c—with different meanings—to the same point. In particular, every point $\omega \in \mathbb{C} \smallsetminus \overline{\mathbb{D}}$ has the form $\omega = \phi_c(c)$ for $c = \phi_M^{-1}(\omega)$.

As in the dynamical plane, the *external ray* of angle $t \in \mathbb{R}/\mathbb{Z}$ in the parameter plane is the set

$$\mathcal{R}(t) = \Phi_M^{-1}\left(\{r\, e^{2i\pi t},\ r > 1\}\right),$$

and the *equipotential* of level $v > 0$ is the closed curve

$$\mathcal{E}(v) = \left\{c \in \mathbb{C} \mid \log|\Phi_M(c)| = v\right\}.$$

One easily checks that the ray $\mathcal{R}(0)$ is exactly the half-line $]1/4, +\infty[$. For $c \in \mathbb{C} \smallsetminus \overline{\mathcal{R}}(0)$, the ray $R_c(0)$ lands at a repelling fixed point of f_c usually denoted by β_c. The second fixed point α_c of f_c is attracting if and only if the parameter c lies inside the open region $\mathcal{H} \subset M$ bounded by the central cardioid. The multiplier $f_c'(\alpha_c)$ defines a conformal representation from $\mathcal{H}$ to the unit disc $\mathbb{D}$ which extends to a homeomorphism between the boundary curves. We denote by $\gamma \colon \mathbb{R}/\mathbb{Z} \simeq \partial\mathbb{D} \to \partial\mathcal{H}$ the inverse of the latter homeomorphism.

The following two theorems (see [DH1, M2]) describe the regions of $\mathbb{C} \smallsetminus \mathcal{H}$ over which the external rays landing at α_c have fixed angles.

Theorem 1.2 [DH1].
a) *Every external ray $\mathcal{R}(t)$ of rational angle t converges to some parameter in M.*
b) *A parameter $c \in M$ is the landing point of some external ray $\mathcal{R}(p/q)$ with $(p,q) = 1$ and q even if and only if the critical point of f_c is eventually periodic. Moreover, for such a parameter c, the external ray $\mathcal{R}(r/s)$ converges to c if and only if, in the dynamical plane of f_c, the external ray $R_c(r/s)$ lands at the critical value c.*
c) *For any rational p/q with $(p,q) = 1$ and $q > 1$, the parameter $\gamma(p/q) \in \partial\mathcal{H}$ is the landing point of exactly two external rays $\mathcal{R}(t^-)$ and $\mathcal{R}(t^+)$, where the angles $t^\pm = t^\pm(p/q)$ belong to the same orbit under multiplication by two and satisfy $0 < t^- < t^+ < 1$.*

The proper arc $\mathcal{R}(t^-) \cup \mathcal{R}(t^+) \cup \gamma(p/q)$ divides the plane into two connected components. We denote by $\mathcal{W}_{p/q}$ the component which does not contain the domain $\mathcal{H}$.

Theorem 1.3 [GM]. *For any parameter c in $\mathcal{W}_{p/q}$, there are exactly q external rays that land at the fixed point α_c, namely the rays $R_c(2^i t^+)$, where $t^+ = t^+(p/q)$ and $i = 0, \ldots, q-1$. Moreover, if we cut out the plane along these rays, the critical value c lies in the region bounded by $R_c(t^-)$ and $R_c(t^+)$.*

1.3 The Yoccoz puzzle

Finally, we define the Yoccoz puzzle (see [M2]) for every parameter c belonging to a wake $\mathcal{W}_{p/q}$ with $(p,q) = 1$ and $q > 1$ (every point of $M \smallsetminus \overline{\mathcal{H}}$ belongs to such a $\mathcal{W}_{p/q}$).

Definition 1.4. Let $c \in \mathcal{W}_{p/q}$, $t^+ = t^+(p/q)$ and $X^c = \{z \in \mathbb{C} \mid G_c(z) < 1\}$. The *Yoccoz puzzle* of f_c is given by the following sequence of graphs:

$$I_0^c = \partial X^c \cup \left(X^c \cap \bigcup_{i \geq 0} \overline{R}_c(2^i t^+) \right),$$

$$I_n^c = f_c^{-n}(I_0^c) \quad \text{for all } n \geq 1.$$

The *puzzle pieces of depth* 0 are the q connected components of $X^c \smallsetminus I_0^c$. Each of them contains exactly one point $f_c^i(0)$ with $0 \leq i \leq q-1$ and will be denoted by $P_0^c(f_c^i(0))$ accordingly. The *puzzle pieces of depth* $n \geq 1$ are the connected components of $f_c^{-n}(X^c) \smallsetminus I_n^c = f_c^{-n}(X^c \smallsetminus I_0^c)$. The puzzle piece containing a given point z will be denoted by $P_n^c(z)$. The puzzle pieces containing the critical value c will be simply denoted by $P_0^c, \ldots, P_n^c, \ldots$.

For $c \in \mathcal{W}_{p/q}$, a theorem of Yoccoz [M2] asserts that, if f_c is non-renormalizable and combinatorially recurrent, then its Julia set is locally connected, and in particular $\bigcap_{n=1}^{\infty} P_n^c = \{c\}$.

The map f_c is *combinatorially recurrent* if, for every n, there exists $m > 0$ such that $f_c^m(0)$ belongs to $P_n^c(0)$. If the critical point 0 is recurrent then f_c is obviously combinatorially recurrent. Conversely, if f_c is non-renormalizable and combinatorially recurrent then it follows from the theorem of Yoccoz that the point 0 is recurrent.

Milnor's paper [M2] gives further information about the puzzle pieces P_n^c that will be needed in Section 3. In particular, it provides a sequence of annuli $A_{n_i}^c$ having the following properties:

- $A_{n_i}^c = P_{n_i}^c \smallsetminus \overline{P_{n_i+1}^c}$ surrounds the critical value c of f_c and is *non-degenerate*—meaning that $P_{n_i}^c \supset \overline{P_{n_i+1}^c}$;

- $f_c^{n_i - n_0}$ induces an unramified covering map from $A_{n_i}^c$ onto $A_{n_0}^c$; and

- the sum $\sum_{i \geq 0} \operatorname{mod} A_{n_i}^c$ is infinite, where $\operatorname{mod} A_{n_i}^c$ denotes the modulus of $A_{n_i}^c$.

2 Puzzles in the parameter plane

As in the previous section, we will use the following conventions: objects in the parameter plane will be denoted by calligraphic capital letters (such as $\mathcal{P}$, $\mathcal{R}$), while those in dynamical planes will be denoted by roman capitals (such as P, R). Symbols associated with objects from the dynamical plane of f_c will also carry a subscript or superscript c.

2.1 Yoccoz para-puzzles

Since the combinatorics of the Yoccoz puzzle of f_c depends on the wake $\mathcal{W}_{p/q}$ that contains c, we will construct one para-puzzle for each $\mathcal{W}_{p/q}$, using the same angles as in the dynamical plane and replacing the Böttcher coordinate by the Douady-Hubbard conformal representation. Fix two relatively prime integers p, q with $q > 1$.

Definition 2.1. Let $t^+ = t^+(p/q)$ and $\mathcal{X}_n = \{ c \in \mathcal{W}_{p/q} \mid \log|\phi_M(c)| < 2^{-n} \}$, $n \geq 0$. The *Yoccoz para-puzzle* is given by the following sequence of graphs defined in the closure of $\mathcal{W}_{p/q}$:

$$\mathcal{I}_n = \partial \mathcal{X}_n \cup \left(\bigcup_{t \in J_n} \overline{\mathcal{R}}(t) \cap \mathcal{X}_n \right)$$

where $J_n = \left\{ t \in \mathbb{R}/\mathbb{Z} \mid 2^n t \in \{t^+, \dots, 2^{q-1} t^+\} \right\}$. The *para-puzzle pieces of depth n* are the connected components of the $\mathcal{X}_n \smallsetminus \mathcal{I}_n$. A general para-puzzle piece of depth n will be denoted by $\mathcal{P}_n$ and the one containing a given parameter c by $\mathcal{P}_n(c)$.

Remark 2.2. For any parameter c in $\mathcal{W}_{p/q}$, Theorem 1.1 shows that I_0^c consists of converging rays and equipotentials. The graph I_1^c has the same property provided the critical value c does not lie on I_0^c. By induction, if c does not belong to $I_{n-1}^c \cup \dots \cup I_0^c$ then $I_n^c \cup \dots \cup I_0^c$ is a union of converging rays and closed equipotentials. We will now characterize the set of parameters c for which this occurs.

Lemma 2.3. *A parameter $c \in \overline{\mathcal{W}}_{p/q}$ belongs to $\mathcal{I}_n$ if and only if the critical value c of f_c lies on I_n^c. As a consequence, if c belongs to a para-puzzle piece $\mathcal{P}_n$ of depth n, then c is outside of $I_n^c \cup \dots \cup I_0^c$.*

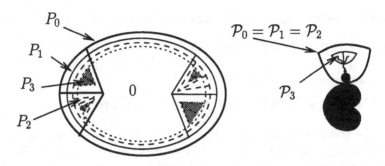

Figure 1: Shape of dynamical and parameter pieces of depth ≤ 3 for $\mathcal{W}_{1/3}$.

Proof. This is clear for $n = 0$. Now recall that if c does not belong to K_c—or equivalently if c does not belong to M—the Böttcher coordinate ϕ_c is defined on a disc U_c containing the critical value. For parameters c in $\overline{\mathcal{W}_{p/q}} \smallsetminus M$, the equality $\Phi_M(c) = \phi_c(c)$ implies that the argument of $\phi_c(c)$ belongs to $[t^-, t^+]$ (Theorem 1.3). Hence, by definition of graphs and para-graphs, a parameter $c \in \overline{\mathcal{W}_{p/q}}$ is in $\mathcal{I}_n \smallsetminus M$ if and only if the critical value c is in $I_n^c \smallsetminus K_c$. Then Theorem1.2-b) completes the proof for $c \in M$. Indeed, the rays of $\mathcal{I}_n \smallsetminus \mathcal{I}_0$ land at parameters c such that $f_c^n(c) = \alpha_c$, and each of these parameters c is the landing point of q external rays corresponding to the q dynamical rays landing at the critical value c (c is a preimage of α_c). The second part of the statement follows readily. $\square$

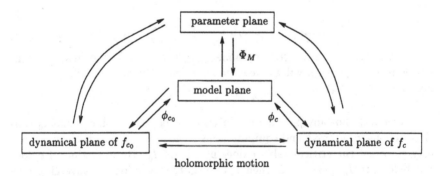

Figure 2: Relations between dynamical, parameter and reference planes.

Remarks 2.4.

a) Let n be the minimal index such that I_n^c contains the critical value. Then the graphs $I_0^c, \ldots, I_n^c$ do not contain the critical point, and therefore consist of rays and equipotentials. This holds in particular when the parameter c belongs to $\mathcal{P}_{n-1}$.

b) All para-puzzle pieces are simply-connected since each graph $\mathcal{I}_n$ is made connected by the equipotential it involves.

c) The unique para-puzzle piece $\mathcal{P}_0$ of depth 0 is bounded by $\mathcal{R}(t^+)$, $\mathcal{R}(t^-)$ and $\mathcal{E}(1)$. In the dynamical plane of f_c for $c \in \mathcal{P}_0$, the curves $R_c(t^+)$, $R_c(t^-)$ and $E_c(1)$ are exactly the rays and equipotential that bound the piece P_0^c containing the critical value (Theorem 1.3). Thus, using the canonical parameterizations of rays and equipotentials, we get a natural homeomorphism from $\partial \mathcal{P}_0$ to ∂P_0^c for any $c \in \mathcal{P}_0$.

2.2 Relations between puzzles and para-puzzles

We want now to generalize Remark 2.4-c) to deeper graphs, *i.e.*, to construct a "canonical" homeomorphism from $\partial \mathcal{P}_n$ to $\partial P_n^{c_0}$ for any $n \geq 0$ and any parameter $c_0 \in \mathcal{P}_n$. Canonical here means the homeomorphism extends to a quasiconformal map from $\mathcal{P}_n$ to $P_n^{c_0}$. The construction proceeds by induction and relies on the fact that, when c moves within a para-puzzle piece $\mathcal{P}_n$, the graph I_{n+1}^c moves holomorphically.

Recall that a *holomorphic motion* of a given subset $Z \subset \widehat{\mathbb{C}}$ parameterized by a complex manifold $\mathcal{C}$ is a map $\mathcal{C} \times Z \to \mathcal{C} \times \widehat{\mathbb{C}}$ of the form $(c, z) \mapsto (c, h^c(z))$ which is holomorphic on each slice $\mathcal{C} \times \{z\}$, $z \in Z$, injective on each slice $\{c\} \times Z$, $c \in \mathcal{C}$, and satisfies $h^{c_0}(z) = z$ for every $z \in Z$ and a given $c_0 \in \mathcal{C}$.

Lemma 2.5. *Let c_0 be a parameter in a para-puzzle piece $\mathcal{P}_n$ of depth $n \geq 0$. There exists a holomorphic motion*

$$h_n: \begin{cases} \mathcal{P}_n \times I_{n+1}^{c_0} \longrightarrow \mathcal{P}_n \times \widehat{\mathbb{C}} \\ (c, z) \longmapsto h_n(c, z) = (c, h_n^c(z)) \end{cases}$$

such that $I_{n+1}^c = h_n^c(I_{n+1}^{c_0})$ for every $c \in \mathcal{P}_n$.

Moreover, if $n \geq 1$, then h_n coincides with h_{n-1} on $\mathcal{P}_n \times (I_n^{c_0} \cap I_{n+1}^{c_0})$ (where $\mathcal{P}_{n-1} = \mathcal{P}_{n-1}(c_0) \supset \mathcal{P}_n$) and for every $c \in \mathcal{P}_n$ the diagram

$$\begin{array}{ccc} I_{n+1}^{c_0} & \xrightarrow{\ h_n^c\ } & I_{n+1}^c \\ {\scriptstyle f_{c_0}}\big\downarrow & & \big\downarrow{\scriptstyle f_c} \\ I_n^{c_0} & \xrightarrow[\ h_{n-1}^c\ ]{} & I_n^c \end{array}$$

is commutative.

Proof. Let $V_n = \{z \in \mathbb{C} \mid \log^+ |z| \geq 2^{-(n+1)}\}$. Since $\phi_c^{-1}(z)$ is holomorphic in the two variables (c, z) and is defined for $(c, z) \in \mathcal{P}_n \times V_n$, the map

$$h_n : \begin{cases} \mathcal{P}_n \times \phi_{c_0}^{-1}(V_n) \longrightarrow \mathcal{P}_n \times \widehat{\mathbb{C}} \\ (c, z) \longmapsto \left(c, \phi_c^{-1} \circ \phi_{c_0}(z)\right) \end{cases}$$

is a holomorphic motion of $\phi_{c_0}^{-1}(V_n)$ (note that this set contains the equipotential of $I_{n+1}^{c_0}$). Moreover, for $c \in \mathcal{P}_n$, the critical point of f_c remains outside of I_{n+1}^c (Lemma 2.3), so that f_c admits local inverses on I_{n+1}^c. Therefore, h_n extends uniquely over $\mathcal{P}_n \times (I_{n+1}^{c_0} \setminus K_{c_0})$ to a holomorphic motion— still denoted by h_n—satisfying $f_c \circ h_n^c = h_n^c \circ f_{c_0}$. One easily checks that $h_n^c(I_{n+1}^{c_0} \setminus K_{c_0}) = I_{n+1}^c \setminus K_c$. The final extension

$$h_n : \mathcal{P}_n \times \left(\phi_{c_0}^{-1}(V_n) \cup I_{n+1}^{c_0}\right) \longrightarrow \mathcal{P}_n \times \widehat{\mathbb{C}}$$

is given by the Λ-Lemma of Mañé-Sad-Sullivan (see [D]) which shows that any holomorphic motion of a given subset in $\widehat{\mathbb{C}}$ extends to a holomorphic motion of the closure. Since the extended map h_n^c is continuous, $h_n^c(I_{n+1}^{c_0}) = I_{n+1}^c$. The others desired properties are satisfied by construction. $\square$

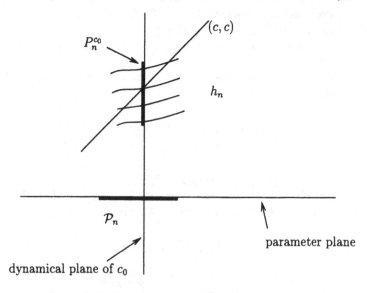

Figure 3: Holomorphic motion of $I_{n+1}^{c_0}$ over $\mathcal{P}_n$.

Notations 1. From now on, we fix a parameter $c_0 \in M$ and we suppose that c_0 lies in some para-puzzle piece $\mathcal{P}_n$ at each depth n. For every c in $\mathcal{P}_n$, we denote by D_{n+1}^c the puzzle piece of depth $n+1$ in the dynamical plane of f_c which is bounded by $h_n^c(\partial P_{n+1}^{c_0})$, where h_n is the holomorphic motion given by Lemma 2.5.

Lemma 2.6. *There exists a unique homeomorphism*

$$H: \bigcup_{n \geq 0} (\overline{\mathcal{P}_n} \cap \mathcal{I}_{n+1}) \longrightarrow \bigcup_{n \geq 0} (\overline{P_n^{c_0}} \cap I_{n+1}^{c_0})$$

that coincides with $(\phi_{c_0})^{-1} \circ \Phi_M$ outside of M. Moreover, H maps $\partial \mathcal{P}_n$ onto $\partial P_n^{c_0}$ and is given, on $\mathcal{P}_n \cap \mathcal{I}_{n+1}$, by the inverse holomorphic motion of the critical value: $H(c) = (h_n^c)^{-1}(c)$.

Proof. We will prove the following assertion $\mathfrak{P}(k)$ by induction on $k \geq 0$:

$\mathfrak{P}(k)$: *The map $(\phi_{c_0})^{-1} \circ \Phi_M$, restricted to*

$$\left(\partial \mathcal{P}_0 \cup (\overline{\mathcal{P}_0} \cap \mathcal{I}_1) \cup \cdots \cup (\overline{\mathcal{P}_{k-1}} \cap \mathcal{I}_k) \right) \setminus M,$$

extends to a homeomorphism H from $\partial \mathcal{P}_0 \cup (\overline{\mathcal{P}_0} \cap \mathcal{I}_1) \cup \cdots \cup (\overline{\mathcal{P}_{k-1}} \cap \mathcal{I}_k)$ to $\partial P_0^{c_0} \cup (\overline{P_0^{c_0}} \cap I_1^{c_0}) \cup \cdots \cup (\overline{P_{k-1}^{c_0}} \cap I_k^{c_0})$ which takes $\partial \mathcal{P}_k$ onto $\partial P_k^{c_0}$.

The step $k = 0$ follows from remark 2.4-c) and we now assume that $\mathfrak{P}(k)$ is satisfied.

We first show that $\phi_{c_0}^{-1} \circ \Phi_M$ induces a homeomorphism from $(\overline{\mathcal{P}_k} \cap \mathcal{I}_{k+1}) \setminus M$ onto $(\overline{P_k^{c_0}} \cap I_{k+1}^{c_0}) \setminus J_{c_0}$. In fact, $\phi_{c_0}^{-1} \circ \Phi_M: \mathbb{C} \setminus M \to \mathbb{C} \setminus J_{c_0}$ maps the point whose Douady-Hubbard coordinate is $re^{2i\pi t}$ to the point whose Böttcher coordinate is the same. Moreover, by induction, the homeomorphism H takes $\partial \mathcal{P}_k$ onto $\partial P_k^{c_0}$. Therefore, the image of $(\overline{\mathcal{P}_k} \cap \mathcal{I}_{k+1}) \setminus M$ under $\phi_{c_0}^{-1} \circ \Phi_M$ is exactly $(\overline{P_k^{c_0}} \cap I_{k+1}^{c_0}) \setminus J_{c_0}$.

Note that on $(\mathcal{P}_k \cap \mathcal{I}_{k+1}) \setminus M$, the map H satisfies

$$H(c) = \phi_{c_0}^{-1} \circ \Phi_M(c) = \phi_{c_0}^{-1}(\phi_c(c)) = (h_k^c)^{-1}(c).$$

Thus we extend H by setting $H(c) = (h_k^c)^{-1}(c)$ for $c \in \mathcal{P}_k \cap \mathcal{I}_{k+1}$. To see that this extension is still injective, we have to show that, if two distinct external rays $\mathcal{R}(t_1), \mathcal{R}(t_2) \subset \mathcal{I}_{k+1}$ land at points $c_1, c_2 \in \mathcal{P}_k$ with the same image under H, then $c_1 = c_2$. Since c_1, c_2 belong to $\mathcal{P}_k$ with $k \geq 1$, they are Misiurewicz parameters (*i.e.*, parameters for which the critical point is eventually periodic). This follows readily from Theorem 1.2-b) and the fact that t_1, t_2 have even denominators since they both differ from $t^{\pm}$. Then Theorem 1.2-b) implies that the two dynamical rays $R_{c_1}(t_1), R_{c_2}(t_2)$ land at c_1 and c_2 respectively. Since $h_k^{c_j}(R_{c_0}(t_j)) = R_{c_j}(t_j)$ for $j \in \{1, 2\}$, the rays

$R_{c_0}(t_1)$ and $R_{c_0}(t_2)$ both land at $H(c_1) = H(c_2)$. In turn, this implies that $R_{c_1}(t_1)$ and $R_{c_1}(t_2)$ land at c_1. By Theorem 1.2-b), $\mathcal{R}(t_2)$ lands at c_1, and therefore $c_1 = c_2$.

Finally, we have to show that H maps $\partial\mathcal{P}_{k+1}$ onto $\partial P^{c_0}_{k+1}$. Since H is a homeomorphism from $\partial\mathcal{P}_k \cup (\overline{\mathcal{P}_k} \cap \mathcal{I}_{k+1})$ to $\partial P^{c_0}_k \cup (\overline{P^{c_0}_k} \cap I^{c_0}_{k+1})$, it maps $\partial\mathcal{P}_{k+1}$ to ∂Q_{k+1} for some piece Q_{k+1} of depth $k+1$. On the other hand, for $c \in \mathcal{P}_{k+1}$, the critical value c of f_c is in D^c_{k+1} because it moves continuously with c, belongs to D^c_{k+1} for $c = c_0$ and does not meet I^c_{k+1} (Lemma 2.3). Hence, if $c \in \partial\mathcal{P}_{k+1}$ lies inside some equipotential segment of $\partial\mathcal{P}_{k+1}$, then the critical value c belongs to ∂D^c_{k+1}. Therefore, $H(c) = (h^c_k)^{-1}(c) \in \partial P^{c_0}_{k+1}$ and lies inside some equipotential segment of $\partial P^{c_0}_{k+1}$. Since $P^{c_0}_{k+1}$ is the only piece of depth $k+1$ whose boundary contains this point, $Q_{k+1} = P^{c_0}_{k+1}$.   □

Corollary 2.7. *The para-puzzle piece $\mathcal{P}_{k+1}$ containing c_0 is the set of parameters c for which the two puzzle pieces P^c_{k+1} and D^c_{k+1} coincide. Thus c belongs to $\mathcal{P}_k \setminus \overline{\mathcal{P}_{k+1}}$ if and only if the critical value c lies in $D^c_k \setminus \overline{D^c_{k+1}}$.*

Proof. As in Lemma 2.6, for $c \in \mathcal{P}_{k+1}$, the critical value c of f_c is in D^c_{k+1}.

Conversely, suppose c belongs to $\mathcal{P}_k \setminus \overline{\mathcal{P}_{k+1}}$. Let c_t, $t \in [0,1]$, be a continuous path in $\mathcal{P}_k$ joining c_0 to $c_1 = c$, crossing $\partial\mathcal{P}_{k+1}$ at exactly one point c_{t_0}, and avoiding $\mathcal{I}_{k+1} \cap M$ (such a path can be constructed using equipotentials and rays). Then for $t < t_0$, the critical value $f_{c_t}(0) = c_t$ belongs to $D^{c_t}_{k+1}$. Moreover, since the parameter path c_t crosses $\mathcal{I}_{k+1}$ at $t = t_0$, the critical value $f_{c_t}(0)$ gets out of $D^{c_t}_{k+1}$ when t passes over t_0. Then Lemma 2.6 insures that, for $t > t_0$, the critical value does not cross $\partial D^{c_t}_{k+1}$ again (because the path does not cross $\mathcal{P}_{k+1}$). Hence $c_1 = c$ lies outside of D^c_{k+1}.   □

Lemma 2.8. *If $\overline{D^{c_0}_{n+1}} \subset D^{c_0}_n$ then $\overline{\mathcal{P}_{n+1}} \subset \mathcal{P}_n$.*

Proof. The corollary 2.7 shows that the homeomorphism H of Lemma 2.6 maps $\partial\mathcal{P}_{n+1} \cap \mathcal{P}_n$ into $\partial D^{c_0}_{n+1}$. If $\overline{D^{c_0}_{n+1}} \subset D^{c_0}_n$, then $H^{-1}(\partial D^{c_0}_{n+1})$ is compact, so $\partial\mathcal{P}_{n+1} \cap \mathcal{P}_n$ is also compact (since it is closed in $\mathcal{P}_n$). On the other hand, $\partial\mathcal{P}_{n+1}$ is connected, and it is not entirely included in $\partial\mathcal{P}_n$ because of the piece of equipotential it contains. Therefore $\overline{\mathcal{P}_{n+1}} \subset \mathcal{P}_n$.   □

3   Proof of the Main Theorem

3.1   Extending the holomorphic motions

Let $c_0 \in M$ be a parameter satisfying the hypotheses of the Main Theorem *i.e.*, for which $f_{c_0}(z) = z^2 + c_0$ is a non-renormalizable map whose fixed points are repelling and whose critical point is recurrent. Since c_0 does not belong to the central cardioid, a well-known result (see for example [GM]) asserts that c_0 lies in some wake $\mathcal{W}_{p/q}$ with $(p,q) = 1$ and $q > 1$; moreover, c_0 is not a Misiurewicz parameter. Thus c_0 belongs to a complete sequence of pieces $\mathcal{P}_n$, $n \geq 0$, of the para-puzzle constructed in Section 2. As in the dynamical planes, we consider the sequence of (possibly degenerate) annuli $\mathcal{A}_n = \mathcal{P}_n \smallsetminus \overline{\mathcal{P}_{n+1}}$. We also denote by A_n^{co} the annulus $P_n^{co} \smallsetminus \overline{P_{n+1}^{co}}$ which surrounds the critical value of f_{c_0}.

The key ingredient for constructing a quasiconformal homeomorphism $\mathcal{A}_n \to A_n^{co}$ is the following theorem of Z. Słodkowski [S, D]:

Theorem 3.1 (Słodkowski). *Every holomorphic motion H of $Z \subset \widehat{\mathbb{C}}$ parameterized by the open unit disc $\mathbb{D}$ and inducing the inclusion map on $\{0\} \times Z$ extends to a holomorphic motion $\tilde{h}\colon \mathbb{D} \times \widehat{\mathbb{C}} \to \mathbb{D} \times \widehat{\mathbb{C}}$. Moreover, for every parameter $c \in \mathbb{D}$, the map $\tilde{h}^c\colon \widehat{\mathbb{C}} \to \widehat{\mathbb{C}}$ is a quasiconformal homeomorphism whose dilatation constant K_c is bounded by $\frac{1+|c|}{1-|c|}$.*

Applying this theorem, we can extend each h_n to a holomorphic motion $\tilde{h}_n$ of $\mathbb{C}$ parameterized by $\mathcal{P}_n$. Then, as in Lemma 2.6, we consider the map $\mathcal{A}_n \to A_n^{co}$ which assigns to each parameter c the point $(\tilde{h}_n^c)^{-1}(c)$. This map is a quasiconformal homeomorphism (see the proof of Proposition 3.2 below). However, without further care, its dilatation will depend on n. We will solve this problem by restricting to the sequence of non-degenerate annuli $A_{n_i}^{co}$ described at the end of Section 1, whose modulus sum is infinite and for which $f^{n_i - n_0}$ induces an unramified covering map from $A_{n_i}^{co}$ to $A_{n_0}^{co}$.

3.2 Comparing the moduli

The Main Theorem is now a straightforward consequence of the following:

Proposition 3.2. *There exists a constant $K > 1$ such that, for every integer $n \in \{n_i, \ i \geq 0\}$,*

$$\frac{1}{K} \operatorname{mod} A_n^{co} \leq \operatorname{mod} \mathcal{A}_n \leq K \operatorname{mod} A_n^{co}.$$

Proof. Let $n \in \{n_i, \ i \geq 0\}$ and d_n be the degree of the covering map $f_{c_0}^{n-n_0}\colon A_n^{co} \to A_{n_0}^{co}$. To get the required inequality, it suffices to construct a quasiconformal homeomorphism $\tilde{H}_n\colon \mathcal{A}_n \to A_n^{co}$ whose dilatation is bounded independent of n.

For every $c \in \mathcal{P}_n$ and $m \in \{n_0, n\}$, let A_m^c denote the annulus $D_m^c \smallsetminus \overline{D_{m+1}^c}$. Then the map $f_c^{n-n_0}$ induces an unramified covering map of degree d_n from A_n^c to $A_{n_0}^c$. Indeed, Lemma 2.5 and the definition of D_m^c, D_{m+1}^c show that the diagram

$$
\begin{array}{ccc}
\partial A_n^{co} & \xrightarrow{\;h_n^c\;} & \partial A_n^c \\[2pt]
f_{c_0}^{n-n_0} \downarrow & & \downarrow f_c^{n-n_0} \\[2pt]
\partial A_{n_0}^{co} & \xrightarrow[\;h_{n_0}^c\;]{} & \partial A_{n_0}^c
\end{array}
$$

is commutative, so $f_c^{n-n_0}$ maps A_n^c onto $A_{n_0}^c$. Lemma 2.3 ensures that no critical point of $f_c^{n-n_0}$ can enter A_n^c while c remains in $\mathcal{P}_n$, so the map $f_c^{n-n_0}\mid_{A_n^c}$ has no ramification points and its degree is d_n.

Now Słodkowski's Theorem gives a holomorphic motion

$$
\widetilde{h}_{n_0} \colon \mathcal{P}_{n_0} \times \widehat{\mathbb{C}} \longrightarrow \mathcal{P}_{n_0} \times \widehat{\mathbb{C}}
$$

which extends h_{n_0} and consists of K_c-quasiconformal homeomorphisms $\widetilde{h}_{n_0}^c$. For every $c \in \mathcal{P}_n$ the homeomorphism $\widetilde{h}_{n_0}^c \colon \overline{A_{n_0}^{co}} \to \overline{A_{n_0}^c}$ lifts—via the holomorphic covering maps $f_{c_0}^{n-n_0}$ and $f_c^{n-n_0}$—to a homeomorphism

$$
\widetilde{h}_n^c \colon \overline{A_n^{co}} \longrightarrow \overline{A_n^c}
$$

which is again K_c-quasiconformal, coincides with h_n^c on ∂P_{n+1}^{co}, and depends holomorphically on c—because $f_c^{n-n_0} \circ \widetilde{h}_n^c = \widetilde{h}_{n_0}^c \circ f_{c_0}^{n-n_0}$. Hence the map

$$
\widetilde{h}_n \colon \mathcal{P}_n \times \overline{A_n^{co}} \longrightarrow \mathcal{P}_n \times \widehat{\mathbb{C}}, \quad (c, z) \longmapsto \left(c, \widetilde{h}_n^c(z)\right),
$$

is a holomorphic motion.

We now recall that, according to Corollary 2.7 and Lemma 2.6, a parameter c is in $\mathcal{A}_n \cup \partial \mathcal{P}_{n+1}$ if and only if the critical value c belongs to $A_n^c \cup \partial D_{n+1}^c$, or equivalently if and only if $(\widetilde{h}_n^c)^{-1}(c)$ lies in $A_n^{co} \cup \partial P_{n+1}^{co}$. Let $\widetilde{H}_n$ be the map

$$
\widetilde{H}_n \colon \begin{cases} \mathcal{A}_n \cup \partial \mathcal{P}_{n+1} \longrightarrow A_n^{co} \cup \partial P_{n+1}^{co}, \\[4pt] \quad\quad c \quad\quad \longmapsto \widetilde{H}_n(c) = (\widetilde{h}_n^c)^{-1}(c). \end{cases}
$$

On $\partial \mathcal{P}_{n+1}$, the map $\widetilde{H}_n$ coincides with H and hence is a homeomorphism onto ∂P_{n+1}^{co} (Lemma 2.6).

Assertion. *The map $\widetilde{H}_n$ is a quasiconformal homeomorphism with dilatation*

$$
K_n = \sup\{K_c,\ c \in \mathcal{P}_n\} \le \sup\{K_c,\ c \in \overline{\mathcal{P}_{n_0+1}}\} = K < +\infty.
$$

We prove this assertion below assuming that all maps are sufficiently smooth. The general case can be derived by approximation arguments, using compactness of quasiconformal homeomorphisms (see [DH2, Lemma IV.3] and [LV, Theorem 2.2 in Section VI.2]). Note that the proof of Theorem 3.1

(in [D]) yields an explicit approximation of class C^∞ for the extended holomorphic motion.

Consider the Beltrami coefficient $\mu = \overline{\partial}\widetilde{H}_n / \partial \widetilde{H}_n$. Differentiating the relation $\widetilde{h}_n^c \circ \widetilde{H}_n(c) = c$, we get the identity

$$\overline{\partial \widetilde{h}_n^c}(\widetilde{H}_n(c))\, \overline{\partial \widetilde{H}_n}(c) + \partial \widetilde{h}_n^c(\widetilde{H}_n(c))\, \overline{\partial}\widetilde{H}_n(c) = 0$$

which shows that

$$|\mu(c)| = \left| \frac{\overline{\partial \widetilde{h}_n^c}(\widetilde{H}_n(c))}{\partial \widetilde{h}_n^c(\widetilde{H}_n(c))} \right| \le \frac{K_n - 1}{K_n + 1} < 1 .$$

The Ahlfors-Bers Theorem then yields a K_n-quasiconformal homeomorphism $\phi \colon \mathcal{A}_n \to \phi(\mathcal{A}_n)$ satisfying $\overline{\partial}\phi = \mu\, \partial\phi$. By a straightforward calculation of $\overline{\partial}(\widetilde{H}_n \circ \phi^{-1})$, the map $\Phi_n = \widetilde{H}_n \circ \phi^{-1}$ is holomorphic. One can easily see that $\widetilde{H}_n$ is a proper map, so Φ_n is a holomorphic ramified covering map from $\phi(\mathcal{A}_n)$ to $A_{n_0}^{co}$. By Lemma 2.8 and the Riemann-Hurwitz formula, Φ_n has no ramification points, so its degree is given by the degree of H on the inner boundary component of $\mathcal{A}_n$. Using the fact that $\widetilde{H}_n$ induces a homeomorphism from $\partial \mathcal{P}_{n+1}$ to ∂P_{n+1}^{co}, we conclude that Φ_n is a conformal homeomorphism. The result follows. $\qquad\square$

In fact the above proof gives the following more precise result:

Corollary 3.3. *For each $n \in \{n_i,\ i \ge 0\}$, there exists a number $K_n > 1$ such that*

$$\frac{1}{K_n}\, \mathrm{mod}\, A_n^{co} \le \mathrm{mod}\, \mathcal{A}_n \le K_n\, \mathrm{mod}\, A_n^{co}$$

and $K_n \longrightarrow 1$ as n tends to infinity.

Proof. Since the sum $\sum_{i>0} \mathrm{mod}\, A_{n_i}^{co}$ is infinite, Proposition 3.2 shows that $\sum_{i\ge 0} \mathrm{mod}\, \mathcal{A}_{n_i}$ is also infinite, so that the intersection of the para-puzzle pieces $\mathcal{P}_n$ reduces to $\{c_0\}$. On the other hand, the preceding proof bounds the dilatation of $\widetilde{H}_n$ by $K_n = \sup\{K_c,\ c \in \mathcal{P}_n\}$. To see that the sequence K_n tends to 1, choose a conformal representation $\chi \colon \mathcal{P}_{n_0} \to \mathbb{D}$ which maps c_0 to 0. Then Słodkowski's Theorem 3.1 asserts that $K_c \le \frac{1+|\chi(c)|}{1-|\chi(c)|}$. This implies the result since the pieces $\mathcal{P}_n$ shrink to c_0. $\qquad\square$

To conclude this paper, we characterize the parameters rays $\mathcal{R}(t)$ which converge to a given Yoccoz parameter c.

Corollary 3.4. *Let $c \in M$ be a parameter satisfying the hypotheses of the Main Theorem. The angles $t \in \mathbb{R}/\mathbb{Z}$ for which the parameter ray $\mathcal{R}(t)$ converges to c are exactly those for which the dynamical ray $R_c(t)$ converges to the critical value c of f_c.*

Proof. Observe that if a ray $\mathcal{R}(t)$ or $R_c(t)$ enters some (para-) puzzle piece of depth n then all its points of potential less than 2^{-n} lie in that piece. Now Lemma 2.6 shows that $\mathcal{R}(t)$ enters $\mathcal{P}_n(c)$ if and only if $R_c(t)$ enters the piece P_n^c containing the critical value of f_c. Since $\bigcap \mathcal{P}_n(c) = \bigcap P_n^c = \{c\}$, the result follows.   □

References

[A]   AHLFORS L. — *Lectures on Quasiconformal Mappings*, Wadsworth & Brook/Cole, Advanced Books & Software, Monterey 1987.

[B]   BRANNER B. — *Puzzles and para-puzzles of quadratic and cubic polynomials*, Proc. of Symp. in Appl. Math. **49** (1994), 31–69.

[D]   DOUADY A. — *Prolongement de mouvements holomorphes [d'après Slodkowski et autres]*, Séminaire Bourbaki n° 775, Astérisque **227** (1995), 7–20.

[DH1]   DOUADY A., HUBBARD J. H. — *Étude dynamique des polynômes complexes I & II*, Publ. math. d'Orsay 1984.

[DH2]   DOUADY A., HUBBARD J. H. — *On the dynamics of polynomial-like mappings*, Ann. Sci. Éc. Norm. Sup. **18** (1985), 287–343.

[GM]   GOLDBERG L. R., MILNOR J. — *Fixed points of polynomial maps, part II: fixed point portraits*, Ann. Sci. Éc. Norm. Sup. **26** (1993), 51–98.

[H]   HUBBARD J. H. — *Local connectivity of Julia sets and bifurcation loci: three theorems of J.-C. Yoccoz*, in *Topological Methods in Modern Mathematics*, 467–511, Goldberg and Phillips eds, Publish or Perish 1993.

[LV]   LEHTO O., VIRTANEN K. I. — *Quasiconformal Mappings in the Plane*, Springer-Verlag, 1973.

[M1]   MILNOR J. — *Dynamics in One Complex Variable: Introductory Lectures*, IMS Preprint, SUNY Stony Brook 1990.

[M2]   MILNOR J. — *Local connectivity of Julia Sets: Expository lectures*, in this volume.

[R]   ROESCH P. — *Topologie locale des méthodes de Newton cubiques*, thèse, ENS de Lyon 1997.

[S]   SŁODKOWSKI Z. — *Extensions of holomorphic motions*, Ann. Scuola
      Norm. Sup. Cl. Sci. **22** (1995), 185–210.

Pascale ROESCH
Université des Sciences et Technologies de Lille
UFR de Mathématiques
59655 Villeneuve d'Ascq CEDEX, FRANCE
email:roesch@agat.univ-lille1.fr

Local properties of the Mandelbrot set at parabolic points

Tan Lei

Abstract

We formulate the technique of parabolic implosion into an easy-to-use result: Orbit correspondence, and apply it to show that for c_0 a primitive parabolic point, the Mandelbrot set M outside the wake of c_0 is locally connected at c_0. This, combined with known results inside the wake, shows that M is locally connected at c_0. The appendices contain sketches of relative results and their proofs.

1 Introduction

Denote by Q_c the map $z \mapsto z^2 + c$. The Mandelbrot set M is defined to be the set of c such that $\lim_{n \to \infty} Q_c^n(0) \not\to \infty$.

We say that c_0 is a *primitive* (resp. *non-primitive*) parabolic point if Q_{c_0} has a periodic point of multiplier 1 (resp. $e^{2\pi i \theta}$ with $\theta \in \mathbb{Q} \smallsetminus \mathbb{Z}$).

For precise definition of the following notations, see §3.2. Denote by $\varphi_M : \mathbb{C} \smallsetminus \overline{\mathbb{D}} \to \mathbb{C} \smallsetminus M$ the conformal isomorphism fixing ∞ and tangent to the identity there. If two M-rays $R_M(\theta^\pm)$ land at the same point c_0 we denote by $wake(\theta^\pm)$ the component of $\mathbb{C} \smallsetminus R_M(\theta^+) \cup R_M(\theta^-) \cup \{c_0\}$ which does not contain 0. Denote by $R_{c_0}(\theta)$ the external ray of Q_{c_0} of angle θ.

Theorem 1.1 *Let $c_0 \neq 1/4$ be a primitive parabolic point. More precisely Q_{c_0} has a k-periodic point of multiplier 1, with $k > 1$.*

a) (landing property) For the two angles $\theta^\pm$ for which $R_{c_0}(\theta^\pm)$ land at a common parabolic point z_0 and are adjacent to the Fatou component containing c_0, the M-rays $R_M(\theta^\pm)$ land at c_0.

b) (parametrisation) A germ of $R_M(\theta^\pm)$ can be written as $\bigcup_{n \geq n_0} c_n^\pm([0,1])$ for some $n_0 > 0$ satisfying: c_n injective, $c_{n+1}^\pm(0) = c_n^\pm(1)$, and, for $T(z) = z^2$ and some $C, C' > 0$,

$$\varphi_M \circ T^k \circ \varphi_M^{-1}(c_n^\pm(t)) = c_{n-1}^\pm(t) \quad \text{and} \quad \frac{C'}{n^2} \leq |c_n^\pm(t) - c_0| \leq \frac{C}{n^2} . \quad (1)$$

c) (transversality) The local solution of $Q_c^k(z) - z = 0$ can be written as $(c_0 + v^2, g_\pm(v))$ with $g_\pm$ holomorphic in v, $g_\pm(0) = z_0$, $g'_\pm(0) \neq 0$. Moreover the multiplier map $\lambda_\pm(v) = (Q_{c_0+v^2}^k)'(g_\pm(v))$ satisfies $\lambda_\pm(0) = 1$ and $\lambda'_\pm(0) \neq 0$.

d) (local connectivity) The set $M \smallsetminus wake(\theta^\pm)$ is locally connected at c_0.

e) (no other rays involved) For any other ray $R_M(\zeta)$ outside of the wake, the impression of its corresponding prime end does not contain c_0 (in particular the ray does not land at c_0) and $M \smallsetminus (wake(\theta^\pm) \cup \{c_0\})$ is connected.

The last statement says that M has no ghost umbilical cords at c_0, in the terminology of D. Sørensen. We will formulate techniques of parabolic implosion into a parabolic orbit correspondence result and then apply it to construct a sequence of puzzle pieces P_n in $\mathbb{C} \smallsetminus wake(\theta^\pm)$, with c_0 on the boundary, such that $\varphi_M \circ T^k \circ \varphi_M(\partial P_n) = \partial P_{n-1}$ and diam$(P_n) \asymp 1/n^2$.

Together with known results inside the wake, see [Hu, Mi, Sø] and Appendix D, we can conclude that M is locally connected at every primitive parabolic points.

A similar statement for non-primitive parabolic points is included in [DH2], [Mi] and [Sø]. In Appendix A and D we give a short account of these results, together with interesting consequences.

Our treatment follows very closely the original approach of Douady and Hubbard ([DH2]), who invented the theory of parabolic implosion while proving landing properties of M-rays.

J. Hubbard in [Hu, Theorem 14.6] (see also Appendix E in the present paper) sketched a proof of the local connectivity of M at a primitive parabolic point, using the theory of Mandelbrot-like families together with the fact that M is locally connected at the cusp $c = 1/4$. This proof is less elementary than the one in the present paper, but has an advantage that it turns the local connectivity property at a cusp into a semi-direct consequence of the existence of sub-Mandelbrot-like families (semi-direct means that extra combinatorial information is needed), and thus can be adopted to more general settings.

We show in Appendix B that the similar statement for repelling orbit correspondence can be applied to show that M is locally connected at Misiurewicz points (i.e. c values such that $Q_c(0)$ is strictly preperiodic).

For a more combinatorial approach of landing properties and local connectivity of M at various points, see [Sc1] and [Sc2].

Acknowledgements. Discussions with A. Douady and M. Shishikura have been very fruitful. J. Milnor has helped to largely improve both the global structure and local details.

2 Orbit correspondence in parabolic implosion

We will use the terminology and results in [S1].

Definition. By a *Fatou coordinate*, associated with a rational map f on a region U, we will mean a univalent map $\Phi : U \to \mathbb{C}$ which satisfies the following condition for every $z \in U$,

$$f(z) \in U \iff \Phi(z) + 1 \in \Phi(U) \iff \Phi(f(z)) = \Phi(z) + 1 ,$$

and moreover satisfies the convexity condition that whenever both w and $w + n$ belong to $\Phi(U)$ the intermediate points $w + 1$, $w + 2$, $\ldots$, $w + (n-1)$ must also belong belong to $\Phi(U)$. Note that if Φ exists f is also univalent in U.

The set up. Fix $r > 0$ small. Let $\Delta_r \subset \mathbb{C}$ be a bounded connected open set with 0 on the boundary, U an open neighbourhood of 0 and $f_s(z)$ be a family of holomorphic maps on U satisfying:
1. $(s, z) \mapsto f_s(z)$ is well defined, continuous on $\overline{\Delta_r} \times U$ and holomorphic in z, with $f_s(z) = \lambda(s)z + O(z^2)$;
2. $(s, z) \mapsto f_s(z)$ is holomorphic on $\Delta_r \times U$;
3. $f_0''(0) \neq 0$;
4. Set $\sigma(s) = (\lambda(s) - 1)/(2\pi i)$. Then σ is continuous on $\overline{\Delta_r}$ with $\sigma(0) = 0$, and σ maps Δ_r univalently onto $\{z \mid |z| < r, \ |\arg z| < \pi/4\}$.

Remark. In practice $f_s(z)$ and $\sigma(s)$ are often defined in a larger sector neighbourhood of $s = 0$ than Δ_r. But we are only interested in the subsector of parameters which is mapped by $\sigma(s)$ bijectively onto $\{z \mid |z| < r, \ |\arg z| < \pi/4\}$. Here the width $\pi/4$ is not really relevant, it can be any angle in the interval $(0, \pi)$.

Theorem 2.1 (Douady-Lavaurs-Shishikura): *If r is sufficiently small, then there exist the followings (see Figure 1):*

For $s = 0$ there are open Jordan domains $\Omega_{-,0}$ and $\Omega_{+,0}$ which are respectively an attracting and a repelling petal for f_0, satisfying $\overline{\Omega_{-,0}} \cap \overline{\Omega_{+,0}} = \{0\}$, with associated Fatou coordinates $\Phi_{\pm,0} : \Omega_{\pm,0} \to \mathbb{C}$.

For $s \in \overline{\Delta_r} \smallsetminus \{0\}$ there is a Jordan domain Ω_s containing two fixed points for f_s on its boundary and two Fatou coordinates $\Phi_{\pm,s} : \Omega_s \to \mathbb{C}$ with

$$\Phi_{-,s}(z) - \frac{1}{\sigma(s)} = \Phi_{+,s}(z) . \tag{2}$$

Taking Ω_0 to be $\Omega_{-,0} \cup \Omega_{+,0}$, both the closure $\overline{\Omega_s}$ and the complement $\hat{\mathbb{C}} \smallsetminus \Omega_s$ depend continuously on $s \in \overline{\Delta_r}$, using the Hausdorff metric for the space of compact subsets of the Riemann sphere, and the correspondence $(s, z) \mapsto \Phi_{\pm,s}(z)$ is continuous in both variables for either choice of sign, wherever it is defined. Moreover $\Phi_{\pm,s}(z)$ is holomorphic in s for $s \in \Delta_r$.

Proof. Let $\alpha(s) \in \mathbb{C}$ be defined so that $\lambda(s) = e^{2\pi i \alpha(s)}$ and $|\Re(\alpha(s))| < 1/2$. This theorem was proved in [S1], Proposition 3.2.2 and Appendix, with $\frac{1}{\alpha(s)}$ in place of $\frac{1}{\sigma(s)}$, and with $|\arg(\alpha(s))| < \pi/4 + \varepsilon$ in place of $s \in \Delta_r$. For $s \in \Delta_r$

we have $|\arg \alpha(s)| \approx |\arg \sigma(s)| < \frac{\pi}{4}$. Using the Taylor series expansion of e^x one can see easily that the difference $\frac{1}{\sigma(s)} - \frac{1}{\alpha(s)}$ is continuous on $\overline{\Delta_r}$. Since the Fatou coordinates are defined up to addition of constants (relative to z), one can add the above difference to $\Phi_{-,s}$ in Shishikura's normalisation without alerting the continuous dependence on (s, z). Therefore (2) holds together with the rest part of the theorem.   ∎

Using this result, we can restate and prove the "Tour de Valse" theorem of Douady and Sentenac ([DH2], exposé n°XI) as follows.

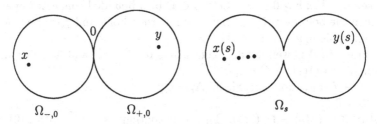

Figure 1 : Parabolic orbit correspondence

Proposition 2.2 (Parabolic orbit correspondence). *Let* $X \subset \Omega_{-,0}$ *and* $Y \subset \Omega_{+,0}$ *be compact sets. If* r *is sufficiently small, then given continuous functions* $x : \overline{\Delta_r} \times [0, 1] \to X$ *and* $y : \overline{\Delta_r} \times [0, 1] \to Y$, *holomorphic in* s *for* $s \in \Delta_r$, *the equations*

$$f_s^n(x(s, t)) = y(s, t) \quad \text{and} \quad \Phi_{-,s}(x(s, t)) + n - \frac{1}{\sigma(s)} = \Phi_{+,s}(y(s, t)) \quad (3)$$

have at least one common solution $s = s_n(t)$ *in* Δ_r, *which depends continuously on* t, *provided that* n *is sufficiently large. Moreover*

$$\lim_{n \to \infty} n - \frac{1}{\sigma(s_n(t))} = C_0(t) \quad \text{and} \quad \lim_{n \to \infty} \left| 1 - \frac{1}{n\sigma(s_n(t))} \right| = 0 \quad \text{as} \quad n \to \infty \quad (4)$$

where $C_0(t)$ *is finite and continuous on* t, *and the convergences are uniform on* t *(see Figure 1).*

Before giving the proof we give a quick application showing that the Julia sets move discontinuously:

Application I. Let $f_s(z) = (1+s)z + z^2$. For Δ_r a sector neighbourhood of 0 containing the direction of $i\mathbb{R}$ the family satisfies all the required conditions. The Julia set $J(f_0)$ of f_0 is a cauliflower (see the left picture in Figure 1 of [S1], in this volume). Let x be any point in $\Omega_{-,0}$.

So $x \notin J(f_0)$. We will show that there is a sequence $s_n \in \Delta_r$ converging to 0 such that $x \in J(f_{s_n})$ for all n. It follows that for any converging subsequence $J(f_{s_{n_k}})$, $\lim J(f_{s_{n_k}}) \neq J(f_0)$.

In order to find s_n, we define at first an appropriate function $y(s)$. On $J(f_0)$ there is a repelling periodic cycle of period two. Let $y(0)$ be a preimage of this cycle such that $y(0) \in \Omega_{+,0}$. For any s near 0, there is a unique $y(s)$ which is holomorphic on s and is mapped by some iterates of f_s onto a repelling periodic cycle of period two. Clearly $y(s) \in J(f_s)$.

Now we can apply Proposition 2.2 for $x(s) \equiv x$ to conclude that there is a sequence $s_n \in \Delta_r$, $s_n \to 0$, such that $f_{s_n}^n(x) = y(s_n)$. As a consequence $x \in J(f_{s_n})$ for every large n.

Proof of Proposition 2.2. First note that we only need to find a solution for the right equation in (3) , since combining it with (2) we get $\Phi_{+,s}(x(s,t)) + n = \Phi_{+,s}(y(s,t))$. By definition of Fatou coordinates we have $\Phi_{+,s}(f_s^n(x(s,t))) = \Phi_{+,s}(x(s,t)) + n = \Phi_{+,s}(y(s,t))$. Since $\Phi_{+,s}$ is univalent, we have $f_s^n(x(s,t)) = y(s,t)$. This is the left equation in (3) .

To simplify the situation, assume at first $x(s,t) = x(s)$ and $y(s,t) = y(s)$ (i.e. they are independent of t). Set $\tilde{x}(s) = \Phi_{-,s}(x(s))$ and $\tilde{y}(s) = \Phi_{+,s}(y(s))$. Then $\tilde{x}(s)$ and $\tilde{y}(s)$ are continuous in $\overline{\Delta_r}$. For n large, we are seeking for solutions of the following:

$$\tilde{x}(s) + n - \frac{1}{\sigma(s)} = \tilde{y}(s) \,. \tag{5}$$

Let $\tilde{x}(s) - \tilde{y}(s) = C_0 + H(s)$, with $C_0 = \tilde{x}(0) - \tilde{y}(0)$, $H(s)$ continuous in $\overline{\Delta_r}$ and $H(0) = 0$. We may assume r is small so that $|H(s)|_{\overline{\Delta_r}} \leq 1/4$.

Set $F(s) = C_0 + n - \frac{1}{\sigma(s)} = \frac{1}{z_n} - \frac{1}{\sigma(s)}$ with $z_n = 1/(C_0 + n)$, and $G(s) = F(s) + H(s)$. Then (5) becomes $G(s) = 0$. We will at first find a solution for $F(s) = 0$ and then use it to approximate a solution of $G = 0$.

For n large enough,

$$z_n \in \overline{D}(z_n, |z_n|^2) \subset \{z \mid |z| < r, \; |\arg(z)| < \frac{\pi}{4}\} \,.$$

Let $\overline{D} = \sigma^{-1}(\overline{D}(z_n, |z_n|^2)) \subset \Delta_r$. By univalence of σ, the set $\overline{D}$ is a closed topological disc. Moreover there is a unique $s_n' \in D$ such that $\sigma(s_n') = z_n$, i.e. $F(s_n') = 0$. Now

$$|F(s)|_{\partial D} = \left| \frac{1}{z_n} - \frac{1}{\sigma(s)} \right|_{\partial D} = \frac{|\sigma(s) - z_n|}{|z_n \sigma(s)|}\bigg|_{\partial D} \geq \frac{|z_n|^2}{2|z_n|^2} \geq \frac{1}{2} \quad \text{and}$$

$$|F(s) - G(s)|_{\partial D} = |H(s)|_{\partial D} \leq \frac{1}{4} \,.$$

Therefore $|F - G|_{\partial D} < |F|_{\partial D}$. By Rouché's theorem we conclude that there is a unique $s_n \in D$ such that $G(s_n) = 0$. (Here is another argument showing the existence of s_n: We have $|\arg(F(s)) - \arg(G(s))|_{s \in \partial D} < \pi/4$. Since $\arg(F)|_{\partial D}$ is of degree one so is $\arg(G)|_{\partial D}$. Now G must have at least one zero in D for otherwise $\arg(G(s))$ would extend to a continuous, homotopically non-trivial map from $\overline{D}$ to $\mathbb{R}/2\pi\mathbb{Z}$. This is impossible.)

Therefore s_n is a common solution of the equations in (5) and (3). Furthermore, by definition of C_0 and (5),

$$\lim_{n \to \infty} n - \frac{1}{\sigma(s_n)} = C_0.$$

This is part of (4), the other part of (4) easily follows.

Finally assume that $x(s,t)$ and $y(s,t)$ do depend on t. Set as above $\tilde{x}(s,t) = \Phi_{-,s}(x(s,t))$ and $\tilde{y}(s,t) = \Phi_{+,s}(y(s,t))$. Then $\tilde{x}(s,t) - \tilde{y}(s,t) = C_0(t) + H(s,t)$, with $C_0(t) = \tilde{x}(0,t) - \tilde{y}(0,t)$, $H(s,t)$ continuous on $\overline{\Delta_r} \times [0,1]$ and $H(0,t) = 0$. By uniform continuity of $H(s,t)$ we may choose r small such that $|H(s,t)| \le 1/4$ on $\overline{\Delta_r} \times [0,1]$. The same argument above would find a unique $s_n(t)$ as zero of $G(s,t)$. This implies that $s_n(t)$ is continuous in t. ∎

3 Proof of Theorem 1.1

In this section let (c_0, z_0) be such that z_0 is a k-periodic point ($k > 1$) of Q_{c_0} with multiplier 1 and it is on the boundary of the Fatou component of Q_{c_0} containing c_0. We would like to find an appropriate family $f_s(z)$ and two appropriate functions $x(s,t)$ and $y(s,t)$ in order to apply Proposition 2.2.

3.1 The family f_s relative to $\vec{w}$

We will at first define holomorphic functions $g_\pm(v), \sigma_\pm(v)$ for $v = \sqrt{c - c_0}$, define a holomorphic map $I(c)$, choose a vector $\vec{w}$ with certain properties and then define a family of maps f_s relative to $\vec{w}$ satisfying the set up of §2.

Lemma 3.1 *Set* $F(c,z) = Q_c^k(z) - z$. *Then* $(Q_{c_0}^k)''(z_0) \ne 0$ *and* $F(c,z) = 0$ *has two local solutions which can be written as* $(c_0 + v^2, g_\pm(v))$ *with* $g_\pm(v)$ *holomorphic.*

Proof. By assumption $Q_{c_0}^k$ is the first return map about z_0 with derivative 1 at z_0. It follows that each parabolic basin attached to the point z_0 is fixed by $Q_{c_0}^k$. In other words, different parabolic basins attached to z_0 belong to different orbits under the iteration of Q_{c_0}. On the other hand, Q_{c_0} has only one critical point so there can be at most one periodic cycle of parabolic basins. It follows that there is exactly one such basin attached to z_0. As a consequence $(Q_{c_0}^k)''(z_0) \ne 0$ (by local theory of Fatou flowers).

Now consider $F(c, z)$. We have

$$F(c_0, z_0) = 0, \quad \left.\frac{\partial F}{\partial z}\right|_{(c_0, z_0)} = 0, \quad \left.\frac{\partial^2 F}{\partial^2 z}\right|_{(c_0, z_0)} = (Q_{c_0}^k)''(z_0) \neq 0 \ .$$

By Weierstrass preparation theorem the local solutions of $F(c, z) = 0$ coincide with the solution of

$$(z - z_0)^2 + (c - c_0) \cdot A(c - c_0) \cdot (z - z_0) + (c - c_0) \cdot B(c - c_0) = 0 \qquad (6)$$

where $A(*), B(*)$ are holomorphic functions. Moreover the analytic set $\{c \mid F(c, z) \text{ has multiple solutions}\}$ can not have dimension one, so it has dimension zero, hence is equal to the singleton $\{c_0\}$ in a neighbourhood of c_0.

Now replace $c - c_0$ by v^2 in (6), we get

$$(z - z_0)^2 + v^2 A(v^2) \cdot (z - z_0) + v^2 B(v^2) = 0 \ .$$

As the discriminant is not constantly 0, it is in the form $v^{2m}H(v^2)$ with $H(0) \neq 0$. Therefore $\sqrt{H(v^2)}$ is a locally well defined holomorphic function. So the solutions of (6) (and of $F(c, z) = 0$) are the two holomorphic functions $g_\pm(v) = -v^2 A(v^2) \pm v^m \sqrt{H(v^2)}$. (We will show later that $g'_\pm(0) \neq 0$ therefore $m = 1$.)   ∎

Set

$$\sigma_\pm(v) = \frac{(Q_{c_0+v^2}^k)'(g_\pm(v)) - 1}{2\pi i} \ .$$

Lemma 3.2 *For $c = c_0 + v^2$ set $I(c) = \sigma_+(v)\sigma_-(v)$. The correspondence $c \mapsto I(c)$ is holomorphic throughout a neighbourhood of c_0 with $I(c_0) = 0$. Furthermore, given a direction $\vec{w}$ such that $I(c_0 + hw)/|I(c_0 + hw)| \to -1$ as $h \searrow 0$, and a small number $r > 0$, there is an open connected set Δ_r containing $\{hw, \ h > 0 \text{ small}\}$ and with 0 on the boundary, such that σ_+ maps a choice Δ of $\sqrt{\Delta_r}$ univalently onto $\{z \mid |z| < r, |\arg(z)| < \pi/4\}$, and maps $\{h\sqrt{w}, \ h > 0 \text{ small}\}$ onto an arc tangent to $\mathbb{R}^+$ at 0 (see Figure 2).*

Proof. Since $z - Q_{c_0+v^2}^k(z) = 0$ has only two solutions $g_\pm(v)$, we have (by residue theorem)

$$\int_{|z-z_0|=t} \frac{1}{Q_{c_0+v^2}^k(z) - z} dz = \frac{2\pi i}{(Q_{c_0+v^2}^k)'(g_+(v)) - 1} + \frac{2\pi i}{(Q_{c_0+v^2}^k)'(g_-(v)) - 1}$$

$$= \frac{1}{\sigma_+(v)} + \frac{1}{\sigma_-(v)} \ .$$

But the left hand side is holomorphic on v and has finite value at $v = 0$. So

$$\lim_{v \to 0} \frac{1}{\sigma_+(v)} + \frac{1}{\sigma_-(v)} \quad \text{is finite.} \qquad (7)$$

Set $\sigma_{\pm}(v) = r_{\pm}(v)e^{i\eta_{\pm}(v)}$. Then (7) gives $\eta_{-}(v) - \eta_{+}(v) \to \pi$ as $v \to 0$ (exchange g_{+} and g_{-} if necessary).

Now we turn to $I(c)$. As a symmetric function of the roots of (6), $I(c)$ is a function of the coefficients of (6). So $c \mapsto I(c)$ is holomorphic in a neighbourhood of c_0. Clearly $I(c_0) = 0$. It has then a power series expansion $I(c_0 + s) = as^{\nu} + \cdots$ with $\nu \geq 1$. Thus there are ν distinct directions $\vec{w}$ such that $I(c_0 + hw)/|I(c_0 + hw)| \to -1$ as $h \searrow 0$.

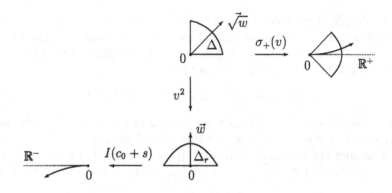

Figure 2 : Finding $\vec{w}$, Δ and Δ_r, in the case $\sigma'_{+}(0) \neq 0$

Fix a choice of w and a choice of $\sqrt{w}$. Let $v = v(h) = h \cdot \sqrt{w}$. Then $I(c_0 + v^2) = \sigma_{+}(v)\sigma_{-}(v)$ and $\frac{I}{|I|} = e^{i(\eta_{+}(v)+\eta_{-}(v))}$. So $\eta_{+}(v) + \eta_{-}(v) \to \pi$ as $h \searrow 0$.

Combining the two equations on $\eta_{\pm}(v)$ we conclude that, as $v = h\sqrt{w}$, $\lim_{h \searrow 0} \eta_{+}(v)$ exists and is equal to 0. In other words, $\sigma_{+}(v)$ maps $\{h\sqrt{w}, h > 0 \text{ small}\}$ onto an arc tangent to $\mathbb{R}^+$ at 0. (Similarly $\sigma_{-}(v)$ maps it onto an arc tangent to $\mathbb{R}^-$ at 0.)

As a holomorphic function fixing 0, $\sigma_{+}(v) = b(v)^p$ in a neighbourhood of 0, with b univalent, $b(0) = 0$ and $p > 0$ some integer. There is therefore a (almost) sector neighbourhood Δ of 0 containing the direction of $\sqrt{w}$ which is mapped by $\sigma_{+}(v)$ univalently onto $\{z \mid |z| < r, |\arg(z)| < \pi/4\}$. Since Δ has an opening angle close to $\pi/(4p) \leq \pi/4$, the map v^2 is univalent on Δ, whose image will be the required set Δ_r. ∎

Definition of f_s relative to $\vec{w}$. Fix a choice of $\vec{w}$ such that $I(c_0+hw)/|I(c_0+hw)| \to -1$ as $h \searrow 0$. Lemma 3.2 provides two sets $\overline{\Delta}$ and $\overline{\Delta_r}$. For $v \in \overline{\Delta}$ and $s = v^2 \in \overline{\Delta_r}$, set $g(s) = g_{+}(v)$ and define $f_s(z)$ to be $Q^k_{c_0+s}(z + g(s)) - g(s)$.

Corollary 3.3 *The family f_s satisfies the hypothesis in the set up of §2 with $(s, \sigma(s))$ parametrised by $(v^2, \sigma_{+}(v))$.*

Proof. The conditions 1 and 2 about regularities of $f_s(z)$ are easily verified, with $\lambda(s) = f'_s(0) = (Q^k_{c_0+s})'(g(s)) = (Q^k_{c_0+s})'(g_+(v))$. So $\sigma(s) := \frac{\lambda(s)-1}{2\pi i} = \sigma_+(v)$. For condition 3, we have $f''_0(0) = (Q^k_{c_0})''(z_0) \neq 0$ (Lemma 3.1). Condition 4 follows from Lemma 3.2. ∎

3.2 The angles $\theta^\pm$, $\eta_n^\pm$ and the functions $x(s,t)$, $y(s,t)$

Here we need to understand more about the dynamics of Q_{c_0} and of Q_c for nearby c. We will at first recall the classic theory of Riemann representations, external rays, state two results about the dynamics of Q_{c_0}, and then use perturbations to construct $x(s,t)$ and $y(s,t)$.

We consider a *Model plane* as a complex plane together with the action of $q(= Q_0) : z \mapsto z^2$. This plane is used to give (external) coordinates in the dynamical plane of Q_c as well as in the parameter plane in the following sense:

For $c \in \mathbb{C}$ define φ_c from a subset of the model plane onto a subset of the dynamical plane of Q_c such that φ_c conjugates q to Q_c, fixes ∞ and is tangent to the identity there (this is often called the Böttcher coordinates). It is known that the maximal domain of definition of φ_c is $\overline{\mathbb{C}} \setminus X_c$, with X_c a bounded star-like compact set containing the closed unit disc. Moreover, if $c \in M$, the set X_c coincides with the closed unit disc, and if $c \notin M$, there is $z_c \in \mathbb{C} \setminus X_c$ such that $\varphi_c(z_c) = c$.

Define φ_M from $\{|z| > 1\}$ (a subset of the model plane) onto $\mathbb{C} \setminus M$ (a subset of the parameter plane) by $\varphi_M(z_c) = c$. Due to a result of Douady-Hubbard, and independently Sibony, we know that φ_M is a well defined conformal isomorphism, fixing ∞ and tangent to the identity there.

The rays $R_M(\theta)$, $R_c(\theta)$ are defined to be $\varphi_M(\{e^{\mu+2\pi i\theta} \mid 0 < \mu\})$ and $\varphi_c(\{e^{\mu+2\pi i\theta} \mid 0 < \mu\})$ respectively. Such a ray is said to *land* at a point z if the limit, as $\mu \searrow 0$, exists and is equal to z. We say also that z has θ as *external angle*.

The great importance of these rays are:

1. they are preserved by the dynamics;

2. when two or more rays land at the same point, they provide a clean cut of the plane, and, together with equipotentials, provide *puzzle pieces* which are candidates of local connected neighbourhoods;

3. they often depend analytically on the parameter.

Concerning our polynomial Q_{c_0}, we have the following known result:

Lemma 3.4 *(Douady-Hubbard) The Julia set of Q_{c_0} is locally connected. The convex hull of the critical orbit forms a topologically finite tree (Hubbard tree) with the critical value c_0 as an extremity. The point z_0, as the unique parabolic* *point on the boundary of the Fatou component U containing c_0, has two external angles adjacent to U. These angles are denoted by $\theta^{\pm}$ (with $\theta^{\pm} \in (0,1)$ and $\theta^- < \theta^+$). The period of them by angle doubling is equal to $\operatorname{per}(z_0)$.*

For a proof, see [DH2], pages 51-52. From this one derives easily

Lemma 3.5 *In the same setting, there is a sequence of prerepelling (i.e. eventually periodic repelling) points z_n in the Hubbard tree such that $z_n \to z_0$, z_n has two external angles $\eta_n^{\pm}$ with $\eta_n^- \nearrow \theta^-$, $\eta_n^+ \searrow \theta^+$, as $n \to \infty$, and, for k the period of z_0, $Q_{c_0}^k(z_n) = z_{n-1}$, $2^k\theta^+ = \theta^+$ (mod 1) and $2^k\theta^- = \theta^-$ (mod 1).*

Proof. In the segment $[\alpha, c_0]$ of the Hubbard tree (where α is the fixed point of Q_{c_0} with a non-zero rotation number), choose a point x so that in $[x, z_0] \smallsetminus \{z_0\}$ there is no branching points of the tree (this is possible since there are only finitely many branching points), and no points in the basin of attraction of the parabolic orbit (this is possible as we have assumed that c_0 is primitive). By expansion property on the Julia set, there is a minimal N such that $Q_{c_0}^N$ restricted to $[x, z_0]$ is no more injective. This implies that there is $z' \in [x, z_0]$ such that $Q_{c_0}^{N-1}(z') = 0$ and $Q_{c_0}^N(z') = c_0$. By minimality of N, $Q_{c_0}^N$ restricted to $[z', z_0]$ is a homeomorphism onto its image. As $Q_{c_0}^N(z_0)$ is a point in the parabolic orbit, distinct from z_0, and is on the Hubbard tree, by the choice of x we have $[Q_{c_0}^N(z_0), z_0] \supset [x, z_0] \supset [z', z_0]$. Therefore

$$Q_{c_0}^N([z_0, z']) = [Q_{c_0}^N(z_0), Q_{c_0}^N(z')] = [Q_{c_0}^N(z_0), c_0] \supset [z', z_0] \ .$$

As $Q_{c_0}^N$ reverses the orientation, there is a point $z_1 \in [z', z_0]$ fixed by $Q_{c_0}^N$, i.e. z_1 is a periodic point. It must be repelling as it is not in the parabolic orbit. It must have exactly two external rays, as it is in the Hubbard tree, and is not a branching or extremal point. By expansion again the local inverse branch g of $Q_{c_0}^k$ mapping z_0 to z_0 maps $[z_1, z_0]$ into itself. Denote by z_n the orbit of z_1 under g. They satisfy the desired property of the Lemma. See Figure 3. ∎

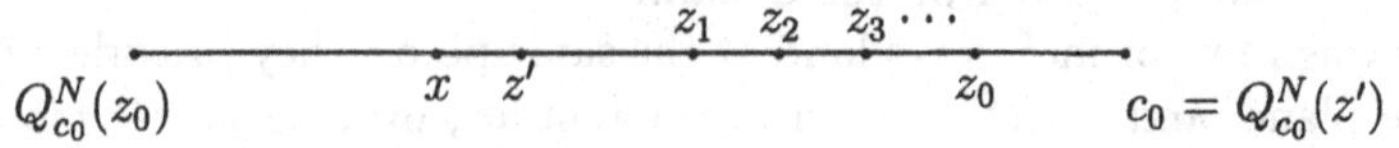

Figure 3 : The segment $[Q_{c_0}^N(z_0), c_0]$ in the Hubbard tree of Q_{c_0}

A proof of a similar statement in the case that 0 is preperiodic for Q_c can be found in [T], Lemma 2.3(2), and in the case 0 is periodic for Q_c can be found in [Sc2], Lemma 5.1.

In order to define an appropriate $y(s,t)$, we define at first $\Gamma_0(t)$ for $t \in [0,1]$ in the log-model plane as follows (we have chosen deliberately $\log e^{2\pi i\theta^+}$ to be $2\pi i\theta^+ - 2\pi i$ for the purpose to give a similar picture as in the dynamical plane and in the parameter plane, Figures 4, 5 and 6): Fix $\hat{\mu} > 0$ small and n_1 large, the map $t \mapsto \Gamma_0(t)$ should be piecewise linear and continuous on $[0, 1/2)$ and $(1/2, 1]$, with

$$
\begin{aligned}
\Gamma_0(0) &= \hat{\mu}/2^k + 2\pi i\theta^+ - 2\pi i , & \Gamma_0(1/6) &= \hat{\mu} + 2\pi i\theta^+ - 2\pi i , \\
\Gamma_0(1/3) &= \hat{\mu} + 2\pi i\eta_{n_1}^+ - 2\pi i , & \Gamma_0(\tfrac{1}{2}^-) &= 2\pi i\eta_{n_1}^+ - 2\pi i , \\
\Gamma_0(\tfrac{1}{2}^+) &= 2\pi i\eta_{n_1}^- , & \Gamma_0(2/3) &= \hat{\mu} + 2\pi i\eta_{n_1}^- , \\
\Gamma_0(5/6) &= \hat{\mu} + 2\pi i\theta^- , & \Gamma_0(1) &= \hat{\mu}/2^k + 2\pi i\theta^- .
\end{aligned}
$$

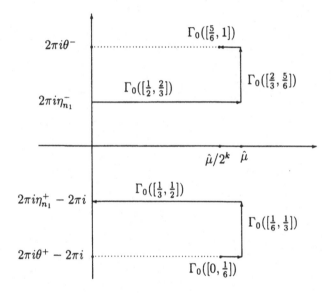

Figure 4 : The curve $\Gamma_0(t)$ and its direction of parametrisation

We then define $y(0, t) = \varphi_{c_0}(e^{\Gamma_0(t)})$. Although Γ_0 has a discontinuity at $t = 1/2$, by the choice of $\eta_{n_1}^\pm$, the map $y(0, t)$ is continuous everywhere and parametrises a Jordan arc.

The function $y(0, t)$ is stable in a neighbourhood of c_0 (cf. [DH1], last chapter), in other words, $y(s, t) = \varphi_{c_0+s}\varphi_{c_0}^{-1}(y(0, t)) = \varphi_{c_0+s}(e^{\Gamma_0(t)})$ is well defined and is an analytic continuation of $y(0, t)$. One may choose $\hat{\mu}$ small enough and n_1 large enough so that $y(s, t)$ belongs to a compact set Y in $\Omega_{+,0}$ for all $(s, t) \in \overline{\Delta_r} \times [0, 1]$.

Define $x(s, t) = x(s)$ to be simply $Q_{c_0+s}^{kl}(c_0 + s)$, where l is the minimal integer such that $x(0) \in \Omega_{-,0}$. We may choose a small enough neighbourhood of c_0 so that $x(s)$ belongs to a compact set X in $\Omega_{-,0}$.

We now define $\Gamma_n(t)$ for $n \geq 0$ inductively as follows: choose $\Theta \in (\theta^-, \theta^+)$. Let τ be the map from $[\Theta - 2\pi i, \Theta)$ to its self mapping z to $2z$ (mod $2\pi i\mathbb{Z}$).

Then $2\pi i\theta^-$ and $2\pi i\theta^+ - 2\pi i$ are fixed points of τ^k. Set

$$\Gamma_n(0) = \frac{\hat{\mu}}{2^{n(k+1)}} + 2\pi i\theta^+ - 2\pi i \ , \quad \Gamma_n(1) = \frac{\hat{\mu}}{2^{n(k+1)}} + 2\pi i\theta^- \ ,$$

and define $\Gamma_n([0, 1/2))$ (resp. $\Gamma_n((1/2, 1])$) to be the lift by τ^{-k} of the arc $\Gamma_{n-1}([0, 1/2))$ (resp. $\Gamma_{n-1}((1/2, 1])$) with the above initial condition. Note that

$$\Gamma_n(\tfrac{1}{6}) = \Gamma_{n-1}(0), \quad \Gamma_n(\tfrac{5}{6}) = \Gamma_{n-1}(1),$$

$$\lim_{t \to \frac{1}{2}^-} \Gamma_n(t) = 2\pi i\eta_{n_1+n}^+ - 2\pi i \text{ and } \lim_{t \to \frac{1}{2}^+} \Gamma_n(t) = 2\pi i\eta_{n_1+n}^- \ .$$

Moreover

$$\bigcup_{n \geq 0} \Gamma_n([0, \tfrac{1}{6}]) = \{\mu + 2\pi i\theta^+ - 2\pi i \mid 0 < \mu \leq \hat{\mu}\}$$

$$\bigcup_{n \geq 0} \Gamma_n([\tfrac{5}{6}, 1]) = \{\mu + 2\pi i\theta^- \mid 0 < \mu \leq \hat{\mu}\} \ .$$

Note also that for $q : z \mapsto z^2$ in the model plane we have $q^k(e^{\Gamma_n(t)}) = e^{\Gamma_{n-1}(t)}$.

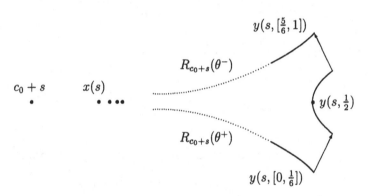

Figure 5 : The dynamical plane of Q_{c_0+s}

3.3   Application II of Proposition 2.2 for a fixed $\vec{w}$

Fix a choice of $c_0 + \vec{w}$ along which $I(c)/|I(c)| \to -1$, as in Lemma 3.2.

We can now apply Proposition 2.2 to the family f_s relative to $\vec{w}$ and the functions $x(s) - g(s)$ and $y(s,t) - g(s)$ defined in the previous sub-sections. There exists then $n_0 > 0$ and, for every $n \geq n_0$, $t \in [0, 1]$, a point $s_n(t) \in \Delta_r$ such that $f_{s_n(t)}^n(x(s_n(t)) - g(s_n(t))) = y(s_n(t), t) - g(s_n(t))$. In other words,

$$Q_{c_0+s_n(t)}^{kn}(x(s_n(t)) = y(s_n(t), t) \ .$$

Lemma 3.6 *We have $c_n(t) := c_0 + s_n(t) = \varphi_M(e^{\Gamma_{kl+n}(t)})$ for $n \geq n_0$ and $t \in [0,1]$.*

Proof. Fix $n \geq n_0$ so that $s_n(t)$ exists.

We claim at first that $x(s_n(t)) = \varphi_{c_0+s_n(t)}(e^{\Gamma_n(t)})$ for $t \in [0,1]$ (cf. Figure 5). Set $Q = Q_{c_0+s_n(t)}$ and $\varphi = \varphi_{c_0+s_n(t)}$. Define $y_j(t) = Q^{k(n-j)}(x(s_n(t)))$ for $j = 0, \cdots, n$. We want to show $y_j(t) = \varphi \circ \exp(\Gamma_j(t))$ for $t \in [0,1]$ and $j \in \{0,1,\cdots,n\}$, i.e. that the following diagram commutes for $j = 0,1,\cdots,n-1$:

$$
\begin{array}{ccc}
\Gamma_{j+1}([0,1]) & \xrightarrow{\varphi \circ \exp} & y_{j+1}([0,1]) \\
{\scriptstyle \tau^k}\downarrow & & \downarrow{\scriptstyle Q^k} \\
\Gamma_j([0,1]) & \xrightarrow{\varphi \circ \exp} & y_j([0,1]).
\end{array}
$$

Since Q^k is univalent where the Fatou coordinates are defined and maps both $y_1(1/6)$ and $y_0(0)$ to $y_0(1/6)$, we must have $y_1(1/6) = y_0(0) = \varphi \circ \exp(\Gamma_0(0)) = \varphi \circ \exp(\Gamma_1(1/6))$. So the diagram commutes for $t = 1/6$, $j = 0$. The rest follows easily by lifting and by induction.

Let $c_n(t) = c_0 + s_n(t)$. Recall that $x(s) = Q^{kl}_{c_0+s}(c_0 + s)$. We want to show now $c_n(t) = \varphi_{c_n(t)}(e^{\Gamma_{kl+n}(t)})$ (cf. Figure 5).

One can choose $c = c_0 + s$ close enough to c_0 so that Ω_s (given by Theorem 2.1) is disjoint from $c, Q_c(c), \cdots, Q^{kl-1}_c(c)$. This last set is the set of critical values of Q^{kl}_c. As a consequence, any arc γ in Ω_s from $Q^{kl}_c(x(s))$ to $x(s)$ would lift by Q^{-kl}_c to an arc starting from $x(s)$ and ending at a point $w(c)$ which is independent of the choice of γ. We claim now $w(c) = c$: On one hand, $w(c_0) = c_0$, which is an easy consequence of the pulling back of the Fatou coordinates $\Phi_{-,0}$. On the other hand, $w(c)$ is continuous on c, and is an element of $Q^{-kl}_c(Q^{kl}_c(c))$. As a consequence, $w(c) = c$ provided that c is close enough to c_0.

We can show that $w(c_n(t)) = \varphi_{c_n(t)}(e^{\Gamma_{kl+n}(t)})$ using the lifting argument as above. Therefore $c_n(t) = \varphi_{c_n(t)}(e^{\Gamma_{kl+n}(t)})$.

Hence $c_n(t) = \varphi_M(e^{\Gamma_{kl+n}(t)})$ for $t \in [0,1]$. ∎

3.4 Conclusions

Lemma 3.7 *(= Theorem 1.1.a)) The M-ray of angle $\theta^\pm$ lands at c_0 and is tangent to $c_0 + \vec{w}$ at c_0, for each direction $c_0 + \vec{w}$ along which $I(c)/|I(c)|$ converges to -1,*

Proof. Note that for $c_n(t)$ defined in Lemma 3.6,

$$\gamma^+ := \bigcup_{n \geq n_0} c_n([0,1/6]) = \varphi_M(\exp(\{\mu + 2\pi i\theta^+ \mid 0 < \mu \leq \hat{\mu}\}))$$

is a germ of $R_M(\theta^+)$. By Lemma 3.2 and Corollary 3.3, the arc $\sigma(\{hw, \ h > 0 \text{ small}\})$ is tangent to $\mathbb{R}^+$ at 0. On the other hand, since $\frac{1}{n}\sigma(s_n(t)) \to 1$ uniformly (by (4) of Proposition 2.2), the arc $\sigma(T_{-c_0}\gamma^+) = \sigma(\bigcup_{n \geq 0} s_n([0, 1/6]))$ is also tangent to $\mathbb{R}^+$ at 0 (where T_{-c_0} denotes the translation $z \mapsto z - c_0$). Therefore $\sigma(T_{-c_0}\gamma^+)$ and $\sigma(\{hw, \ h > 0 \text{ small}\})$ are tangent. So $R_M(\theta^+)$ lands at c_0 and is tangent to $c_0 + \vec{w}$ at c_0.

Similarly $R_M(\theta^-)$ lands at c_0 and is tangent to $c_0 + \vec{w}$. ∎

Corollary 3.8 *(= Theorem 1.1.c)) The functions $g_\pm(v)$ defined in Lemma 3.1 and their multiplier functions $\lambda_\pm(v)$ satisfy $(g_\pm)'(0) \neq 0$ and $\lambda'_\pm(0) \neq 0$.*

Proof. Relying on the fact that φ_M is a conformal isomorphism we know that there is only one M-ray of angle θ^+. Combining with Lemma 3.7 we conclude that there is a unique $\vec{w}$ such that $I(c)/|I(c)|$ converges to -1 along $c_0 + \vec{w}$. It follows that $I(c)$ is univalent.

If $\sigma_\pm(v)$ were holomorphic on c, the function $I(c) = \sigma_+(v)\sigma_-(v)$ could not be univalent (since $\sigma_\pm(0) = 0$). Therefore none of $\sigma_\pm(v)$ is holomorphic on c. It follows that none of $g_\pm(v)$ can be holomorphic on c.

Since $I(c)$ is univalent, $I(c_0 + v^2)$ as a function of v has a double zero at 0. It follows that each $\sigma_\pm(v)$ has a simple zero at 0, i.e. $\sigma'_\pm(0) \neq 0$. It follows easily that $\lambda'_\pm(0) \neq 0$, as $\lambda_\pm(v) = 1 + 2\pi i \sigma_\pm(v)$. Finally

$$\lambda'_\pm(0) = \frac{d}{dv}(Q_{c_0+v^2}^k)'(g_\pm(v))\Big|_{v=0} =$$

$$\frac{\partial}{\partial c}(Q_c^k)'(z)\Big|_{(c_0, z_0)} \cdot \frac{d}{dv}(c_0 + v^2)\Big|_{v=0} + \frac{\partial}{\partial z}(Q_c^k)'(z)\Big|_{(c_0, z_0)} \cdot \frac{d}{dv}g_\pm(v)\Big|_{v=0} =$$

$$= (Q_{c_0}^k)''(z_0) \cdot g'_\pm(0)$$

so $g'_\pm(0) \neq 0$. ∎

Corollary 3.9 *There exist constants $C, C' > 0$ independent of n, t such that $C'/n^2 \leq |c_n(t) - c_0| \leq C/n^2$, for all $t \in [0, 1]$ and all $n \geq n_0$.*

Proof. The curve $(s, \sigma(s))$ is parametrised by $(v^2, \sigma_+(v))$. By the left limit in (4) of Proposition 2.2 and the fact that $C_0(t)$ is continuous on t, we know that $\sigma(s_n(t))/\frac{1}{n} \to 1$ uniformly. Now $\sigma(s_n(t)) = \sigma_+(v_n(t))$, where $v_n(t) = \sqrt{s_n(t)} \in \Delta$, and σ_+ is univalent by the above corollary, so $\sigma_+(v_n(t))/(b \cdot v_n(t)) \to 1$ (for some $b \neq 0$) uniformly. Therefore $b \cdot v_n(t)/\frac{1}{n} \to 1$ uniformly. Hence

$$\frac{c_n(t) - c_0}{1/(bn)^2} = \frac{v_n(t)^2}{1/(bn)^2} \to 1$$

uniformly. In particular the constants C, C' required in the Corollary exist.

∎

Proof of Theorem 1.1.b), d) and e).

b) To get $c_n^+([0,1])$ and $c_n^-([0,1])$ for $n \geq n_0$ we just need to reparametrise $c_n([0, 1/6])$ and $c_n([5/6, 1])$ respectively. The rest follows from Lemma 3.6 and Corollary 3.9.

d) By construction and the fact that $R_M(\theta^\pm)$ land at c_0, for $n \geq n_0$, the set

$$c_n([\tfrac{1}{6}, \tfrac{5}{6}]) \cup \bigcup_{m \geq n} c_m([0, \tfrac{1}{6}]) \cup \bigcup_{m \geq n} c_m([\tfrac{5}{6}, 1]) \cup \{c_0\}$$

bounds a Jordan domain (called a puzzle piece, Figure 6). Let us denote it by P_n. We have $\mathrm{diam}(P_n) \leq C/n^2$ due to the above corollary. Moreover $\partial P_n \cap M$ has only two points c_0 and $c_n(1/2)$. So $M \cap \overline{P_n}$ is connected (this is a general fact, cf. for example Lemma 2.1 of [T]). Therefore $M \smallsetminus wake(\theta^\pm)$ is locally connected at c_0 and $M \smallsetminus (wake(\theta^\pm) \cup \{c_0\})$ is connected.

e) Given any $R_M(\zeta)$ outside the wake, there is n such that $\zeta > \eta_n^+$ or $\zeta < \eta_n^-$. So the corresponding impression is contained in $(M \smallsetminus wake(\eta_n^\pm)) \cup \{c_n(1/2)\}$, hence is disjoint from c_0.

∎

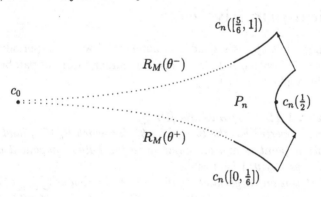

$c_n([\tfrac{5}{6}, 1])$

$R_M(\theta^-)$

c_0

P_n

$c_n(\tfrac{1}{2})$

$R_M(\theta^+)$

$c_n([0, \tfrac{1}{6}])$

Figure 6 : The puzzle piece P_n

4   Application III of Proposition 2.2

Here is a qualitative version and the most common form of parabolic implosion:

Corollary 4.1 *In the set up of §2, for any* $x \in \Omega_{-,0}$, $y \in \Omega_{+,0}$, *there are* $s_n \to 0$ *and* $k_n \to \infty$ *such that* $f_{s_n}^{k_n}$ *converges uniformly on compact sets of* $\Omega_{-,0}$ *to a holomorphic map* g *with* $g(x) = y$.

Proof. Apply Proposition 2.2 to $x(s,t) \equiv x$ and $y(s,t) \equiv y$, we find a sequence $s_n \to 0$ such that $f_{s_n}^n(x) = y$. We will see that $f_{s_n}^n|_{\Omega_{-,0}}$ converges uniformly on compact sets to a function g satisfying $g(x) = y$ and $g \circ f_0 = f_0 \circ g$.

Recall from the proof of Theorem 2.1 that for $\alpha(s)$ defined as $\lambda(s) = e^{2\pi i \alpha(s)}$ with $\alpha(s)$ close to 0, the function $\frac{1}{\alpha(s)} - \frac{2\pi i}{\lambda(s)-1} = \frac{1}{\alpha(s)} - \frac{1}{\sigma(s)}$ is continuous in $\overline{\Delta_r}$, in particular has a finite value at $s = 0$.

In [S2], 3.4, Shishikura proved that for any sequence $s_n \to 0$ such that $\frac{1}{\alpha(s_n)} = k_n + \beta_n$ with $k_n \in \mathbb{N}$, $k_n \to +\infty$ and $\beta_n \to \beta \in \mathbb{C}$, the sequence $f_{s_n}^{k_n}|_{\Omega_{-,0}}$ converges uniformly to a limit function.

Now assume that s_n is a solution of (5) . We have

$$\frac{1}{\alpha(s_n)} = n + \tilde{x}(s_n) - \tilde{y}(s_n) + \left(\frac{1}{\alpha(s_n)} - \frac{1}{\sigma(s_n)}\right)$$

with $\tilde{x}(s_n) - \tilde{y}(s_n) \to \tilde{x}(0) - \tilde{y}(0)$ as $n \to \infty$. So the hypothesis of Shishikura's theorem is satisfied with $k_n = n$. Therefore $f_{s_n}^n|_{\Omega_{-,0}}$ converges uniformly to a limit function g. Clearly $g(x) = y$, g is holomorphic and $g \circ f_0 = f_0 \circ g$. ■

A   Non-primitive case

Assume now that Q_{c_0} is a quadratic polynomial with a k-periodic point of multiplier $e^{2\pi i p/q}$ with $q > 1$. In this case a similar theory of parabolic implosion can be applied to show:

Theorem A.1 *(Douady-Hubbard, [DH2])*
a) (landing property) For the two angles $\theta^\pm$ for which $R_{c_0}(\theta^\pm)$ land at a common parabolic point z_0 and are adjacent to the Fatou component containing c_0, the M-rays $R_M(\theta^\pm)$ land at c_0.
b) (parametrisation) A germ of $R_M(\theta^\pm)$ can be written as $\bigcup_{n \geq n_0} c_n^\pm([0,1])$ for some $n_0 > 0$ satisfying: c_n injective, $c_{n+1}^\pm(0) = c_n^\pm(1)$, and, for $T(z) = z^2$ and some $C_1, C_2 > 0$,

$$\varphi_M \circ T^{kq} \circ \varphi_M^{-1}(c_n^\pm(t)) = c_{n-1}^\pm(t) \quad \text{and} \quad \frac{C_1}{n} \leq |c_n^\pm(t) - c_0| \leq \frac{C_2}{n} . \quad (8)$$

c) (transversality) The local solution of $Q_c^k(z) - z = 0$ can be written as $(c, h(c))$ with h holomorphic, $h(c_0) = z_0$ and $h'(c_0) \neq 0$. Moreover the multiplier function $\lambda(c) = (Q_c^k)'(h(c))$ satisfies $\lambda(c_0) = e^{2\pi i p/q}$ and $\lambda'(c_0) \neq 0$.

The inequalities in (1) and (8) can be used to prove a result of the degree of tangency. See Appendix C.

B Misiurewicz case

A parallel theory for repelling periodic points can be sketched as follows:

Theorem B.1 *Let $f_s(z) = \lambda(s)z + O(z^2)$ be a holomorphic family of maps such that $|\lambda(0)| > 1$. Then for s in a neighbourhood of 0, there are neighbourhood U_s of 0 and linearising coordinates Φ_s such that for any z, n with $f_s^j(z) \in U_s$, $j = 0, 1, \cdots, n$, we have*

$$\Phi_s(f_s^n(z)) = \lambda(s)^n \Phi_s(z)$$

and $\Phi_s(z)$ depends holomorphically on (s, z).

Proposition B.2 (Repelling orbit correspondence). *In the same setting, let $x(s)$, $y(s)$ be two holomorphic functions with $x(0) = 0$, $x(s) \not\equiv 0$ and $y(0) \in U_0 \smallsetminus \{0\}$. Then there is a sequence $s_n \to 0$ for $n > N$ such that $f_{s_n}^n(x(s_n)) = y(s_n)$ and $f_{s_n}^j(x(s_n)) \in U^{s_n}$ for $j = 0, 1, \cdots, n$. Moreover, if $\lambda'(0) \neq 0$, there are constants C_1, C_2 such that $\frac{C_1}{|\lambda(0)|^n} < |s_n| < \frac{C_2}{|\lambda(0)|^n}$.*

Proof. Set $\tilde{x}(s) = \Phi_s(x(s))$, $\tilde{y}(s) = \Phi_s(y(s))$. They are holomorphic functions with $\tilde{x}(0) = 0$, $\tilde{x}(s) \not\equiv 0$ and $\tilde{y}(0) \neq 0$. Given n, we seek for solution of

$$\lambda(s)^n \tilde{x}(s) = \tilde{y}(s) .$$

With the help of power series expansions one can carry out a similar proof as the one of Proposition 2.2. ∎

One can find extensive applications of Proposition B.2 in [DH3] and [Mc]. We present here an application to the Mandelbrot set (Parts 1) and 2) are due to Douady and Hubbard, Parts 3) and 4) are due to Tan Lei):

Theorem B.3 *Let c_0 be a Misiurewicz point, more precisely there are $k \geq 0$, $l \geq 0$ minimal such that $Q_{c_0}^{k+l}(0) = Q_{c_0}^l(0)$ and $l > 0$. Then*
1) The angles of M-rays landing at c_0 coincide with the angles of Q_{c_0}-rays landing at c_0
2) $\frac{d}{dc}(Q_c^{k+l}(0) - Q_c^l(0))|_{c=c_0} \neq 0$.
3) M is locally connected at c_0.
4) The impression of the prime end of any other M-ray does not contain c_0.

Details of proofs can be found in [T].

C degree

The following illustrates a quick application of the inequalities in Theorems 1.1, A.1, although the statement is not in its sharpest form.

Theorem C.1 *For $c_0 \neq 1/4$ a primitive parabolic point, there is a unique hyperbolic component W attached to c_0, and for the two rays $R_M(\theta^\pm)$ landing at c_0, we have that the degree of tangency at c_0 of $R_M(\theta^+)$ and $R_M(\theta^-)$ is 2^+ (i.e. is larger than or equal to 2, but smaller than $2 + \mu$ for any $\mu > 0$), as well as the degree of tangency at c_0 of $R_M(\theta^i)$ and ∂M, $i = +$ or $-$. For c_0 a non-primitive parabolic point, there are exactly two hyperbolic components W_0 and W_1 attached to c_0 and the two rays $R_M(\theta^\pm)$ landing at c_0 separate them, moreover the degree of tangency at c_0 of $R_M(\theta^i)$ and ∂W_j is 2^+ , $i = +$ or $-$, $j = 0, 1$, as well as the degree of tangency at c_0 of ∂W_0 and ∂W_1.*

Proof. For the existence of such hyperbolic components see Corollary D.4 below.

Here we only prove the part about degree of tangency, which is due to Tan Lei.

Assume at first c_0 is non-primitive. We need an inequality between hyperbolic metric and Euclidean metric: Let U be a simply connected domain in $\mathbb{C}$. Then for $x, y \in U$, we have

$$|y - x| \leq e^{2d_U(x,y)} \cdot d(x, \partial U) , \tag{9}$$

where d_U denotes the hyperbolic metric of U. For a proof, see Appendix A of [ST].

Let W_0, W_1 be the two hyperbolic components attached to c_0. We are going to show for n large enough, and $c_n^\pm = c_n^\pm(0)$ (in the notation of Theorem A.1),

$$|c_n^+ - c_{n+1}^+| \leq B \cdot d(c_n^+, W_0 \cup W_1), \quad |c_n^- - c_{n+1}^-| \leq B \cdot d(c_n^-, W_0 \cup W_1) , \tag{10}$$

where B is a constant.

To prove this estimate, we need the relationship between $c_n^\pm$ in (8) and the help of hyperbolic metric. We will only work on the $+$ case. The other one is similar. Take a simply connected domain V in the model plane so that $T^{-kq}(V)$ has a component V' contained in V and that the two sets $U = \varphi_M(V)$, and $U' = \varphi_M(V')$ both contain c_n^+ for $n \geq N$. Therefore, for V_n the component of $T^{-kqn}(V)$ with $e^{2\pi i \theta^+}$ on the boundary, and $U_n = \varphi_M(V_n) \subset U$,

$$d_U(c_{n+N}^+, c_{n+N+1}^+) \leq d_{U_n}(c_{n+N}^+, c_{n+N+1}^+) = d_U(c_N^+, c_{N+1}^+) = A .$$

Now apply (9) we get

$$|c_{n+N}^+ - c_{n+N+1}^+| \leq e^{2d_U(c_{n+N}^+, c_{n+N+1}^+)} \cdot d(c_{n+N}^+, \partial U) \leq B \cdot d(c_{n+N}^+, \partial U) .$$

In particular, since $(W_0 \cup W_1) \cap U = \emptyset$,

$$|c_n^+ - c_{n+1}^+| \leq B \cdot d(c_n^+, W_0 \cup W_1)$$

for any large n. This is (10).

Assume by contradiction that the θ^+-ray is tangent to ∂W_0 with a degree of tangency at least $2 + \mu$, $\mu > 0$. Then

$$d(c_n^+, W_0 \cup W_1) \le d(c_n^+, W_0) \le C \cdot |c_n^+ - c_0|^{2+\mu} . \tag{11}$$

So

$$\frac{C_1}{n} \le |c_n^+ - c_0| \le \sum_{k \ge n} |c_k^+ - c_{k+1}^+| \le B \cdot \sum_{k \ge n} d(c_k^+, W_0 \cup W_1)$$

$$\le B' \cdot \sum_{k \ge n} |c_k^+ - c_0|^{2+\mu} \le B'' \cdot \sum_{k \ge n} \frac{1}{k^{2+\mu}}$$

where B', B'' are constants, the first and the last inequalities is due to (8), the second one is trivial, the third one is due to (10) and the fourth one is due to the assumption (11). Therefore

$$\frac{C_1}{B''} \le \sum_{k \ge n} \frac{n}{k^{2+\mu}} \le \sum_{k \ge n} \frac{1}{k^{1+\mu}} .$$

This is a contradiction to the fact that $\sum \frac{1}{k^{1+\mu}}$ converges.

This shows that the rays have degree 2^+ tangency to each ∂W_0 and ∂W_1. A similar calculation using (1) would show that in the primitive case, the two rays $R_M(\theta^\pm)$ are tangent to each other at c_0 with degree 2^+ tangency.∎

D A short account of relative results

(One can find in [Mi], [Sc1] and [Sc2] radically different approaches of results in this appendix).

Here we give a short exposition of a proof for

Theorem D.1 *For c Misiurewicz or on the boundary of a hyperbolic component, M is locally connected at c.*

The part concerning Misiurewicz points is already established in Theorem B.3. We will only deal with the remaining cases.

We will sketch a proof of a stronger result (Theorem D.3) and show how to deduce from it the above result. We will also give a list of important corollaries.

For $c_0 \ne 1/4$ a parabolic point, define $Ang(c_0)$ to be the set of two angles such that the corresponding dynamical rays for Q_{c_0} land at the parabolic point z_0 on the boundary of the Fatou component containing c, and adjacent to this Fatou component. For $c_0 = 1/4$ we set by convention $Ang(c_0) = \{0, 1\}$.

Our starting point is the following combination of Theorems 1.1.a and A.1.a, proved by parabolic implosion technique in the case $c_0 \neq 1/4$ (in the case $c_0 = 1/4$ it can be checked by hand):

Theorem D.2 *Let c_0 be a parabolic point. Denote by $\theta^\pm$ the two angles in $Ang(c_0)$. Then the M-rays $R_M(\theta^\pm)$ land at c_0.*

Now define $wake(c_0)$ to be the connected component not containing 0 of $\mathbb{C} \smallsetminus (R_M(\theta^+) \cup R_M(\theta^-) \cup \{c_0\})$.

Given any hyperbolic component W of M it is quite easy to see (with the help to quasi-conformal surgery) that the multiplier $\rho(c)$ of an attracting periodic point $z(c)$ for each $c \in W$ realizes a conformal homeomorphism from W onto the unit disc $\mathbb{D}$, and extends to a homeomorphism of the closure (see [DH2], pp.108-109). Denote by $\rho^{-1}(1)$ the root of W. The parabolic points on ∂W are exactly $\rho^{-1}(e^{2\pi i \mathbb{Q}})$.

Theorem D.3 *Let c_0 be a parabolic point with $Ang(c_0) = \{\theta^-, \theta^+\}$.*

a) There is a holomorphic function $\beta_{c_0}(c)$ defined on $wake(c_0)$ such that $\beta_{c_0}(c)$ is the common landing point of $R_c(\theta^+)$ and of $R_c(\theta^-)$.

b) There is a hyperbolic component $W(c_0)$ of M with root c_0 and $W(c_0) \subset wake(c_0)$. Moreover for the common landing point z_0 of $R_{c_0}(\theta^\pm)$,

$$\lim_{c \in wake(c_0),\, c \to c_0} \beta(c) = z_0 \ .$$

c) Structure of sub-wakes of $W(c_0)$ (no ghost limbs):

$$M \cap \overline{wake(c_0)} \smallsetminus \bigcup_{c' \in \partial W \smallsetminus \{c_0\},\, \text{parabolic}} wake(c') = \overline{W(c_0)} \ . \tag{12}$$

d) M is locally connected at c for any $c \in \partial W(c_0)$ which is non-parabolic; $M \cap \overline{wake(c_0)}$ is locally connected at c_0; and $M \smallsetminus wake(c')$ is locally connected at c' for any parabolic $c' \in \partial W(c_0) \smallsetminus \{c_0\}$.

Proof.

a) can be found in [Mi], Theorem 3.1. It is proved by an elementary argument. A sketch is included in Theorem D.8 below.

b) is due to a parabolic perturbation argument (in the non-imploded direction). See [Mi], section 4.

c) We will follow essentially the route of D. Sørensen ([Sø]).

At first we apply a) and b) to every parabolic point on the boundary of $W(c_0)$. Combining with the tuning algorithm (cf. for example [Ha1, §7] and [Sø]), we obtain the following: for any parabolic $c' \in \partial W(c_0) \smallsetminus \{c_0\}$ and any $c \in wake(c')$, the common landing point $\beta_{c'}(c)$ of $R_c(\theta^\pm(c'))$ has a constant and non-trivial combinatorial rotation number $\hat{\rho}(c')$. Moreover, if we choose a converging non-constant sequence c'_n of such parabolic points, $\hat{\rho}(c'_n) \to \infty$. This allows us to apply the powerful Yoccoz inequality (see for example [Hu] and [Sø]), to show that

$$\text{diam}(M \cap \overline{wake(c'_n)}) \to 0 \quad \text{as } n \to \infty. \tag{13}$$

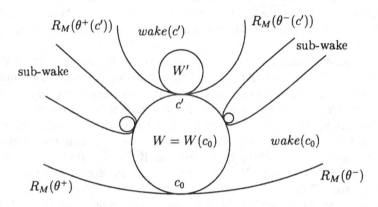

Figure 7 : The sub-wake structure of $W(c_0)$

Two more ingredients are needed to end the proof:
A. The parabolic points are dense in $\partial W(c_0)$, moreover, a consequence of the tuning algorithm gives $\sum_{c' \in \partial W(c_0) \smallsetminus \{c_0\}, \text{ parabolic}} \theta^+(c') - \theta^-(c') = \theta^+ - \theta^-$;
B. Every point of ∂M is accumulated by points in the hyperbolic components (by a normal family argument, cf. Branner [Br]).

   Denote by M' the difference of the left set in (12) to the right set. Assume it is not empty. Let $c \in \partial M' \smallsetminus \overline{W(c_0)}$. Part A says that there is no room left for M' to contain hyperbolic components. Therefore by Part B the point c must be the limit of some $c_n \in wake(c'_n)$, for some non-constant converging parabolic sequence $c'_n \in \partial W(c_0) \smallsetminus \{c_0\}$. Applying (13) , we get $c \in \partial W(c_0)$. A contradiction.

   d) This is a trivial topological consequence of c): Let $c \in \partial W(c_0)$ be non-parabolic. Choose c'_n, c''_n two sequences of parabolic points on $\partial W(c_0)$ tending to c from opposite sides, with $\theta^+(c'_n) < \theta^-(c''_n)$. Then the short arc in $\partial W(c_0)$ connecting c'_n and c''_n together with the two M-rays $R_M(\theta^+(c'_n))$, $R_M(\theta^-(c''_n))$ bounds a simply connected region U_n. Let $M_n = M \cap \overline{U_n}$. Then $\{M_n\}$ is a nested sequence of connected neighbourhoods in ∂M of c. As $\bigcap_n M_n$ is disjoint from any of the sub-wakes, and $\bigcap_n M_n \subset M$ (as M is closed) we conclude from (12) that $\bigcap_n M_n = \{c\}$.

For $c = c_0$, the points c'_n, c''_n can be chosen such that $\theta^+(c'_n) \nearrow \theta^+$ and $\theta^-(c''_n) \searrow \theta^-$, and U_n the union of two simply connected regions bounded by the short arc in $\partial W(c_0)$ connecting c'_n and c''_n and the four M-rays $R_M(\theta^+(c'_n))$, $R_M(\theta^-(c''_n))$ and $R_M(\theta^\pm)$. The rest follows.

Similar argument can be carried out for $c = c' \in \partial W(c_0)$ parabolic. ∎

Proof of Theorem D.1. Let W be a hyperbolic component with root c_0.

Assume at first that c_0 is primitive. Theorem D.3.d) applied to $wake(c_0)$ shows that $M \cap \overline{wake(c_0)}$ is locally connected at c_0. Theorem 1.1.d shows that $M \smallsetminus wake(c_0)$ is locally connected at c_0. As a consequence M is locally connected at c_0.

Another consequence of this is that $W = W(c_0)$, for $W(c_0)$ given by Theorem D.3.b). In other words $W(c_0)$ is the unique hyperbolic component with root c_0. We claim that this is also true whether c_0 is primitive or not.

Proof. We proceed by induction on the period of W. For period 1, there is a unique hyperbolic component which is the central cardioid. Its root $1/4$ is primitive and the assertion is true. Assume it is true for all hyperbolic components of period up to k. Assume W is of period $k + 1$. We want to show $W = W(c_0)$. If c_0 is primitive we are done. Assume c_0 is non-primitive. Then by a perturbation argument as in [Mi], section 4 (see also [Ha2]), one can show that there is a hyperbolic component W_1 with root c_1, with $c_0 \in \partial W_1 \smallsetminus \{c_1\}$ and with period less than or equal to k. By induction $W_1 = W(c_1)$. If $W \neq W(c_0)$, Theorem D.3.c) applied to $wake(c_0)$ shows that $W \cap wake(c_0) = \emptyset$, and applied to $wake(c_1)$ shows that $W \cap (M \smallsetminus wake(c_0)) = \emptyset$. This is a contradiction. ∎

Coming back to the proof of Theorem D.1, we may already conclude that M is locally connected at any $c \in \partial W$ which is non-parabolic, by applying Theorem D.3.d) to $W = W(c_0)$.

We may assume now that c_0 is non-primitive. By the above argument $c_0 \in \partial W(c_1)$ for some different parabolic point c_1. Theorem D.3.d) applied to $wake(c_0)$ shows that $M \cap \overline{wake(c_0)}$ is locally connected at c_0; and then applied to $wake(c_1)$ shows that $M \smallsetminus wake(c_0)$ is locally connected at c_0. As a consequence M is locally connected at c_0.

For any $c' \in \partial W$ parabolic, it is the root of a hyperbolic component W' by Theorem D.3.b). We can then repeat the same argument above with W replaced by W', and conclude that M is locally connected at c'. ∎

Let us restate a few important corollaries in the proof:

Corollary D.4 *Every parabolic point is the root of a unique hyperbolic component. Moreover c_0 is on the boundary of precisely two (resp. one) hyperbolic components if c_0 is non-primitive (resp. primitive).*

Corollary D.5 *For c_0 a parabolic point $M \smallsetminus \{c_0\}$ has exactly two connected components.*

Corollary D.6 *For c_0 a parabolic point and θ any angle not in $Ang(c_0)$. The impression of the prime end corresponding to $R_M(\theta)$ does not contain c_0. In particular c_0 is the landing point of precisely two M-rays, i.e. those with angle in $Ang(c_0)$.*

Note that the same result as Corollary D.6 for c_0 a Misiurewicz point is stated in Theorem B.3.

The following important result comes as a corollary: For any $\theta \in \mathbb{T}$, define $L(\theta)$ to be the landing point of $R_M(\theta)$ if it exists. For any c a Misiurewicz point, define $Ang(c)$ to be the set of angles such that the corresponding dynamical ray for Q_c lands at c.

Corollary D.7 (Douady-Hubbard-landing-theorem)

1. The map L is well defined on the set of rationals with even denominators and maps it onto the set of Misiurewicz parameters. Moreover for any c Misiurewicz $L^{-1}(c) = Ang(c)$, and $Ang(c)$ consist of finitely many rationals with even denominators.

2. The map L is well defined on the set of rationals with odd denominators and maps it onto the set of parabolic parameters. Moreover for any c parabolic $L^{-1}(c) = Ang(c)$.

Proof. By an elementary argument it is quite easy to see that L is well defined on the set of rationals and maps it into the set of Misiurewicz and parabolic parameters (cf. [PR] in this volume or [GM]).

Dynamical properties tell us that $Ang(c)$ consists of non-empty but finitely many rationals with even denominators for c Misiurewicz; and of two rationals with odd denominators for c parabolic.

Our theorem B.3 establishes $L^{-1}(c) = Ang(c)$ for c Misiurewicz, and Corollary D.6 establishes $L^{-1}(c) = Ang(c)$ for c parabolic. ∎

Remark. One can find a more combinatorial approach of Corollary D.7 in [Mi] and [Sc1]. One hidden difficulty in proving Corollary D.7.1 is that for θ a rational with even denominator, $L(\theta)$ must be a Misiurewicz point, or equivalently can not be a parabolic point. Our proof, as well as the original proof of Douady-Hubbard, relies on results about parabolic points. However, there are ways to avoid this, for example one can use Thurston's algorithm to construct a parameter c such that $\theta \in Ang(c)$ (a rather involved method), or one can apply the pretty elementary method given by Petersen and Ryd (cf. [PR], in this volume).

The following is a refinement of Theorem D.3.a) and gives a dynamical description of $wake(c_0)$:

Theorem D.8 *Let c_0 be a parabolic point with $Ang(c_0) = \{\theta^-, \theta^+\}$. Then $wake(c_0)$ coincides with the set of $c \in \mathbb{C}$ such that $R_c(\theta^-)$ and $R_c(\theta^+)$ both land, and land at the same repelling periodic point.*

Proof. The main ideas are the following: At first this is true for $c \in \mathbb{C} \setminus M$, by a combinatorial study conducted by Milnor ([Mi], Lemmas 2.6, 2.9 and 2.11). An argument of P. Haïssinsky ([Ha1], Proposition A.1) shows that the landing points $z_\pm(c)$ of $R_c(\theta^\pm)$ are holomorphic in c, on any connected open set where they are both defined, and moreover on this connected open set either $z_+(c) \equiv z_-(c)$ or $z_+(c) \neq z_-(c)$ for any c. We can then conclude easily by combining these two arguments. For details, see [Mi], Theorem 3.1.   ∎

The following function is as important as the function $\beta_{c_0}(c)$ defined in Theorem D.3.a).

Theorem D.9 *Let W be a hyperbolic component with root c_0. Let $\alpha_{c_0}(c)$ be an attracting periodic point for $c \in W$ depending holomorphically on c. Then $\alpha_{c_0}(c)$ extends to a holomorphic map on the entire $wake(c_0)$, is a periodic point of Q_c with constant period, repelling for $c \in wake(c_0) \setminus \overline{W}$, and co-incides, in each sub-wake of W, i.e. $wake(c')$, $c' \in \partial W$ parabolic, with the holomorphic function $\beta_{c'}(c)$ (which is the landing point of $R_c(\theta^\pm(c'))$).*

Proof. This is basically a corollary of Theorem D.3. By the property of the multiplier function $\rho(c)$ and the implicit function theorem, $\alpha_{c_0}(c)$ extends holomorphically to a neighbourhood of $\overline{W} \setminus \{c_0\}$ in $wake(c_0)$. Now fix $c' \in \partial W \setminus \{c_0\}$ parabolic. Let W' be the unique hyperbolic component with root c' (Corollary D.4). Theorem D.3.b) applied to $W' = W(c')$ says $\alpha_{c_0}(c) = \beta_{c'}(c)$ on W'. Since $\beta_{c'}(c)$ is defined and holomorphic on the entire sub-wake $wake(c')$, so is $\alpha_{c_0}(c)$, and they coincide throughout the subwake.

We need now the help of a third holomorphic function $\xi(c)$ which is defined to be $\alpha_{c_0}(c_1)$ for some $c_1 \in wake(c') \setminus M$ (with c' as above), and to be its analytic continuation throughout $\mathbb{C} \setminus (M \cup R_M(0))$ following the holomorphic motion of the Cantor Julia set. Clearly $\alpha_{c_0}(c) = \xi(c)$ on a open set. So $\alpha_{c_0}(c)$ extends to $\mathbb{C} \setminus (M \cup R_M(0))$ and coincides with $\xi(c)$ there. In particular $\alpha_{c_0}(c)$ extends to $wake(c_0) \setminus M$.

Finally by (12) every point $c \in wake(c_0)$ is either outside of M, or in one of the sub-wakes, or in $\overline{W}$. Therefore the theorem.   ∎

Generally speaking $\alpha_{c_0}(c)$, as a periodic point of Q_c, can be analytically continued along any path in $\mathbb{C} \setminus \{$finitely many parabolic points$\}$. But we may loose tract of the rays landing at $\alpha_{c_0}(c)$, it may become non-repelling, and we do not know a priori whether these parabolic points present real branch points or not. Theorem D.9 answers these questions at least in $wake(\theta^\pm(c_0))$.
Remark. In case W is the central cardioid of M, the proof of D.3, D.9 can be made more elementary and thus lead to a simpler proof of local connectivity of M at $c_0 = 1/4$. See [GM], Appendix C.

Our final corollary is the following Lavaurs continuity result ([DH2], Chapter XVII: Une propriété de continuité), which is a stronger version of Theorem D.3.b):

Theorem D.10 (Lavaurs) *For θ a rational angle, denote by $\gamma_\theta(c)$ the landing point of $R_c(\theta)$ if it exists. Then $\gamma_\theta(c)$ is continuous on $\mathbb{C} \setminus \bigcup_{k \geq 1} R_M(2^k\theta)$. Here the rays $R_M(\eta)$ do not include their landing points. This in particular says that $\gamma_\theta(c)$ is continuous at the landing points of these rays.*

Proof. We will only treat the case θ has odd denominator. The other case is a much easier exercise.

On the open set

$$\mathbb{C} \setminus \overline{\bigcup_{k \geq 1} R_M(2^k\theta)}$$

$\gamma_\theta(c)$ is well defined (Douady-Hubbard) and holomorphic ([Ha1], appendix). So we just need to show that $\gamma_\theta(c)$ is continuous at $L(\theta) = c_0$, the landing point of $R_M(\theta)$ (as this will imply that $\gamma_{2^l\theta}(c)$ for $l \geq 1$ is continuous at $L(\theta)$ and as a consequence $\gamma_\theta(c)$ is continuous at $L(2^m\theta)$ for $m \geq 1$).

Let U be a neighbourhood of c_0 such that $U \setminus R_M(\theta)$, $U \cap wake(c_0)$ and $U \setminus wake(c_0)$ are all simply connected. Set $\gamma(c) = \gamma_\theta(c)$. Denote as before, by z_0 the parabolic point of Q_{c_0} on the boundary of the Fatou component of c_0.

Case 1. The point c_0 is primitive. In this case the other angle θ' in $Ang(c_0)$ is not in the form $2^l\theta$. So $\gamma(c)$ is well defined in $U \setminus R_M(\theta)$. But $\gamma(c) = \beta_{c_0}(c)$ for $c \in U \cap wake(c_0)$ (Theorem D.3.a)) and $\beta_{c_0}(c) = g_i(v)$, $i = +$ or $-$, where $v = \sqrt{c - c_0}$ and $g_\pm(v)$ is defined in Section 3.1 (they are the two local periodic points perturbed from z_0). As $U \setminus R_M(\theta)$ is simply connected, $\gamma(c) \equiv g_i(\sqrt{c - c_0})$ on $U \setminus R_M(\theta)$ and is continuous at c_0 (as $g_i(\sqrt{c - c_0})$ is).

Case 2. The point c_0 is non-primitive. This is the case where the two angles θ, θ' in $Ang(c_0)$ are in the same orbit by angle doubling and $\gamma(c)$ is defined on $U \setminus (R_M(\theta) \cup R_M(\theta'))$, and is holomorphic on $U \setminus (R_M(\theta) \cup R_M(\theta') \cup \{c_0\})$ ([Ha1], appendix).

In $U \cap wake(c_0)$ we have $\gamma(c) = \beta_{c_0}(c)$ and $\gamma(c_0) = z_0$ (Theorem D.3.a) and b)). So γ is continuous at c_0 on $(U \cap wake(c_0)) \cup \{c_0\}$.

In $U \setminus wake(c_0)$, let W_1 the hyperbolic component containing c_0 on the boundary. As $c \in U \cap W_1$ tending to c_0 radically, the same perturbative argument of Milnor ([Mi], section 4) shows that $\gamma(c)$ is repelling for Q_c and tends to z_0. Let $z(c)$ be this repelling periodic point. It has analytic continuation along any path in $U \setminus \{c_0\}$ (shrinking U if necessary) and is periodic of constant period for Q_C. It is therefore a well defined holomorphic function in $U \setminus wake(c_0)$ as this last set is simply connected. As $\gamma(c)$ is holomorphic, we have $\gamma(c) \equiv z(c)$ on $(U \setminus wake(c_0)) \cup \{c_0\}$ and continuous at c_0, as $z(c)$ is. ∎

E   Local connectivity via Mandelbrot-like families (following Hubbard)

Here we sketch a proof of the local connectivity of M at a primitive parabolic point using the theory of Mandelbrot-like families and the fact that M is locally connected at $1/4$. We follow essentially the route indicated by Hubbard ([Hu]).

Let c_0 be a primitive parabolic point, more precisely Q_{c_0} has a k periodic point of multiplier 1. In [Ha1] in this volume, it is shown that there are open discs U_0, U_1 containing c_0, open discs B_c (resp. B'_c) whose boundary undergo a holomorphic motion, such that $\mathbf{f} = \{Q_c^k : B'_c \to B_c,\ c \in U_0\}$ forms a Mandelbrot-like family, with $\overline{U_1} \subset U_0$, U_1 is the set of c such that the critical point of Q_c^k in B'_c does not escape B'_c after one iteration, $\overline{U_1} \cap M$ is connected (it is in fact a puzzle piece in the sense of Yoccoz), and, for $M_{\mathbf{f}}$ the connectedness locus and $\chi : M_{\mathbf{f}} \to M$ the straightening map, $M_{\mathbf{f}} \subset M \cap U_0$ and χ is a homeomorphism with $\chi(c_0) = 1/4$.

Let U_n to be the set of $c \in U_0$ such that the critical point of Q_c^k in B'_c does not escape B'_c after n iterations. Then U_n is again a puzzle piece and $\bigcap_n U_n = M_{\mathbf{f}}$.

Denote by W the unique hyperbolic component with c_0 as root (Corollary D.4). Choose c'_n, c''_n two sequences of parabolic points on ∂W converging to c_0 such that $\theta^+(c'_n) < \theta^+$ and $\theta^-(c''_n) > \theta^-$. Denote by V_n the simply connected domain containing $R_M(\theta^+)$ bounded by the short arc in ∂W connecting c'_n and c''_n and the two rays $R_M(\theta^+(c'_n))$ and $R_M(\theta^-(c''_n))$ (here we need Theorem A.1 for the landing of these rays). Denote by M_n the set $\chi(\overline{V_n} \cap M_{\mathbf{f}})$.

Set $L_n = \overline{V_n} \cap U_n$. Then $L_n \subset L_{n-1}$ and $L_n \cap \partial M$ is a connected neighbourhood (in ∂M) of c_0. We now show that $\bigcap_n L_n = \{c_0\}$:

$$c_0 \in \bigcap_n L_n = \left(\bigcap_n L_n\right) \cap M_{\mathbf{f}} = \bigcap_n (L_n \cap M_{\mathbf{f}}) \subset \bigcap_n (\overline{V_n} \cap M_{\mathbf{f}}) =$$

$$\bigcap_n \chi^{-1}(M_n) = \chi^{-1}(\{\tfrac{1}{4}\}) = \{c_0\}\ ,$$

where the first equality is due to $\bigcap_n L_n \subset \bigcap_n U_n = M_{\mathbf{f}}$, the second is trivial, the set inclusion is because $L_n \subset \overline{V_n}$, the next equality is by definition of M_n and the fact that χ is a homeomorphism, the following is due to the facts that $\lim_{n\to\infty} \chi(c'_n) = \lim_{n\to\infty} \chi(c''_n) = 1/4$ (as χ is continuous) and that M is locally connected at $1/4$ (see [GM] Appendix C for an elementary proof), and the final inequality is by properties of χ.

References

[Br] B. Branner, *The Mandelbrot set*, Proc. Symp. Appl. Math., AMS 39 (1989), 75-105.

[DH1] A. Douady & J. Hubbard, *Etudes dynamique des polynômes complexes*, I, Pub. Math. d'Orsay, 84-02.

[DH2] A. Douady & J. Hubbard, *Etudes dynamique des polynômes complexes*, II, Pub. Math. d'Orsay, 85-05.

[DH3] A. Douady & J. Hubbard, *On the dynamics of polynomial-like mappings*, Ann. Sci. ENS, vol. 18, pp. 287-343, 1985.

[GM] L. Goldberg & J. Milnor, *Fixed points of polynomial maps*, part II, Ann. Sci. Ens. 26 (1993), 51-98.

[Ha1] P. Haïssinsky, *Modulation dans l'ensemble de Mandelbort*, in this volume.

[Ha2] P. Haïssinsky, *Déformation J-équivalente de polynômes géométriquement finis - Pincements de polynômes*, ENS Lyon, preprint no. 235, 1998.

[Hu] Hubbard, *Local connectivity of Julia sets and bifurcation loci*, in Topological Methods in Modern Mathematics, ed. by L. Goldberg and A. Phillips, Publish or Perish, Inc. 1993.

[Mc] C. McMullen, *The Mandelbrot set is universal*, in this volume.

[Mi] J. Milnor, *Periodic Orbits, External Rays and the Mandelbrot set: An expository account*, Stony Brook IMS preprint #1999/3, Astérisque, to appear.

[PR] C. Petersen & G. Ryd, *Convergence of rational rays in parameter spaces*, in this volume.

[S1] M. Shishikura, *Bifurcation of parabolic fixed points*, in this volume.

[S2] M. Shishikura, *The boundary of the Mandelbrot set has Hausdorff dimension two*, S.M.F. Astérisque 222, 1994.

[Sc1] D. Schleicher, *Rational external rays of the Mandelbrot set*, Astérisque, to appear.

[Sc2] D. Schleicher, *On fibers and local connectivity of Mandelbrot and multibrot sets*, Stony Brook IMS preprint #1998/13.

[Sø]   D. Sørensen, *Complex Dynamical Systems: Rays and Non-Local Connectivity*, Ph.D. thesis, DTU Denmark, 1994.

[ST]   M. Shishikura & Tan L., *An alternative proof of Mañé's theorem on non-expanding Julia sets*, in this volume.

[T]   Tan L, *Voisinages connexes des points de Misiurewicz*, Ann. Inst. Fourier, Grenoble, 42, 4, 707-735, 1992.

DEPARTMENT OF MATHEMATICS, UNIVERSITY OF WARWICK, COVENTRY CV4 7AL, UNITED KINGDOM
e-mail: tanlei@maths.warwick.ac.uk

Convergence of rational rays in parameter spaces

Carsten Lunde Petersen and Gustav Ryd[*]

Abstract

We give an elementary proof of the landing Theorem for rational external rays of the Mandelbrot set and related connectedness loci for the one-parameter families of polynomials $\{P_c(z) = z^d + c\}_{c \in \mathbb{C}}$, $d \geq 2$. The proof is quite general and applies with marginal changes to numerous other families with one free critical point.

1 The setting

Fix an integer $d \geq 2$ and consider the family $P_c(z) = P_c = z^d + c$, $c \in \mathbb{C}$, of monic polynomials of degree d with a degree $(d-1)$ critical point at the origin. For each c let J_c denote the Julia set for P_c and define the domain of attraction to infinity

$$\Lambda_c = \Lambda_c(\infty) = \{z \in \mathbb{C} | P_c^n(z) \longrightarrow \infty, \text{as } n \to \infty\}.$$

Let $\phi_c : \Omega_c \longrightarrow \overline{\mathbb{C}} \setminus \overline{\mathbb{D}(R)}$ be a Böttcher coordinate for P_c at ∞ normalised by being tangent to the identity at ∞. Then ϕ_c^{-1} has a unique maximal radial extension to a subset $(\mathbb{C} \setminus \overline{\mathbb{D}})_c$ of $\mathbb{C} \setminus \overline{\mathbb{D}}$. This radial extension terminates at a point w with $|w| > 1$ if and only if ϕ_c^{-1} extends continuously to w and $\phi_c^{-1}(w)$ is a (pre)critical point for P_c. Thus $(\mathbb{C} \setminus \overline{\mathbb{D}})_c = \mathbb{C} \setminus \overline{\mathbb{D}}$, if and only if $c \notin \Lambda_c$. Let $\widetilde{\Lambda}_c = \phi_c^{-1}((\overline{\mathbb{C}} \setminus \overline{\mathbb{D}})_c)$, then $\phi_c : \widetilde{\Lambda}_c \longrightarrow (\mathbb{C} \setminus \overline{\mathbb{D}})_c$ is biholomorphic and depends holomorphically on the parameter c. That is if $z_0 \in \widetilde{\Lambda}_{c_0}$ then there exists neighbourhoods U of z_0 and V of c_0 such that the map $(z, c) \mapsto \phi_c(z)$ is analytic on $U \times V$.

Define $M_d = \{c \in \mathbb{C} | c \notin \Lambda_c\}$ and let $\Phi_d : \overline{\mathbb{C}} \setminus M_d \longrightarrow \overline{\mathbb{C}} \setminus \overline{\mathbb{D}}$ denote the unique Riemann map tangent to the identity at infinity, then $\Phi_d(c) = \phi_c(c)$. Given $\theta \in \mathbb{T} = \mathbb{R}/\mathbb{Z}$ and $c \in \mathbb{C}$: The dynamical (external) ray $R_c(\theta)$ of J_c is the analytic arc $\phi_c^{-1}(\{\exp(s + i2\pi\theta) | s > 0\} \cap (\overline{\mathbb{C}} \setminus \overline{\mathbb{D}})_c)$. It starts at ∞ and either ends at a precritical point $z_0 \in \cup_{n \geq 0} P_c^{-n}(c)$ or ends by accumulating on some subset of the Julia set J_c. The parameter (external) ray $R_{M_d}(\theta)$ of M_d is the analytic arc $\Phi_d^{-1}(\{\exp(s + i2\pi\theta) | s > 0\})$. A ray is called a rational ray if θ is rational, i.e. $\theta \in \mathbb{Q}/\mathbb{Z}$. A ray R is said to land or converge, if the

---

[*]Research undertaken at Dept. of Mathematics, KTH, 100 44 Stockholm, Sweden.

accumulation set $\overline{R} \smallsetminus R$ is a singleton subset of J (if it is a dynamical ray) or M_d (if it is a parameter ray).

On the boundary of M_d we distinguish two particular and different types of parameters, parabolic parameters and Misiurewicz parameters. A parameter c_0 is called parabolic, if P_{c_0} has a parabolic cycle, i.e., there exists a positive integer q and a periodic point z_0 such that $P_{c_0}^q(z_0) = z_0$ and $(P_{c_0}^q)'(z_0) = 1$. A parameter c_0 is called a Misiurewicz parameter, if the orbit of c_0 is finite and ends in a repelling periodic orbit, i.e. there exist positive integers l, q such that $P_{c_0}^{l+q}(c_0) = P_{c_0}^l(c_0)$ and $|(P_{c_0}^q)'(P_{c_0}^l(c_0))| > 1$.

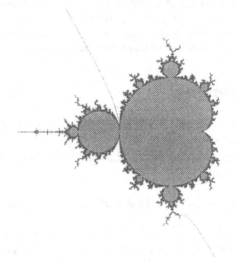

Figure 1: The Mandelbrot set with a periodic (under angle doubling) and a strictly preperiodic ray.

In this paper we give an elementary proof of the landing Theorem for rational external rays of M_d, for any integer $d \geq 2$:

Theorem 1 (Douady-Hubbard) *Given $\theta \in \mathbb{T}$ with $d^l\theta \equiv d^{l+q}\theta \mod 1$, for some minimal integers $l \geq 0$ and $q \geq 1$. Then the parameter ray $R_{M_d}(\theta)$ lands on a parameter $c_0 \in \partial M_d$. Furthermore,*

1. *if $l > 0$ then c_0 is a Misiurewicz parameter, $P_{c_0}^l(c_0) = P_{c_0}^{l+q}(c_0)$ and the dynamic ray $R_{c_0}(\theta)$ land at c_0.*

2. *if $l = 0$ then c_0 is a parabolic parameter, the (unique) parabolic orbit for P_{c_0} has one cycle of immediate attracted basins. This cycle contains both 0 and c and its exact period is q. Moreover the dynamic ray $R_{c_0}(\theta)$*

lands on the parabolic point to which c_0 is attracted under iteration by $P_{c_0}^q$.

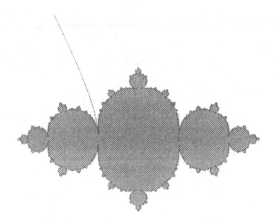

Figure 2: The Julia set for a quadratic polynomial with a period-2 ray landing at a parabolic fixed point.

This Theorem is well-known, at least for $d = 2$, since [DH]. The original proof is however rather involved. There are several other proofs, such as [HS], which uses iteration on Teichmüller spaces, [Mil], [Sch] and [K]. In [CG] this theorem is claimed, but there seems to be a part missing. This paper has evolved from [Pet] and [Ryd], which both were extending Theorem 1 to different settings using the same techniques, but formulated in terms of Caratheodory Kernel convergence and hyperbolic distance on the one side and harmonic functions on the other side. Here we have chosen to formulate the proof in terms of normal families.

The idea behind our proof for landing of the parameter ray $R = R_{M_d}(\theta)$ with $d^{l+q}\theta \equiv d^l\theta \mod 1$ is the following simple one: Let $c_0 \in M_d$ be an accumulation point of R. If $c_0 \in J_{c_0}$ then $l > 0$ and $P_{c_0}^{l+q}(c_0) = P_{c_0}^l(c_0)$ and the dynamical ray $R_{c_0}(\theta)$ lands at c_0. Thus c_0 is a Misiurewicz parameter, by the snail-Lemma, [DH] (alternatively the repellingness of $P_{c_0}^l(c_0)$ is proved directly, also in [DH] by showing P_{c_0} is expanding in the orbifold metric). If $c_0 \in K_{c_0} \smallsetminus J_{c_0}$, then $l = 0$ and c_0 is contained in a immediate basin of exact period q for a parabolic point. Thus c_0 is a parabolic parameter. The Douady-Hubbard landing Theorem then easily follows.

Our proof differs from the original proof in [DH] mainly in our direct approach to convergence and uses only elementary analytical means. Our

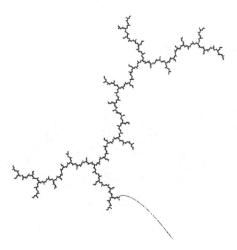

Figure 3: The Julia set for a quadratic Misiurewicz polynomial with a preperiodic ray landing at the critical value c_0.

proof has easy generalisations (beyond the plenty of the original proof) to many other settings.

2   The central tool

We will frequently use $g_c : \mathbb{C} \longrightarrow \mathbb{R}_+$, the Green's function for J_c with pole at ∞. It is the sub-harmonic function, which is harmonic on Λ_c, coincides with $\log |\phi_c(z)|$ on $\widetilde{\Lambda}_c$ and which is identically zero on $\mathbb{C} \smallsetminus \Lambda_c$. Similarly we use $G_{M_d} : \mathbb{C} \longrightarrow \mathbb{R}_+$, the Green's function for M_d with pole at ∞. It is the sub-harmonic function, which coincides with $\log |\Phi_{M_d}|$ on $\mathbb{C} \smallsetminus M_d$ and which is 0 on M_d. It satisfies the relation $G_{M_d}(c) = g_c(c)$ for all $c \in \mathbb{C}$. Note that an external ray is a gradient line for the appropriate Green's function.

Our central tool is a sequences ψ_n of holomorphic maps, which relates the dynamics in the filled Julia set, K_{c_0}, with the dynamics in Λ_{c_0}. The maps ψ_n are obtained as limit maps of sequences of inverse branches of iterates of P_c with c in the appropriate parameter ray. The underlying idea is the same as the idea behind parabolic implosion.

Given a parameter $c \in \mathbb{C}$, an angle $\eta \in \mathbb{T}$, and a potential level $p \geq g_c(c)$ we define 'sectors' $V(c, \eta, p) = \{z | p2/3 < g_c(z)\} \smallsetminus R_c(\eta + \frac{1}{2})$ and $V'(c, \eta, p)$ as the sub-sector of $V(c, \eta, p)$, which maps univalently onto $V(c, \eta, pd^q)$ by P^q and which intersects the ray $R_c(\eta)$. Given $\theta \in \mathbb{T}$ with $\eta = d^l\theta \equiv d^{l+q}\theta \mod 1$, for some minimal $l \geq 0$ and $q \geq 1$. Let $c_0 \in \partial M_d$ be any limit point of $R_{M_d}(\theta)$ and let $\{c_j\}_{j \in \mathbb{N}} \subset R_{M_d}(\theta)$ be a sequence converging to c_0. Define

$$R_j = (R_{c_j}(\theta) \cap \{z | g_{c_j}(z) \geq g_{c_j}(c_j)\}) \cup \{\infty\} \qquad (2.1)$$

Then the sequence $\{R_j\}_{j \in \mathbb{N}}$ is a sequence of compact sets in $\overline{\mathbb{C}}$. Passing to a subsequence, if necessary, we can suppose this sequence converges to some compact subset $R \subset \overline{\mathbb{C}}$ containing c_0 and ∞. Then $R \cap \Lambda_{c_0} = R_{c_0}(\theta) \cup \{\infty\}$, because the Böttcher coordinates ϕ_c depends continuously on the parameter c on any set of the form $\{z | g_c(z) \geq p\}$, where $p > g_c(c)/d$. Moreover $c_0 \in R$ and $P^{l+q}(R) \subseteq P^l(R)$.

Notation 2.1 *During the rest of this article we use without further notice the symbols θ, l, q, η, c_0, c_j, R and R_j for $j \in \mathbb{N}$ in the sense introduced in the above paragraph.*

Proposition 2.2 *Let $z_0 \in R \cap K_{c_0}$ be arbitrary. There exist $w = w_{z_0}$ in $R_{c_0}(\eta)$, a 'sector' $V = V_{z_0} = V(c_0, \eta, g_{c_0}(w))$ and $\{\psi_n = \psi_{n,z_0} : V \longrightarrow K_{c_0}\}_{n \geq 0}$, a sequence of holomorphic maps such that.*

1. *For all $n \geq 0$: $\psi_n(w) = P_{c_0}^n(z_0)$ and $\psi_{n+1} = P_{c_0} \circ \psi_n$.*

2. *For all $n \geq l$: $\psi_n \circ P_{c_0}^q = P_{c_0}^q \circ \psi_n$ on $V' = V'(c_0, \eta, g_{c_0}(w))$ and in particular $\psi_n(P_{c_0}^q(w)) = P_{c_0}^{n+q}(z_0) = \psi_{n+q}(w)$.*

3. *One of the following mutually exclusive cases occurs*

   (a) *The ψ_n are univalent with $\psi_n(V) \subseteq (K_{c_0} \smallsetminus J_{c_0})$ and for all $n, m \geq 0$ such that either $n < l$ or $(n - m) \notin q\mathbb{Z}$: $\psi_n(V) \cap \psi_m(V) = \emptyset$.*

   (b) *All ψ_n are constants.*

Moreover if $z_0 \notin \psi_{0,c_0}(V_{c_0})$, then each ψ_{n,z_0} extends holomorphically to the unrestricted sector $\widehat{V} = V(c_0, \eta, 0) = \Lambda_{c_0} \smallsetminus R_{c_0}(\eta + \frac{1}{2})$, the extensions still enjoying the above properties.

Notation 2.3 *We shall refer to the above items as Property 1, Property 2, Property 3a and Property 3b respectively.*

Proof : We prove the Proposition only in the extreme case $z_0 = c_0$, and leave the analogous general case to the reader.
We can suppose $G_{M_d}(c_j) = g_{c_j}(c_j) < d^{-l}$. For each j let N_j be given by $1 \leq d^{N_j} \cdot g_{c_j}(c_j) < d^q$ and $P_{c_j}^{N_j}(c_j) \in R_{c_j}(\eta)$. By relative compactness, passing to a subsequence if necessary, we have $w_j := P_{c_j}^{N_j}(c_j) \longrightarrow w \in R_{c_0}(\eta)$ with $1 \leq g_{c_0}(w) \leq d^q$.

Define $V = V(c_0, \eta, g_{c_0}(w))$ and note that V is a simply connected domain. We can assume that V does not contain any critical values for $P_{c_j}^{N_j}$, discarding possibly finitely many initial values of j. Also we can assume that $R_j \cap \{z|g_{c_j}(z) \geq g_{c_j}(w_j)\} \subseteq V$ for all j. For each j define $\psi_{0,c_j} : V \longrightarrow \mathbb{C}$ as the unique branch of $P_{c_j}^{-N_j}$, which maps w_j onto c_j (see Figure 4). Moreover for each n with $0 < n \leq N_j$ define $\psi_{n,c_j} : V \longrightarrow \mathbb{C}$ by $\psi_{n,c_j} = P_{c_j}^n \circ \psi_{0,c_j}$.

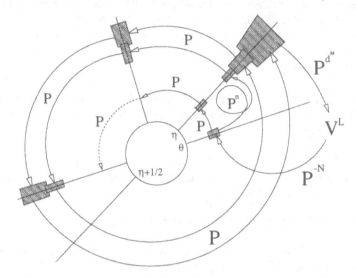

Figure 4: Image domains $\psi_{n,c_j}(V^L)$ viewed in the Böttcher coordinate at infinity (we have taken $L \simeq g_{c_0}(w)d^q$ and M to be a multiple of q).

The maps ψ_{n,c_j} are univalent and satisfy for each fixed j :

$$\forall n, 0 \leq n < N_j : \qquad \psi_{n,c_j}(w_j) = P_{c_j}^n(c_j) \quad \text{and} \quad \psi_{n+1,c_j} = P_{c_j} \circ \psi_{n,c_j} \quad (2.2)$$

$$\forall n, l \leq n \leq N_j - q : \qquad \psi_{n,c_j} \circ P_{c_j}^q = P_{c_j}^q \circ \psi_{n,c_j}. \qquad (2.3)$$

$\forall n, m$ with $0 \leq n, m \leq N_j - M$ and either $n < l$ or $|n - m| \notin q\mathbb{Z}$:

$$\psi_{n,c_j}(V) \cap \psi_{m,c_j}(V) = \emptyset \qquad (2.4)$$

Where $M = M(\theta)$ is any integer which satisfies

$$\frac{1}{d^M} < \min\{\text{dist}_\mathbb{T}(d^n\theta, d^m\theta)|0 \leq n < m < l + q\}.$$

Note that $N_j \longrightarrow \infty$ as $j \to \infty$. Moreover for any $L > d^q$ and any $n \in \mathbb{N}$ the restrictions of ψ_{n,c_j} to $V^L = V \cap \{z|g_{c_0}(z) < L\}$ are uniformly bounded.

We shall obtain the maps ψ_n as locally uniform limits of the maps ψ_{n,c_j} as $j \to \infty$. By Montel's Theorem there exists a holomorphic map $\psi_0 : V \longrightarrow \overline{\mathbb{C}}$

such that, extracting a subsequence if necessary, the maps ψ_{0,c_j} converge locally uniformly to ψ_0. Define for each $n \in \mathbb{N}$ a holomorphic map $\psi_n : V \longrightarrow \overline{\mathbb{C}}$ by $\psi_n = P_{c_0}^n \circ \psi_0$. Then $\psi_{n,c_j} \underset{j \to \infty}{\overrightarrow{}} \psi_n$, because $P_{c_j}^n$ converge locally uniformly to $P_{c_0}^n$ as $j \to \infty$. By the Hurwitz Theorem the maps ψ_n are either univalent or constant. Moreover $\psi_n(V) \subseteq K_{c_0}$ or equivalently $g_{c_0} \equiv 0$ on $\psi_n(V)$, because $g_{c_j} \to 0$ uniformly on $\psi_{0,c_j}(V^L)$ for each fixed L.

By locally uniform convergence the properties (2.2) and (2.3) also holds for the ψ_n (with P_{c_0}), so Properties 1 and 2 follows for ψ_n. Similarly (2.4) holds for ψ_n and ψ_m if not both ψ_n and ψ_m are constants. Moreover if $J_{c_0} \cap \psi_n(V) \neq \emptyset$, then ψ_n is constant, by the openness of non-constant holomorphic maps. By (2.2) if one ψ_n is constant they all are, so the Property 3 follows.

Suppose $z_0 \notin \psi_{0,c_0}(V_{c_0})$ and let $\{z_j \in R_j\}_{j \in \mathbb{N}}$ be a sequence converging to z_0. Then the ratio $g_{c_j}(z_j)/g_{c_j}(c_j)$ diverges to infinity, since if not, then $z_0 \in \psi_{0,c_0}(V_{c_0})$ by the locally uniform convergence of ψ_{0,c_j} to ψ_{0,c_0}. But then for any $p > 0$ there exists $N \in \mathbb{N}$, such that $\forall j \geq N$ the maps ψ_{n,z_j} are defined and univalent on the sector $V(c_0, \eta, p)$. Thus the domain of ψ_{n,z_0} can be taken to include $\widehat{V} = V(c_0, \eta, 0)$. **q.e.d.**

We obtain some immediate corollaries.

Corollary 2.4 *For $z_0 \in R \cap K_{c_0}$ we have*

$$P_{c_0}^l(z_0) = P_{c_0}^{l+q}(z_0) \quad \Leftrightarrow \quad \forall n \in \mathbb{N} \text{ the map } \psi_{n,z_0} \text{ is constant.}$$

Proof : The map $\psi_l = \psi_{l,z_0} : V_{z_0} \longrightarrow K_{c_0}$ is either univalent or constant by Property 3. Moreover $\psi_l(w) = P_{c_0}^l(z_0)$, and $\psi_l(P_{c_0}^q(w)) = P_{c_0}^{l+q}(z_0)$ by Property 1 and Property 2 respectively. As $P_{c_0}^q(w) \neq w$ the Corollary follows.

q.e.d.

Corollary 2.5 *Let $z \in (R \cap K_{c_0}) \setminus J_{c_0}$ and suppose $P_{c_0}^l(z) \neq P_{c_0}^{l+q}(z)$. Then $P_{c_0}^l(z)$ belongs to a q-periodic immediate basin of some parabolic point α.*

Proof : The map $\psi_l = \psi_{l,z} : V_z \longrightarrow K_{c_0}$ is univalent by Property 3 and Corollary 2.4. Hence the image $\psi_l(V_z)$ is contained in some q-periodic Fatou component U. A periodic Fatou component is either a hyperbolic, a parabolic or a rotation domain. The Fatou domain U can not be hyperbolic, because hyperbolicity is an open condition. It can not be a rotation domain either, because any point on $\psi_l(R_{c_0}(\eta))$ is non recurrent. **q.e.d.**

Corollary 2.6 *For $z_0 \in R \cap K_{c_0}$ we have*

$$P_{c_0}^{l+q}(z_0) = P_{c_0}^l(z_0) \quad \Leftrightarrow \quad z_0 \in J_{c_0}.$$

Proof : If $z_0 \in J_{c_0}$, then ψ_{0,z_0} is constant, because it can not be open. Thus $P_{c_0}^l(z_0) = P_{c_0}^{l+q}(z_0)$ by Corollary 2.4.

Next assume $P_{c_0}^l(z_0) = P_{c_0}^{l+q}(z_0)$ and $z_0 \notin J_{c_0}$. Then $P_{c_0}^l(z_0)$ is a q periodic Siegel point, because the alternative, attracting, is an open condition. However the set of centres of Siegel disks is discrete and R is a continuum. Thus we get a contradiction with Corollary 2.5. More detailed let U denote the Siegel disk with centre $P_{c_0}^l(z_0)$ and let $z' \in R$ be a point with $P_{c_0}^l(z') \in U \smallsetminus \{P_{c_0}^l(z_0)\}$. Then $P_{c_0}^l(z') \neq P_{c_0}^{l+q}(z')$, because a Siegel disk contains precisely one periodic point. But then U is a parabolic domain and not a Siegel domain, by Corollary 2.5, hence $P_{c_0}^l(z_0) \neq P_{c_0}^{l+q}(z_0)$. This contradiction completes the proof.   **q.e.d.**

For $w \in R_c(\eta)$ with $g_c(c) < d \cdot g_c(w)$ define a real analytic parametrisation $\Gamma_{c,w} : \mathbb{R}_+ \longrightarrow R_c(\eta)$ by $\Gamma_{c,w}(t) = \phi_c^{-1}(\exp(g_c(w) \cdot d^{qt} + i2\pi\eta))$ of the sub-arc of the dynamic ray $R_c(\eta)$ from w to ∞. The map $\Gamma_{c,w}$ satisfies

$$\forall t \geq 0 : P_c^q \circ \Gamma_{c,w}(t) = \Gamma_{c,w}(t+1).$$

Moreover $\Gamma_{c,w}$ has a unique real analytic extension to the interval $] - t_0, \infty[$, where $t_0 = \frac{1}{q \log d} \log(g_c(c)/(d \cdot g_c(w)))$. In particular for $c \in M_d$ the parametrisation $\Gamma_{c,w}$ extends to $\widehat{\Gamma}_{c,w} : \mathbb{R} \longrightarrow R_c(\eta)$, a parametrisation of $R_c(\eta)$. In the light of Corollary 2.6 we define arcs:

Definition 2.7 *For $z_0 \in (R \cap K_{c_0}) \smallsetminus J_{c_0}$ let $\Gamma_{z_0} : \mathbb{R}_+ \longrightarrow (K_{c_0} \smallsetminus J_{c_0})$ be the real analytic arc given by $\Gamma_{z_0} = \psi_{0,z_0} \circ \Gamma_{c_0, w_{z_0}}$. Then for all $t \geq 0$ ($t \in \mathbb{R}$ if $z_0 \notin \psi_{0,c_0}(V_{c_0})$):*

$$P_{c_0}^{l+q} \circ \Gamma_{z_0}(t) = P_{c_0}^l \circ \Gamma_{z_0}(t+1).$$

By Corollary 2.5 the arc $P_{c_0}^l(\Gamma_{z_0})$ is contained in a q-periodic immediate basin of a parabolic point α'_{z_0} and $\Gamma_{z_0}(t) \xrightarrow[t \to \infty]{} \alpha_{z_0}$, where $\alpha'_{z_0} = P_{c_0}^l(\alpha_{z_0}) = P_{c_0}^{l+q}(\alpha_{z_0})$.

3   The Douady-Hubbard Landing Theorem

Proposition 3.1 *If $c_0 \in J_{c_0}$, then $l > 0$ and c_0 is a Misiurewicz parameter with $P_{c_0}^l(c_0) = P_{c_0}^{l+q}(c_0)$. Moreover the dynamic ray $R_{c_0}(\theta)$ lands on c_0.*

Proof :   By Corollary 2.6 we have $P_{c_0}^l(c_0) = P_{c_0}^{l+q}(c_0)$. If $l = 0$ then c_0 is q-periodic, but then also the critical point 0 is q-periodic, because it is the only preimage of c_0. This contradicts however the fact that $c_0 \in \partial M_d$, thus $l > 0$.

Let us next prove that the dynamical ray $R_{c_0}(\theta) \subseteq R$ lands on c_0. however this follows from Corollary 2.6, because $R_{c_0}(\theta) \cap J_{c_0} \subseteq R \cap J_{c_0}$ and the set of points z_0 with $P_{c_0}^l(z_0) = P_{c_0}^{l+q}(z_0)$ is discrete. Hence it follows from the snail Lemma, [DH, Exp. VIII, p. 71, 1e. P], that $P_{c_0}^l(c_0)$ is repelling or parabolic. however it can not be parabolic, because the unique critical point is preperiodic. Alternatively by [DH, Exp. V.4 p. 40-42, 1e. P] $P_{c_0}^l(c_0)$ is repelling.

q.e.d.

Proposition 3.2 *Suppose $c_0 \in U \subseteq K_{c_0}$, where U is some Fatou component. Then c_0 is a parabolic parameter, U is an immediate parabolic basin of exact period q and the preperiod $l = 0$.*

Proof : Define $U_l = P_{c_0}^l(U)$, then U_l is a q-periodic immediate parabolic basin, by Corollary 2.6 and Corollary 2.5. But then so is U, because the cycle of such immediate basins should contain a critical point and value. Let the (exact) period of U be p. For $n \geq 0$ define $\gamma_n = P_{c_0}^n(\Gamma_{c_0})$. Then $\gamma_p \in U$. Let $\Phi : U \longrightarrow \mathbb{C}$ be a Fatou coordinate for $P_{c_0}^p$ on U normalised by $\Phi(c_0) = 0$ and $\Phi(P_{c_0}^p(c_0)) = 1$, i.e. $\Phi \circ P_{c_0}^p = 1 + \Phi$. The map Φ has a univalent inverse branch

$$\Psi : \mathbb{H}_{-1} = \{z = x + iy | y > -1\} \longrightarrow \Omega$$

with $\Psi(0) = c_0$ and with $P_{c_0}^p(\Omega) \subset \Omega$.

There exists $n \geq 0$ such that $\gamma_{pn}, \gamma_{p(n+1)} \subseteq \Omega$, because $\gamma_m([0,1])$ is compact and $P_{c_0}^q(\gamma_m(t)) = \gamma_m(t+1) = \gamma_{m+q}(t)$ for all $m \geq l$ and $t \geq 0$. Define $\tilde{\gamma}_{pn} = \Phi(\gamma_{pn})$ and $\tilde{\gamma}_{p(n+1))} = \Phi(\gamma_{p(n+1)})$, so that $\tilde{\gamma}_{pn}(t+1) = \tilde{\gamma}_{pn}(t) + q/p$ and similarly for $\tilde{\gamma}_{p(n+1)}$. The map $z \to \exp(2\pi i z p/q)$ maps these two arcs onto two simple closed curves, which are rotations (around the origin) of each other. But the later intersect, being rotations of each other, and so do $\tilde{\gamma}_{pn}$ and $\tilde{\gamma}_{p(n+1)}$. Since γ_{pn} and $\gamma_{p(n+1)}$ are contained in Ω they also intersect. But then p is a multiple of q by Property 3a and thus $p = q$.

Next we prove that $l = 0$. Choose n with $\gamma_{nq} \subset \Omega$ so that $\gamma_{(n+m)q} \subset \Omega$ for all $m \geq 0$. Then $\tilde{\gamma}_{qn} = \Phi(\gamma_{nq})$ satisfies the hypothesis of the isotopy Lemma 3.3 below. Let $\tilde{\Gamma}_n$ be the isotopy, whose existence is granted by Lemma 3.3. Increasing n, if necessary, we can assume $\tilde{\Gamma}_n(\mathbb{R}_+ \times [0,1]) \subset \mathbb{H}_{-1}$, because $\tilde{\gamma}_{q(n+1)} = 1 + \tilde{\gamma}_{qn}$ and $\tilde{\Gamma}_n(t+1, s) = 1 + \tilde{\Gamma}_n(t,s)$. Define an isotopy $\Gamma_n : \mathbb{R}_+ \times [0,1] \longrightarrow \Omega$ by $\Gamma_n = \Psi \circ \tilde{\Gamma}$. Then the image $\Gamma_n(\mathbb{R}_+ \times [0,1])$ does not contain any of the critical values for $P_{c_0}^{nq}$. Thus there exists a unique lift Γ_0 of Γ_n to $P_{c_0}^{nq}$ with $\Gamma_0(0,0) = c_0$. But then by construction

$$\Gamma_0(n,0) = \Gamma_0(n,1) = \gamma_0(n) = \gamma_{nq}(0).$$

which contradicts Property 3a, when $l > 0$.   **q.e.d.**

Note that the above isotopy argument is already given in [DH, Exp. XII.2 p. 35-36 2e. P]

Lemma 3.3 *Let* $\gamma_1 : \mathbb{R}_+ \longrightarrow \mathbb{C}$ *be an arc with* $\gamma_1(0) = n$ *and such that for all* $t \geq 0$ $\gamma_1(t+1) = \gamma_1(t) + 1$. *Then there exists* $\Gamma : \mathbb{R}_+ \times [0,1] \longrightarrow \mathbb{C}$, *an isotopy of arcs with* $\Gamma(t,1) = \gamma_1(t)$, $\Gamma(t,0) = n + t := \gamma_0(t)$, $\Gamma(0,[0,1]) = n$ *and for all* $t \geq 0, s \in [0,1] : \Gamma(t+1,s) = 1 + \Gamma(t,s)$.

Proof :   Extend γ_1 to an arc $\gamma_1 : \mathbb{R}_+ \longrightarrow \mathbb{C}$ by use of the periodicity $\gamma_1(t+1) = \gamma_1(t) + 1$. For some $r > 0$ sufficiently big the vertical translate $\gamma_1 + ir$ is disjoint from the arc $\mathbb{R}$, the real axis, and those two arcs bound a 'horizontal strip B. Let $h > 0$ be such that there exists a biholomorphic map $\phi : \{z = x + iy | 0 < y < h\} \longrightarrow B$ with $\phi(z+1) = 1 + \phi(z)$. Then ϕ extends to a homeomorphism of the closed sets (the boundaries being analytic arcs) and we can suppose ϕ normalised by $\phi(0) = n$. Let $w = \phi^{-1}(ir)$, then a isotopy Γ with the required properties is obtained as

$$\Gamma(s,t) = \begin{cases} (1-4s)(n+t) + 4s \cdot \phi(t) & \text{if } 0 \leq s \leq \frac{1}{4} \\ \phi(t + 2(s - \frac{1}{4})w) - \phi(2(s - \frac{1}{4})w) + n & \text{if } \frac{1}{4} \leq s \leq \frac{3}{4} \\ \gamma_1\left(4(s - \frac{3}{4})t + 4(1-s) \cdot \gamma_1^{-1}(\phi(t+w) - ir)\right) & \text{if } \frac{3}{4} \leq s \leq 1 \end{cases}$$

q.e.d.

The following Lemma implies that even in the periodic case the set $R \cap J_{c_0}$ is a singleton.

Lemma 3.4 *Suppose* θ *is* q *periodic and let* U *be a Fatou component for* P_{c_0} *with* $R_U = U \cap R \neq \emptyset$. *If* R_U *connects two distinct boundary points* $\alpha \neq \beta$ *of* U, *i.e.* $\alpha, \beta \in \overline{R}_U$. *Then the restriction* $P_{c_0}^q : U \longrightarrow U$ *has two distinct critical points in* U.

Proof :   The Fatou component U is a q-periodic immediate basin for a parabolic periodic point, which we can assume to be α, by Corollary 2.5 and the remark following Definition 2.7.

Let $\phi : U \longrightarrow \mathbb{D}$ be a Riemann map, say mapping α to 1. Let $\widehat{L} = \phi(R_U)$ and define $\widehat{P}$ as the conjugate Blaschke product $\widehat{P} = \phi \circ P_{c_0}^q \circ \phi^{-1}$. Then L connects the parabolic point 1 for $\widehat{P}$ with a repelling fixed point $\widehat{\beta} \neq 1$. The conclusion of the Lemma is then equivalent to: $\widehat{P}$ has at least two distinct critical points in $\mathbb{D}$. To reduce indexing define for $x \in \widehat{L} : z = z(x) = \phi^{-1}(x)$.

Choose a round disk ω around $\widehat{\beta}$ such that $\widehat{P}$ is univalent on ω and $\overline{\omega} \subset \widehat{P}(\omega)$. We can choose a possibly smaller round disk $\omega' \subset \omega$ around $\widehat{\beta}$

such that for any $x \in \omega' \cap \widehat{L}$: $z(x) \notin \Gamma_{c_0}$ and $\phi(\Gamma_z([0,1])) \subset \omega$. The first because $\overline{\Gamma}_{c_0}$ only intersects J_{c_0} in one point. And the second, because for any $z \in (R \cap K_{c_0}) \smallsetminus (J_{c_0} \cup \Gamma_{c_0})$ the map $\psi_{0,z}$ extends univalently to the sector $\widehat{V} = V(c_0, \eta, 0)$ and $\psi_{0,z}^{-1}(\Gamma_z([0,1]))$ is contained in the compact subset $\{w \in R_{c_0}(\eta)|1 \leq g_{c_0}(w) \leq d^{2q}\}$ of $\widehat{V}$. Hence by the Köebe Distortion Theorem for univalent maps the diameter of $\phi(\Gamma_z([0,1]))$ tends to 0 as the distance of x to $\mathbf{S}^1 = \partial \mathbb{D}$ tends to zero.

Let $x \in \omega' \cap \widehat{L}$ be arbitrary and let $\gamma = \phi \circ \Gamma_z$. Then $\gamma : \mathbb{R} \longrightarrow \mathbb{D}$ is an analytic arc with $\gamma(0) = x$ and $\widehat{P} \circ \gamma(s) = \gamma(s+1)$. But then $\gamma(s) \to \beta$ as $s \to \infty$, because $\widehat{P}$ univalently covers ω by $\widehat{P}(\omega)$.

Let $U_1, U_2 \subset \mathbb{D}$ be the two complimentary components of $\mathbb{D} \smallsetminus \gamma$. Consider say U_1. It contains a unique connected component U_1' of $\widehat{P}^{-1}(U_1)$, because the dynamics of P on γ is diffeomorphically conjugate to translation by 1. Moreover the degree of the restriction $\widehat{P} : U_1' \longrightarrow U_1$ is at least 2, because $\overline{U_1} \cap \mathbf{S}^1$ is covered at least twice by $\overline{U_1'} \cap \mathbf{S}^1$. Thus U_1 contains a critical point for $\widehat{P}$. And similarly so do U_2. **q.e.d.**

Remark 3.5 *Consider the family $Q_c(z) = z^4 - 2z^3 + z^2 + z + c$, $c \in \mathbb{C}$ of quartic polynomials. For each polynomial Q_c the real axis contains one critical point $\omega \in]-\frac{1}{3}, 0[$ with critical value v_c. For Q_0 the critical value v_0 is contained in the same interval and is attracted to the parabolic fixed point 0. The two other critical points are complex conjugate and are attracted to the parabolic fixed point 1. Moreover the external ray of argument 0 equals $[1, \infty]$. For any sequence $\{c_n\}_{n \in \mathbb{N}} \subset]0, \infty[$ converging to 0 the Hausdorff limit R of the segments of external rays $[v_n, \infty] \subset \mathbb{R}$ equals $[v_0, \infty]$. In this case the segment $]0, 1[$ is like the arc R_U in the proof of the above Lemma.*

Proposition 3.6 *Suppose $c_0 \in U \subseteq K_{c_0}$, where U is some Fatou component. Then the dynamic ray $R_{c_0}(\theta)$ lands on the parabolic point on the boundary of U.*

Proof : It was proved in Proposition 3.2 that c_0 is a parabolic parameter, U is an immediate parabolic basin of exact period q and θ is q-periodic ($l = 0$). Let α denote the parabolic point to which c_0 is iterated by $P_{c_0}^q$. Then $\alpha \in R$ by closeness and forward invariance of R. We claim that $R \cap J_{c_0} = \{\alpha\}$.

The set $R \cap J_{c_0}$ is a finite collection of q-periodic repelling or parabolic points, by Corollary 2.6 and analytic continuation. If this set was to contain any point other than α, then there should be a Fatou component U' such that $R \cap U'$ connects two different points of $R \cap J_{c_0}$. But no such component exist by Lemma 3.4 and the fact that P_{c_0} has only one critical point. **q.e.d.**

Proof of Theorem 1:   The cluster set $Cl = \overline{R_{M_d}(\theta)} \smallsetminus R_{M_d}(\theta) \subseteq \partial M_d$ is a
continuum and is thus connected. Let $c_0 \in Cl$. By Propositions 3.1 and 3.2,
c_0 has either parabolic or Misiurewicz dynamics. Since the period q and the
preperiod l is fixed, the parameter c_0 varies in a finite set and is thus unique,
which proves convergence. The dichotomy follows from Proposition 3.1 and
Proposition 3.2. Finally the statements about dynamic rays is contained
in Proposition 3.1 for the Misiurewicz case $l \neq 0$, and follows by combining
Proposition 3.2 and Proposition 3.6 for the parabolic case $l = 0$.   **q.e.d.**

References

[CG]   L. Carleson and T. Gamelin. *Complex Dynamics*. Springer-Verlag, 1993.

[DH]   A. Douady and J.H. Hubbard. Etude Dynamique des Polynômes Complexes. Première Partie, Publications Mathématique d'Orsay 84-02, 1984 and Deuxième Partie Publications Mathématique d'Orsay, 85-04, 1985.

[HS]   J. H. Hubbard and D Schleicher. The Spider Algorithm. In *Proceedings of Symposia in Applied Mathematics*, volume 49, 1994.

[K]   J. Kiwi. Rational rays and critical portraits of complex polynomials. 1997, IMS Preprint 1997/15 SUNY at Stony Brook.

[Mil]   J. Milnor. Periodic orbits, external rays and the Mandelbrot set; an expository account. Preprint, 1995.

[Pet]   C. L. Petersen. Convergence of (pre)periodic rays in parameter spaces. Manuscript, 1997.

[Ryd]   G. Ryd. *Iterations of one parameter families of complex polynomials.* PhD thesis, Department of Mathematics, KTH Stockholm, Sweden, 1997.

[Sch]   D. Schleicher. Rational Parameter Rays of the Mandelbrot Set. November 1997, IMS Preprint 1997/13 SUNY Stony Brook.

C.L. PETERSEN,IMFUFA, ROSKILDE UNIVERSITY, POSTBOX 260, DK-4000 ROSKILDE, DENMARK. e-mail: lunde@ruc.dk
GUSTAV RYD, RESEARCH & TRADE, 103 95 STOCKHOLM, SWEDEN
e-mail: gustav@rat.se

Bounded recurrence of critical points and Jakobson's Theorem

Stefano Luzzatto

Abstract

We prove that the set of quadratic interval maps which satisfy a strong bounded recurrence condition on the orbit of the critical point has positive Lebesgue measure. This implies classical Theorems of Jakobson and Benedicks-Carleson.

1 Introduction

1.1 Statement of results

In this paper we will be concerned with the dynamics of maps belonging to the quadratic family

$$f_a(x) = x^2 - a$$

for $a \in \Omega_\varepsilon = [2 - \varepsilon, 2]$, $\varepsilon > 0$ and $x \in I = [-2, 2]$. Let $c_0 = c_0(a) = f_a(0)$ denote the critical value of f_a and for $i \geq 0$, $c_i = c_i(a) = f^i(c_0)$. Notice that this differs slightly from the usual convention of letting c_1 denote the critical value.

Definition (Bounded recurrence of the critical point). For $\delta > 0, \alpha > 0$ and $a \in \mathbb{R}$ we say that f_a satisfies the bounded recurrence condition $(BR)_n = (BR)_n(\delta, \alpha)$ if for all $k \in [1, n]$:

$$\prod_{\substack{c_i \in (-\delta, \delta) \\ 0 \leq i \leq k}} |c_i| \geq e^{-\alpha k}. \qquad (BR)_n$$

We say that f satisfies (BR) if it satisfies $(BR)_n$ for all n and we let

$$\Omega_\varepsilon^* = \Omega_\varepsilon^*(\alpha, \delta) = \{a \in \Omega_\varepsilon : f_a \text{ satisfies } (BR)(\delta, \alpha)\}$$

Theorem. *For every $\alpha > 0$ there exists $\delta > 0$ such that*

$$\frac{|\Omega_\varepsilon^*|}{|\Omega_\varepsilon|} \to 1 \quad as \quad \varepsilon \to 0.$$

Here $|\cdot|$ denotes Lebesgue measure.

The proof also shows that the complement of Ω_ε^* in Ω_ε contains an open and dense set so that in particular Ω_ε^* is a nowhere dense set of positive Lebesgue measure. In section 4.1 we shall show that for small ε and any $\alpha < \log \sqrt{2}$, all parameters in Ω_ε^* have an exponentially growing derivative along the critical orbit. By standard results this implies that the corresponding maps f_a admit an absolutely continuous invariant probability measure μ. In fact quite minor modifications of the constructions and estimates given here can be used to obtain the same conclusion. In any case we have as a Corollary the following classical

Theorem (Jakobson, 1981). *The parameter value 2 is a (one-sided) full Lebesgue density point of the parameters for which the corresponding maps f_a admit an absolutely continuous invariant probability measure.*

This result is considered a milestone in the theory of dynamical systems, both for the intrinsic interest of the statement as well as for the ideas and techniques introduced. It has also motivated a large amount of deep and difficult research, although the full potential of the concepts and methods involved is still far from being fully grasped or realized. The main purpose of this paper is to take advantage of the substantial progress which has been made through the application and generalizations of the original ideas to more complicated examples, and to revisit the arguments in their original and simplest setting.

The formulation of the main result in terms of condition (∗) is new in this setting and is motivated by several factors. Amongst these are the fact that it appears naturally in the study of maps with infinite derivatives, where the same condition is imposed on the recurrence near the critical points and near the singularities; it is extremely simple to state and it is not hard to develop an intuitive feeling for it; a simple application of the Ergodic Theorem actually shows that it is satisfied quite generally, i.e. typical points satisfy (∗) in relation to their recurrence with respect to any other given typical point, for any maps admitting an absolutely continuous invariant measure with a density in $L^p, p > 1$.

The proof is divided into 5 sections, in which the different kinds of arguments required are developed: e.g. the combinatorial structure, the probabilistic argument, the distortion estimates. Throughout the proof I have tried to identify the exact assumptions used at each step and to bring these to the reader's attention. In connection with this I have also tried to show how most of the arguments apply in much more general situations than the one considered here and to give some indication as to how they have been or could be generalized.

Acknowledgements. These notes would never have been written without the enthusiastic support of the participants in the Non-Hyperbolic Dynamics Seminar at Warwick during the spring term 1997 : James Gallagher, Oleg Koslovsky, Lambros Lambrou, Hans Henrik Rugh, Sebastian van Strien and Tan Lei. Special thanks also to Hans Henrik Rugh, Samual Senti and Warwick Tucker for reading preliminary versions of this paper and making useful suggestions, to Sebastian van Strien for many discussions, his encouragement and his excellent advice, and to Jacob Palis and Marcelo Viana for their continuing friendship and support and expert guidance which has led me to the problems and techniques developed in this paper. This work was supported by a grant of the British Engineering and Physical Sciences Research Council (EPSRC) No. GR/K86329

1.2 Idea of the proof

We fix $\varepsilon > 0$ and restrict our attention to the parameter interval $\Omega = \Omega_\varepsilon = [2-\varepsilon, 2]$ (from now on we shall usually omit the subscript ε). We begin with a combinatorial construction leading to the definition of the set Ω^*. The details are slightly technical but the basic idea is easy to explain. Fix a partition $\mathcal{I}$ of the critical neighbourhood $\Delta = (-\delta, \delta)$ which is fine enough so that two points which are close enough together and always fall in the same element of the partition can be said to have essentially the same recurrence. We then construct a sequence of partitions $\mathcal{P}^{(1)}, \mathcal{P}^{(2)}, \ldots$ of Ω so that points in the same element of the partition of order n "stay together" up to time n; let $\Omega^{(n)}$ denote the union of elements which satisfy a certain condition which will be shown to imply $(BR)_n$.

It then remains to show that $\Omega^* = \cap_n \Omega^{(n)}$ has positive measure and $|\Omega^*|/\varepsilon \to 1$ as $\varepsilon \to 0$. For this it is sufficient to show (i) that the measure $|\Omega^{(n-1)} \setminus \Omega^{(n)}|$ of parameters which need to be excluded at each stage is exponentially small in n and (ii) that no exclusions have to be made before some time $N \to \infty$ as $\varepsilon \to 0$. The second claim follows immediately from the fact that the critical point is non-recurrent for the parameter value $a = 2$ while the first is non-trivial and constitutes the heart of the proof. In probabilistic language it corresponds to saying that the conditional probability of *not satisfying* $(BR)_n$ given that $(BR)_{n-1}$ *is satisfied*, is exponentially small in n.

Before going into the details let's consider a particularly simple example which illustrates some of the mechanisms involved. Let $T : I \to I$ be an expanding interval map with a Markov partition $\mathcal{P}$. For simplicity we assume that $T(\omega) = I$ for all elements $\omega \in \mathcal{P}$. Then by standard results there is a sequence $\mathcal{P}_1, \mathcal{P}_2, \ldots$ of refinements of $\mathcal{P}$ such that T^n maps each $\omega_n \in \mathcal{P}_n$ to I with uniformly bounded distortion. Now fix an arbitrary point $c \in I$ and consider the set $\Gamma^* = \cap_n \Gamma^{(n)}$ where

$$\Gamma^{(n)} = \{x \in I : |f^n(x) - c| \geq \delta e^{-\alpha n}\}$$

for some $\alpha > 0$ and $\delta > 0$. I claim that for arbitrarily small α and sufficiently small δ, $|\Gamma^*| > 0$. Indeed, it follows from the bounded distortion property that the proportion of each interval ω_n which falls into $\Delta_n = (c - \delta e^{-\alpha n}, c + \delta e^{-\alpha n})$ at time n is of the order of $\delta e^{-\alpha n}$. Thus the same is true of the entire interval and the total measure of points which fall into Δ_n at time n for some n, is bounded by some constant multiple of $\sum \delta e^{-\alpha n}$ which can be made arbitrarily small by taking δ small.

This simple argument can be interpreted in the following way. The distortion bounds (which in this case follow from the expansivity of the map) and the Markov property imply that small intervals ω_n grow sufficiently large sufficiently fast and in a sufficiently regular way that their distribution in the original interval I at time n is essentially random. Therefore the probability of falling into Δ_n is proportional to the size of Δ_n and the fact that this tends to zero exponentially fast in n then yields the conclusion. Notice that the construction shows that Γ^* must be nowhere dense.

Our condition $(BR)_n$ is very similar to the one given above although somewhat stronger since it takes into account the past orbit of points. Since we are interested in the recurrence of a single (critical) point for different parameters we shall consider the image in the dynamical interval I of the iterates of the critical point for different parameter values. It is easy to see that for any $n \geq 0$ the images $f_a^n(c)$ for an interval of parameter values in form an interval in I. To simplify the terminology we shall usually refer to these sets as *iterates* of an interval of parameter values. We shall show below that intervals of good parameters are growing in size at exponential speeds and that our construction of the partitions in parameter space yields uniform distortion estimates. The main difficulty here compared to the situation in the toy model described above is the absence of a Markov structure and even of a uniform lower bound on the images of partition elements. To overcome this problem we develop a statistical argument to show (and make precise the notion) that most partition elements have uniformly large images and therefore can be considered to have a sufficiently random distribution. More specifically we prove two key technical estimates which link the *size* and *number* of elements in $\mathcal{P}^{(n)}$ to their recurrence: elements with strong recurrence are *small* and *not too many*. Therefore the *average* recurrence of the critical points of maps in $\Omega^{(n-1)}$ up to time n is very low and it follows by a large deviation argument that for exponentially most parameters it will be sufficiently low so that they will belong to $\Omega^{(n)}$ (i.e. the ratio $|\Omega^{(n-1)} - \Omega^{(n)}|/|\Omega^{(n-1)}| \to 0$ exponentially fast).

In section 2 we give the complete combinatorial definition of the sets $\Omega^{(n)}$

and their partitions $\mathcal{P}^{(n)}$. In section 3 we assume two technical lemmas (the ones mentioned above) and carry out the large deviation argument. In section 4 we prove three fundamental lemmas which constitute the basis of the proof. These deal respectively with expansion estimates outside the critical neighbourhood, parameter dependence, and the notion of binding. In section 5 we prove the bounded distortion property for elements of $\mathcal{P}^{(n)}$ and show that all points of $\Omega^{(n)}$ actually satisfy condition $(BR)_n$. Finally, in section 6 we prove the technical lemmas used in the large deviations argument.

1.3 Technical remarks

Remarks about specific aspects of the proof will be interspersed throughout the text. I will limit myself here to some remarks of a more general nature.

Persistence of hyperbolicity. On one level, Jakobson's Theorem as well as the Theorem given here and the many other generalizations which will be mentioned below are about the general question of *persistence of hyperbolicity*. While uniform hyperbolicity is essentially an open condition, this is not the case for the weaker forms of hyperbolicity which occur in systems with critical points or (in higher dimensions) homoclinic and heteroclinic tangencies. However, these Theorems do say that in certain cases non-uniform hyperbolicity is persistent in a strong measure-theoretical sense. In the present setting the proof makes it quite clear that the uniform hyperbolicity in most of the phase space, in fact outside the critical neighbourhood Δ, which moreover persists under perturbations, is crucial. However the fact that these estimates do not hold everywhere in a persistent way means that parameter choices have to be made and is reflected in the fact that Ω^* contains no open sets. On the other hand the fact that we can take $|\Delta| \to 0$ without affecting the uniform expansion estimates is reflected in the fact that 2 is a full Lebesgue density point of Ω^*.

Connected to this point is that of characterizing dynamically the Lebesgue density points of all parameters for which the corresponding maps have absolutely continuous invariant probability measures (*acip*'s); an interesting statement would be something like: a^* is a (possibly one-sided) full Lebesgue density point of parameters for which an *acip* exists (such measures always reflect a certain degree of hyperbolicity, see next paragraph) if and only if it satisfies certain dynamical conditions, e.g. it has an *acip* itself. Some very partial results are known (see references below). Notice that maps without *acip*'s can be partial density points of maps with *acip*'s (i.e. the ratio of the measure of parameters for which an *acip* exists in an ε neighbourhood of such a map tend to a number strictly between 0 and 1 as ε tends to 0. For example it follows from the hyperbolicity of the renormalization operator that this is true for infinitely renormalizable maps in the quadratic family. There are no known examples of maps which are in the closure of the set of maps with *acip*'s but are *zero* Lebesgue density points of this set, or even for which such density is not defined. Such examples would be very interesting.

It is very likely that questions related to the degree of persistence of hyperbolicity should be naturally formulated and connected to questions related to the the very notion of hyperbolicity and to the degrees of hyperbolicity which can occur, even though this notion is not yet well deinfed ! All maps in the quadratic family (and even in much more general families) with an absolutely continuous invariant probability measure μ are *non-uniformly hyperbolic with respect to μ*: there exists $\lambda > 0$ such that $\log |(f^n)'(x)|^{1/n} \to \lambda$ for μ-almost every x, i.e. the derivative is growing asymptotically at a (uniform) exponential rate. However it is clear that not all such maps exhibit the same *degree of hyperbolicity*: for example the rate of decay of correlations, which reflects some fairly intrinsic properties of the map, can be exponential in certain cases and subexponential in others [NS98, BLvS99].

Construction of the invariant measure. The construction and the estimates given here can be used for parameters in Ω^* to prove the existence of an absolutely continuous invariant probability measure. Indeed they easily give rise to an induced map with very strong hyperbolicity properties which in fact implies exponential decay of correlations for Hölder continuous observables. Indeed, a version of Lemma 4.3 shows immediately that there exists a simple partition $\mathcal{P}$ of the phase space, a function $P : \mathcal{P} \to \mathbf{N}$ and an induced uniformly expanding map F with uniformly bounded distortion given by $F|\omega = f^{P(\omega)}|\omega$. This map however does not satisfy $\inf_{\omega \in \mathcal{P}} |P(\omega)| > 0$, i.e. elements of $\mathcal{P}$ do not grow to a fixed size under F. Nevertheless Proposition 3.1 implies that F can be iterated and $\mathcal{P}$ refined to give a new induced map $\hat{F}$, a refinement $\hat{\mathcal{P}}$ of $\mathcal{P}$ and a function $\hat{P} : \hat{\mathcal{P}} \to \mathbf{N}$ with $\hat{F}|(\hat{\omega}) = f^{\hat{P}}|(\omega)$ such that all $\hat{\omega} \in \hat{\mathcal{P}}$ grow to fixed size under $\hat{F}$. Moreover $\hat{P}$ is summable and has an exponentially decaying tail: $|\{x : \hat{P}(x) \geq n\}|$ is exponentially small in n. The existence of an *acip* μ for f then follows by classical arguments. By recent results of Lai-Sang Young [You98], the exponentially decaying tail of $\hat{N}$ immediately implies that if μ is mixing it also satisfies strong stochastic properties such as exponential decay of correlation and the central limit theorem. See [BLvS99] for a generalization to a larger class of maps satisfying some very weak conditions on the critical orbits.

Comparison with other approaches. All proofs and generalizations of Jakobson's Theorem must essentially rely on the construction of some sequence of partitions with some combinatorial information which keeps track of the certain aspects of the *itinerary* of partition elements. The partition needs to be constructed in such a way as to guarantee uniform distortion bounds and sufficiently strong analytic estimates on the size of partition elements in terms of their combinatorics. These ingredients are then all plugged into a statistical argument to obtain the necessary estimates on the conditional probabilities of satisfying the required conditions.

There are two main reasons for which this essential similarity in the various proofs is not always easily recognizable. One is that different approaches differ in the specifics and the order in which these constructions and estimates are carried out. In this paper for example, the partitions are constructed following [BC91] by fixing a partition in dynamical space near the critical point and essentially pulling it back via a special family of maps from parameter to dynamical space. In other approaches, such as recent notes of Yoccoz [Yoc], the partition is constructed topologically using preimages of a fixed point. Both constructions do the job but require different approaches to the distortion bounds and to the way the combinatorial information is defined. This can also lead to an (apparently) different structure to the overall argument as definitions and estimates need to be introduced at different stages.

Probably the most confusing aspect of trying to compare different proof is related to the choice of the (essentially geometric) conditions which are needed as an *intermediate step* in the construction. The existence of an *acip* is a relatively abstract notion and therefore a more concrete condition is introduced and is then shown on the one side to occur with positive probability in parameter space and on the other to imply the existence of an *acip*. Depending on what kind of condition is chosen, the proportion of work required to prove its persistence in parameter space and the fact that it implies the existence of an *acip* can vary considerably. In this paper a bounded recurrence condition on the critical orbit is used which I would say is about half way between the two: the proof of the existence of an *acip* given condition (∗) involves the construction of an induced Markov map and the estimates involved are almost identical to those required in the proof of the persistence of condition (∗) in parameter space. Yoccoz, on the other hand, defines a priori a set of parameters which by definition have induced Markov maps. The existence of an *acip* is then almost automatic and most of the work goes into proving the persistence of such structures in parameter space. Interestingly this involves also defining a condition on the critical orbits in terms of their itineraries with respect to the given partition, which is strongly reminiscent of our condition (∗).

1.4 Historical Remarks

The existence of quadratic maps with absolutely continuous invariant probabilities was first noticed by Ulam and von Neumann [UvN45] for the case

$a = 2$; they used the very special fact that f_2 is a Chebyshev polynomial smoothly conjugate to the angle-doubling map on the circle to obtain an explicit formula for the density of the invariant measure μ. The extensive work carried out during the 50's, 60's and 70's on uniformly expanding maps, see [Ren57, Wil79, LY73, Bow79] formed the basis for the more sophisticated techniques required to treat more general maps with critical points. The existence of an *acip* is now known to follow from some very weak conditions on the critical orbit, see [NvS88] and [BLvS99]. Jakobson [Jak81] was the first to show the existence of *acip*'s could be *typical* in some sense, even for systems with critical points. Benedicks and Carleson [BC85] gave an alternative proof based on controlling the recurrence of the critical point and other proofs include [Ryc88, BC91, Tsu93a, Yoc] and the one given in the present paper. The proof in [Tsu93a] is particularly remarkable for its conciseness as it takes full advantage of the special properties of maps close to the top quadratic map. It has recently been proved that in the special case of the quadratic family almost all parameters have either an attracting periodic orbit or admit an absolutely continuous invariant probability measure [GS97, LMN97].

There are many interesting generalizations of Jakobson's Theorem. These are usually limited to proving the persistence of some geometric and/or analytic condition (analogous to the bounded recurrence condition given above) but do not go into the details of the ergodic properties of systems satisfying such conditions. In general however, the methods used can be adapted to show that invariant measures with certain hyperbolicity properties exist. In [dMvS93] a generalization is proved to show that under a mild transversality condition, maps satisfying a *non-recurrence* condition on the critical orbit are full Lebesgue density points of maps with *acip*'s. In [TTY92] a sketch of the proof is given for one-parameter families of maps with finitely many critical points. In [Tsu93b] it is shown that under similar transversality assumptions as in [dMvS93] maps with a finite number of critical points and satisfying some exponential growth and bounded recurrence conditions along forward and backward critical orbits are Lebesgue density points of similar maps. Such transversality conditions have been shown to be satisfied for all maps with exponentially growing derivatives along the critical orbit in the quadratic family [Tsu] and for maps with non-recurrent critical points in rather general polynomial families [vS]. In [Ree86] a version of Jakobson's Theorem is proved in the context of rational maps; in [Rov93] "contracting Lorenz maps" are studied in which the map has a critical point coinciding with a discontinuity; in [Thu98]families of maps with completely degenerate ("flat") critical points are studied; in [LVb, LT] maps in which both singularities with infinite derivatives and critical points coexist and in [PRV98] a family of maps with an infinite number of critical points is analysed.

Of particular interest are generalizations to higher dimensional situations.

The ideas and techniques required for such a major breakthrough were first introduced in [BC91] in the context of the Hénon family of maps. These were generalized in [MV93, Via93] to quadratic-like maps in the context of homoclinic bifurcations. In [DRV96, Cos98] similar techniques were applied to study the dynamics near saddle-node bifurcation points. In [Via97] a class of systems with many expanding directions is shown to be persistent under perturbations. In [PR97] and [PRV] a class of flows with an equilibrium point with complex eigenvalues is considered and in [LVa] a generalized model of the Lorenz equations is developed to include parameters for which the return map contains folds as well as an unbounded derivative. These results do not study ergodic properties of the systems but rather concentrate on proving the persistence of certain geometrical/analytical structures which embody some degree of hyperbolicity. The ergodic properties in some cases have also been studied, see [BY93, BY98, BVb, BVa] for the Hénon maps, [Alv97, AV] for the systems studied in [Via97] and [HL] for generalized Lorenz-like attractors of [LVa].

2 The combinatorial structure

For the rest of the paper we fix constants

$$\log \sqrt{2} > \hat{\lambda} \gg \alpha \gg \delta \gg \iota \gg \varepsilon > 0$$

in the order given, where the symbol $\gg$ means that the constant on the right is chosen sufficiently small after the constant on the left has been fixed. To simplify the notation we shall also introduce the constants

$$\lambda = \hat{\lambda} - 2\alpha \quad \text{and} \quad \beta = \alpha/\lambda.$$

The constant $\hat{\lambda}$ gives the expansion rate outside a critical neighbourhood. It can be chosen arbitrarily close to $\log \sqrt{2}$ and we shall take advantage of this fact to simplify certain estimates. The constants α and δ are those used in the definition of condition (*) and in the formulation of the results. Notice that taking α smaller *strengthens* the statement of the Theorem. The constant ι is an incidental constant used in the construction to define a critical neighbourhood $\Delta^+ = (-\delta^\iota, \delta^\iota)$ which strictly contains the critical neighbourhood $\Delta = (-\delta, \delta)$. The fact that ε is chosen last is no problem as our Main Theorem is an asymptotic statement about what happens as $\varepsilon \to 0$ once the other constants have been fixed.

2.1 Inductive assumptions

Let $\Omega^{(0)} = \Omega = \Omega_\varepsilon$ and $\mathcal{P}^{(0)} = \{\Omega^{(0)}\}$ denote the trivial partition of Ω. Given $n \geq 1$ suppose that for each $k \leq n - 1$ there exists a set $\Omega^{(k)} \subseteq \Omega$ and a partition $\mathcal{P}^{(k)}$ of $\Omega^{(k)}$ into intervals such that each $\omega \in \mathcal{P}^{(k)}$ has an associated *itinerary* constituted by the following information (for the moment this is given as abstract data, the geometrical meaning of this data will become clear in the next section). A sequence

$$0 = \eta_0 < \eta_1 < \cdots < \eta_s \leq k, \qquad s = s(\omega) \geq 0$$

of *escape times*. Between any two escape times η_{i-1} and η_i (and between η_s and k) there is a sequence

$$\eta_{i-1} < \nu_1 < \cdots < \nu_t < \eta_i \qquad t = t(\omega, i) \geq 0$$

of *essential return times* (or *essential returns*) and between any two essential returns ν_{j-1} and ν_j (and between ν_t and η_i) there is a sequence

$$\nu_{j-1} < \mu_1 < \cdots < \mu_u < \nu_j \qquad u = u(\omega, i, j) \geq 0$$

of *inessential return times* (or *inessential returns*). Following every essential (resp. inessential) return there is a time interval

$$[\nu_j + 1, p_j] \quad (\text{resp. } [\mu_j + 1, p_j])$$

with $p_j > 0$ called the *binding period* associated to the return time ν_j (resp. μ_j). A binding period cannot contain any essential or inessential return times although it may contain a sequence of *bound return times* (or *bound returns*). Associated to each essential, inessential and bound return time is a positive integer r called the *return depth* (which is roughly the (norm of the) logarithm of the distance from the critical point) . We define the functions

$$\mathcal{R}^{(k)} : \Omega^{(k)} \to \mathbf{N} \quad \text{and} \quad \mathcal{E}^{(k)} : \Omega^{(k)} \to \mathbf{N}$$

which associate to each $a \in \Omega^{(k)}$ respectively the total sum of all return depths and the total sum of essential return depths only of the element $\omega \in \mathcal{P}^{(k)}$ containing a in its itinerary up to and including time k. By construction both functions are constant on elements of $\mathcal{P}^{(k)}$.

We also assume that the following two analytic properties are satisfied.

Bounded recurrence: Each $a \in \Omega^{(k)}$ satisfies $(BR)_k$.

Bounded distortion: For every $\omega \in \mathcal{P}^{(k)}$, the map

$$c_j : \omega \to \omega_j = \{c_j(a) : a \in \omega\}$$

is a diffeomorphism with uniformly bounded distortion $\forall \, j \leq \nu + p + 1$ where ν is the last essential or inessential return of ω before time k and

p is the binding period. More precisely, there exists a constant $D > 0$ independent of k and $\omega \in \mathcal{P}^{(k)}$ such that

$$\frac{|c_j'(a)|}{|c_j'(b)|} \leq D \qquad \forall\, a, b \in \omega,\ \forall\, j \leq \nu + p + 1$$

Moreover if $\nu+p+1 < k$ the same statement holds for all $\nu+p+1 < j \leq k + 1$ replacing ω by any subinterval $\omega' \subseteq \omega$ which satisfies $\omega_j' \subseteq \Delta^+$.

2.2 Definition of $\Omega^{(n)}$ and $\mathcal{P}^{(n)}$

For $r \in \mathbb{N}$ let

$$I_r = [e^{-r}, e^{-r+1}) \quad \text{and } I_{-r} = -I_r.$$

Then

$$\Delta^+ = \{0\} \cup \bigcup_{|r| \geq r_{\delta+} - 1} I_r \quad \text{and} \quad \Delta = \{0\} \cup \bigcup_{|r| \geq r_\delta - 1} I_r.$$

where

$$r_\delta = \log \delta^{-1} \quad \text{and} \quad r_{\delta+} = \iota \log \delta^{-1}$$

and we can suppose without loss of generality that $r_\delta, r_{\delta+} \in \mathbb{N}$. Finally subdivide each I_r into r^2 subintervals of equal length. This defines partitions $\mathcal{I}, \mathcal{I}^+$ of Δ^+ with $\mathcal{I} = \mathcal{I}^+|_\Delta$. An interval belonging to either one of these partitions is of the form $I_{r,m}$ with $m \in [1, r^2]$ Let $I_{r,m}^\ell$ and $I_{r,m}^\rho$ denote the elements of $\mathcal{I}^+$ adjacent to $I_{r,m}$ and let $\hat{I}_{r,m} = I_{r,m}^\ell \cup I_{r,m} \cup I_{r,m}^\rho$. If $I_{r,m}$ happens to be one of the extreme subintervals of $\mathcal{I}^+$ then let $I_{r,m}^\ell$ or $I_{r,m}^\rho$, depending on whether $I_{r,m}$ is a left or right extreme, denote the intervals $(-2\delta^\iota, -\delta^\iota]$ or $[\delta^\iota, 2\delta^\iota)$ respectively.

Remark 1. The reason for the subdivision of each I_r into r^2 subintervals will become clear in the last part of the argument in the proof of the distortion estimates, see (12) and Sublemma 5.2.2. It is related to the fact that the distortion of a single iterate on each I_r is of the order of 1 (which is "not summable") whereas the distortion of a single iterate of $I_{r,m}$ is of the order of $1/r^2$ (which is "summable").

We now define a refinement $\hat{\mathcal{P}}^{(n)}$ of $\mathcal{P}^{(n-1)}$. Let $\omega \in \mathcal{P}^{(n-1)}$ and $\omega_n = \{c_n(a), a \in \omega\}$ denote its "image" at time n. We distinguish different cases.

Non-chopping times. We say that n is a non-chopping time for ω if one (or more) of the following situations occur:

1. $\omega_n \cap \Delta^+ = \emptyset$;

2. n belongs to the binding period associated to some (essential or inessential) return time $\nu < n$ of ω;

3. $\omega_n \cap \Delta_+ \neq \emptyset$ but ω_n does not intersect more than two elements of the partition $\mathcal{I}^+$.

In all three cases we let $\omega \in \hat{\mathcal{P}}^{(n)}$. If $\omega_n \cap (\Delta \cap I_{\pm r_\delta}) \neq \emptyset$ in cases $b)$ or $c)$ we say that n is a bound return time or an inessential return time respectively for $\omega \in \hat{\mathcal{P}}^{(n)}$ and we define the corresponding return depth by

$$r = \max\{|r| : \omega_n \cap I_r \neq \emptyset\}.$$

Notice that we are allowing the possibility of intervals having inessential or bound returns not actually intersecting Δ, but just the partition elements of $\mathcal{I}^+$ adjacent to (but outside) Δ. This is for minor technical reasons related to the estimates in Section 5.1.

Chopping times. In all remaining cases, i.e. if $\omega_n \cap \Delta^+ \neq \emptyset$ and ω_n intersects at least three elements of $\mathcal{I}^+$, we say that n is a chopping time for $\omega \in \mathcal{P}^{(n-1)}$. We define a natural subdivision

$$\omega = \omega^\ell \cup \bigcup_{(r,m)} \omega^{(r,m)} \cup \omega^\rho.$$

so that each $\omega_n^{(r,m)}$ fully contains a unique element of $\mathcal{I}_+$ (though possibly extending to intersect adjacent elements) and ω_n^ℓ and ω_n^ρ are components of $\omega_n \setminus (\Delta^+ \cup \omega_n)$ with $|\omega_n^\ell| \geq \delta^\iota$ and $|\omega_n^\rho| \geq \delta^\iota$. If the connected components of $\omega_n \setminus (\Delta^+ \cup \omega_n)$ fail to satisfy the above condition on their length we just glue them to the adjacent interval of the form $w_n^{(r,m)}$. By definition we let each of the resulting subintervals of ω be elements of $\hat{\mathcal{P}}^{(n)}$. The intervals ω^ℓ, ω^ρ and $\omega^{(r,m)}$ with $|r| < r_\delta$ are called *escape components* and are said to have an *escape at time* n. All other intervals are said to have an *essential return* at time n and the corresponding values of $|r|$ are the associated *essential return depths*. We make the same remark here as above: partition elements $I_{\pm r_\delta}$ do not belong to Δ but we still say that the associated intervals $\omega^{(\pm r_\delta, m)}$ have a return rather than an escape.

This defines the partition $\hat{\mathcal{P}}^{(n)}$ of $\Omega^{(n-1)}$. In particular the functions $\mathcal{E}^{(n)}$ and $\mathcal{R}^{(n)}$ are defined on $\Omega^{(n-1)}$ and constant on elements of $\hat{\mathcal{P}}^{(n)}$. Thus we can define

$$\Omega^{(n)} = \{\omega \in \hat{\mathcal{P}}^{(n)} : \mathcal{E}^{(n)}(\omega) \leq \alpha n/10\}$$

and

$$\mathcal{P}^{(n)} = \hat{\mathcal{P}}^{(n)} | \Omega^{(n)}.$$

Notice that an itinerary of escape times, return times, etc. is well defined for all elements of $\mathcal{P}^{(n)}$ up to time n. In Section 5 we shall prove that the two analytic assumptions on the recurrence and the distortion are also satisfied:

Proposition 2.1. *The inductive assumptions are satisfied for $\Omega^{(n)}$ and $\mathcal{P}^{(n)}$.*

In particular all parameters in $\Omega^{(n)}$ satisfy $(BR)_n$. Let

$$\Omega^* = \bigcap_{n \geq 0} \Omega^{(n)}.$$

2.3 Renormalization properties of the combinatorics

We give an alternative combinatorial description of the partition elements defined above. This description is crucial to the probabilistic argument which will be applied in the next section and highlights some remarkable renormalization properties of the construction. To each $\omega \in \hat{\mathcal{P}}^{(n)}$ is associated a sequence $0 = \eta_0 < \eta_1 < \cdots < \eta_s \leq n$, $s = s(\omega) \geq 0$ of escape times and a corresponding sequence of escaping components

$$\omega \subseteq \omega^{(\eta_s)} \subseteq \cdots \subseteq \omega^{(\eta_0)} \quad \text{with} \quad \omega^{(\eta_i)} \subseteq \Omega^{(\eta_i)} \text{ and } \omega^{(\eta_i)} \in \mathcal{P}^{(\eta_i)}.$$

Letting

$$\omega^{(\eta_i)} = \omega \quad \text{for all } s + 1 \leq i \leq n$$

gives a well defined parameter interval $\omega^{(\eta_i)}$ associated to $\omega \in \hat{\mathcal{P}}^{(n)}$ for each $0 \leq i \leq n$. Notice that for two intervals $\omega, \tilde{\omega} \in \hat{\mathcal{P}}^{(n)}$ and any $0 \leq i \leq n$, the corresponding intervals $\omega^{(\eta_i)}$ and $\tilde{\omega}^{(\eta_i)}$ are either disjoint or coincide. Then we define

$$Q^{(i)} = \bigcup_{\omega \in \hat{\mathcal{P}}^{(n)}} \omega^{(\eta_i)}$$

and let

$$\mathcal{Q}^{(i)} = \{\omega^{(\eta_i)}\}$$

denote the natural partition of $Q^{(i)}$ into intervals of the form $\omega^{(\eta_i)}$. Notice that

$$\Omega^{(n-1)} = Q^{(n)} \subseteq \cdots \subseteq Q^{(0)} = \Omega^{(0)}$$

and

$$Q^{(n)} = \hat{\mathcal{P}}^{(n)}$$

since the number s of escape times is always strictly less than n and therefore in particular $\omega^{(\eta_m)} = \omega$ for all $\omega \in \hat{\mathcal{P}}^{(n)}$. For a given $\omega = \omega^{(\eta_i)} \in Q^{(i)}$, $0 \le i \le n-1$ we let

$$Q^{(i+1)}(\omega) = \{\omega' = \omega^{(\eta_{i+1})} \in Q^{(i+1)} : \omega' \subseteq \omega\}$$

denote all the elements of $Q^{(i+1)}$ which are contained in ω and $Q^{(i+1)}(\omega)$ the corresponding partition. Then we define

$$\Delta\mathcal{E}^{(i)} : Q^{(i+1)}(\omega) \to \mathbf{N} \quad \text{where} \quad \Delta\mathcal{E}^{(i)}(a) = \mathcal{E}^{(\eta_{i+1})}(a) - \mathcal{E}^{(\eta_i)}(a)$$

gives the total sum of all essential return depths associated to the itinerary of the element $\omega' \in Q^{(i+1)}(\omega)$ containing a, between the escape at time η_i and the escape at time η_{i+1}. Finally we let

$$Q^{(i+1)}(\omega, R) = \{\omega' \in Q^{(i+1)} : \omega' \subseteq \omega, \Delta\mathcal{E}^{(i)}(\omega') = R\}.$$

3   The probabilistic argument

Modulo the proof of Proposition 2.1 which will be given in section 5, it remains to prove that

$$|\Omega^*| > 0$$

and that 2 is a Lebesgue density point of Ω^*. As mentioned above our main strategy is to estimate the average value of the recurrence of the critical point for parameters in $\Omega^{(n-1)}$ taking into account the orbit up to time n. This is the heart of the proof, where the combinatorial structure and the analytical estimates really need to work together.

A relatively easy argument can be used to show that intervals in $\tilde{\mathcal{P}}^{(n)}$ are exponentially small in terms of their recurrence, i.e. in terms of the total sum of their return depths. However this in itself does not imply that the total measure of points belonging to intervals with strong recurrence is small, we also need to estimate the *cardinality* of the set of intervals in $\tilde{\mathcal{P}}^{(n)}$ with a given recurrence. The easiest way to do this is to use the sequence of return depths as a way of indexing the elements of $\tilde{\mathcal{P}}^{(n)}$ and to use standard combinatorial techniques to estimate the possible number of sequences whose terms add up to a given total. Unfortunately (unboundedly) many intervals may have the same sequence of returns depths and therefore we need to divide the combinatorial structure into *blocks* for which an indexing with bounded multiplicity can be obtained. This is precisely the purpose of the construction of section 2.3. These blocks have in an intrinsic interest as they contain certain scale-invariant properties of the construction. Recall that $\beta = \alpha/\lambda \ll 1$.

Proposition 3.1. *For all $i \leq n - 1$, $\omega \in \mathcal{Q}^{(i)}$ and $R \geq 0$ we have*

$$\sum_{\omega' \in \mathcal{Q}^{(i+1)}(\omega,R)} |\omega'| \leq e^{(10\beta-1)R}|\omega|.$$

The statement in the this Proposition may be considered *the* key estimate in the entire paper. Starting with a "large" interval (the notion of escape time has the purpose precisely of fixing the concept of large in this context) we can construct a subdivision into subintervals which are again large after a certain history of returns to the critical neighbourhood, quantified through the total return depth function R, . The *tail* of R on one escaping component, i.e. the set of points belonging to intervals which have not yet achieved large scale after a total number of returns with return depths R, is exponentially small in R.

Proposition 3.1 will be proved in Section 6. For the moment we assume this estimate and complete the proof of our main Theorem. In section 3.1 we show how the estimate of Proposition 3.1 which applies to a single *block* leads to a bound for the overall average recurrence up to time n. In section 3.2 we complete the proof with the large deviations argument.

3.1 Average recurrence

Lemma 3.1.

$$\int_{\Omega^{(n-1)}} e^{\mathcal{E}^{(n)}/2} = \sum_{\omega \in \mathcal{Q}^{(n)}} e^{\mathcal{E}^{(n)}(\omega)/2}|\omega| \leq e^{3n/r_\delta}|\Omega|.$$

Proof. The equality follows immediately by the definitions. For the inequality, let $0 \leq i \leq n - 1$, and $\omega \in \mathcal{Q}^{(i)}$ and write

$$\sum_{\omega' \in \mathcal{Q}^{(i+1)}(\omega)} e^{\Delta\mathcal{E}^{(i)}(\omega')/2}|\omega'| = \sum_{\omega' \in \mathcal{Q}^{(i+1)}(\omega,0)} |\omega'| + \sum_{R \geq r_\delta} e^{R/2} \sum_{\omega' \in \mathcal{Q}^{(i+1)}(\omega,R)} |\omega'|.$$

Proposition 3.1 implies

$$\sum_{\omega' \in \mathcal{Q}^{(i+1)}(\omega,0)} |\omega'| + \sum_{R \geq r_\delta} e^{R/2} \sum_{\omega' \in \mathcal{Q}^{(i+1)}(\omega,R)} |\omega'| \leq \left(1 + \sum_{R \geq r_\delta} e^{(10\beta-\frac{1}{2})R}\right)|\omega|$$

and therefore

$$\sum_{\omega' \in \mathcal{Q}^{(i+1)}(\omega)} e^{\Delta\mathcal{E}^{(i)}(\omega')/2}|\omega'| \leq (1 + e^{-r_\delta/3})|\omega| \leq e^{3/r_\delta}|\omega|. \tag{1}$$

Since $\mathcal{E}^{(n)} = \Delta\mathcal{E}^{(0)} + \cdots + \Delta\mathcal{E}^{(n-1)}$ and $\Delta\mathcal{E}^{(i)}$ is constant on elements of $\mathcal{Q}^{(i)}$ we can write $\sum_{\omega \in \mathcal{Q}^{(n)}} e^{\mathcal{E}^{(n)}(\omega)/2}|\omega|$ as

$$\sum_{\omega^{(1)} \in \mathcal{Q}^{(1)}(\omega^{(0)})} e^{\Delta\mathcal{E}^{(0)}(\omega^{(1)})/2} \sum_{\omega^{(2)} \in \mathcal{Q}^{(2)}(\omega^{(1)})} e^{\Delta\mathcal{E}^{(1)}(\omega^{(2)})/2} \quad \cdots$$

$$\sum_{\omega^{(n-1)} \in \mathcal{Q}^{(n-1)}(\omega^{(n-2)})} e^{\Delta\mathcal{E}^{(n-1)}(\omega^{(n-1)})/2} \sum_{\omega^{(n)} \in \mathcal{Q}^{(n)}(\omega^{(n-1)})} e^{\Delta\mathcal{E}^{(n-1)}(\omega^{(n)})/2}|\omega^{(n)}|\,.$$

Notice the *nested* nature of the expression. Applying (1) repeatedly gives

$$\sum_{\omega \in \mathcal{Q}^{(n)}} e^{\mathcal{E}^{(n)}(\omega)/2}|\omega| \le e^{3n/r_\delta}|\Omega|.$$

$\square$

3.2 Large deviations

The definition of $\Omega^{(n)}$ gives

$$|\Omega^{(n-1)}| - |\Omega^{(n)}| = |\Omega^{(n-1)} \setminus \Omega^{(n)}| = |\{\omega \in \hat{\mathcal{P}}^{(n)} = \mathcal{Q}^{(n)} : e^{\mathcal{E}^{(n)}/2} \ge e^{\alpha n/20}\}|$$

and therefore using Chebyshev's inequality $(M\,|\{x \in I : f(x) \ge M\}| \le \int_I f)$ and Lemma 3.1 we have

$$|\Omega^{(n-1)}| - |\Omega^{(n)}| \le e^{-\alpha n/20} \int_{\Omega^{(n-1)}} e^{\mathcal{E}^{(n)}/2} \le e^{\left(\frac{3}{r_\delta} - \frac{\alpha}{20}\right)n}|\Omega| \le e^{-\alpha n/30}|\Omega|$$

which implies

$$|\Omega^{(n)}| \ge |\Omega^{(n-1)}| - e^{-\alpha n/30}|\Omega|$$

Thus using the fact that for $\varepsilon > 0$ small there exists a $N > 0$ such that $\Omega^{(j)} = \Omega^{(j+1)}$ for all $j \le N$ (this follows easily from the fact that $c_j(2) \equiv -2, \forall j \ge 2$) we get

$$|\Omega^{(n)}| \ge \left(1 - \sum_{i=N}^{n} e^{-\alpha i/30}\right)|\Omega| \quad \text{and} \quad |\Omega^*| \ge \left(1 - \sum_{i=N}^{\infty} e^{-\alpha i/30}\right)|\Omega|.$$

Moreover it is easy to see that $N(\varepsilon) \to +\infty$ as $\varepsilon \to 0$. Therefore

$$|\Omega_\varepsilon^*|/|\Omega_\varepsilon| \to 1 \quad \text{as} \quad \varepsilon \to 0$$

which implies that 2 is a one sided Lebesgue density point of Ω^*.

4 Fundamental tools

We state and prove three preliminary Lemmas which are not difficult but whose importance cannot be overestimated. They provide the foundations for the estimates of Sections 5 and 6 and are also of independent interest. Generalizations of each of these results have been successfully developed and applied.

The first is an estimate on the expansion outside a given critical neighbourhood $\Delta = (-\delta, \delta)$. As a corollary we will show that all parameters in Ω_ε^* satisfy the Collet-Eckmann condition. The second is a basic result about the dependence of iterates of the critical points on the parameter. The third formalizes the intuitive idea that when an iterate of a critical point lands in Δ it then retraces part of its initial orbit. Therefore information on the past is useful to study the future.

4.1 Expansion outside Δ

Lemma 4.1. *For all $0 < \hat{\lambda} < \log 2$ and $\delta > 0$ small, there exist constants $\varepsilon > 0$ and $C_\delta > 0$ such that for any $a \in \Omega_\varepsilon$, $f = f_a$, $n \geq 1$ and $x \in I$ such that when $x, fx, \ldots, f^{n-1}x \notin \Delta$ we have*

$$|(f^n)'x| \geq C_\delta e^{\hat{\lambda} n}. \tag{UE1}$$

If moreover $f^n x \in \Delta^+$ we have

$$|(f^n)'x| \geq e^{\hat{\lambda} n} \tag{UE2}$$

Proof. Consider the map $h : [0,1] \to [-2,2]$ given by $h(\theta) = 2\cos \pi\theta$ and define a family of maps on $[0.1]$ by

$$g_a = h^{-1} \circ f_a \circ h.$$

Notice that for $a = 2$, h defines a conjugacy between $f_2 = x^2 - 2$ and the angle doubling map $g_2(\theta) = 2\theta \mod 1$. Indeed using the standard equality $cos^2\psi - 1 = cos 2\psi$ we have

$$f_2(h(\theta)) = (2\cos \pi\theta)^2 - 2 = 2\cos 2\pi\theta = h(g_2(\theta)).$$

Now by the chain rule

$$
\begin{aligned}
|(f_a^n)'(x)| &= |h'(g_a^n(h^{-1}(x))) \cdot (g_a^n)'(h^{-1}(x)) \cdot (h^{-1})'(x)| \\
&= |(g_a^n)'(h^{-1}(x))| \cdot \left| \frac{h'(g_a^n(h^{-1}(x)))}{h'(h^{-1}(x))} \right|.
\end{aligned}
$$

Using the fact $h'(\theta) = \sin \pi \theta$ is close to zero only near the extreme points it is easy to see that

$$\left| \frac{h'(g_a^n(h^{-1}(x)))}{h'(h^{-1}(x))} \right| \geq \begin{cases} C_\delta & \text{if } f^{n-1}(x) \notin (-\delta, \delta) \\ 1 & \text{if } |f^n(x)| \leq |x| \end{cases}$$

for some constant $C_\delta > 0$ depending on δ. Since $|(g_2^n)'(\theta)| = 2^n$ for any θ this clearly implies the statement with $\hat{\lambda} = \log 2$ for $a = 2$. For $a < 2$ an explicit calculation shows that

$$|g_a'(\theta) - g_2'(\theta)| \leq C_\delta' |2 - a| \leq \varepsilon C_\delta'$$

for some (large) C_δ' and therefore

$$|(g_a^n)'(\theta)| \geq (2 - \varepsilon C_\delta')^n = e^{\log(2 - \varepsilon C_\delta')} \geq e^{\hat{\lambda}}$$

for any given $\hat{\lambda} < \log 2$ and $\delta > 0$ as long as ε is small enough.   □

Remark 2. Estimates analogous to (UE1) with $\hat{\lambda}$ depending on δ hold in very great generality, e.g. for any C^2 unimodal map without periodic attractors [Mis81, Man85]. The remarkable aspects of the statement given here are 1) that the constant $\hat{\lambda}$ can be chosen independent of δ, and 2) that the constant C_δ can be dropped if $f^n(x)$ is closer to the criticalpoint than x is. The proof given here takes full advantage of the special nature of the parameter value $a = 2$, namely that it is smoothly conjugate to a uniformly expanding map. A similar result holds for perturbation of more general maps with non-recurrent critical orbits [dMvS93] or even for maps with certain exponential growth (Collet-Eckmann) conditions on the forward and backward orbits of the critical points [Tsu93b], however it is harder there to obtain good estimates on the dependence of constants such as C_δ, C_δ' and ε in terms of δ. Sharp estimates (or indeed any estimates at all) for $\hat{\lambda}$ and C_δ in terms of δ in more general situations would be extremely interesting. These issues are related to the questions mentioned above about the characterization of maps which are Lebesgue density points of maps with *acip*'s.

Recall that $\hat{\lambda}, \alpha$ and δ are assumed to have been fixed.

Corollary 4.1.1. *For all $\varepsilon > 0$ sufficiently small, $a \in \Omega_\varepsilon$ and $k \geq 1$ such that f_a satisfies $(BR)_k$ we have*

$$|(f^{k+1})'(c_0)| \geq e^{(\hat{\lambda} - 2\alpha)(k+1)} = e^{\lambda(k+1)}.$$

Proof. Let N be a large constant such that $c_i(a) \notin \Delta$ and $(f_a^i)'(c_0(a)) \geq 3^i$ for all $i \leq N$ and all $a \in \Omega_\varepsilon$. N can be chosen arbitrarily large if ε is chosen small, this follows from the fact that for $a = 2$ the critical point is in the preimage of a fixed point with derivative equal to 4. Let $0 < N < \nu_1 < \cdots < \nu_s \leq k$ denote all the iterates such that $c_{\nu_j} \in \Delta$. By that chain rule we have, for any fixed $a \in \Omega_\epsilon$ and $f = f_a$,

$$(f^{k+1})'(c_0) = (f^N)'(c_0) \cdot (f^{\nu_1 - N})'(c_N) \cdots (f^{\nu_s - \nu_{s-1}})'(c_{\nu_{s-1}}) \cdot (f^{k+1-\nu_s})'(c_{\nu_s}).$$

Moreover each term in the product, except for the first one, can be further split up into a piece of orbit outside Δ for which (UE2) holds, and the last iterate inside Δ. Thus rearranging the terms be separating iterates outside Δ and inside Δ gives

$$|(f^{k+1})'(c_0)| \geq 3^N e^{\hat{\lambda}(k+1-N-s)} \prod_{j=1}^{s} |f'(c_{\nu_j})|$$

and using the definition of $(BR)_k$ yields

$$|(f^{k+1})'(c_0)| \geq 3^N e^{\hat{\lambda}(k+1-N-s)} e^{-\alpha k}.$$

Finally notice that that whenever $c_{\nu_j} \in \Delta$ we have $\log|c_{\nu_j}|^{-1} \geq \log \delta^{-1}$ and therefore condition $(BR)_k$ also implies in particular a bound on the number of returns:

$$s \log \delta^{-1} \leq \sum_{\nu_i \leq n} \log|c_{\nu_i}|^{-1} \leq \alpha k.$$

Substituting this into the last display above completes the proof. $\qquad \square$

4.2 Parameter dependence

Lemma 4.2. *For all $a \in \Omega_\epsilon$ and all $k \geq 1$ such that f_a satisfies $(BR)_k$,*

$$\frac{1}{2} \leq \frac{|c_k'(a)|}{|(f_a^k)'c_0|} \leq 2.$$

Proof. Let $F : \Omega_\epsilon \times I \to I$ be the function of two variables defined inductively by

$$F(a, x) = f_a(x) \quad \text{and} \quad F^k(a, x) = F(a, f_a^{k-1}x).$$

Then, for $x = c_0$,

$$c_k'(a) = \partial_a F^k(a, c_0) = \partial_a F(a, f_a^{k-1} c_0) = -1 + f_a'(c_{k-1}) c_{k-1}'(a).$$

Iterating this argument gives

$$-c_k'(a) = 1 + f_a'(c_{k-1}) + f_a'(c_{k-1}) f_a'(c_{k-2}) + \dots$$
$$\dots + f_a'(c_{k-1}) f_a'(c_{k-2}) \cdots f_a'(c_1) f_a'(c_0)$$

192   *Stefano Luzzatto*

and dividing both sides by $(f^k)'(c_0) = f_a'(c_{k-1})f_a'(c_{k-2})\ldots f_a'(c_1)f_a'(c_0)$ gives

$$-\frac{c_k'(a)}{(f_a^k)'(c_0)} = 1 + \sum_{i=1}^{k} \frac{1}{(f^i)'(c_0)}. \tag{2}$$

Now choose N so that

$$\sum_{i=N}^{\infty} e^{-\lambda i} \leq \frac{1}{10}$$

and $\varepsilon > 0$ sufficiently small so that for all $a \in \Omega_\varepsilon$,

$$(f_a^i)'(c_0) \leq -(3.5)^i \qquad \forall\, i \leq N. \tag{3}$$

Then, by (2),

$$\frac{|c_k'(a)|}{|f_a^k(c_0)|} \geq 1 - \sum_{i=1}^{N}(3.5)^{-i} - \sum_{N+1}^{\infty} e^{-\lambda i} \geq 1 - \frac{2}{5} - \frac{1}{10} \geq \frac{1}{2}.$$

The upper bound is obtained in the same way. $\square$

Remark 3. A similar remark applies here as in the previous section. The proof is relatively straightforward in this case where we take advantage of (3) which follows from the fact that the critical point is preperiodic repelling. Moreover, in this case transversality of the family has a clear geometric meaning: a preperiodic repelling critical point is a codimension one condition in the space of maps and the statement in the Lemma is saying that the quadratic family crosses this subspace transversally and with positive speed. Similar statements are usually assumed in more general versions of Jakobson's Theorem; a geometric interpretation is not obvious in such cases but one can work with the analytic condition that the expression in (2) is bounded away from 0. Tsujii [Tsu] has recently shown that the quadratic family does indeed satisfy such a transversality condition at Collet-Eckmann parameters and van Strien [vS] has proved the same for rather more general polynomial families at Misiurewicz parameters (non-recurrent critical points).

The results just mentioned certainly have an intrinsic interest, although in this context it is not clear why any form of transversality at all should be required. Transversality conditions, in the theory of local bifurcations for example, are often imposed to guarantee that one moves *away from a bad situation* with positive speed; the opposite is true here: we begin with a map with an *acip* and want to show that this phenomenon *persists*. Probably some smoothness condition (maybe even a very weak one) on the dependence of the maps with respect to the parameter is all that is really needed. In fact we shall not use the statement of Lemma 4.2 directly but rather an immediate consequence stated in the following Corollary. The rest of the proof could most likely be adapted to work even if the constants 4 and 1/4 below were replaced by some other constants possibly even depending on ε, which would probably be necessary if the transversality conditions were not satisfied.

Corollary 4.2.1. *Let* $\omega \subset \Omega$ *and suppose that all* $a \in \omega$ *satisfy* $(BR)_k$. *Then for all* $1 \leq i \leq j \leq k$ *there exists* $\xi \in \omega$ *such that*

$$|(f_\xi^{j-i})'(\xi_i)|/4 \leq \frac{|\omega_i|}{|\omega_k|} \leq 4|(f_\xi^{j-i})'(\xi_i)|$$

Proof. Consider the map $\varphi : \omega_i \to \omega_j$ given by $\varphi(c_i(a)) = c_j \circ c_i^{-1}(a)$. By the mean value theorem there exists $\xi \in \omega$ such that $|\omega_j|/|\omega_i| = |\varphi'(\xi_i)|$. Then by the chain rule and Proposition 4.2 we get the desired statement. $\square$

4.3 Binding

Lemma 4.3. *Let* $0 \leq k \leq n-1$, $\omega \in \mathcal{P}^{(k)}$ *and suppose that* k *is an essential return time for* ω *with return depth* r. *Then there exists an integer*

$$p \leq 2r$$

such that

$$|\omega_{k+p+1}| \geq |\omega_k|^{8\beta} \quad \text{with } \beta = \alpha/\lambda < 1/8.$$

Proof. For $a \in \omega$ let $\gamma = \gamma(a)$ be the interval $[c_k(a), 0]$ (or $[0, c_k(a)]$) and for $j \geq 0$, let $\gamma_j = f_a^{j+1}(\gamma)$. We define the *binding period* of $c_k(a)$ as

$$p(a, k) = \sup\{m \in \mathbb{N} : |\gamma_j| \leq e^{-2\alpha j} \quad \forall j \leq m - 1\}$$

and the *binding period* of ω as

$$p = p(\omega, k) = \min\{p(a, k) : a \in \omega\}$$

Intuitively $p(a, k)$ is essentially constant on ω but strictly speaking there can be some small variation. An additional argument is required at the end of the proof to deal with this problem.

For the next two sublemmas we fix a parameter $a \in \omega$ and let $\hat{p} = \min\{p - 1, k\}$. We shall show below that $p \ll k$ but for the moment we want to make sure that our bounded recurrence condition holds during the time interval under consideration.

Sublemma 4.3.1 (Distortion during the binding period). *For all y_0 and z_0 in γ_0 and $0 \le i \le \hat{p} + 1$ we have*

$$\left| \frac{(f^i)'(z_0)}{(f^i)'(y_0)} \right| \le e^{(1-e^{-\alpha})^{-2}} := D_\alpha$$

Proof. By the chain rule

$$\left| \frac{(f^i)'(z_0)}{(f^i)'(y_0)} \right| = \left| \prod_{j=0}^{i-1} \frac{f'(z_j)}{f'(y_j)} \right| = \left| \prod_{j=0}^{i-1} \left(1 + \frac{f'(z_j) - f'(y_j)}{f'(y_j)} \right) \right|.$$

By the definition of f, $f'(z) = 2|z| \; \forall \, z \in I$ and $f''(z) \equiv 2$. Then by the mean value theorem, $|f'(z_j) - f'(y_j)| \le |\gamma_j| \sup_{z \in I} f''(z) = 2|\gamma_j|$. Then, using the fact that $\log(1 + x) < x$,

$$\left| \frac{(f^i)'(z_0)}{(f^i)'(y_0)} \right| \le e^{\log \prod_{j=0}^{i-1} \left(1 + \frac{|\gamma_j|}{|y_j|} \right)} = e^{\sum_{j=0}^{i-1} \log \left(1 + \frac{|\gamma_j|}{|y_j|} \right)} \le e^{\sum_{j=0}^{i-1} \frac{|\gamma_j|}{|y_j|}}$$

and thus it is sufficient to show that

$$\sum_{j=0}^{i-1} \frac{|\gamma_j|}{|y_j|} \le (1 - e^{-\alpha})^{-2}.$$

Since $|\gamma_j| \le e^{-2\alpha j}$ this follows immediately if we show that

$$|y_j| \ge (1 - e^{-\alpha}) e^{-\alpha j}$$

which in turn follows if $|c_j| \ge e^{\alpha j}$ since

$$|y_j| \ge |c_j| - |\gamma_j| \ge e^{-\alpha j} - e^{-2\alpha j} \ge e^{-\alpha j}(1 - e^{-\alpha j}) \ge e^{-\alpha j}(1 - e^{-\alpha}).$$

Let N be the smallest integer such that $e^{-\alpha N} \le \delta$ (i.e. N is the earliest time at which condition (BR) allows a return to Δ; notice that $N \sim \log \delta^{-1}/\alpha$). Then for $j \ge N$ we have clearly have $|c_j| \ge e^{-\alpha j}$ by condition $(BR)_j$ and for $j \le N$ we have the basic estimate $|c_j| \ge 1 - \varepsilon 4^j$ (since 4 is the maximum expansion rate). Therefore we just choose ε small enough (recall that it can be chosen arbitrarily small after fixing α and δ). $\qquad \square$

Sublemma 4.3.2 (Expansion at the end of the binding period). *Let $x = c_k(a)$. We have an estimate for the length of the binding period:*

$$\log |x|^{-1} \le p \le \frac{2}{\lambda} \log |x|^{-1} \ll k;$$

and two alternative estimates for the growth of the derivative at the end of the binding period:

$$|(f^{p+1})'(x)| \ge |x|^{5\beta - 1} \quad \text{and} \quad |(f^{p+1})'(x)| \ge e^{\frac{\lambda(p+1)}{6}}.$$

Notice that the derivatives are calculated at $x = c_k(a)$, thus there is an overall expansion during the binding period which more than compensates the small derivative at the return.

Proof. By assumption $(BR)_{\hat{p}}$ is satisfied and therefore $|(f_a^j)'(c_0)| \geq e^{\lambda j}$ for all $j \leq \hat{p}$. By the mean value theorem, the definition of binding and the bounded distortion estimate above,

$$e^{-2\alpha\hat{p}} \geq |\gamma_{\hat{p}}| \geq |(f^{\hat{p}})'(c_0)||\gamma_0|/D_\alpha \geq e^{\lambda(\hat{p}-1)}|x|^2/D_\alpha.$$

Taking logarithms and rearranging yields

$$\begin{aligned}
\hat{p} &\leq \frac{2\log|x|^{-1} + 2\log D_\alpha + \lambda}{\lambda + 2\alpha} \\
&\leq \frac{2}{\lambda}(\log|x|^{-1} + \log D_\alpha) + 1 \leq \frac{2\log|x|^{-1}}{\lambda} - 1
\end{aligned} \tag{4}$$

as long as $|x| \leq 2\delta^\iota$ is sufficiently small. Thus $\hat{p} < k$ and in particular $p = \hat{p} + 1$. Similarly, using the fact that $|f'(z)| \leq 4$, $\forall z$ and that $|\gamma_p| \geq e^{-2\alpha p}$, we have

$$4^p|x|^2 D_\alpha \geq |(f^p)'(c_0)||\gamma_0|D_\alpha \geq |\gamma_p| \geq e^{-2\alpha p}$$

and therefore $p \geq (2\log|x|^{-1} - \log D_\alpha)/(2\alpha + \log 4) \geq \log|x|^{-1}$. This completes the estimates on the size of p. To get the derivative estimates, notice that by the mean value theorem there exists $\xi_0 \in \gamma_0$ such that $|(f^p)'\xi_0| = |\gamma_p|/|\gamma_0|$. Thus by Sublemma 4.3.1, the fact that $|f'(x)| = 2|x|$ and $|\gamma_0| = x^2$ and the chain rule, we have

$$|(f^{p+1})'(x)| = |f'(x)|\,|(f^p)'(x_0)| \geq \frac{1}{D_\alpha}\frac{|\gamma_p|}{|\gamma_0|} \geq \frac{2}{D_\alpha}\frac{e^{-2\alpha p}}{|x|}.$$

The two required estimates for the expansion are now easily obtained by substituting a bound for p in terms of x in one case and a bound for x in terms of p in the other. More precisely (4) gives $e^{-2\alpha p} \geq e^{-4\alpha\log|x|^{-1}/\lambda} \geq |x|^{4\alpha/\lambda}$ and therefore $|(f^{p+1})'(x)| \geq \frac{e^{-2\alpha p}}{D_\alpha|x|} \geq |x|^{\frac{5\alpha}{\lambda}-1}$. Similarly (4) also gives $\log|x|^{-1} \geq \lambda p/4$ i.e. $|x|^{-1} \geq e^{\lambda p/4}$ and therefore $|(f^{p+1})'(x)| \geq \frac{e^{-2\alpha p}}{D_\alpha|x|} \geq e^{-2\alpha p}e^{\lambda p/4}/D_\alpha \geq e^{(\lambda-2\alpha)(p+1)/5}$. $\square$

We can now complete the proof of the Lemma. Suppose first of all that $p(a)$ is constant on ω. Then by Sublemma 4.3.2 and using the fact that $|c_k(a)| \geq e^{-2r}$ by assumption, we have $|(f^{p+1})'(c_k(a))| \geq |c_k(a)|^{5\beta-1} \geq e^{(6\beta-1)r}$ and thus by Corollary 4.2.1 $|\omega_{k+p+1}| \geq e^{(7\beta-1)r}|\omega_k|$. The result then follows from the assumption that k is an essential return for ω implying that $e^{-r}/r^2 \leq |\omega_k| \leq 2e^{-r}$. If $p(a)$ is not constant on ω then for all $\tilde{a} \in \omega$ we have

$$|(f_{\tilde{a}}^p)'(c_{k+1}(\tilde{a}))| \sim |(f_{\tilde{a}}^p)'(c_0(\tilde{a}))| \sim |(f_a^p)'(c_0(a))| \sim |(f_a^p)'(c_{k+1}(a))|.$$

Here the symbol $\sim$ is used to indicate that the quantities on either side are the same up to a multiplicative constant which does not depend on a, ω or k. The middle $\sim$ follows from the inductive hypotheses and the other two from Sublemma 4.3.1. Therefore

$$|(f_{\tilde{a}}^{p+1})'(c_k(\tilde{a}))| \sim |(f_a^{p+1})'(c_k(a))|$$

and the rest of the argument proceeds as above.   □

5   Distortion and recurrence bounds

In this section we prove the general inductive step in the definition of $\Omega^{(n)}$ and $\mathcal{P}^{(n)}$. Notice that the combinatorial properties follow by construction thus we only need to prove the two analytic conditions: bounded distortion and bounded recurrence. We begin with the latter.

5.1   Condition $(BR)_n$

Lemma 5.1. *All points in $\Omega^{(n)}$ satisfy $(BR)_n$.*

Recall that functions $\mathcal{R}^{(k)}$ and $\mathcal{E}^{(k)}$ are defined on $\Omega^{(n)}$ and constant on elements of $\mathcal{P}^{(n)}$ for all $k \leq n$ and most importantly we have $\mathcal{E}^{(k)}(a) \leq \lambda \alpha k/10$ for all $k \leq n$ and all $a \in \Omega^{(n)}$. We claim that it is sufficient to prove that for all $a \in \Omega^{(n)}$ and $0 \leq k \leq n$ we have

$$\mathcal{R}^{(k)}(a) \leq 5\mathcal{E}^{(k)}(a) \tag{5}$$

by virtue of the following

Sublemma 5.1.1. *Let $a \in \Omega^{(n)}$ and suppose that $\mathcal{R}^{(k)}(a) \leq \alpha k/2$ for all $k \leq n$. Then a satisfies $(BR)_n$.*

Proof. Notice first of all that by construction $c_\nu(a) \in \Delta$ only if a has some non-zero return depth associated to a return at time ν. In other words all the returns to Δ which contribute to the sum in the definition of condition (BR) are accounted for by considering all the return times and return depths associated to a. By construction, if $\nu \leq n$ is a return for ω and the associated return depth is r, then all points in ω satisfy $|c_\nu(a)| \geq e^{-2r}$ (in fact they satisfy $|\varphi^\nu a| \sim e^{-r}$ but the weaker estimate is sufficient for our purposes). Therefore $\log |c_\nu(a)|^{-1} \le 2r$ and

$$\sum_{\nu_j \le n} \log |c_{\nu_j}(a)|^{-1} \le \sum_{\nu_j \le n} 2r_j \le 2\mathcal{R}^{(n)}(\omega) \le \alpha n.$$

$\square$

Equation 5 says that the at least one fifth of the total sum of return depths associated to the orbit of the critical point of a given parameter a is formed by essential return depths. Thus all we need to do is prove that the inessential and bound returns occurring between two essential returns have a total sum of return depths which can be bounded in terms of the return depths of the preceding essential return. We divide the argument into two parts; first we consider the essential returns and then the bound returns.

Sublemma 5.1.2. *Let* $\omega \in \mathcal{P}^{(\nu)}$, $\nu \in [1, n]$ *and suppose that all points in* ω *satisfy* $(BR)_\nu$ *and that* ν *is an essential return for* ω. *Set* $\nu := \mu_0$ *and let* $\nu < \mu_1 < \cdots < \mu_u$ *be a maximal sequence of inessential returns occurring after time* ν *and before any subsequent chopping time. Let* $r_0, r_1, \ldots, r_u$ *be the corresponding inessential return depths. Then*

$$\sum_{r=0}^{u} r_i \le \frac{3}{2} r_0.$$

Proof. The idea of the proof is simple. Firstly, there are no chopping times between time μ_0 and time μ_u and the intervals are growing exponentially, in fact the growth can be estimates in terms of the return depths using the binding period estimates. Secondly, large intervals cannot fall too close to the critical point or else they would give rise to a chopping time. Thus we can bound the return depths at each return in terms of the size of the interval. More formally, let $p_0, \ldots, p_u$ be the lengths of the binding periods corresponding to the returns $\mu_0, \ldots, \mu_u$. By the bounded recurrence condition and Sublemma 4.3.2 we have for each $a \in \omega$ and $i = 0, \ldots, u$,

$$|(f_a^{p_i+1})'(c_{\mu_i}(a))| \ge |c_{\mu_i}(a)|^{\frac{5\alpha}{\lambda}-1} \ge e^{\left(1-\frac{6\alpha}{\lambda}\right)r}.$$

By Lemma 4.1 we also have

$$|(f_a^{\mu_{i+1}-(\mu_i+p_i+1)})'(c_{\mu_i+p_i+1}(a))| \ge 1$$

and therefore combining these two estimates and using the chain rule gives

$$|(f_a^{\mu_u+p_u+1-\mu_0})'(c_{\mu_0}(a))| \ge e^{\left(1-\frac{6\alpha}{\lambda}\right)\sum_{i=0}^{u} r_i}.$$

Stefano Luzzatto

Thus by Corollary 4.2.1,

$$4 \geq 2|\omega_{\mu_u+p_u+1}| \geq |\omega_{\mu_0}|e^{\left(1-\frac{6\alpha}{\lambda}\right)\sum_{i=0}^{u} r_i}.$$

Now since μ_0 is an essential return we have $|\omega_{\mu_0}| \geq e^{-r_0}/r_0^2 \geq 8e^{-(1+\tilde{\alpha})r_0}$ for some $\tilde{\alpha} > 0$ which can be chosen small and so

$$e^{-(1+\tilde{\alpha})r_0} \leq |\omega_{\mu_0}| \leq 4e^{-\left(1-\frac{6\alpha}{\lambda}\right)\sum_{i=0}^{u} r_i}$$

which immediately implies

$$\sum_{i=0}^{u} r_i \leq \frac{1+\tilde{\alpha}}{1-\frac{6\alpha}{\lambda}}r_0 \leq \frac{3}{2}r_0.$$

$\square$

Sublemma 5.1.3. *Let* $\omega \in \mathcal{P}^{(\mu)}$, $\mu \in [1, n]$ *and suppose that* μ *is an essential or inessential return for* ω *with return depth* ρ. *Let* $p > 0$ *be the length of the binding period following time* μ *and let* $\mu < \zeta_1 < \cdots < \zeta_v \leq \mu+p$ *be the bound return times of* ω, *i.e. those times for which* $\omega_{\zeta_i} \cap \Delta \neq \emptyset$, *and* $\rho_1, \ldots, \rho_v$ *the corresponding bound return depths:* $\rho_i = \max\{|r| : \omega_{\zeta_i} \cap I_r \neq \emptyset\}$. *Then*

$$\sum_{i=1}^{q} \rho_i \leq \rho.$$

Proof. For simplicity we fix $a \in \omega$ which has the deepest sequence of bound returns. By the definition of binding the points $c_{\mu+j}(a)$ is close to $c_{j-1}(a)$ for all $j \leq p$ and thus, intuitively, a bound return ζ_i for c_μ occurs exactly when $c_0(a)$ has a return at time $\zeta_i - \mu + 1$. We assume for the moment that this is indeed the case. The main idea is that by the definition of binding period c_{ζ_i} and $c_{\zeta_i-\mu+1}$ are very close (of the order of $e^{-2\alpha\zeta_i-\mu+1}$) and by condition $(BR)_{\zeta_i-\mu+1}$, $|c_{\zeta_i-\mu+1}|$ is of the order of $e^{-\alpha\zeta_i-\mu+1}$ and therefore the return depth ρ_i corresponding to the bound return at time ζ_i is almost the same as the associated return depth r_i corresponding to the return at time $\zeta_i - \mu + 1$. Thus the sum of the bound return depths is uniformly almost the same as the sum of all the return depths associated to $c_0(a)$ up to time p. By condition $(BR)_p$ and the estimates on the length of p this is of the order of $\alpha p \sim \alpha r$. The details are left to the reader.

It can happen however that $\zeta_i - \mu + 1$ is not a return time but $c_{\zeta_i-\mu+1}$ falls very close to Δ so that c_{ζ_i} falls inside Δ and gives rise to a bound return time. This situation has no real relevance to the dynamics and could to all effects and purposes be ignored by modifying slightly the definitions. To be formally correct however there are at least a couple of ways to deal with it. One way is to change slightly the definition of condition $(BR)_n$ so that not all returns to

Δ are counted but those to a neighbourhood which shrinks slowly from Δ to Δ minus the two extreme subintervals of the partition. In this way we could guarantee that the above situation does not occur. Alternatively, notice that the bound return depths corresponding to such situations are necessarily very small, essentially r_δ. The techniques used for the binding estimates can be applied here to show that there is a binding period of length $\sim r_\delta$ following time $\zeta_i - \mu + 1$ during which this situation cannot happen, and is therefore very rare. Once again the details are straightforward and are left to the reader. □

5.2 Bounded distortion

In this section we shall prove a strong version of the distortion bounds.

Lemma 5.2. *Let* $\omega \in \mathcal{P}^{(n)}$. *Then*

$$\frac{|c'_k a|}{|c'_k b|} \leq e^{1/r_\delta} = D$$

for all $a, b \in \omega$ *and all* $k \leq \nu_q + p_q + 1$ *where* ν_q *is the last essential or inessential return of* ω *before time* n *and* p_q *is the length of the corresponding binding period. If* $n > \nu_q + p_q + 1$ *then the same statement holds for all* $k \leq n$ *restricted to any subinterval* $\bar{\omega} \subseteq \omega$ *such that* $|\bar{\omega}_k| \subseteq \Delta^+$.

Notice that the distortion bound *improves* as δ tends to zero. This is only true under the additional assumptions stated that $k \leq \nu_q + p_q + 1$ or restricting to $\bar{\omega} \subseteq \omega$ such that $|\bar{\omega}_k| \subseteq \Delta^+$. Without any assumptions the result still holds for any $k \leq n$ with a constant D_δ independent of n but depending on δ.

By Lemma 4.2 it is sufficient to show that

$$\frac{|(f_a^k)'(c_0)|}{|(f_b^k)'(c_0)|} \ll e^{1/r_\delta} = D, \tag{6}$$

i.e. critical orbits with the same combinatorics satisfy the same derivative estimates. By standard arguments (see also the proof of Lemma 4.3), (6) follows if we can show that

$$\sum_{j=0}^{k-1} D_j \ll 1/r_\delta \quad \text{where} \quad D_j = \sup_{a \in \omega} \frac{|\omega_j|}{|c_j(a)|}.$$

Let $0 < \nu_1 < \cdots < \nu_q < k$ be the essential and inessential returns of ω up to time k. By construction there is a unique element I_{ρ_i, m_i} in $\mathcal{I}^+$ associated to each ν_i. Let p_i be the length of the binding period associated to ν_i. For notational convenience define ν_0 and p_0 so that $\nu_0 + p_0 + 1 = 0$. We suppose first that $k \leq \nu_q + p_q + 1$. Then write

$$\sum_{j=0}^{\nu_q+p_q} D_j = \sum_{i=0}^{q-1} \sum_{\nu_i+p_i+1}^{\nu_{i+1}+p_{i+1}} D_j.$$

Sublemma 5.2.1. *For each* $i = 0, \ldots, q-1$ *we have*

$$\sum_{\nu_i+p_i+1}^{\nu_{i+1}+p_{i+1}} D_j \leq D_\alpha^3 |\omega_{\nu_{i+1}}| e^{\rho_{i+1}}.$$

Proof. Write

$$\sum_{\nu_i+p_i+1}^{\nu_{i+1}+p_{i+1}} D_j = \sum_{\nu_i+p_i+1}^{\nu_{i+1}-1} D_j + D_{\nu_{i+1}} + \sum_{\nu_{i+1}+1}^{\nu_{i+1}+p_{i+1}} D_j. \tag{7}$$

We shall estimate each of the three terms on the right hand side in separate arguments.

For the first notice that since $\omega_{\nu_{i+1}} \subseteq \hat{I}_{\rho_i,m}$, condition (UE2) implies that $|(f^{\nu_{i+1}-j})' c_{\nu_{i+1}-j}(a)| \geq e^{\lambda(\nu_{i+1}-j)}$ and therefore by Corollary 4.2.1 we have $|\omega_j| \leq 4 e^{-\lambda(\nu_{i+1}-j)} |\omega_{\nu_{i+1}}|$. Moreover $|c_{\nu_{i+1}-j}(a)| \geq \delta^\iota$ and so letting $\Lambda = \sum_{i \geq 0} e^{-\lambda i}$ we have

$$\sum_{j=\nu_i+p_i+1}^{\nu_{i+1}-1} D_j \leq 4 \sum_{j=\nu_i+p_i+1}^{\nu_{i+1}-1} e^{-\lambda j} |\omega_{\nu_{i+1}}| \delta^{-\iota} \leq 4\Lambda |\omega_{\nu_{i+1}}| e^{\rho_{i+1}}. \tag{8}$$

The last inequality follows from the fact that $\delta^{-\iota} < \delta^{-1} \leq e^{\rho_{i+1}}$. For the second term we immediately have

$$D_{\nu_{i+1}} \leq \frac{1}{2} |\omega_{\nu_{i+1}}| e^{\rho_{i+1}} \tag{9}$$

since $D_{\nu_{i+1}}$ is the supremum of $|\omega_{\nu_{i+1}}|/|c_{\nu_{i+1}}|$ over all points $c_{\nu_{i+1}}$ in $\omega_{\nu_{i+1}}$ and $|c_{\nu_{i+1}}| \geq 2e^{-\rho_{i+1}}$ by definition.

To estimate the third term we argue as follows. By Corollary 4.2.1 and the bounded distortion during binding periods we have for all $j \in [\nu_{i+1} + 1, \nu_{i+1} + p_{i+1}]$,

$$|\omega_j| \leq 8 D_\alpha |\omega_{\nu_{i+1}}| \sup_{a \in \omega} |(f_a^{j-\nu_{i+1}})'(c_{\nu_{i+1}})(a)|. \tag{10}$$

Now fix $a \in \omega$ and let $\gamma_j, j \geq 0$ be as in section 4.3. By the definition of binding periods we have $|\gamma_j| \leq e^{-2\alpha(j-\nu_{i+1})}$ and $|\gamma_0| \geq e^{-2\rho_{i+1}}$ and therefore

$$|(f^{j-\nu_{i+1}-1})'(c_{\nu_{i+1}+1}(a))| \leq D_\alpha |\gamma_j|/|\gamma_0| \leq D_\alpha e^{-2\alpha(j-\nu_{i+1})} e^{-2\rho_{i+1}}.$$

Moreover $|f'_a(c_{\nu_{i+1}}(a))| \leq 2e^{-\rho_{i+1}}$ and therefore

$$\sup_{a\in\omega} |(f_a^{j-\nu_{i+1}})'(c_{\nu_{i+1}})(a)| \leq 2D_\alpha e^{-2\alpha(j-\nu_{i+1})}e^{-\rho_{i+1}}.$$

Substituting in (10) gives

$$\sum_{j=\nu_{i+1}+1}^{\nu_{i+1}+p_{i+1}} D_j \leq 16D_\alpha|\omega_{\nu_{i+1}}|e^{\rho_{i+1}} \sum \frac{e^{-2\alpha(j-\nu_{i+1})}}{e^{-\alpha(j-\nu_{i+1})}} \leq 16D_\alpha^2|\omega_{\nu_{i+1}}|e^{\rho_{i+1}}. \quad (11)$$

Using the fact that Λ depends only on λ and D_α is large when α is small, equations (8)-(11) complete the proof.   □

Sublemma 5.2.1 gives immediately

$$\sum_{j=0}^{\nu_q+p_q} D_j = \sum_{i=0}^{q-1} \sum_{\nu_i+p_i+1}^{\nu_{i+1}+p_{i+1}} D_j \leq D_\alpha^3 \sum_{i=0}^{q-1} |\omega_{\nu_i}|e^{\rho_i}. \quad (12)$$

Since $|\omega_{\nu_i}| \lesssim e^{-\rho_i}/\rho_i^2$ we have almost obtained our estimate except for the fact that we have no control over the multiplicity with which each ρ_i occurs. To overcome this problem we show that essentially only the *last* return with a given return depth needs to be considered in the sum (12) since the lengths of the preceeding ones form a decreasing geometric sequence. First of all subdivide the sum on the right hand side into partial sums corresponding to return times with the same return depth:

$$\sum_{i=0}^{q-1} |\omega_{\nu_i}|e^{\rho_i} = \sum_{r\geq r_{\delta^+}} e^r \sum_{i:\rho_i=r} |\omega_{\nu_i}|. \quad (13)$$

Sublemma 5.2.2. *For any $r \geq r_{\delta^+}$,*

$$\sum_{i:\rho_i=r} |\omega_{\nu_i}| \leq 10e^{-r}/r^2.$$

Proof. Let $\mu_j = \nu_{i_j}, j = 1, \ldots, m$ be the subsequence of returns with return depths equal to r. By construction $|\omega_{\mu_m}| \leq 5e^{-r}/r^2$. Using the binding period estimates and (UE2) we have for all $a \in \omega$ and $j = 1, \ldots, m-1$,

$$|(f_a^{\mu_{i+1}-\mu_i})'c_{\mu_i}a| \geq |(f_a^{\mu_i+p_i+1})'c_{\mu_i}a| \geq e^{(1-5\beta)r} \geq e^{(1-5\beta)r_{\delta^+}} = \delta^{-\iota(1-5\beta)}$$

Therefore by Corollary 4.2.1, $|\omega_{\mu_i}| \geq \delta^{\iota(1-5\beta)}|\omega_{\mu_{i+1}}|/4$ and

$$\sum_{j=1}^{m} |\omega_{\mu_i}| \leq \sum_{j=0}^{m-1} \delta^{-\iota(1-5\beta)j}|\omega_{\mu_m}| \leq 2|\omega_{\mu_m}|.$$

□

Substituting the estimate of Sublemma 5.2.2 into (12) and (13) we get

$$\sum_{j=0}^{\nu_q+p_q} D_j \leq D_\alpha^3 \sum_{r \geq r_{\delta^+}} e^r \sum_{i:\rho_i=r} |\omega_{\nu_i}| \leq D_\alpha^4 \sum_{r \geq r_{\delta^+}} \frac{1}{r^2} \leq \frac{2D_\alpha^4}{r_{\delta^+}^2} \ll 1/r_\delta.$$

This completes the proof of the Lemma for $k \leq \nu_q + p_q + 1$. If $k > \nu_q + p_q + 1$ we consider the additional terms D_j restricting ourselves to some subinterval $\bar{\omega} \subset \omega$ with $\bar{\omega}_k \subseteq \Delta^+$. Clearly the preceding estimates are unaffected by this restriction. Using (UE2) we have

$$|\bar{\omega}_j| \leq 4e^{-\hat{\lambda}(k-j)} |\bar{\omega}_k| \leq 2e^{-\hat{\lambda}(k-j)} \delta^\iota$$

and therefore using also the fact that $|f'(c_j(a))| \geq \delta^\iota$ since $\omega_j \cap \Delta^+ = \emptyset$ we get

$$\sum_{j=\nu_q+p_q+1}^{k-1} D_j \leq \sum_{j=\nu_q+p_q+1}^{k-1} e^{-\lambda(k-j)} \leq \sum_{i=0}^{\infty} e^{-\lambda i}.$$

6   Combinatorial and metric estimates

The aim of this section is to prove Proposition 3.1: for all $i \leq n - 1$, $\omega \in \mathcal{Q}^{(i)}$ and $R \geq 0$ we have

$$\sum_{\omega' \in \mathcal{Q}^{(i+1)}(\omega,R)} |\omega'| \leq e^{(10\beta-1)R} |\omega|.$$

We use a counting argument to show that the cardinality of $\mathcal{Q}^{(i+1)}(\omega, R)$ is bounded above by $e^{\beta R}$ and an analytic argument to show that the size of each $\omega' \in \mathcal{Q}^{(i+1)}(\omega, R)$ is bounded above by $e^{(9\beta-1)R}|\omega|$.

6.1   The counting argument

Lemma 6.1. *For all* $0 \leq i \leq n - 1$, $\omega \in \mathcal{Q}^{(i)}$ *and* $R \geq 0$, *we have*

$$\#\mathcal{Q}^{(i+1)}(\omega, R) \leq e^{\beta R}.$$

We begin with a standard abstract counting argument. Let $\mathcal{S}_R$ denote the the set of all sequences of pairs of integers $(\pm r_1, m_1), (\pm r_2, m_2), \ldots, (\pm r_s, m_s)$ with $s \geq 1$, $|r_1| + \cdots + |r_s| = R$, $|r_j| \geq r_\delta$ and $m_j \in [1, r_j^2]$ for all $j = 1, \ldots, s$.

Sublemma 6.1.1.

$$\#\mathcal{S}_R \leq e^{\beta R/2}.$$

Proof. We estimate first of all the cardinality of the set of possible sequences $r_1, \ldots, r_s$ of positive integers $r_j \geq r_\delta$ with $r_1 + \cdots + r_s = R$ for a fixed $s \geq 1$. Notice that the number of of possible sequences as prescribed corresponds exactly to the number of ways of partitioning R objects into s disjoint non-empty subsets and can be bounded above by the number of ways of choosing s balls from a row of $R + s$ balls since this will determine a partition of the remaining R balls into at most s disjoint subsets (the fact that each of these subsets must contain at least r_δ elements will be used in the calculation below). Using Stirling's approximation formula for factorials $k! \in [1, 1 + 1/4k]\sqrt{2\pi k}k^k e^{-k}$ we have

$$\binom{R+s}{s} \leq \frac{(R+s)!}{R!s!} \leq \frac{(R+s)^{R+s}}{R^R s^s} \leq \left(\frac{R+s}{R}\right)^R \left(\frac{R+s}{s}\right)^s.$$

To estimate the second term we use the fact that $sr_\delta \leq R$ and that $\log(1+x) < x$ for all $x > 0$ and we get

$$\left(\frac{R+s}{R}\right)^R \leq \left(\frac{R(1+\frac{1}{r_\delta})}{R}\right)^R \leq \left(1 + \frac{1}{r_\delta}\right)^R \leq e^{R\log(1+\frac{1}{r_\delta})} \leq e^{R/r_\delta}.$$

To estimate the first term we write first of all

$$\left(\frac{R+s}{s}\right)^s = \left[\left(\frac{s}{R+s}\right)^{-\frac{s}{R}}\right]^R$$

$$\leq \left[\left(\frac{s}{R(1+\frac{1}{r_\delta})}\right)^{-\frac{s}{R}}\right]^R \leq \left[\left(\frac{s}{R}\right)^{-\frac{s}{R}}\left(1 + \frac{1}{r_\delta}\right)^{s/R}\right]^R$$

Since $s/R \leq 1/r_\delta$ and $x^{-x} \to 1$ as $x \to 0$ both terms inside the parenthesis tend to 1 as $\delta \to 0$ and it is sufficient to choose a sufficiently small δ to get

$$\binom{R+s}{s} \leq \left(\frac{R+s}{R}\right)^R \left(\frac{R+s}{s}\right)^s \leq e^{R/r_\delta}e^{\beta R/6} \leq e^{\beta R/5}$$

Next we need to allow each r_j to occur with multiplicity r_j^2 and this gives a bound

$$e^{\beta R/5}\sum_{j=1}^{s} r_j^2 \leq e^{\beta R/5}R^2 \leq e^{\beta R/5}e^{2\log R} \leq e^{\beta R/4}.$$

Also, each r_i can have a positive or negative sign. This gives an additional multiplicative factor of 2^s and thus a bound of

$$2^s e^{\beta R/4} \leq e^{s\log 2}e^{\beta R/4} \leq e^{(R\log 2)/r_\delta}e^{\beta R/4} \leq e^{\beta R/3}.$$

Finally, keeping in mind that for fixed R we have $s \le R/r_\delta$ we can sum over all possible sequence lengths s to get

$$\sum_{s \le R/r_\delta} e^{\beta R/3} = \frac{R}{r_\delta} e^{\beta R/3} = e^{\log(R/r_\delta)} e^{\beta R/3} \le e^{\beta R/2}.$$

$\square$

The statement in Lemma 6.1 now follows from the above and the following estimate about the actual number of partition elements which can share exactly the same sequence.

Sublemma 6.1.2. *For $\omega \in \mathcal{Q}^{(i-1)}$ and any sequence $(r_1, m_1), \dots, (r_s, m_s)$ with $s \ge 1, |r_j| \ge r_\delta, m_j \in [1, r_j^2]$ there exists at most r_δ^3 elements $\tilde\omega \in \mathcal{Q}^{(i)}(\omega)$ with a maximal sequence of essential returns $\eta_{i-1} < \nu_1 < \cdots < \nu_s \le \eta_i$ such that $\tilde\omega_{\nu_j} \cap I_{r_j, m_j} \ne \emptyset$ for $j = 1, \dots, s$.*

Proof. The proof is an immediate consequence of the construction described above. The first time, after time η_{i-1}, that ω intersects Δ^+ in a chopping time, every subinterval which arises out of the 'chopping' has either an escape time, in which case the sequence above is empty and the lemma does not apply or an essential return time in which case a unique pair of integers r_1 and m_1 are associated to it. Thus no two elements created up to this time can share the same sequence. Fixing one of these subintervals which has an essential return we consider higher iterates until the next time that it intersects Δ^+. At this time it is further subdivided into subintervals. Those which have essential returns at this time all have another uniquely defined pair of integers r_2 and m_2 associated to them. However multiplicity can occur for those which have escape times: all the subintervals which fall in $\Delta^+ \setminus \Delta$ have an escape at this time, and therefore belong to $\mathcal{Q}^{(i)}$ and we do not consider further iterates, but all share the same first (and only) term of the associated sequence of return depths. However the number of such subintervals can be estimated by the number of elements of the partition $\mathcal{I}^+|_{\Delta^+ \setminus \Delta}$ plus at most two elements which can escape by falling outside Δ^+. The number of such intervals is then $\le 2(r_{\delta^+} - r_\delta) r_\delta^2 + 2 \le r_\delta^3$. In the case of the intervals which have two or more returns we repeat the same reasoning to get the result. $\square$

6.2 The metric estimates

Lemma 6.2. *For all $0 \le i \le n - 1$, $\omega \in \mathcal{Q}^{(i)}$, $R \ge 0$ and $\tilde\omega \in \mathcal{Q}^{(i+1)}(\omega, R)$ we have*

$$|\tilde\omega| \le e^{(9\beta - 1)R} |\omega|.$$

Proof. From the construction of $\tilde{\omega}$ there is a nested sequence of intervals

$$\tilde{\omega} \subseteq \omega^{(\nu_s)} \subseteq \cdots \subseteq \omega^{(\nu_1)} \subseteq \omega^{(\nu_0)} = \omega$$

such that each $\omega^{(\nu_j)}$ has an essential return at time ν_j (intuitively $\omega^{(\nu_j)}$ is created as a consequence of the intersection of $\omega^{(\nu_{j-1})}$ with Δ at time ν_j). Write

$$\frac{|\tilde{\omega}|}{|\omega|} = \frac{|\omega^{(\nu_1)}|}{|\omega^{(\nu_0)}|} \frac{|\omega^{(\nu_2)}|}{|\omega^{(\nu_1)}|} \cdots \frac{|\omega^{(\nu_s)}|}{|\omega^{(\nu_{s-1})}|} \frac{|\tilde{\omega}|}{|\omega^{(\nu_s)}|}. \tag{14}$$

Letting $r_0 = r_\delta$ we have

Sublemma 6.2.1. *For $j = 0, \ldots, s - 1$ we have*

$$\frac{|\omega^{(\nu_{j+1})}|}{|\omega^{(\nu_j)}|} \leq e^{-r_{j+1} + 9\beta r_j}$$

Proof. We consider first the cases $j = 1, \ldots, s - 1$. The bounded distortion property holds up to time $\nu_j + p_j + 1$ and so

$$\frac{|\omega^{(\nu_{j+1})}|}{|\omega^{(\nu_j)}|} \leq D \frac{|\omega^{(\nu_{j+1})}_{\nu_j + p_j + 1}|}{|\omega^{(\nu_j)}_{\nu_j + p_j + 1}|}.$$

To estimate the numerator notice that by definition $\omega^{(\nu_{j+1})}_{\nu_{j+1}} \subset \Delta_+$ and therefore by (UE2) and Corollary 4.2.1 $|\omega^{(\nu_{j+1})}_{\nu_{j+1}}| \geq e^{\hat{\lambda}(\nu_{j+1} - \nu_j - p_j - 1)} |\omega^{(\nu_{j+1})}_{\nu_{j+1}}|/4$ and therefore

$$|\omega^{(\nu_{j+1})}_{\nu_j + p_j + 1}| \leq 4e^{-\lambda(\nu_{j+1} - \nu_j - p_j - 1)} |\omega^{(\nu_{j+1})}_{\nu_{j+1}}| \leq |\omega^{(\nu_{j+1})}_{\nu_{j+1}}| \leq 5e^{-r_{j+1}}.$$

To estimate the denominator, observe that by construction $\omega^{(\nu_j)}$ has an essential return at time ν_j, and therefore by Lemma 4.3 $|\omega^{(\nu_j)}_{\nu_j + p_j + 1}| \geq |\omega^{(\nu_j)}_{\nu_j}|^{8\beta} \geq e^{-8\beta r_j}/r_j^{16\beta}$.

For $j = 0$, recall that $\omega^{(\nu_0)} = \omega \in Q^{(i-1)}$ which means that $\omega^{(\nu_0)}$ has an escape at time $\eta_{i-1} = \nu_0$. Suppose first that $\omega^{(\nu_0)}_{\nu_1} \subseteq \Delta^+$. The the bounded distortion holds up to time ν_1 and we have

$$\frac{|\omega^{(\nu_1)}|}{|\omega^{(\nu_0)}|} \leq D \frac{|\omega^{(\nu_1)}_{\nu_1}|}{|\omega^{(\nu_0)}_{\nu_1}|} \leq De^{-r_1} |\omega^{(\nu_0)}_{\nu_1}|^{-1}$$

and we just need to show that $|\omega^{(\nu_0)}_{\nu_1}| \geq De^{-9\beta r_\delta}$. There are two cases to consider according to the position of $\omega^{(\nu_0)}_{\nu_0}$. If $\omega^{(\nu_0)}_{\nu_0} \cap \Delta \neq \emptyset$ then $I_{r,m} \subseteq$

$\omega_{\nu_0}^{(\nu_0)} \subseteq \hat{I}_{r,m}$ for some $r_{\delta^+} \geq r \geq r_\delta$ and we can repeat the binding arguments of section 4.3 and together with (UE2) and Corollary 4.2.1 these give

$$|\omega_{\nu_0}^{(\nu_1)}| \geq |\omega_{\nu_0}^{(\nu_0+p_0+1)}|/4 \geq |\omega_{\nu_0}^{(\nu_0)}|^{8\beta}/4 \geq e^{-8\beta r}/4r^{16\beta} \geq De^{-9\beta r_\delta}.$$

On the other hand if $\omega_{\nu_0}^{(\nu_0)} \cap \Delta^+ = \emptyset$ then by construction $|\omega_{\nu_0}^{(\nu_0)}| \geq \delta^\iota$ and using (UE2) again and the fact that the critical point is quadratic

$$|\omega_{\nu_1}^{(\nu_0)}| \geq |\omega_{\nu_0+1}^{(\nu_0)}|/4 \geq \delta^{2\iota}/4 \geq \delta^{8\beta} \geq e^{-8\beta r_\delta}$$

as long as $\iota < 4\beta$. This completes the proof in the case $\omega_{\nu_0}^{(\nu_1)} \subseteq \Delta^+$.

If $\omega_{\nu_0}^{(\nu_1)}$ is not completely contained in Δ^+ we cannot apply the bounded distortion property to the whole of $\omega^{(\nu_0)}$ up to time ν_1. However, restricting ourselves to a maximal subinterval $\bar{\omega}^{(\nu_0)} \subset \omega^{(\nu_0)}$ with $\bar{\omega}_{\nu_1}^{(\nu_0)} \subseteq \Delta^+$ we do have

$$\frac{|\omega^{(\nu_1)}|}{|\omega^{(\nu_0)}|} \leq \frac{|\omega^{(\nu_1)}|}{|\bar{\omega}^{(\nu_0)}|} \leq D\frac{|\omega_n^{(\nu_1)} u_1|}{|\omega_{\nu_1}^{(\nu_0)}|} \leq De^{r_1}|\bar{\omega}_{\nu_1}^{(\nu_0)}|^{-1}.$$

Therefore it is sufficient to show that $|\bar{\omega}_{\nu_1}^{(\nu_0)}| \geq De^{-9\beta r_\delta}$. This is clearly true since by hypothesis $|\bar{\omega}_{\nu_1}^{(\nu_0)}|$ intersects Δ (since it contains $|\omega_{\nu_1}^{(\nu_1)}|$ which is contained in Δ) and also extends all the way to the boundary of Δ^+ (since we have assumed that $|\omega_{\nu_1}^{(\nu_0)}|$ is not properly contained in Δ^+). Therefore

$$|\bar{\omega}_{\nu_1}^{(\nu_0)}| \sim \delta^\iota \gg \delta^{8\beta} \geq e^{-8\beta r_\delta}.$$

$\square$

The statement in the Lemma now follows immediately. Substituting into (14) and using the fact that $|\tilde{\omega}|/|\omega^{(\nu_s)}| \leq 1$ since $\tilde{\omega} \subseteq \omega^{(\nu_s)}$ gives

$$\frac{|\tilde{\omega}|}{|\omega|} \leq e^{-\sum_1^s r_j + 9\beta \sum_1^{s-1} r_j} e^{8\beta r_\delta} \leq e^{(8\beta-1)R} e^{-8\beta r_s + 8\beta r_\delta}.$$

The result follows since $r_s \geq r_\delta$. $\square$

References

[Alv97] J. Alves. *SRB measures for nonhyperbolic systems with multidimensional expansion*. PhD thesis, IMPA, Rio de Janeiro, 1997.

[AV] J. Alves and M. Viana. Statistical properties of some multidimensional attractors. In preparation.

[BC85] M. Benedicks and L. Carleson. On iterations of $1 - ax^2$ on $(-1, 1)$. *Annals of Math.*, 122:1–25, 1985.

[BC91] M. Benedicks and L. Carleson. The dynamics of the Hénon map. *Annals of Math.*, 133:73–169, 1991.

[BLvS99] H. Bruin, S. Luzzatto, and S. van Strien. Decay of correlations in one dimensional dynamics. *preprint* 1999. http://www.maths.warwick.ac.uk/ luzzatto/papers.html.

[Bow79] R. Bowen. Invariant measures for Markov maps of the interval. *Comm. Math. Phys.*, 69:1–17, 1979.

[BVa] M. Benedicks and M. Viana. Random perturbations and statistical properties of some Hénon-like maps. In preparation.

[BVb] M. Benedicks and M. Viana. Solution of the basin problem for certain non-uniformly hyperbolic attractors. In preparation.

[BY93] M. Benedicks and L.-S. Young. SBR-measures for certain Hénon maps. *Invent. Math.*, 112:541–576, 1993.

[BY98] M. Benedicks and L.-S. Young. Markov extensions and decay of correlations for certain Hénon maps. to appear in *Astèrisque*, 1999.

[CE83] P. Collet and J.-P. Eckmann. Positive Lyapunov exponents and absolute continuity for maps of the interval. *Ergod. Th. & Dynam. Sys.*, 3:13–46, 1983.

[Cos98] M. J. Costa. *Global strange attractors after collision of horseshoes with periodic sinks*. PhD thesis, IMPA, Rio de Janeiro, 1998.

[dMvS93] W. de Melo and S. van Strien. *One-dimensional dynamics*. Springer Verlag, Berlin, 1993.

[DRV96] L. J. Díaz, J. Rocha, and M. Viana. Strange attractors in saddle-node cycles: prevalence and globality. *Invent. Math.*, 125:37–74, 1996.

[GS97] J. Graczyk and G. Swiatek. Generic hyperbolicity in the logistic family. *Annals of Math.*, 146:1–52, 1997.

[HL] M. Holland and S. Luzzatto. Invariant measures and decay of correlations for systems with criticalities and singularities. Work in progress, 1999

[Jak81] M.V. Jakobson. Absolutely continuous invariant measures for one-parameter families of one-dimensional maps. *Comm. Math. Phys.*, 81:39–88, 1981.

[JN96]   M.V. Jakobson and S. Newhouse A two dimensional version of a folklore theorem. *Amer. Math. Soc. Transl. (2)*, 171: 89–105, 1996.

[LMN97]   M. Lyubich, M. Martens, and T. Nowicki. Almost every real quadratic map is either regular or stochastic. Preprint Stony Brook, 1997.

[LT]   S. Luzzatto and W. Tucker. Non-uniformly expanding dynamics in maps with criticalities and singularities. preprint http://www.maths.warwick.ac.uk/ luzzatto/papers.html.

[LVa]   S. Luzzatto and M. Viana. Lorenz-like attractors without invariant foliations. In preparation.

[LVb]   S. Luzzatto and M. Viana. Positive Lyapunov exponents for Lorenz-like maps with criticalities. to appear in *Astérisque*.

[LY73]   A. Lasota and J.A. Yorke. On the existence of invariant measures for piecewise monotonic transformations. *Trans. Amer. Math. Soc.*, 186:481–488, 1973.

[Man85]   R. Mañé. Hyperbolicity, sinks and measure in one dimensional dynamics. *Commun. Math. Phys.*, 100:495-524, 1985.

[Mis81]   M. Misiurewicz. Absolutely continuous invariant measures for certain maps of the interval. *Publ. Math. IHES*, 53:17–51, 1981.

[MV93]   L. Mora and M. Viana. Abundance of strange attractors. *Acta Math.*, 171:1–71, 1993.

[NS98]   T. Nowicki and D. Sands Non-uniform hyperbolicity and universal bounds for S-unimodal maps. *Invent. Math.*, 132:633-680, 1998

[NvS88]   T. Nowicki and S. van Strien. Absolutely continuous measures under a summability condition. *Invent. Math.*, 93:619–635, 1988.

[PR97]   A. Pumariño and A. Rodriguez. *Persistence and coexistence of strange attractors in homoclinic saddle-focus connections*, volume 1658 of *Lect. Notes in Math.* Springer Verlag, Berlin, 1997.

[PRV]   M. J. Pacifico, A. Rovella, and M. Viana. Persistence of global spiraling attractors. In preparation.

[PRV98]   M. J. Pacifico, A. Rovella, and M. Viana. Infinite-modal maps with global chaotic behaviour. *Annals of Math.*, pages 1–44, 1998.

[Ree86]   M. Rees. Positive measure sets of ergodic rational maps. *Ann. Sci. Ecole Norm. Sup.*4⁰ *Série*, 19:383–407, 1986.

[Ren57]   A. Renyi. Representations for real numbers and their ergodic properties. *Acta Math. Acad. Sci. Hungar.*, 8:477–493, 1957.

[Rov93]   A. Rovella. The dynamics of perturbations of the contracting Lorenz attractor. *Bull. Braz. Math. Soc.*, 24:233–259, 1993.

[Ryc88]   M. Rychlik. Another proof of Jakobson's theorem and related results. *Erg. Th. & Dyn. Sys.*, 8:93–109, 1988.

[Thu98]   H. Thunberg. Positive Lyapunov exponents for maps with flat critical points. *to appear in Erg. Th. & Dyn. Sys.*, 1998.

[Tsu]   M. Tsujii. A simple proof of the monotonicity conjecture in the quadratic family. *Erg. Th. & Dyn. Sys.* to appear.

[Tsu93a] M. Tsujii. A proof of Benedicks-Carleson -Jakobson Theorem. *Tokyo J. Math.*, 16:295–310, 1993.

[Tsu93b] M. Tsujii. Positive Lypaunov exponents in families of one-dimensional dynamical systems. *Invent. Math.*, 111:113–137, 1993.

[TTY92]   P. Thieullen, C. Tresser, and L.-S. Young. Exposant de Lyapunov positif dans des familles à un paramètre d'applications unimodales. *C. R. Acad. Sci. Paris*, 315, Série I:69–72, 1992.

[UvN45]   S. Ulam and von Neumann. *Bull. AMS*, 1945.

[Ves87]   A.P. Veselov. Integrable mappings and Lie algebras. *Soviet Math. Dokl.*, 35:211–213, 1987.

[Ves91]   A.P. Veselov. Integrable maps. *Russian Math. Surveys*, 46(5):1–51, 1991.

[Via93]   M. Viana. Strange attractors in higher dimensions. *Bull. Braz. Math. Soc.*, 24:13–62, 1993.

[Via97]   M. Viana. Multidimensional nonhyperbolic attractors. *Publ. Math. IHES*, 85:69–96, 1997.

[vS]   S. van Strien. Generic unfoldings of rational maps. Work in progress.

[Wil79]   R.F. Williams. The structure of the Lorenz attractor. *Publ. Math. IHES*, 50:321–347, 1979.

[Yoc] J.-C. Yoccoz. *Weakly Hyperbolic Dynamics*. Birkhäuser, Basel, in preparation.

[You98] L.-S. Young. Statistical properties of dynamical systems with some hyperbolicity. *Ann. of Math* 147:585–650, 1998.

MATHS DEPT, WARWICK UNIVERSITY, COVENTRY CV4 7AL, UK
E-mail: luzzatto@maths.warwick.ac.uk
Url: http://www.maths.warwick.ac.uk/~luzzatto

The Herman-Światek Theorems with applications

Carsten Lunde Petersen

Abstract

This article first presents a proof of the following Theorem of Herman, [Her] : An analytic circle homeomorphism f with a critical point and irrational rotation number θ is quasi-symmetrically conjugate to the corresponding rigid rotation, if and only if θ is of constant type. The proof is presented so as to clearly exhibit, that it only uses the hypotheses on Df through the fact that f satisfies the Światek Cross-ratio distortion inequality [Swi], and that there is at least one critical point a_0 on the circle, where f contracts (quasisymmetrically) like a power x^ν with $\nu > 1$.

Secondly the Theorem is applied, using an idea of Herman, to prove the existence of quadratic Siegel polynomials, $P_\theta(z) = e^{i2\pi\theta}z + z^2$, with θ of non-constant type, for which the Julia set is locally connected.

1 Introduction

In his survey at Séminaire Bourbaki, February 1987, A. Douady [Dou] presented a conjectural procedure to produce synthetically a quadratic polynomial P_θ with a Siegel disk, whose rotation number is θ and whose boundary is a quasi-circle containing the critical point. The idea was to use quasi-conformal surgery on a certain cubic Blaschke product f_θ, whose restriction to S^1 is a real analytic circle homeomorphism with rotation number θ and with a critical point, an inflection point. (see also page 12). However in order to perform the surgery, the restriction of f_θ to S^1 has to be quasi-symmetrically conjugate to the rigid rotation with rotation number θ. This lead Douady to pose the question: Does there exist any irrational θ for which the conjugation on S^1 of f_θ to the corresponding rigid rotation is quasi-symmetric? Herman, who attended the seminar, responded positively only a few days later. Using Światek's newly developed Cross-ratio distortion inequality [Swi] (see also Theorem 1.3), he proved the following result, which in the folklore is known as the Herman-Światek Theorem:

Theorem 1.1 ([Her]) *Let f be an analytic circle homeomorphism with irrational rotation number θ. Then f is c-quasi-symmetrically conjugate to the rigid rotation (of angle θ), if and only if θ is of constant type. Moreover the constant $c > 1$ depends only on Df and the constant type bound N for θ.*

I shall give a proof of this Theorem with the analytic assumption replaced by a technical hypothesis to be presented below. The proof is an edited version of Yoccoz proof, in his unpublished manuscript [Yoc, Section 3].

We shall identify the circle with $\mathbb{T} = \mathbb{R}/\mathbb{Z}$ and give $\mathbb{T}$ the induced orientation. Moreover we shall not distinguish a circle map f and lifts of f to $\mathbb{R}$. Let f be an orientation preserving circle homeomorphism. Call a quadruple (a, b, c, d) admissible if $a < b < c < d < a + 1$ in $\mathbb{R}$ and define the cross-ratio

$$[a, b, c, d] = \frac{b - a}{c - a}\frac{d - c}{d - b}$$

and the cross-ratio distortion

$$D(a, b, c, d, f) = \frac{[f(a), f(b), f(c), f(d)]}{[a, b, c, d]}.$$

We shall be working under the following technical assumptions on f

Hypothesis 1 *The circle homeomorphism f satisfies*

a. *There exists a constant $C > 1$ such that for any family $(a_i, b_i, c_i, d_i)_{i \in I}$ of admissible quadruples with $\max_{x \in \mathbb{T}}(\#\{i | x \in \,]a_i, d_i[\}) \leq m$:*

$$\prod_{i \in I} D(a_i, b_i, c_i, d_i, f) \leq C^m$$

b. *There exist constants $\nu, C' > 1$ and (at least) one 'critical point' $a_0 \in \mathbb{T}$, such that for all $x, y \in \mathbb{R}$ with $|x - a_0| \leq |y - a_0|$*

$$\left| \frac{f(x) - f(a_0)}{f(y) - f(a_0)} \right| \leq C' \cdot \left| \frac{x - a_0}{y - a_0} \right|^\nu$$

Note that for any circle homeomorphism f and for any $t \in \mathbb{R}$ the circle homeomorphism $f_t = f + t$ satisfies Hypothesis 1 if and only if f does and that with the same constants. Though not explicitly formulated the following slightly stronger version of Theorem 1.1 is implicit in both [Her] and [Yoc].

Theorem 1.2 *Let f be a circle homeomorphism with irrational rotation number θ and which satisfies Hypothesis 1. Then f is c-quasi-symmetrically conjugate to the rigid rotation (of angle θ), if and only if θ is of constant type. Moreover the constant $c > 1$ depends only on the constant type bound N for θ and the bounds $C, \nu, C' > 1$ in Hypothesis 1.*

The connection between Theorem 1.1 and Theorem 1.2, that is the connection between Hypothesis 1 and conditions on the derivative Df is made through the following Theorem of Świątek:

Theorem 1.3 ([Swi]) *Let f be a C^3 orientation preserving circle homeomorphism and suppose f has finitely many critical points $a_0, a_1, \ldots a_n$ such that there exist neighbourhoods $W_i \subseteq \mathbb{T}$ of a_i for each i and such that :*

1. *f has negative Schwarzian derivative Sf on $W_i \smallsetminus a_i$.*

2. *There exist constants $A_i > 0$ and $l_i \in \mathbb{N}$ such that*

$$\forall x \in W_i : A_i |x - a_i|^{2l_i} < f'(x) < 2A_i |x - a_i|^{2l_i}$$

3. *The variation of $\log Df$ on $\mathbb{T} \smallsetminus \cup_{i=0}^n W_i$ is bounded by $\rho > 0$.*

Then f satisfies Hypothesis 1 with C, ν, C' given by the A_i, l_i, and ρ.

Proof : For Hypothesis 1.a. this is the *Cross-Ratio Inequality* of Świątek [Swi, p.112]. Hypothesis 1.b. follows by integration of 2.   **q.e.d.**

Here Sf denotes the Schwarzian derivative of f and is given by

$$Sf = D^2 \log Df - \frac{1}{2}(D \log Df)^2$$
$$= (D^3 f Df - \frac{3}{2}(D^2 f)^2)(Df)^{-2}$$
$$= -2(Df)^{-\frac{1}{2}} \cdot D^2 (Df)^{-\frac{1}{2}}.$$

Any real analytic circle homeomorphism with a critical point is easily seen to satisfy the hypotheses and thus the conclusion of Theorem 1.3. Thus Theorem 1.1 follows by combining Theorem 1.2 and Theorem 1.3.

The fact that c in Theorem 1.1 depends essentially only on the constant type bound N for θ, enables the proof of a result for quadratic polynomials, with Siegel disks of certain non constant type rotation numbers:

For $|\theta| = 1$ let c_θ be the complex number for which the quadratic polynomial $Q_{c_\theta}(z) = z^2 + c_\theta$ has a fixed point with multiplier $\exp(i2\pi\theta)$. Herman has proved the following Theorem :

Theorem 1.4 (Herman, unpublished) *There exist irrational θ of non-constant type, such that the boundary of the Siegel disk for the quadratic polynomial Q_{c_θ} is a Jordan curve containing the critical point.*

Here I take his proof one step further and prove:

Theorem 1.5 *There exist irrational θ of non-constant type, such that the Julia set of the quadratic polynomial Q_{c_θ} is locally connected.*

2   Notation and prerequisites

Following the established conventions of the subject, we shall freely use the letter c to denote any constant, which depends only on Df or as here the constants $C, \nu, C' > 1$ of Hypothesis 1. In particular the constants do not depend on the irrational rotation number θ of f. That is given any f satisfying Hypothesis 1 and any $\eta \in \mathbb{T}$ such that $f_\eta = f + \eta$ has irrational rotation number, the constants c for f_η only depend on f.

Notation 2.1 *We shall use freely the following notations:*

1. *f will denote a critical circle map with a (not necessarily unique) critical point a_0 and irrational rotation number θ.*

2. *The irrational $\theta \in [0,1]$ has (the unique) continued fraction expansion with integer coefficients b_n and with convergents $\frac{p_n}{q_n}$ for $n \in \mathbb{N}$:*

$$\theta = \cfrac{1}{b_1 + \cfrac{1}{b_2 + \cfrac{1}{\ddots \cfrac{}{b_n + \cfrac{1}{\ddots}}}}}, \qquad \frac{p_n}{q_n} = \cfrac{1}{b_1 + \cfrac{1}{b_2 + \cfrac{1}{\ddots \cfrac{}{b_{n-2} + \cfrac{1}{b_{n-1}}}}}},$$

where $p_n = b_n p_{n-1} + p_{n-2}$ and $q_n = b_n q_{n-1} + q_{n-2}$.

3. *For each $i \in \mathbb{Z}$ define $a_i = f^{-i}(a_0)$, (note that the critical orbit is indexed time reversely).*

4. *For $n \geq 0$ we define*

$$I_n(x) = \begin{cases} [x, f^{-q_n}(x)], & n \text{ odd}, \\ [f^{-q_n}(x), x], & n \text{ even}. \end{cases}$$
$$m_n(x) = |f^{-q_n}(x) - x| = |I_n(x)|,$$
$$K_n = \{a_i \mid 0 \leq i \leq q_n\}, \qquad K_n^* = K_n \setminus \{a_{q_n}\}$$

We recall that the irrational number θ is of constant type with bound N, if and only if the sequence $\{b_n\}$ of coefficients is bounded by N, and more generally of constant type, if the coefficients are bounded by some unspecified N. Moreover θ is of constant type, if and only if, it is Diophantine of exponent

2, where in general θ is Diophantine of exponent $p \geq 2$, if and only if there exists $C > 0$ such that

$$\forall q \geq 1, \forall p \in \mathbb{Z} : |\theta - \frac{p}{q}| \geq \frac{C}{q^p}.$$

The most basic property of circle homeomorphisms is captured by the Poincaré semi-conjugation Theorem:

Theorem 2.2 (Poincaré semi-conjugation Theorem) *Let f be any orientation preserving circle homeomorphism with irrational rotation number θ. Then f is semi-conjugate to the rigid rotation $R_\theta(z) = z + \theta$ with angle θ, i.e. there exists a continuous map $\phi : \mathbb{T} \longrightarrow \mathbb{T}$ such that $\phi \circ f = R_\theta \circ \phi$.*

The Poincaré semi-conjugation Theorem implies that the combinatorial orbit structure of any orientation preserving circle homeomorphism is the same as that for the corresponding rigid rotation, in particular:

Remark 2.3 *The circle homeomorphism f has the following properties:*

1. *For even n (respectively for odd n) the point of K_n^* following (respectively preceding) a point a_i for $0 \leq i < q_n$ is a_j with*

$$j = \begin{cases} i + q_{n-1}, & \text{if } 0 \leq i < q_n - q_{n-1} \\ i + q_{n-1} - q_n, & \text{if } q_n - q_{n-1} < i < q_n \end{cases}$$

2. *For any $x \in \mathbb{T}$ and any $n \geq 1$ all the intervals $I_n(f^j(x))$ for $0 \leq j < q_{n+1}$ and $I_{n+1}(f^j(x))$ for $0 \leq j < q_n$ have mutually disjoint interiors.*

3. *For any $x \in \mathbb{T}$*

$$\mathbb{T} = \bigcup_{0 \leq j < q_{n+1}+q_n} I_n(f^j(x)) = \bigcup_{0 \leq j < q_{n+1}} I_n(f^j(x)) \cup \bigcup_{0 \leq j < q_n} I_{n+1}(f^j(x))$$

3   Theorem 1.2 and the ' à priori ' bounds

In this section we prove Theorem 1.2. We shall assume throughout this section that f is a circle homeomorphism with irrational rotation number θ and which satisfies Hypothesis 1. Moreover the letters a_0, C, ν, C' shall always denote respectively the critical point and the constants of Hypothesis 1. We begin with a fundamental, but elementary consequence of Hypothesis 1.a.:

Lemma 3.1 *For any admissible quadruple (a, b, c, d) and any $N \in \mathbb{N}$ one has*

$$[f^N(a), f^N(b), f^N(c), f^N(d)] \leq C^m \cdot [a, b, c, d],$$

where $m \in \mathbb{N}$ is the covering number

$$m = \max_{x \in \mathbb{T}} \#\{j | 0 \le j < N \text{ and } x \in \,]f^j(a), f^j(d)[\}.$$

Proof :   By Hypothesis 1.a. we have

$$\frac{[f^N(a), f^N(b), f^N(c), f^N(d)]}{[a, b, c, d]} = \prod_{0 \le j < N} D(f^j(a), f^j(b), f^j(c), f^j(d), f) \le C^m.$$

q.e.d.

Let us fix $n \ge 0$, say even to fix the ideas. Set $q = q_n$, $m(x) = m_n(x)$, $I(x) = I_n(x)$ for $x \in \mathbb{T}$ (see Notation 2.1) and define for $x \in \mathbb{T}$

$$\Delta(x) = [f^{-2q}(x), f^{-q}(x), x, f^q(x)]$$
$$= \frac{m_n(f^{-q}(x))}{m_n(f^{-q}(x)) + m_n(x)} \cdot \frac{m_n(f^q(x))}{m_n(f^q(x)) + m_n(x)}.$$

Lemma 3.2 *There exists a constant $c > 1$ such that $\forall x \in \mathbb{T}$ and for all $z \in I(x)$*

$$c^{-1} \cdot \min\{m(z), m(f^q(z))\} \le m(x) \le m(z) + m(f^q(z)).$$

Proof :   The right-hand inequality is a useful but trivial observation (see the figure). Regarding the left hand inequality, let us prove that, there exists a constant $c_1 > 1$ such that

$$\forall x \in \mathbb{T} \qquad m(x) \ge c_1^{-1} \cdot \min\{m(f^q(x)), m(f^{-q}(x))\}. \qquad (1)$$

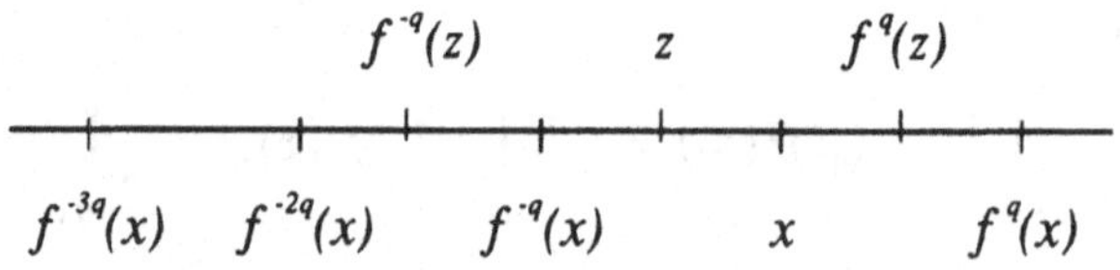

Figure 1: Relative positions of the relevant points.

Given any $x \in \mathbb{T}$ write $\min\{m(f^q(x)), m(f^{-q}(x))\} = r(x) \cdot m(x)$, we shall prove $r(x)$ is bounded independent of x. We can assume $r(x) \ge 1$ so that $m(x) \le \min\{m(f^q(x)), m(f^{-q(x)})\}$, which implies $\Delta(x) \ge \frac{1}{4}$. We estimate

$$\Delta(f^{-q}(x)) = \frac{m(f^{-2q}(x))}{m(f^{-2q}(x)) + m(f^{-q}(x))} \cdot \frac{m(x)}{m(x) + m(f^{-q}(x))}$$
$$\le \frac{m(x)}{m(x) + m(f^{-q}(x))} \le \frac{1}{1 + r(x)}.$$

The intervals $f^j(]f^{-3q}(x), x[)$, $0 \leq j < q$ cover any point $y \in \mathbb{T}$ at most 3 times. Thus by Lemma 3.1

$$\frac{1}{4} \leq \Delta(x) \leq C^3 \cdot \Delta(f^{-q}(x)) \leq \frac{C^3}{1 + r(x)},$$

hence $r(x) \leq 4 \cdot C^3 - 1$.

Let $c = 4 \cdot C^3$, so that the constant $c_1 = c - 1$ applies in (1), by arbitrariness of $x \in \mathbb{T}$. Let $x \in \mathbb{T}$ and $z \in I(x)$ be arbitrary. Then by (1) either

$$m(z) < m(f^{-q}(x)) + m(x) \leq (c_1 + 1) \cdot m(x) = c \cdot m(x) \qquad \text{or}$$
$$m(f^q(z)) < m(f^q(x)) + m(x) \leq (c_1 + 1) \cdot m(x) = c \cdot m(x).$$

q.e.d.

Proposition 3.3 *There exists a constant $c > 1$ such that for all $x \in \mathbb{T}$ and $z \in I(x)$ one has*

$$c^{-1} \cdot m(x) \leq m(z) \leq c \cdot m(x).$$

Proof : By the above Lemma 3.2 it suffices to prove that there exists $c_1 > 1$ such that for all $z \in \mathbb{T}$

$$c_1^{-1} \cdot m(z) \leq m(f^q(z)) \leq c_1 \cdot m(z) \qquad (2)$$

Let $y \in \mathbb{T}$ be a point where the continuous function $z \mapsto m(z)$ attains its minimum, so that $\Delta(y) \geq 1/4$, and let $y_j = f^{-j}(y)$ for all $j \in \mathbb{Z}$. Given any $z \in \mathbb{T}$ let j be the minimal non negative integer with $y_j \in I_n(f^{2q}(z))$. Then $0 \leq j < q_{n+1} + q_n$ by Remark 2.3.3. Thus by Lemma 3.1 there exists $c_2 > 1$ such that $c_2^{-1} \leq \Delta(y_j)$, because the intervals $]y_{i+2q}, y_{i-q}[$, $0 < i \leq j$ cover any point of the circle at most three times. Furthermore again by the Lemma 3.1 we can suppose $\Delta(y_{j+q}), \Delta(y_{j+2q}) \geq c_2^{-1}$ increasing c_2 if necessary. Writing out the definitions of $\Delta(y_j)$, $\Delta(y_{j+q})$ and $\Delta(y_{j+2q})$ we obtain for $c_3 = c_2 + 1$ that:

$$c_2^{-1} \leq \Delta(y_j) \Rightarrow m(y_j) \leq c_3 m(y_{j+q})$$
$$c_2^{-1} \leq \Delta(y_{j+q}) \Rightarrow \begin{cases} m(y_{j+q}) \leq c_3 m(y_j) \\ m(y_{j+q}) \leq c_3 m(y_{j+2q}) \end{cases}$$
$$c_2^{-1} \leq \Delta(y_{j+2q}) \Rightarrow m(y_{j+2q}) \leq c_3 m(y_{j+q})$$

So that

$$c_3^{-1} m(y_{j+q}) \leq m(y_j), m(y_{j+2q}) \leq c_3 m(y_{j+q}) \qquad (3)$$

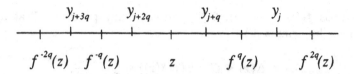

Figure 2: Relative positions of the relevant points.

Let c_4 be the constant from Lemma 3.2 then

$$c_4^{-1}\min\{m(y_{j+2q}), m(y_{j+q})\} \le m(z) \le m(y_{j+2q}) + m(y_{j+q}) \qquad (4)$$
$$c_4^{-1}\min\{m(y_j), m(y_{j+q})\} \le m(f^q(z)) \le m(y_j) + m(y_{j+q}) \qquad (5)$$

Combining the estimates (3) with the pair of estimates (4) and (5) we obtain (2), and the Proposition follows.   **q.e.d.**

Corollary 3.4 *The circle homeomorphism f is minimal, i.e every orbit is dense and any Poincaré semi-conjugacy is a homeomorphism and true conjugacy.*

Proof :   Suppose to the contrary that f is a Denjoy counter example with invariant Cantor set L. Let z be a left endpoint of some (wandering) interval I of the complement of L. Then

$$\lim_{n \to \infty} m_{2n}(z) = 0,$$
$$\lim_{n \to \infty} m_{2n}(f^{q_{2n}}(z)) = |I| > 0,$$

which contradicts Proposition 3.3. Here $|I|$ denotes the length of the interval I.   **q.e.d.**

Theorem 3.5 *There exists a constant $c > 1$ such that*

$$\forall x \in K_n : m_{n-1}(x) \le c \cdot m_n(x).$$

Proof :   To fix the ideas let us suppose n is even and define for $i \in \mathbb{Z}$

$$\Sigma_i = [a_{i+q_n}, a_i, a_{i+q_{n-1}}, a_{i+2q_{n-1}}]$$
$$= \frac{m_n(a_i)}{m_n(a_i) + m_{n-1}(a_i)} \cdot \frac{m_{n-1}(a_{i+q_{n-1}})}{m_{n-1}(a_i) + m_{n-1}(a_{i+q_{n-1}})}$$

and

$$u_i = \frac{m_n(a_i)}{m_{n-1}(a_i)} = \frac{a_i - a_{i+q_n}}{a_{i+q_{n-1}} - a_i}.$$

Then by Proposition 3.3 there exists $c_1 > 1$ such that

$$c_1^{-1}\frac{u_i}{1+u_i} \leq \Sigma_i \leq u_i \tag{6}$$

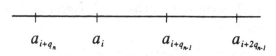

Figure 3: Relative positions of the points defining Σ_i.

Thus it suffices to prove that there exists $c > 1$ such that

$$c^{-1} \leq \Sigma_i \quad \text{for } 0 \leq i \leq q_n. \tag{7}$$

Moreover by Lemma 3.1 it suffices to prove (7) only in the case $i = 0$, because the intervals $]a_{i+q_n}, a_{i+2q_{n-1}}[$, $0 < i \leq q_n$ cover any point of the circle at most three times.

Comparing interval by interval and using Proposition 3.3 we find that there exists $c_2 > 1$ such that

$$\Sigma_0 \leq c_2 \cdot \Sigma_{-q_n}. \tag{8}$$

We can suppose $u_0 < 1$, as we want to bound u_0 away from 0. Thus by Hypothesis 1.b.

$$\Sigma_{-1} \leq u_{-1} \leq C' \cdot u_0^\nu.$$

Invoking Lemma 3.1 again (from Σ_{-1} to Σ_{-q_n}) we get

$$\Sigma_{-q_n} \leq C' \cdot C^3 \cdot u_0^\nu. \tag{9}$$

As we can suppose $u_0 < 1$, we obtain from (6), (8) and (9)

$$\frac{u_0}{2c_1} \leq \Sigma_0 \leq c_2 \cdot C' \cdot C^3 \cdot u_0^\nu$$

This shows that both u_0 and Σ_0 are bounded from below and completes the proof. **q.e.d.**

The above Theorem and/or the following Corollary is often in the literature referred to as *à-priori* real bounds for critical circle maps. They are used as a bootstrap to control the geometry and topology of holomorphic maps, which restricts to critical circle homeomorphisms. See for example [dF], [dFdM], [dFaWdM], [GS], [Pet] and [Yam]. The papers [GS] and [dFdM] also contain proofs of the *à-priori* real bounds.

Corollary 3.6 (à-*priori* real bounds) *There exists a constant $c > 1$ such that*

$$\forall\, n \geq 2 \; : \; |a_0 - a_{q_{n-1}}| \leq c \cdot |a_{q_n} - a_0|.$$

Now Theorem 1.2 follows as a Corollary of Theorem 3.5. We need however an initial Corollary first:

Corollary 3.7 *Suppose θ is of constant type. Then there exists a constant $c > 1$ depending only on C, ν, C' and on the bound N on the coefficients b_n so that*

$$\forall z \in \mathbb{T}, \forall n \in \mathbb{N} : \frac{m_n(z)}{m_{n+1}(z)} \leq c.$$

Proof :   Let $z \in \mathbb{T}$ be arbitrary. Given $n \in \mathbb{N}$ there exists $x \in K_n$ such that either $z \in I_{n-1}(x)$ or $z \in I_{n-1}(f^{-q_{n-1}}(x))$ by Remark 2.3.3. For this x there exists $k \leq N + 1$ such that $z \in I_n(f^{\pm k q_n}(x))$. Let c_1 be the constant from Proposition 3.3 and let c_2 be the constant from Proposition 3.5 then we obtain

$$\frac{m_{n-1}(z)}{m_n(z)} \leq \frac{m_{n-1}(x)}{m_n(x)} \cdot c_1^{k+1} \leq c_2 c_1^{N+2}.$$

q.e.d.

For $n \in \mathbb{N}$ let $M_n = d(\mathbb{Z}, q_n\theta) = |z - R_\theta^{q_n}(z)|$, where the last equality holds for any $z \in \mathbb{T}$ exactly because R_θ is a rigid rotation. Then for any $n \in \mathbb{N}$

$$b_{n+1}M_n < M_{n-1} < (1 + b_{n+1})M_n.$$

Corollary 3.8 *Suppose θ is of constant type. Then there exists a constant $c > 1$ depending only on C, ν, C' and on the bound N on the coefficients b_n so that f is c-quasi-symmetrically conjugate to the rigid rotation R_θ.*

Proof :   By Corollary 3.4 there exists a homeomorphism $h : \mathbb{T} \longrightarrow \mathbb{T}$ conjugating the rigid rotation R_θ to f. Let $z \in \mathbb{T}$ and $0 < \delta < \frac{1}{2}$ be arbitrary. Choose $n \in \mathbb{N}$ such that $M_{n+1} < \delta \leq M_n$. Let us suppose n is even to fix the

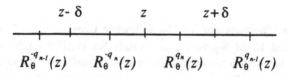

Figure 4: Relative positions of the relevant points.

ideas (n odd being similar) then

$$m_{n+1}(h(z)) \leq |h([z, z + \delta])| \leq m_n(f^{q_n}(h(z))) \tag{10}$$
$$m_{n+1}(f^{q_{n+1}}(h(z))) \leq |h([z - \delta, z])| \leq m_n(h(z)). \tag{11}$$

Let c_1 be the constant from Proposition 3.3 and let c_2 be the constant from Proposition 3.7 then we obtain

$$\frac{1}{c_2} \leq \frac{|h([z, z + \delta])|}{|h([z - \delta, z])|} \leq c_1^2 c_2.$$

q.e.d.

Corollary 3.9 *Suppose θ is of non-constant type. Then f is not quasi-symmetrically conjugate to the rigid rotation.*

Proof : It suffices to prove that a homeomorphism $h : \mathbb{T} \longrightarrow \mathbb{T}$ conjugating f to the rigid rotation R_θ is not quasi symmetric. To this end choose a subsequence $\{b_{n_k}\}_{k \in \mathbb{N}}$ diverging to ∞. Then

$$\frac{|h(I_{n_k-1}(a_0))|}{|h(I_{n_k-2}(a_0))|} = \frac{M_{n_k-1}}{M_{n_k-2}} \leq \frac{1}{b_{n_k}} \xrightarrow[k \to \infty]{} 0,$$

which contradicts that h is quasi symmetric, because

$$\frac{|I_{n_k-1}(a_0)|}{|I_{n_k-2}(a_0)|} = \frac{m_{n_k-1}(a_0)}{m_{n_k-2}(a_0)} \geq c,$$

where c is the constant from Corollary 3.6.   q.e.d.

Theorem 1.2 is a collection of Corollary 3.8 and Corollary 3.9

4 Local connectivity of Julia sets

Let $f_0 : \overline{\mathbb{C}} \longrightarrow \overline{\mathbb{C}}$ denote the Blaschke product

$$f_0(z) = z^2 \frac{z - 3}{1 - 3z}.$$

For each irrational $\theta \in]0, 1[$ let λ_θ be the unique unimodular constant for which the restriction $f_\theta = \lambda_\theta \cdot f_0 : \mathbb{S}^1 \longrightarrow \mathbb{S}^1$ is real analytic, has 1 as critical point and has irrational rotation number θ. Thus Theorem 1.1 applies. Let J_{f_θ} denote the Julia set of f_θ and let $J_\theta \subset J_{f_\theta}$ denote the boundary of the immediate attracted basin $\Lambda_\theta(\infty)$ for ∞. It was proved in [Pet] that

Theorem 4.1 *For every irrational θ the subset J_θ and the full Julia set J_{f_θ} are locally connected.*

We shall prove the following Theorem which combined with Theorem 4.1 implies Theorem 1.5:

Theorem 4.2 *There exists irrational θ of non-constant type for which the Julia set J_{c_θ} for Q_{c_θ} is homeomorphic to J_θ.*

The proof of this Theorem is based on the similar result for θ of constant type:

Theorem 4.3 (Douady, Ghys, Herman, Hubbard and Shishikura)
Suppose θ is of constant type with bound N. Then there exists a $K(N)$-quasi conformal homeomorphism $\phi_\theta : \overline{\mathbb{C}} \longrightarrow \overline{\mathbb{C}}$, which is a holomorphic conjugacy between f_θ and P_{c_θ} on the immediate attracted basins of ∞ and which maps J_θ (homeomorphically) onto J_{c_θ}.

The idea of the proof of Theorem 4.2 is to extract a uniform limit of the maps ϕ_θ of Theorem 4.3 for a sequence $\{\theta_n\}$ of irrationals of constant type converging to an irrational of non-constant type.

To facilitate the proof of Theorem 4.2 we will first outline the proof of Theorem 4.3:
Suppose the number θ is of constant type. Let $h : \mathbb{S}^1 \longrightarrow \mathbb{S}^1$ denote the conjugacy, between f_θ and the rigid rotation R_θ. As θ is of constant type, the Herman-Świątek Theorem, Theorem 1.2 states that the map h is c-quasi-symmetric, with a constant $c > 1$, which only depends on the bound N on the coefficients of the continued fraction expansion for θ. Let $\psi_\theta : \overline{\mathbb{D}} \longrightarrow \overline{\mathbb{D}}$ be a K-quasi-conformal extension of h (with a $K > 1$ which only depends on c and hence on N) and define $F_\theta : \overline{\mathbb{C}} \longrightarrow \overline{\mathbb{C}}$ by

$$F_\theta(z) = \begin{cases} f_\theta(z) & \text{if } z \in \overline{\mathbb{C}} \smallsetminus \mathbb{D} \\ \psi_\theta^{-1} \circ R_\theta \circ \psi_\theta & \text{if } z \in \overline{\mathbb{D}}. \end{cases}$$

Let σ_0 denote the standard almost complex structure on $\overline{\mathbb{C}}$, and let σ_θ denote the F_θ invariant almost complex structure given by

$$\sigma_\theta(z) = \begin{cases} \psi^*(\sigma_0)(z) & \text{if } z \in \mathbb{D} \\ ((\psi \circ F_\theta^n)^*(\sigma_0)(z) & \text{if } F^n(z) \in \mathbb{D} \\ \sigma_0(z) & \text{otherwise.} \end{cases}$$

Finally let $\phi_\theta : \overline{\mathbb{C}} \longrightarrow \overline{\mathbb{C}}$ be the integrating map, the quasi-conformal homeomorphism for which $\sigma_\theta = \phi_\theta^*(\sigma_0)$, normalized so that the conjugate map

$\phi_\theta \circ F_\theta \circ \phi_\theta^{-1}$ equals the quadratic polynomial $Q_\theta(z) = z^2 + c_\theta$, with an indifferent fixed point of multiplier $e^{i2\pi\theta}$.

The conjugacy $h : \mathbb{S}^1 \longrightarrow \mathbb{S}^1$ between f_θ and R_θ, normalised by $h(1) = 1$ depends continuously on θ in the C^0-topology (uniform convergence), because f_θ and R_θ depend continuously on θ in the C^0-topology.

Thus by choosing the quasi-conformal extension ψ_θ in some canonical way, say by using the Ahlfors-Beurling extension, [LV, Th. 6.3] or the Douady-Earle extension, [DE] we can suppose that also the quasi-conformal extensions ψ_θ depend continuously on θ in the C^0-topology. To fix the ideas we choose to use say the Douady-Earle extension for every θ.

Lemma 4.4 *Let (θ_n) be a sequence of irrationals converging to some irrational θ_0. Suppose the θ_n are of constant type with a uniform constant type bound N. Then θ_0 is of constant type with bound N and the two sequences (ϕ_{θ_n}) and $\phi_{\theta_n}^{-1}$ converge to ϕ_{θ_0} respectively $\phi_{\theta_0}^{-1}$ in the C^0-topology.*

Proof : The coefficients of θ_n converge to those of θ_0 so that θ_0 is of constant type with bound N. We shall prove that ϕ_{θ_n} converges uniformly (C^0) to ϕ_{θ_0}, from which the Lemma follows.

There exists $K(N) \geq 1$, such that each map ϕ_{θ_n} is K-quasi conformal. Thus extracting a subsequence, if necessary we can assume the sequence converge C^0 to a K quasi-conformal homeomorphism ϕ_0. We shall prove that

$$\phi_0 = \phi_{\theta_0}. \tag{12}$$

As (F_{θ_n}) converges to F_{θ_0} in the C^0-topology, and $(Q_{c_{\theta_n}})$ converges to $Q_{c_{\theta_0}}$ in the C^0-topology, the map ϕ_0 conjugates F_{θ_0} to Q_{θ_0} and the restriction of ϕ_0 to the immediate attracted basin of ∞ is biholomorphic. Thus (12) holds on the closure of the immediate attracted basin of ∞, by uniqueness of the holomorphic conjugacy. Moreover (12) also holds on the 'Siegel-disk' $\mathbb{D}$, because the Douady-Earle extension depends continuously on the boundary data h_θ. Finally it holds on the grand orbit of $\mathbb{D}$, because both ϕ_0 and ϕ_{θ_0} conjugates dynamics. Thus any (C^0) limit function of the ϕ_{θ_n} equals ϕ_{θ_0}. Combining this with the precompactness of the sequence (ϕ_{θ_n}) and the compactness of $\overline{\mathbb{C}}$ completes the proof.   q.e.d.

Proof of Theorem 4.2:   Let (ϵ_n) be a summable sequence of strictly positive numbers. Let (d_n) be any unbounded sequence of natural numbers and let θ_1 be any irrational of constant type. Given a natural number k_1 let for each $n \geq 1$ the irrational number $\theta_{1,n}$ be obtained from θ_1 by replacing the $(n + k_1)$-th coefficient b_{n+k_1} in the continued fraction expansion of θ_1 with d_1. Then the sequence $(\theta_{1,n})$ converges to θ_1. Hence by Lemma 4.4 there exists n such that

$$d_{C^0}(\phi_{\theta_1}, \phi_{\theta_{1,n}}) \leq \epsilon_1 \quad \text{and} \quad d_{C^0}(\phi_{\theta_1}^{-1}, \phi_{\theta_{1,n}}^{-1}) \leq \epsilon_1$$

Let $\theta_2 = \theta_{1,n}$ and let $k_2 = k_1 + n$. Then we may restart the process using θ_2, k_2, d_2 and ϵ_2 to obtain an integer $k_3 > k_2$ and an irrational number θ_3 of constant type, with $b_{k_2} = d_1$, $b_{k_3} = d_2$ and with all other coefficients of the continued fraction expansion of θ_3 equal to those of θ_1, such that ϕ_{θ_2} and ϕ_{θ_3} are ϵ_2 close in the C_0 topology. Proceeding recursively we find a sequence (θ_n) of irrational numbers of constant type converging to some irrational of non constant type θ_0 and a Cauchy sequence (in the C^0-norm) of (quasiconformal) homeomorphisms (ϕ_{θ_n}), such that the sequence of inverse maps is also a Cauchy sequence. Let ϕ_0 denote the limit of the sequence (ϕ_{θ_n}), then ϕ_0 is a homeomorphism conjugating F_{θ_0} to Q_{c_0}. In particular $J_{c_{\theta_0}} = \phi_0(J_{\theta_0})$.

q.e.d.

References

[dF]   E. de Faria. Assymptotic rigidity of scaling ratios for critical circle mappings. *Ergod. Th. & Dyn. Sys.* **17**(1998), 227–260.

[dFdM]   E. de Faria and W. de Melo. Rigidity critical circle mappings I. preprint IMS#1997/16, Institute for Mathematical Sciences, Stony Brook, November 1997.

[dFaWdM]   E. de Faria an W. de Melo. Rigidity critical circle mappings II. preprint IMS#1997/17, Institute for Mathematical Sciences, Stony Brook, November 1997.

[Dou]   A. Douady. Disques de Siegel et anneaux de Herman. *Séminaire Bourbaki, Volume 1986–87, exposé no. 677, Astérisque* **153**(1988), 151–172.

[DE]   A. Douady and C. J. Earle. Conformally natural extension of homeomorphisms of the circle. *Acta Math.* **157**(1986), 25–48.

[GS]   J. Graczyk and G. Świątek. Critical circle maps near bifurcation. *Comm. Math. Phys.* **176**(1996), 227–260.

[Her]   M. R. Herman. Conjugaison quasi symétrique des homéomorphismes analytiques du cercle à des rotations. Version très très préliminaire, 19 pages manuscrit, 1987.

[LV]   O. Lehto and K. I. Virtanen. *Quasiconformal mappings in the plane.* Springer Verlag, 1973.

[Pet]   C .L. Petersen. Local connectivity of some Julia sets containing a circle with an irrational rotation. *Acta Math.* **177**(1996), 163–224.

[Swi]   G. Światek. Rational Rotation Numbers for Maps of the Circle. *Comm. Math. Phys.* **119**(1988), 109–128.

[Yam]   M. Yampolsky. Complex bounds for critical circle maps. *Ergod. Th. & Dyn. Sys.* **19**(1999), 227–257.

[Yoc]   J.-C. Yoccoz. Structure des orbites des homéomorphismes analytiques possédant un point critique. 30 pages manuscript, 1987.

IMFUFA, ROSKILDE UNIVERSITY, POSTBOX 260, DK-4000 ROSKILDE, DENMARK. e-mail: lunde@ruc.dk

PERTURBATION D'UNE FONCTION LINÉARISABLE

Habib Jellouli

Résumé :

Dans cette note, on démontre dans un premier temps que si $P(z) = e^{2\pi i \alpha} z + z^2$ est linéarisable en 0 et si $(\frac{p_n}{q_n})_{n \in \mathbb{N}}$ est la suite des réduites de α dans le développement en fraction continue, alors pour n assez grand les q_n pétales en 0 du polynôme quadratique parabolique $P_n(z) = e^{2\pi i \frac{p_n}{q_n}} z + z^2$, sont proches du bord du disque de Siegel de P. Ensuite on considère $f(z) = e^{2\pi i \alpha} z + \mathcal{O}(z^2)$ une fonction holomorphe linéarisable en 0 et $(f_\theta)_{\theta \in I}$ une famille $\mathbb{R}$-analytique de fonctions holomorphes sur le disque de Siegel maximal Δ de f en 0, avec $f_\alpha = f$, $f_\theta(0) = 0$ et $f_\theta'(0) = e^{2\pi i \theta}$ pour tout $\theta \in I$. On démontre que si α est un nombre diophantien alors on peut construire pour $k \in \mathbb{N}$, une famille $\mathbb{R}$-analytique $(\varphi_{k,\theta})_{\theta \in I}$ de fonctions holomorphes tel que $\varphi_{k,\theta}$ conjugue f_θ à sa partie linéaire avec un terme d'erreur $\mathcal{O}((\theta - \alpha)^k)$ uniformément sur tout compact de Δ. Enfin on donne une réduction d'une conjecture de Douady et on utilise la famille $(\varphi_{k,\theta})_{\theta \in I}$ pour donner quelques propriétés sur la conjecture réduite.

Abstract :

In this note, we first prove that if $P(z) = e^{2\pi i \alpha} z + z^2$ is linearizable at 0 and if $(p_n/q_n)_{n \in \mathbb{N}}$ is the sequence of the approximants of α in the development in continued fraction then, for n sufficiently large, the q_n petals at 0 of the parabolic polynomial $P_n(z) = e^{2\pi i (p_n/q_n)} z + z^2$ come close to the boundary of the Siegel disc of P. Later, we consider a linearizable holomorphic function $f(z) = e^{2\pi i \alpha} z + \mathcal{O}(z^2)$ at 0 and $(f_\theta)_{\theta \in I}$ an $\mathbb{R}$-analytic family of holomorphic functions on the maximal Siegel disc Δ of f at 0, with $f_\alpha = f$, $f_\theta(0) = 0$ and $f_\theta'(0) = e^{2\pi i \theta}$ for any $\theta \in I$. We show that if α is diophantine, then we can construct for $k \in \mathbb{N}$, a family of $\mathbb{R}$-analytic holomorphic functions, say $(\varphi_{k,\theta})_{\theta \in I}$ such that $\varphi_{k,\theta}$ conjugates f_θ to its linear part with an error term $\mathcal{O}((\theta - \alpha)^k)$ uniformly on any compact of Δ. Finally we give a reduction of a conjecture by Douady, and we use the family $(\varphi_{k,\theta})_{\theta \in I}$ to give some properties on the reduced conjecture.

0. Introduction

Considérons $f(z) = e^{2\pi i \alpha} z + \mathcal{O}(z^2)$ une fonction holomorphe linéarisable en 0 telle que $\alpha \in \mathbb{R} - \mathbb{Q}$ et désignons par Δ son disque de Siegel maximal en 0. Alors on sait qu'il existe un isomorphisme φ de Δ sur le disque unité $\mathbb{D} = \{z \in \mathbb{C}/|z| < 1\}$ avec $\varphi(0) = 0$ et conjuguant f à la rotation $R_\alpha(z) = e^{2\pi i \alpha} z$ (i.e. $\varphi(f(z)) = R_\alpha(\varphi(z))$ pour tout $z \in \Delta$). On appelle sous-disque de Siegel pour f en 0 tout sous-ensemble Δ' de Δ de la forme $\Delta' = \varphi^{-1}(\mathbb{D}_r)$ où $r \in]0,1]$ et $\mathbb{D}_r = \{z \in \mathbb{C}$ tel que $|z| < r\}$.

Soit maintenant I un intervalle ouvert de $\mathbb{R}$ contenant α et $(f_\theta)_{\theta \in I}$ une famille $\mathbb{R}$-analytique de fonctions holomorphes sur Δ, vérifiant : $f_\alpha = f$; $f_\theta(0) = 0$ et $f'_\theta(0) = e^{2\pi i \theta}$ pour tout $\theta \in I$. Pour $\mathcal{U} \subset \Delta$ et $(\theta, b) \in I \times \mathbb{N}$, on note $\mathcal{U}(\theta, b) = \bigcap_{i=0}^{b} f_\theta^{-i}(\mathcal{U}) = \{z \in \mathcal{U}$ tel que $f_\theta^i(z) \in \mathcal{U}$ pour $1 \le i \le b\}$.

Enfin dans tout ce qui suit on note $R_\theta(z) = e^{2\pi i \theta} z$ et $(\theta_n = \frac{p_n}{q_n})_{n \in \mathbb{N}}$ désigne la suite des réduites de α dans le développement de α en fraction continue.

Theorème 1. *Soit $\Delta_1 \subset \Delta_2$ deux sous-disques de Siegel distincts pour f en 0 et $(b_n)_{n \in \mathbb{N}}$ une suite d'entiers naturels positifs vérifiant $b_n = o(q_n q_{n+1})$.*

i) *si K est un compact dans Δ_1, alors pour n assez grand on a:*

$$K \subset \Delta_1(\theta_n, b_n)$$

ii) *si K est un compact dans $\Delta_2 - \overline{\Delta}_1$, alors pour n assez grand on a :*

$$K \subset (\Delta_2 - \overline{\Delta}_1)(\theta_n, b_n).$$

Un cas intéressant se présente lorsqu'on étudie la famille des polynômes quadratiques $P_\theta(z) = e^{2\pi i \theta} z + z^2$. A ce propos on sait que chaque polynôme quadratique $P_{\theta_n}(z) = e^{2\pi i \frac{p_n}{q_n}} z + z^2$ admet q_n bassins immédiats paraboliques attachés à 0. Comme conséquence du Théorème 1, on a le corollaire suivant:

Corollaire 1. *Si $\Delta' \subsetneq \Delta$ est un sous-disque de Siegel pour $P_\alpha(z) = e^{2\pi i \alpha} z + z^2$, alors il existe un rang $n_0 \in \mathbb{N}$ tel que pour tout $n \ge n_0$, chaque bassin immédiat parabolique B attaché à 0 pour $P_{\theta_n}(z) = e^{2\pi i \frac{p_n}{q_n}} z + z^2$ vérifie $B \cap (\Delta - \overline{\Delta'}) \ne \phi$.*

On dit que α satisfait à une condition diophantienne d'ordre $\beta \ge 0$ s'il existe une constante $c > 0$ telle qu'on ait $|\alpha - \frac{p}{q}| \ge \frac{c}{q^{2+\beta}}$ pour tout $\frac{p}{q} \in \mathbb{Q}$. On dit que α est un nombre diophantien s'il satisfait à une condition diophantienne d'ordre β pour un certain $\beta \ge 0$.

Il est bien connu depuis Siegel que si α est un nombre diophantien, alors tout germe holomorphe $f(z) = e^{2\pi i \alpha} z + \mathcal{O}(z^2)$ est linéarisable en 0.

Théorème 2. *Supposons que α est un nombre diophantien et soit k un entier naturel positif donné. Alors il existe un intervalle ouvert $J \subset I$ contenant α et une famille $\mathbb{R}$-analytique $(\varphi_{k,\theta})_{\theta \in J}$ de fonctions holomorphes sur leurs domaines de définitions notés $Dom(\varphi_{k,\theta})$ et vérifiant les trois propriétés suivantes :*

i) $\varphi_{k,\alpha} = \varphi; 0 \in Dom(\varphi_{k,\theta}); \varphi_{k,\theta}(0) = 0$ *et* $\varphi'_{k,\theta}(0) = \varphi'(0)$ *pour tout* $\theta \in J$.

ii) si K est un compact de Δ alors pour θ assez voisin de α, on a:

$$K \subset Dom(\varphi_{k,\theta})$$

iii) $\varphi_{k,\theta} \circ f_\theta = R_\theta \circ \varphi_{k,\theta} + \mathcal{O}((\theta - \alpha)^k)$ *uniformément sur tout compact de* Δ.

Ainsi $\varphi_{k,\theta}$ conjugue f_θ à sa partie linéaire avec un terme d'erreur $\mathcal{O}((\theta - \alpha)^k)$ uniformément par rapport à θ sur tout compact de Δ. $(\varphi_{k,\theta})_{\theta \in J}$ est appelée famille de coordonnées suivies à l'ordre $k \in \mathbb{N}$ associée à la famille $(f_\theta)_{\theta \in I}$.

Corollaire 2. *Supposons que α est un nombre diophantien et soit k un entier naturel positif donné. Si $(b_n)_{n \in \mathbb{N}}$ est une suite d'éléments de $\mathbb{N}$ vérifiant $b_n = o(q_n q_{n+1})$ alors $\varphi_{k,\theta_n} \circ f^j_{\theta_n} = R^j_{\theta_n} \circ \varphi_{k,\theta_n} + \mathcal{O}((\theta_n - \alpha)^{k-1})$ uniformément sur tout compact de Δ et uniformément par rapport à j avec $0 \le j \le b_n$.*

Notons $P_\theta(z) = e^{2\pi i \theta} z + z^2$; $K(P_\theta) = $ l'ensemble de Julia rempli de P_θ. Le paragraphe 4 de cet article est une tentative pour aborder la conjecture suivante de Douady :

Conjecture. *Si α est un nombre diophantien et $(\theta_n = p_n/q_n)_{n \in \mathbb{N}}$ est la suite des réduites de α, alors on a :*

$$\lim_{n \to +\infty} mes(K(P_{\theta_n}) \cap \overset{\circ}{K}(P_\alpha)) = mes(\overset{\circ}{K}(P_\alpha))$$

(où mes désigne la mesure de Lebesgue).

Le théorème suivant permet de faire une réduction de cette conjecture :

Théorème 3. *Si $P_\alpha(z) = e^{2\pi i \alpha} z + z^2$ est linéarisable en 0 et si $(\alpha_n)_{n \in \mathbb{N}}$ est une suite de nombres réels vérifiant $\lim_{n \to +\infty} \alpha_n = \alpha$, alors les deux propriétés suivantes sont équivalentes :*

i) $\lim_{n \to +\infty} mes(K(P_{\alpha_n}) \cap \overset{\circ}{K}(P_\alpha)) = mes(\overset{\circ}{K}(P_\alpha))$

ii) il existe une composante connexe Ω de $\overset{\circ}{K}(P_\alpha)$ vérifiant:

$$\lim_{n \to +\infty} mes(K(P_{\alpha_n}) \cap \Omega) = mes(\Omega).$$

Il en résulte que si on a $\lim\limits_{n\to+\infty} \mathrm{mes}\,(K(P_{\theta_n}) \cap \Delta') = \mathrm{mes}\,(\Delta')$ pour tout sous-disque de Siegel $\Delta' = \varphi^{-1}(\mathbb{D}_r)$ avec $r \in]0,1[$, alors on aura $\lim\limits_{n\to+\infty} \mathrm{mes}\,(K(P_{\theta_n}) \cap \Delta) = \mathrm{mes}\,(\Delta)$ et par suite la conjecture sera vraie. Notre tentative consiste à conjuguer $P_{\theta_n}^{q_n}$ par la coordonnée suivie φ_{k,θ_n} sur Δ' (la famille $(\varphi_{k,\theta_n})_{n\in\mathbb{N}}$ est donnée par le théorème 2) puis en passant dans le revêtement universel $\mathbb{H} = \{\omega \in \mathbb{C}/\mathrm{R\acute{e}}\ \omega < 0\}$ de $\mathbb{D}_r^* = \{z \in \mathbb{C}/0 < |z| < r\}$ au moyen de $\Pi(\omega) = re^{\frac{\omega}{q_n}}$, on aboutit à une fonction holomorphe, injective, $G_n : \mathbb{H} \to \mathbb{C}$ de la forme $G_n(\omega) = \omega + \lambda_n e^\omega + h_n(\omega)$.

La fonction G_n semble posséder des propriétés très intéressantes ; quelquesunes sont démontrées au paragraphe 4.

1. Préliminaires

1.1. Développement en fraction continue

Si x est un nombre réel on note $[x]$ sa partie entière et $\{x\} = x - [x]$ sa partie décimale. Soit $\alpha \in \mathbb{R} - \mathbb{Q}$. On pose $a_0 = [\alpha] \in \mathbb{Z}$, $\alpha_0 = \{\alpha\} \in [0,1[$; on définit par induction, pour $n \in \mathbb{N}^*$, un entier a_n et un réel α_n par :

$$a_n = \left[\frac{1}{\alpha_{n-1}}\right] \ \text{et}\ \alpha_n = \left\{\frac{1}{\alpha_{n-1}}\right\}.$$

On pose $p_{-2} = q_{-1} = 0$, $p_{-1} = q_{-2} = 1$, et on définit inductivement pour $n \in \mathbb{N}$ des entiers p_n et q_n par les relations $p_n = a_n p_{n-1} + p_{n-2}$ et $q_n = a_n q_{n-1} + q_{n-2}$. Les rationnels $\frac{p_n}{q_n}$, pour $n \in \mathbb{N}$ s'appellent les réduites du nombre irrationnel α. On a alors les propriétés suivantes :

$$p_n \wedge q_n = 1 \quad , \quad \text{pour } n \geq 1,$$

$$p_n q_{n+1} - p_{n+1} q_n = (-1)^{n+1} \quad , \quad \text{pour } n \geq 0,$$

$$|\alpha - \frac{p_n}{q_n}| < \frac{1}{q_n q_{n+1}} \quad , \quad \text{pour } n \geq 0,$$

$$q_{n+1} > q_n > \left(\frac{1 + \sqrt{5}}{2}\right)^{n-1} \quad , \quad \text{pour } n \geq 2.$$

1.2. Disque de Siegel maximal

Soit $f(z) = e^{2\pi i\alpha} z + \mathcal{O}(z^2)$ une fonction holomorphe définie dans un voisinage de 0 telle que $\alpha \in \mathbb{R} - \mathbb{Q}$.

Définition. f est dite linéarisable on 0 s'il existe un domaine simplement connexe $\mathcal{U}$ de $\mathbb{C}$ tels que $0 \in \mathcal{U}$ et f est un isomorphisme de $\mathcal{U}$ sur lui-même.

Dans ces conditions $\mathcal{U}$ est appelé un disque de Siegel pour f en 0.

Lemme. *Si f est linéarisable en 0 alors l'ensemble des disques de Siegel pour f en 0 est ordonné par inclusion.*

En effet : Si $\mathcal{U}$ et V sont deux disques de Siegel pour f en 0 tel que $\mathcal{U} \not\subset V$, alors il existe un point $z_0 \in \mathcal{U}$ avec $z_0 \notin \overline{V}$. Soit ψ un isomorphisme de $\mathcal{U}$ sur le disque unité $\mathbb{D}$ tel que $\psi(0) = 0$. Il en résulte d'après le lemme de Schwarz que $\psi(f(z)) = e^{2\pi i \alpha}\psi(z)$ pour tout $z \in \mathcal{U}$.

Considérons la courbe de Jordan L image réciproque par ψ du cercle de centre 0 et passant par $\psi(z_0)$. On en déduit que L est invariante par f et que l'orbite directe du point z_0 par f est dense dans L (car $\alpha \in \mathbb{R} - \mathbb{Q}$). Par suite $V \cap L = \phi$, ce qui donne $V \subset \mathcal{U}$ car $0 \in V$.

Définition. Si f est linéarisable en 0 alors d'après le lemme précédent, l'union Δ de tous les disques de Siegel pour f en 0 est aussi un disque de Siegel pour f en 0, appelé disque de Siegel maximal pour f en 0.

Remarque. Chaque isomorphisme φ de Δ sur le disque unité $\mathbb{D}$ avec $\varphi(0) = 0$, conjugue f à la rotation $R_\alpha(z) = e^{2\pi i \alpha}z$.

φ est appelé une application linéarisante de f en 0. Si $\widetilde{\varphi}$ est une autre application linéarisante de f en 0 alors il existe $\lambda \in \mathbb{C}$ avec $|\lambda| = 1$ et $\widetilde{\varphi} = \lambda\varphi$.

2. Démonstration du théorème 1

2.1. Démonstration du théorème 1

On peut supposer sans perte de généralité que $\Delta_2 \neq \Delta$. Soit $r_1, r_2 \in]0,1[$ tel que $\Delta_i = \varphi^{-1}(\mathbb{D}_{r_i})$ où $\mathbb{D}_{r_i} = \{z \in \mathbb{C}/|z| < r_i\}$ pour $i \in \{1,2\}$. Posons $\delta = \inf\{|x - y|, x \in \partial\Delta_2$ et $y \in \partial\Delta\} \cdot \delta > 0$ et il existe $\varepsilon > 0$ tel que $|f_\theta(z) - f(z)| \leq \frac{\delta}{2}$ pour tout $z \in \overline{\Delta}_2$ et pour tout θ vérifiant $|\theta - \alpha| \leq \varepsilon$. Du fait que $f(\overline{\Delta}_2) = \overline{\Delta}_2$, on en déduit que $f_\theta(z) \in \Delta$ pour tout $z \in \overline{\Delta}_2$ et $|\theta - \alpha| \leq \varepsilon$.

Ainsi la fonction $F_\theta(z) = (\varphi \circ f_\theta \circ \varphi^{-1})(z)$ est définie sur $\overline{\mathbb{D}}_{r_2}$ pour tout $|\theta - \alpha| \leq \varepsilon$, $\mathbb{R}$-analytique en θ et holomorphe en z. Il est facile de voir qu'on peut écrire $F_\theta(z) = e^{2\pi i \theta}z + (\theta - \alpha)z^2 H_\theta(z)$ pour $|\theta - \alpha| \leq \varepsilon$ et $z \in \overline{\mathbb{D}}_{r_2}$. Posons $c = \sup\{|z^2 H_\theta(z)|, z \in \overline{\mathbb{D}}_{r_2}$ et $|\theta - \alpha| \leq \varepsilon\}$. Comme $\lim_{n\to+\infty} \theta_n = \alpha$, alors il existe $n_0 \in \mathbb{N}$ tel que $|\theta_n - \alpha| \leq \varepsilon$ pour tout $n \geq n_0$, et en utilisant la propriété $|\theta_n - \alpha| < \frac{1}{q_n q_{n+1}}$, on obtient :

$$|z| - \frac{c}{q_n q_{n+1}} < |F_{\theta_n}(z)| < |z| + \frac{c}{q_n q_{n+1}} \tag{1}$$

pour tout $z \in \overline{\mathbb{D}}_{r_2}$ et pour tout $n \geq n_0$.

i) Si K est un compact de Δ_1 alors il existe $r'_1 \in]0, r_1[$ tel que $\varphi(K) \subset \mathbb{D}_{r'_1}$. Par hypothèse, $b_n = o(q_n q_{n+1})$, donc il existe $n_1 \geq n_0$ tel que :

$$\frac{b_n}{q_n q_{n+1}} \leq \frac{r_1 - r'_1}{c} \qquad \text{pour } n \geq n_1.$$

On en déduit d'après (1) que pour $n \geq n_1$ et $|z| < r'_1 + j\frac{r_1 - r'_1}{b_n}$ (avec $0 \leq j \leq b_n - 1$), on a : $|F_{\theta_n}(z)| < r'_1 + (j+1)\frac{r_1 - r'_1}{b_n}$.

Il en résulte que si $z \in \mathbb{D}_{r'_1}$ alors $F_{\theta_n}^j(z) \in \mathbb{D}_{r_1}$ pour $0 \leq j \leq b_n$ et $n \geq n_1$, d'où $K \subset \Delta_1(\theta_n, b_n)$ pour $n \geq n_1$.

ii) Si K est un compact de $\Delta_2 - \overline{\Delta}_1$, alors il existe $r''_1, r''_2 \in]r_1, r_2[$ avec $r''_1 < r''_2$ et $\varphi(K) \subset \mathbb{D}_{r''_2} - \overline{\mathbb{D}}_{r''_1}$.

Comme $b_n = o(q_n q_{n+1})$ alors il existe $n_2 \geq n_0$ tel que

$$\frac{b_n}{q_n q_{n+1}} \leq \min\left(\frac{r_2 - r''_2}{c}, \frac{r''_1 - r_1}{c}\right) \qquad \text{pour } n \geq n_2.$$

On en déduit d'après la double inégalité (1) que pour $n \geq n_2$ et

$$r''_1 - j\frac{r''_1 - r_1}{b_n} < |z| < r''_2 + j\frac{r_2 - r''_2}{b_n} \qquad (\text{avec } 0 \leq j \leq b_n - 1)$$

on a:

$$r''_1 - (j+1)\frac{r''_1 - r_1}{b_n} < |F_{\theta_n}(z)| < r''_2 + (j+1)\frac{r_2 - r''_2}{b_n}.$$

Ainsi si $z \in \mathbb{D}_{r''_2} - \overline{\mathbb{D}}_{r''_1}$ on aura $F_{\theta_n}^j(z) \in \mathbb{D}_{r_2} - \overline{\mathbb{D}}_{r_1}$ pour $0 \leq j \leq b_n$ et $n \geq n_2$, d'où $K \subset (\Delta_2 - \overline{\Delta}_1)(\theta_n, b_n)$ pour $n \geq n_2$.

Ceci achève la démonstration du théorème 1.

2.2. Démonstration du corollaire 1

Soit $\Delta' \subsetneq \Delta$ un sous-disque de Siegel pour le polynôme quadratique linéarisable $P_\alpha(z) = e^{2\pi i \alpha} z + z^2$. Désignons par $B_n^{(0)}$ le bassin immédiat parabolique attaché à 0 pour $P_{\theta_n}(z) = e^{2\pi i \frac{p_n}{q_n}} z + z^2$, et qui contient le point critique de P_{θ_n}.

Le point $x_n = -e^{2\pi i \frac{p_n}{q_n}}$ (vérifiant $P_{\theta_n}(x_n) = 0$) se trouve sur le bord de $B_n^{(0)}$ et pour n assez grand $x_n \notin \overline{\Delta}$ (car $x = -e^{2\pi i \alpha} \notin \overline{\Delta}$). Fixons une courbe de Jordan invariante L pour P_α dans $(\Delta - \overline{\Delta}')$. On en déduit qu'il existe $n_0 \in \mathbb{N}$ tel que $B_n^{(0)} \cap L \neq \phi$ pour $n \geq n_0$. D'autre part, d'après le théorème 1, ii), en prenant $b_n = q_n, \forall n \in \mathbb{N}$, alors il existe $n_1 \geq n_0$ tel que $L \subset (\Delta - \overline{\Delta}')(\theta_n, q_n)$ pour $n \geq n_1$. Or on sait que les q_n bassins immédiats paraboliques attachés à 0 pour P_{θ_n} sont donnés par $B_n^{(j)} = P_{\theta_n}^j(B_n^{(0)})$ avec

$0 \leq j \leq q_n - 1$. En conclusion $B_n^{(j)} \cap (\Delta - \overline{\Delta}') \neq \phi$ pour $n \geq n_1$ et $0 \leq j \leq q_n - 1$. Ainsi le corollaire 1 se trouve démontré.

2.3. Compléments

1) Si $\Delta_1 \subsetneq \Delta$ est un sous-disque de Siegel pour f en 0 et $(b_n)_{n \in \mathbb{N}}$ une suite d'entiers naturels positifs vérifiant $b_n = o(q_n q_{n+1})$, alors il existe un rang $n_0 \in \mathbb{N}$ tel que pour $n \geq n_0$, toute composante connexe de $\Delta_1(\theta_n, b_n)$ est simplement connexe.

En effet : d'après le théorème 1, il existe $n_0 \in \mathbb{N}$ tel que pour $n \geq n_0$, on a $\overline{\Delta}_1 \subset \Delta(\theta_n, b_n)$. On pose $\mathbb{D}_r = \varphi(\Delta_1) = \{z \in \mathbb{C}/|z| < r\}$.

Ainsi pour $n \geq n_0$ et $0 \leq k \leq b_n$, les fonctions $g_{n,k} = \varphi \circ f_{\theta_n}^k \circ \varphi^{-1}$ sont définies et holomorphes sur $\mathbb{D}_r$.

Soit $\mathcal{C}$ une composante connexe de $\Delta_1(\theta_n, b_n)$ pour $n \geq n_0$, et soit Γ un lacet de Jordan dans $\mathcal{C}$. Notons W la composante connexe bornée de $\mathbb{C} - \Gamma$ et $M = \varphi(\overline{W}) \subset \mathbb{D}_r$. Ainsi pour $0 \leq k \leq b_n$ et $z \in \partial M = \varphi(\Gamma)$ on a $|g_{n,k}(z)| < r$. Il en résulte d'après le principe du maximum que pour tout $z \in M$ et pour tout $k \in \{0, 1, 2, \ldots, b_n\}$, on a $|g_{n,k}(z)| < r$. Ceci exprime que $W \subset \Delta_1(\theta_n, b_n)$ et par suite $\mathcal{C}$ est simplement connexe.

2) Il est facile de voir d'après le théorème 1 qu'au sens de la distance de Hausdorff entre compacts de $\mathbb{C}$, $\partial(\Delta_1(\theta_n, b_n))$ (resp. $\partial((\Delta_2 - \overline{\Delta}_1)(\theta_n, b_n))$) tend vers $\partial \Delta_1$ (resp. $\partial \Delta_1 \cup \partial \Delta_2$) lorsque n tend vers $+\infty$.

3. Démonstration du théorème 2

3.1. Préliminaire

Si Ω est un domaine de $\mathbb{C}$, alors on note par $\mathcal{H}(\Omega)$ l'ensemble de toutes les applications holomorphes sur Ω et à valeurs dans $\mathbb{C}$.

Considérons deux domaines $\mathcal{U}, V$ de $\mathbb{C}$ et $\varphi : \mathcal{U} \to V$ une application holomorphe telle que $\varphi'(z) \neq 0$ pour tout $z \in \mathcal{U}$.

On définit l'opérateur linéaire φ^* de $\mathcal{H}(V)$ dans $\mathcal{H}(\mathcal{U})$ par :

$$\varphi^* g = \frac{g \circ \varphi}{\varphi'} \quad \text{pour } g \in \mathcal{H}(V).$$

Et lorsque φ est un isomorphisme de $\mathcal{U}$ sur V, on définit l'opérateur inverse φ_* de $\mathcal{H}(\mathcal{U})$ dans $\mathcal{H}(V)$ par $\varphi_* = (\varphi^{-1})^* = (\varphi^*)^{-1}$.

Propriétés.

i) Si $\varphi : \mathcal{U} \to V$ et $\psi : V \to W$ sont holomorphes et vérifiant $\varphi'(z) \neq 0$ pour $z \in \mathcal{U}$ et $\psi'(\xi) \neq 0$ pour $\xi \in V$, alors $(\psi \circ \varphi)^* = \varphi^* \circ \psi^*$. Si de plus φ et ψ sont des isomorphismes, alors $(\psi \circ \varphi)_* = \psi_* \circ \varphi_*$.

ii) Soit un diagramme commutatif

$$\begin{array}{ccc} \mathcal{U} & \xrightarrow{\varphi} & \mathcal{U} \\ h \downarrow & & \downarrow h \\ V & \xrightarrow{\psi} & V \end{array}$$

Si φ, ψ et h sont holomorphes et vérifiant $\varphi'(z) \neq 0$ pour $z \in \mathcal{U}$, $\psi'(z) \neq 0$ pour $z \in V$ et $h'(z) \neq 0$ pour $z \in \mathcal{U}$, alors on a aussi un diagramme commutatif pour les opérateurs associés :

$$\begin{array}{ccc} \mathcal{H}(V) & \xrightarrow{\psi^*} & \mathcal{H}(V) \\ h^* \downarrow & & \downarrow h^* \\ \mathcal{H}(\mathcal{U}) & \xrightarrow{\varphi^*} & \mathcal{H}(\mathcal{U}) \end{array}$$

En particulier si $g_0 \in \mathcal{H}(V)$ est un point fixe pour ψ^* (i.e. $\psi^* g_0 = g_0$), alors $f_0 = h^* g_0 \in \mathcal{H}(\mathcal{U})$ est un point fixe pour φ^*.

3.2. Lemme. *Soit α un nombre diophantien et $\lambda \in \mathbb{C}$. Considérons $f_\alpha(z) = e^{2\pi i \alpha} z + \mathcal{O}(z^2)$ une fonction holomorphe dans un voisinage de 0 et désignons par Δ son disque de Siegel maximal en 0.*

Alors, pour toute fonction holomorphe $G : \Delta \to \mathbb{C}$ avec $G(0) = G'(0) = 0$, il existe une seule fonction holomorphe $g : \Delta \to \mathbb{C}$ vérifiant $g(0) = 0$, $g'(0) = \lambda$ et $g - f_\alpha^* g = G$.

Démonstration : Soit $\varphi : \Delta \longrightarrow \mathbb{D} = \{z \in \mathbb{C}/|z| < 1\}$ une application linéarisante de f en 0. On en déduit, d'après la propriété ii) ci-dessus, que le diagramme suivant est commutatif

$$\begin{array}{ccc} \mathcal{H}(\mathbb{D}) & \xrightarrow{R_\alpha^*} & \mathcal{H}(\mathbb{D}) \\ \varphi^* \downarrow & & \downarrow \varphi^* \\ \mathcal{H}(\Delta) & \xrightarrow{f_\alpha^*} & \mathcal{H}(\Delta) \end{array}$$

On pose $H = \varphi_* G \in \mathcal{H}(\mathbb{D})$. On a $H(0) = H'(0) = 0$. Ainsi on se ramène à montrer qu'il existe une seule fonction holomorphe $h : \mathbb{D} \to \mathbb{C}$ telle que $h(0) = 0$, $h'(0) = \lambda$ et $h - R_\alpha^* h = H$. Ecrivons $H(z) = \displaystyle\sum_{n \geq 2} a_n z^n$, $h(z) = \displaystyle\sum_{n \geq 0} c_n z^n$. Donc l'égalité $h - R_\alpha^* h = H$ est équivalente à l'égalité $\displaystyle\sum_{n=0}^{+\infty} c_n (1 -$

$$e^{2\pi i(n-1)\alpha})z^n = \sum_{n=2}^{+\infty} a_n z^n \quad \text{ou encore}$$

$$\begin{cases} c_0 = 0 \\ c_1 \text{ quelconque} \\ c_n = \frac{a_n}{1 - e^{2\pi i(n-1)\alpha}} \quad \text{pour } n \geq 2 \ . \end{cases}$$

Soit m_n l'entier relatif vérifiant $|(n-1)\alpha - m_n| < \frac{1}{2}$ pour $n \geq 2$. On a

$$|1 - e^{2\pi i(n-1)\alpha}| = |e^{\pi i(n-1)\alpha} - e^{-\pi i(n-1)\alpha}| = 2|\sin(\pi(n-1)\alpha)|$$
$$= 2|\sin(\pi[(n-1)\alpha - m_n])| \ .$$

En utilisant l'inégalité de Jordan $\frac{2}{\pi}|x| \leq |\sin x|$ pour $|x| \leq \frac{\pi}{2}$, on obtient

$$|1 - e^{2\pi i(n-1)\alpha}| \geq 4|(n-1)\alpha - m_n| = 4(n-1)|\alpha - \frac{m_n}{(n-1)}| \ .$$

Or α est un nombre diophantien, donc il existe $\beta \geq 0$; il existe $M > 0$ tel que $|\alpha - p/q| \geq \frac{M}{q^{2+\beta}}$ pour tout $p/q \in \mathbb{Q}$. Par conséquent, on aura

$$|1 - e^{2\pi i(n-1)\alpha}| \geq \frac{4M}{(n-1)^{1+\beta}} \quad \text{pour } n \geq 2 \ ;$$

d'où

$$|c_n| = \frac{|a_n|}{|1 - e^{2\pi i(n-1)\alpha}|} \leq \frac{(n-1)^{1+\beta}}{4M}|a_n| \quad \text{pour } n \geq 2 \ .$$

Ainsi la fonction $h(z) = \sum_{n \geq 1} c_n z^n$, où c_1 arbitraire et $c_n = \frac{a_n}{1 - e^{2\pi i(n-1)\alpha}}$ pour $n \geq 2$, est holomorphe sur $\mathbb{D}$ et vérifie $h - R_\alpha^* h = H$.

En conclusion, il existe une seule fonction holomorphe $h \in \mathcal{H}(\mathbb{D})$ vérifiant $h(0) = 0$, $h'(0) = \lambda$ (ainsi $c_1 = \lambda$) et $h - R_\alpha^* h = H$. Par suite, le lemme se trouve démontré en considérant $g = \varphi^* h \in \mathcal{H}(\Delta)$.

3.3. Dans ce qui suit, on se place dans les hypothèses et notations du théorème 2: α étant donc un nombre diophantien, $f(z) = e^{2\pi i\alpha}z + \mathcal{O}(z^2)$ une fonction holomorphe dans un voisinage de 0, Δ son disque de Siegel maximal en 0; φ désigne une application linéarisante de f en 0, et enfin $(f_\theta)_{\theta \in I}$ est une famille $\mathbb{R}$-analytique de fonctions holomorphes sur Δ vérifiant $f_\alpha = f$, $f_\theta(0) = 0$ et $f_\theta'(0) = e^{2\pi i\theta}$ pour $\theta \in I$.

3.3.1. Proposition. *Pour tout $k \in \mathbb{N}$, il existe un intervalle ouvert $J \subset I$ contenant α et une famille $\mathbb{R}$-analytique $(g_{k,\theta})_{\theta \in J}$ de fonctions holomorphes sur Δ vérifiant $g_{k,\theta}(0) = 0$, $g'_{k,\theta}(0) = 1$ et $g_{k,\theta} - f^*_\theta g_{k,\theta} = \mathcal{O}((\theta - \alpha)^k)$ uniformément sur tout compact de Δ.*

Démonstration : On a $\varphi \circ f = R_\alpha \circ \varphi$ et $R^*_\alpha(\mathrm{id}_\mathbb{D}) = \mathrm{id}_\mathbb{D}$. Donc

$$g_\alpha = \varphi^*(\mathrm{id}_\mathbb{D}) \quad \text{vérifie} \quad g_\alpha - f^*_\alpha g_\alpha \equiv 0 \ . \tag{2}$$

La fonction $f^*_\theta g_\alpha$ est définie et holomorphe sur l'ensemble $\Delta^*(\theta) = \{z \in \Delta$ tel que $f_\theta(z) \in \Delta$ et $f'_\theta(z) \neq 0\}$. Remarquons que $\Delta^*(\alpha) = \Delta$ et si K est un compact de Δ, alors $K \subset \Delta^*(\theta)$ pour θ assez voisin de α. On en déduit d'après (2) que $g_\alpha - f^*_\theta g_\alpha = \mathcal{O}((\theta - \alpha))$ uniformément sur tout compact de Δ et par suite la proposition est vraie pour $k = 0$ et $k = 1$ avec $g_{0,\theta} = g_{1,\theta} = g_\alpha = \frac{\varphi}{\varphi'}$. Nous allons continuer la preuve par récurrence sur $k \in \mathbb{N}$.

Supposons qu'on a $g_{k,\theta} - f^*_\theta g_{k,\theta} = (\theta - \alpha)^k G_{k,\theta}$ où $(g_{k,\theta})_{\theta \in J}$ est $\mathbb{R}$-analytique, holomorphe sur Δ, $g_{k,\theta}(0) = 0$, $g'_{k,\theta}(0) = 1$ et $G_{k,\theta} : \Delta^*(\theta) \to \mathbb{C}$, holomorphe sur $\Delta^*(\theta)$, $\mathbb{R}$-analytique en $\theta \in J$.

En particulier, $G_{k,\alpha} : \Delta \to \mathbb{C}$ vérifie $G_{k,\alpha}(0) = G'_{k,\alpha}(0) = 0$; par suite, d'après le lemme 3.2 appliqué à $G = -G_{k,\alpha}$, il existe une seule fonction holomorphe $h_{k,\alpha} : \Delta \to \mathbb{C}$ tel que $h_{k,\alpha}(0) = 0$, $h'_{k,\alpha}(0) = 1$ et $h_{k,\alpha} - f^*_\alpha h_{k,\alpha} = -G_{k,\alpha}$. D'où $G_{k,\theta} + (h_{k,\alpha} - f^*_\theta h_{k,\alpha}) = \mathcal{O}((\theta - \alpha))$ uniformément sur tout compact de Δ. Posons

$$g_{k+1,\theta} = \frac{1}{1 + (\theta - \alpha)^k} \left[g_{k,\theta} + (\theta - \alpha)^k h_{k,\alpha} \right] , \tag{3}$$

pour θ dans un petit intervalle ouvert $J' \subset J$ contenant α. On a

$$g_{k+1,\theta} - f^*_\theta g_{k+1,\theta} = \frac{1}{1 + (\theta - \alpha)^k} \left[g_{k,\theta} - f^*_\theta g_{k,\theta} + (\theta - \alpha)^k (h_{k,\alpha} - f^*_\theta h_{k,\alpha}) \right]$$

$$= \frac{1}{1 + (\theta - \alpha)^k} \left[(\theta - \alpha)^k G_{k,\theta} + (\theta - \alpha)^k (\mathcal{O}((\theta - \alpha)) - G_{k,\theta}) \right]$$

$$= \mathcal{O}((\theta - \alpha)^{k+1}) \text{ uniformément sur tout compact de } \Delta \ .$$

D'autre part $g_{k+1,\theta}(0) = 0$ et $g'_{k+1,\theta}(0) = 1$.
Ceci achève la démonstration de la proposition.

3.3.2. Corollaire. *Soit $k \in \mathbb{N}$ et $(g_{k,\theta})_{\theta \in J}$ la famille $\mathbb{R}$-analytique de fonctions holomorphes sur Δ donnée par la proposition 3.3.1. Si Δ' est un sous-disque de Siegel distinct de Δ, alors il existe un intervalle ouvert*

$J' \subset J$ *contenant* α *tel que, pour tout* $\theta \in J'$, *il existe une seule fonction* $\varphi_{k,\theta} : \Delta' \to \mathbb{C}$, *holomorphe sur* Δ' *et vérifiant:*

i) $\varphi_{k,\theta}(0) = 0$, $\varphi'_{k,\theta}(0) = \varphi'(0)$, $\varphi_{k,\theta}(z) \neq 0$ *pour* $z \in \Delta' - \{0\}$;

ii) $g_{k,\theta}(z)\varphi'_{k,\theta}(z) = \varphi_{k,\theta}(z)$ *pour tout* $z \in \Delta'$.

De plus, la famille $(\varphi_{k,\theta})_{\theta \in J'}$ *est* $\mathbb{R}$-*analytique.*

La démonstration de ce corollaire nécessite le lemme préparatif ci-dessous dont l'énoncé est dans un cadre plus général et dont on aura besoin plus tard dans la démonstration du théorème 2.

3.3.3. Lemme. *Soient* $\mathcal{U}$ *un domaine simplement connexe de* $\mathbb{C}$, $z_0 \in \mathcal{U}$ *et* $\lambda \in \mathbb{C}^*$. *Si* $h : \mathcal{U} \to \mathbb{C}$ *est une fonction holomorphe avec* $h(z_0) = 0$, $h'(z_0) = 1$ *et* $h(z) \neq 0$ *pour* $z \in \mathcal{U} - \{z_0\}$, *alors il existe une seule fonction* $\psi : \mathcal{U} \to \mathbb{C}$ *holomorphe sur* $\mathcal{U}$ *et vérifiant:*

i) $\psi(z_0) = 0$, $\psi'(z_0) = \lambda$, $\psi(z) \neq 0$ *pour* $z \in \mathcal{U} - \{z_0\}$;

ii) $h(z)\psi'(z) = \psi(z)$ *pour tout* $z \in \mathcal{U}$.

De plus, si h *varie dans une famille* $\mathbb{R}$-*analytique* $(h_\theta)_{\theta \in I}$, *alors la famille* $(\psi_\theta)_{\theta \in I}$ *associée est aussi* $\mathbb{R}$-*analytique.*

Démonstration :

— $\mathcal{U}$ étant simplement connexe, par suite la fonction

$$H(z) = \int_{z_0}^{z} \frac{(\xi - z_0) - h(\xi)}{(\xi - z_0)h(\xi)} d\xi$$

est définie, holomorphe sur $\mathcal{U}$ et on vérifie facilement que la fonction ψ définie sur $\mathcal{U}$ par $\psi(z) = \lambda(z - z_0)e^{H(z)}$ est solution du lemme 3.3.3.

— Unicité: supposons que $\widetilde{\psi} : \mathcal{U} \to \mathbb{C}$ est solution du lemme. On pose $h(z) = (z - z_0)h_1(z)$ et $\widetilde{\psi}(z) = (z - z_0)\widetilde{\psi}_1(z)$ donc $h_1(z) \neq 0$ et $\widetilde{\psi}_1(z) \neq 0$ pour tout $z \in \mathcal{U}$. On aura

$$h(z)\widetilde{\psi}'(z) = \widetilde{\psi}(z) \qquad \text{pour tout } z \in \mathcal{U} \qquad\qquad \Leftrightarrow$$

$$(z - z_0)h_1(z)\left[\widetilde{\psi}_1(z) + (z - z_0)\widetilde{\psi}'_1(z)\right]$$
$$= (z - z_0)\widetilde{\psi}_1(z) \qquad \text{pour tout } z \in \mathcal{U} \qquad\qquad \Leftrightarrow$$

$$\frac{\widetilde{\psi}'_1(z)}{\widetilde{\psi}_1(z)} = \frac{1 - h_1(z)}{(z - z_0)h_1(z)} \qquad \text{pour tout } z \in \mathcal{U} - \{z_0\} \qquad \Leftrightarrow$$

$$\frac{\widetilde{\psi}'_1(z)}{\widetilde{\psi}_1(z)} = \frac{(z - z_0) - h(z)}{(z - z_0)h(z)} \qquad \text{pour tout } z \in \mathcal{U} \qquad\qquad \Leftrightarrow$$

$$\frac{\widetilde{\psi}'_1(z)}{\widetilde{\psi}_1(z)} = H'(z) \qquad \text{pour tout } z \in \mathcal{U} \qquad\qquad \Leftrightarrow$$

$$\left(\widetilde{\psi}_1(z)e^{-H(z)}\right)' = 0 \qquad \text{pour tout } z \in \mathcal{U} \qquad\qquad \Leftrightarrow$$

Il existe $c \in \mathbb{C}^* = \mathbb{C} - \{0\}$ tel que

$$\widetilde{\psi}_1(z) = c e^{H(z)} \qquad \text{pour tout } z \in \mathcal{U} \qquad \Leftrightarrow$$

$$\widetilde{\psi}(z) = c(z - z_0) e^{H(z)} \qquad \text{pour tout } z \in \mathcal{U} \ .$$

Mais $\widetilde{\psi}'(z_0) = \lambda$ donne $c = \lambda$, et par suite $\widetilde{\psi} = \psi$.

— Si $(h_\theta)_{\theta \in I}$ est une famille $\mathbb{R}$-analytique de fonctions holomorphes sur $\mathcal{U}$ avec $h_\theta(z_0) = 0$, $h'_\theta(z_0) = 1$ et $h_\theta(z) \neq 0$ pour $z \in \mathcal{U} - \{z_0\}$, alors pour $\theta \in I$

$$\psi_\theta(z) = \lambda(z - z_0) \exp\left(\int_{z_0}^z \frac{(\xi - z_0) - h_\theta(\xi)}{(\xi - z_0) h_\theta(\xi)} d\xi \right) \text{ pour tout } z \in \mathcal{U} \ .$$

Il est clair à partir de cette formule que la famille $(\psi_\theta)_{\theta \in I}$ est aussi $\mathbb{R}$-analytique.

Démonstration du corollaire 3.3.2. D'après la formule (3) dans la démonstration de la proposition 3.3.1, on a:

$$g_{k,\alpha} = g_{k-1,\alpha} = \cdots\cdots = g_{1,\alpha} = g_{0,\alpha} = g_\alpha = \frac{\varphi}{\varphi'} \ .$$

On pose $\delta = \min\limits_{z \in \partial\Delta'} |\frac{\varphi(z)}{\varphi'(z)}| > 0$.

Comme $\lim\limits_{\theta \to \alpha} g_{k,\theta} = g_{k,\alpha} = g_\alpha$ uniformément sur tout compact de Δ, alors il existe un intervalle ouvert $J' \subset J$ contenant α tel que

$$|g_{k,\theta}(z) - g_\alpha(z)| < \delta \leq |g_\alpha(z)| \quad \text{pour } z \in \partial\Delta' \text{ et } \theta \in J' \ .$$

On en déduit, d'après le théorème de Rouché, que pour tout $\theta \in J'$, on a $g_{k,\theta}(z) \neq 0$ pour $z \in \Delta' - \{0\}$.

Maintenant, il suffit d'appliquer le lemme 3.3.3 à $\mathcal{U} = \Delta'$, $z_0 = 0$, $\lambda = \varphi'(0)$ et $h_\theta = g_{k,\theta}$ pour $\theta \in J'$. La fonction $\varphi_{k,\theta} : \Delta' \to \mathbb{C}$ est donc donnée par

$$\varphi_{k,\theta}(z) = \varphi'(0) z \exp\left(\int_0^z \frac{\xi - g_{k,\theta}(\xi)}{\xi g_{k,\theta}(\xi)} d\xi \right) \ .$$

Nous sommes maintenant en mesure de donner la démonstration du théorème 2 énoncé dans l'introduction.

3.4. Démonstration du théorème 2.

— Soit $\Delta' \subset \widetilde{\Delta}'$ deux sous-disques de Siegel distincts de Δ. D'après le corollaire 3.3.2, il existe un intervalle ouvert $J' \subset I$ (resp. $\widetilde{J}' \subset I$) contenant α et une famille $(\varphi_{k,\theta})_{\theta \in J'}$ (resp. $(\widetilde{\varphi}_{k,\theta})_{\theta \in \widetilde{J}'}$) vérifiant les propriétés du corollaire 3.3.2.

Comme $\Delta' \subset \tilde{\Delta}'$, alors d'après l'unicité, on aura $\tilde{\varphi}_{k,\theta}/\Delta' = \varphi_{k,\theta}$ pour $\theta \in J' \cap \tilde{J}'$. En d'autres termes, pour tout compact K de Δ, on a $K \subset \text{Dom}(\varphi_{k,\theta})$ pour θ assez voisin de α. Ceci démontre ii).

— D'après le corollaire 3.3.2, la famille $(\varphi_{k,\theta})_{\theta \in J'}$ vérifie $0 \in \text{Dom}(\varphi_{k,\theta})$, $\varphi_{k,\theta}(0) = 0$ et $\varphi'_{k,\theta}(0) = \varphi'(0)$ pour $\theta \in J'$. D'autre part, on a $g_{k,\alpha} = g_\alpha = \frac{\varphi}{\varphi'}$. Ceci donne $g_{k,\alpha}(z)\varphi'(z) = \varphi(z)$ pour tout $z \in \Delta$. Il en résulte d'après l'unicité dans le corollaire 3.3.2 que $\varphi_{k,\alpha} = \varphi$. Ceci démontre i).

— Pour achever la démonstration du théorème 2, il reste à vérifier iii).

Soit K un compact de Δ. Considérons deux sous-disques de Siegel Δ', Δ'' vérifiant $K \subset \Delta' \subsetneq \Delta'' \subsetneq \Delta$. Alors, il existe $\varepsilon > 0$ tel que, pour tout $\theta \in]\alpha - \varepsilon, \alpha + \varepsilon[\subset J'$, on a:

a) $\Delta'' \subset \text{Dom}(\varphi_{k,\theta})$;

b) $\overline{\Delta}' \subset \{z \in \Delta'' \text{ tels que } f_\theta(z) \in \Delta'' \text{ et } f'_\theta(z) \neq 0\}$;

c) $g_{k,\theta}(z) \neq 0$ pour $z \in \Delta'' - \{0\}$.

Ainsi on a, pour $\theta \in]\alpha - \varepsilon, \alpha + \varepsilon[$, $g_{k,\theta}(z)\varphi'_{k,\theta}(z) = \varphi_{k,\theta}(z)$ pour $z \in \Delta''$; en particulier $g_{k,\theta}(f_\theta(z))\varphi'_{k,\theta}(f_\theta(z)) = \varphi_{k,\theta}(f_\theta(z))$ pour $z \in \Delta'$. Donc, d'après b) ci-dessus, on obtient

$$\frac{g_{k,\theta}(f_\theta(z))}{f'_\theta(z)} \cdot \left(\varphi_{k,\theta} \circ f_\theta\right)'(z) = \left(\varphi_{k,\theta} \circ f_\theta\right)(z) \quad \text{pour } z \in \Delta' \ ;$$

d'où $\left(f_\theta^* g_{k,\theta}\right)(z) \cdot \left(\varphi_{k,\theta} \circ f_\theta\right)'(z) = \left(\varphi_{k,\theta} \circ f_\theta\right)(z)$ pour $z \in \Delta'$. Or la fonction $z \mapsto (f_\theta^* g_{k,\theta})(z)$ est holomorphe sur Δ', et vérifie $\left(f_\theta^* g_{k,\theta}\right)(0) = 0$, $\left(f_\theta^* g_{k,\theta}\right)'(0) = 1$ et $\left(f_\theta^* g_{k,\theta}\right)(z) \neq 0$ pour $z \in \Delta' - \{0\}$. Il en résulte, d'après le lemme 3.3.3, que l'on a pour $z \in \Delta'$

$$\left(\varphi_{k,\theta} \circ f_\theta\right)(z) = e^{2\pi i\theta}\varphi'(0)z\exp\left(\int_0^z \frac{\xi - (f_\theta^* g_{k,\theta})(\xi)}{\xi(f_\theta^* g_{k,\theta})(\xi)} d\xi\right) \ .$$

D'autre part

$$\int_0^z \frac{\xi - (f_\theta^* g_{k,\theta})(\xi)}{\xi(f_\theta^* g_{k,\theta})(\xi)} d\xi = \int_0^z \frac{\xi - g_{k,\theta}(\xi)}{\xi g_{k,\theta}(\xi)} d\xi + \int_0^z \frac{g_{k,\theta}(\xi) - (f_\theta^* g_{k,\theta})(\xi)}{g_{k,\theta}(\xi)(f_\theta^* g_{k,\theta})(\xi)} d\xi \ .$$

D'après la proposition 3.3.1, on peut écrire $g_{k,\theta}(\xi) - (f_\theta^* g_{k,\theta})(\xi) = (\theta - \alpha)^k G_{k,\theta}(\xi)$, où $G_{k,\theta}$ holomorphe en $\xi \in \Delta^*(\theta) = \{\xi \in \Delta/f_\theta(\xi) \in \Delta$ et $f'_\theta(\xi) \neq 0\}$ et $\mathbb{R}$-analytique en θ. On pose

$$F_{k,\theta}(z) = \int_0^z \frac{\xi - g_{k,\theta}(\xi)}{\xi g_{k,\theta}(\xi)} d\xi \ , \quad H_{k,\theta}(z) = \int_0^z \frac{G_{k,\theta}(\xi)}{g_{k,\theta}(\xi)(f_\theta^* g_{k,\theta})(\xi)} d\xi \ .$$

On aura

$$\left(\varphi_{k,\theta} \circ f_\theta\right)(z) = e^{2\pi i \theta}\varphi'(0)z\, e^{F_{k,\theta}(z)} \cdot e^{(\theta-\alpha)^k H_{k,\theta}(z)} \; ;$$

$$\left(\varphi_{k,\theta} \circ f_\theta\right)(z) = \left(R_\theta \circ \varphi_{k,\theta}\right)(z) \cdot e^{(\theta-\alpha)^k H_{k,\theta}(z)} \; ;$$

$$\left(\varphi_{k,\theta} \circ f_\theta\right)(z) = (R_\theta \circ \varphi_{k,\theta})(z) + \left(e^{(\theta-\alpha)^k H_{k,\theta}(z)} - 1\right) \cdot (R_\theta \circ \varphi_{k,\theta})(z) \quad (4)$$

et on a $\displaystyle\lim_{t\to 0}\frac{e^t - 1}{t} = 1$, donc il existe $\mu > 0$ tel que $|t| < \mu$ implique $|e^t - 1| \le 2|t|$. Les fonctions $R_\theta \circ \varphi_{k,\theta}$ et $H_{k,\theta}$ sont uniformément bornées sur le compact K, par suite il existe $\varepsilon' > 0$ ($\varepsilon' < \varepsilon$), il existe une constante $M > 0$ tels que $|(R_\theta \circ \varphi_{k,\theta})(z)| \le M$, $|H_{k,\theta}(z)| \le M$ et $|(\theta-\alpha)^k H_{k,\theta}(z)| < \mu$ pour $z \in K$ et $\theta \in]\alpha - \varepsilon', \alpha + \varepsilon'[$. On en déduit, d'après (4), $\left(\varphi_{k,\theta} \circ f_\theta\right)(z) = \left(R_\theta \circ \varphi_{k,\theta}\right)(z) + \mathcal{O}((\theta-\alpha)^k)$ uniformément sur K. Ainsi, le théorème 2 se trouve démontré.

3.5. Démonstration du corollaire 2. Soit K un compact de Δ. Considérons un sous-disque de Siegel Δ' avec $K \subset \Delta' \subsetneq \Delta$.

D'après le théorème 2, il existe $\varepsilon > 0$ et $M > 0$ tels que, pour $\theta \in]\alpha - \varepsilon, \alpha + \varepsilon[$, on a $\varphi_{k,\theta} \circ f_\theta = R_\theta \circ \varphi_{k,\theta} + (\theta - \alpha)^k H_{k,\theta}$, où $|H_{k,\theta}(z)| \le M$ pour $z \in \overline{\Delta}'$.

D'autre part, d'après le théorème 1 (point i)), il existe $n_0 \in \mathbb{N}$ tel que $K \subset \Delta'(\theta_n, b_n)$ et $|\theta_n - \alpha| < \varepsilon$ pour $n \ge n_0$. On en déduit que, pour $0 \le j \le b_n$ et $n \ge n_0$, on a sur K les égalités suivantes:

$$\varphi_{k,\theta_n} \circ f_{\theta_n} = R_{\theta_n} \circ \varphi_{k,\theta_n} + (\theta_n - \alpha)^k H_{k,\theta_n}$$

$$\varphi_{k,\theta_n} \circ f_{\theta_n}^2 = R_{\theta_n}^2 \circ \varphi_{k,\theta_n} + (\theta_n - \alpha)^k \left[R_{\theta_n} \circ H_{k,\theta_n} + H_{k,\theta_n} \circ f_{\theta_n}\right]$$

$$\dots\dots\dots$$

$$\varphi_{k,\theta_n} \circ f_{\theta_n}^j = R_{\theta_n}^j \circ \varphi_{k,\theta_n} + (\theta_n - \alpha)^k \sum_{i=1}^{j} \left(R_{\theta_n}^{(j-i)} \circ H_{k,\theta_n} \circ f_{\theta_n}^{(i-1)}\right)$$

Du fait que $|H_{k,\theta_n}(z)| \le M$ pour $z \in \overline{\Delta}'$ et $f_{\theta_n}^j(z) \in \Delta'$ pour $0 \le j \le b_n$ et $z \in K$, il en résulte qu'on a

$$\left|\sum_{i=1}^{j} \left(R_{\theta_n}^{(j-i)} \circ H_{k,\theta_n} \circ f_{\theta_n}^{(i-1)}\right)(z)\right| \le jM \le b_n M \; . \quad (5)$$

Or, par hypothèse, $b_n = o(q_n q_{n+1})$; donc il existe $\lambda > 0$ tel que $b_n \leq \lambda q_n q_{n+1}$ pour $n \in \mathbb{N}$. Maintenant, d'après le préliminaire 1.1, on a $q_n q_{n+1} < |\theta_n - \alpha|^{-1}$. Ainsi on aura

$$|\sum_{i=1}^{j} \left(R_{\theta_n}^{(j-i)} \circ H_{k,\theta_n} \circ f_{\theta_n}^{(i-1)} \right)(z)| \leq \lambda M |\theta_n - \alpha|^{-1} \text{ pour } z \in K \text{ et } n \geq n_0 \; .$$

Soit donc $|\varphi_{k,\theta_n} \circ f_{\theta_n}^j - R_{\theta_n}^j \circ \varphi_{k,\theta_n}| \leq \lambda M |\theta_n - \alpha|^{k-1}$ pour $z \in K$, $n \geq n_0$ et $0 \leq j \leq b_n$. Ceci achève la démonstration du corollaire 2.

3.6. Remarque. L'hypothèse $b_n = q_n$ pour tout $n \in \mathbb{N}$ permet d'améliorer légèrement la conclusion du corollaire 2. En effet, on a $|\theta_n - \alpha| < \frac{1}{q_n q_{n+1}} < \frac{1}{q_n^2}$, donc $b_n = q_n < |\theta_n - \alpha|^{-1/2}$. Par suite, d'après l'inégalité (5) ci-dessus, on obtient $\varphi_{k,\theta_n} \circ f_{\theta_n}^j = R_{\theta_n}^j \circ \varphi_{k,\theta_n} + \mathcal{O}((\theta_n - \alpha)^{k-\frac{1}{2}})$ uniformément sur tout compact de Δ et uniformément par rapport à j avec $0 \leq j \leq q_n$.

4. Remarques sur la Conjecture

Rappelons la conjecture de Douady déjà énoncée dans l'introduction :

Conjecture. Si α est un nombre diophantien et si $(\theta_n = p_n/q_n)_{n \in \mathbb{N}}$ est la suite des réduites de α, alors $\lim_{n \to +\infty} \text{mes}(K(P_{\theta_n}) \cap \overset{\circ}{K}(P_\alpha)) = \text{mes}(\overset{\circ}{K}(P_\alpha))$ où on a noté $P_\theta(z) = e^{2\pi i \theta} z + z^2$.

Dans ce qui suit, on démontre le théorème 3 (qui permet de réduire la conjecture ci-dessus), et on donne des motivations pour aborder la conjecture réduite.

4.1. Démonstration du théorème 3
Commençons par établir les deux lemmes préparatifs ci-dessous et dont on aura besoin encore plus tard.

4.1.1. Lemme. *Soient $\mathcal{U}$ un domaine simplement connexe de $\mathbb{C}$ (autre que $\mathbb{C}$), $\omega_0 \in \mathcal{U}$ et $f : \mathcal{U} \to \mathbb{C}$ holomorphe et injective sur $\mathcal{U}$. Si K est un compact dans $\mathcal{U}$, alors il existe une constante $\mu \geq 1$ (indépendante de f) tel que $\frac{1}{\mu} |f'(\omega_0)| \leq |f'(\omega)| \leq \mu |f'(\omega_0)|$ pour $\omega \in K$.*

Démonstration : Considérons un isomorphisme ψ du disque unité $\mathbb{D} = \{z \in \mathbb{C}/|z| < 1\}$ sur $\mathcal{U}$ tel que $\psi(0) = \omega_0$.

Ainsi les deux fonctions ψ et $g = f \circ \psi$ sont holomorphes, injectives sur $\mathbb{D}$, par suite d'après le théorème de distorsion de Koebe ([Po], p. 21), on obtient pour $z \in \mathbb{D}$:

$$\frac{1-|z|}{(1+|z|)^3} \leq |\frac{\psi'(z)}{\psi'(0)}| \leq \frac{1+|z|}{(1-|z|)^3} \quad \text{et} \quad \frac{1-|z|}{(1+|z|)^3} \leq |\frac{g'(z)}{g'(0)}| \leq \frac{1+|z|}{(1-|z|)^3} \; .$$

On en déduit que pour $z \in \mathbb{D}$:

$$\left(\frac{1-|z|}{1+|z|}\right)^4 |f'(\omega_0)| \le |f'(\psi(z))| \le \left(\frac{1+|z|}{1-|z|}\right)^4 |f'(\omega_0)|.$$

Or $\psi^{-1}(K)$ est un compact dans $\mathbb{D}$, donc il existe $r \in [0,1[$ tel que $\psi^{-1}(K) \subset \overline{\mathbb{D}}_r = \{z \in \mathbb{C}/|z| \le r\}$. D'où on aura :

$$\left(\frac{1-r}{1+r}\right)^4 |f'(\omega_0)| \le |f'(\omega)| \le \left(\frac{1+r}{1-r}\right)^4 |f'(\omega_0)| \quad \text{pour } \omega \in K.$$

Il suffit de prendre $\mu = \left(\frac{1+r}{1-r}\right)^4$ et le lemme se trouve démontré.

4.1.2. Lemme. *Soient $\mathcal{U}, \mathcal{V}$ deux domaines simplements connexes de $\mathbb{C}$, ψ un isomorphisme de $\mathcal{U}$ sur $\mathcal{V}$ et (ψ_n) une suite de fonctions holomorphes sur $\mathcal{U}$, qui converge vers ψ uniformément sur tout compact de $\mathcal{U}$.*

Si K est un compact dans $\mathcal{V}$ et $\mathcal{U}_1$ est un domaine simplement connexe délimité par un lacet de Jordan dans $\mathcal{U}$ sans points doubles, vérifiant $K \subset \psi(\mathcal{U}_1)$; alors il existe un rang $n_0 \in \mathbb{N}$ tel que pour $n \ge n_0$, on a :

    i)   *ψ_n est injective sur $\mathcal{U}_1$;*

    ii)   *$K \subset \psi_n(\mathcal{U}_1)$.*

Démonstration : Considérons deux domaines simplement connexes $\mathcal{U}_2$, $\mathcal{U}_3$ de $\mathcal{U}$ avec $\overline{\mathcal{U}}_1 \subset \mathcal{U}_2$; $\overline{\mathcal{U}}_2 \subset \mathcal{U}_3$; $\overline{\mathcal{U}}_3 \subset \mathcal{U}$ et $\mathcal{U}_3$ délimité par un lacet de Jordan sans points doubles. On pose

$$\delta_1 = \inf\{|\psi(z) - \psi(z')|; z \in \overline{\mathcal{U}}_1 \text{ et } z' \in \mathcal{U} - \mathcal{U}_2\}$$
$$\delta_2 = \inf\{|\psi(z) - \omega|; z \in \partial\mathcal{U}_3 \text{ et } \omega \in \psi(\overline{\mathcal{U}}_2)\}$$
$$\delta_3 = \inf\{|\psi(z) - \omega|; z \in \partial\mathcal{U}_1 \text{ et } \omega \in K\}.$$

Alors il existe un rang $n_0 \in \mathbb{N}$ tel que pour tout $n \ge n_0$, on a :

$$|\psi_n(z) - \psi(z)| \le \frac{\delta_1}{2} \quad \text{pour } z \in \overline{\mathcal{U}}_1 \tag{6}$$

$$|\psi_n(z) - \psi(z)| \le \frac{\delta_2}{2} \quad \text{pour } z \in \partial\mathcal{U}_3 \tag{7}$$

$$|\psi_n(z) - \psi(z)| \le \frac{\delta_3}{2} \quad \text{pour } z \in \partial\mathcal{U}_1 \tag{8}$$

Il en découle d'après (7) et (8) que pour $n \ge n_0$, on a :

$$|\psi_n(z) - \psi(z)| < |\psi(z) - \omega| \quad \text{pour } z \in \partial\mathcal{U}_3 \text{ et } \omega \in \psi(\overline{\mathcal{U}}_2);$$
$$|\psi_n(z) - \psi(z)| < |\psi(z) - \omega| \quad \text{pour } z \in \partial\mathcal{U}_1 \text{ et } \omega \in K.$$

On en déduit d'après le théorème de Rouché et le fait que ψ est un isomorphisme que pour $n \geq n_0$, on a :

pour tout $\omega \in \psi(\overline{\mathcal{U}}_2)$, il existe un seul $z \in \mathcal{U}_3$ tel que $\psi_n(z) = \omega$, $\qquad$ (9)

pour tout $\omega \in K$, il existe un seul $z \in \mathcal{U}_1$ tel que $\psi_n(z) = \omega$. $\qquad$ (10)

En particulier (10) implique que $K \subset \psi_n(\mathcal{U}_1)$ pour $n \geq n_0$, d'où ii) est démontré.

Pour $n \geq n_0$, ψ_n est injective sur $\mathcal{U}_1$, en effet :

Soit $z_1, z_2 \in \mathcal{U}_1$ tel que $\psi_n(z_1) = \psi_n(z_2) = \omega \in \psi_n(\mathcal{U}_1)$. D'après (6), on a $\psi_n(\overline{\mathcal{U}}_1) \subset \psi(\mathcal{U}_2)$, et on a $\mathcal{U}_1 \subset \mathcal{U}_3$ par hypothèse. Ainsi on obtient $z_1, z_2 \in \mathcal{U}_3$, $\omega \in \psi(\overline{\mathcal{U}}_2)$ avec $\psi_n(z_1) = \psi_n(z_2) = \omega$. Il en résulte de (9) que $z_1 = z_2$ et donc i) est vérifié.

Ceci achève la démonstration du lemme.

4.1.3. Démonstration du théorème 3.

* Il est immédiat que i) $\Rightarrow$ ii).

* Montrons que ii) $\Rightarrow$ i) :

Désignons par Δ le disque de Siegel maximal pour P_α en 0 et soit Ω une composante connexe de $\overset{\circ}{K}(P_\alpha)$.

On sait qu'il existe $k \in \mathbb{N}$ tel que la restriction g_α de P_α^k à Ω est un isormorphisme de Ω sur Δ. Posons $\omega_0 = g_\alpha^{-1}(0) \in \Omega$. Soit Δ' un sous-disque de Siegel arbitraire dans Δ avec $\Delta' \neq \Delta$ et posons $\Omega' = g_\alpha^{-1}(\Delta') \subset \Omega$.

Considérons deux sous-disques de Siegel auxiliaires Δ'', Δ''' de Δ tel que $\Delta' \subsetneqq \Delta'' \subsetneqq \Delta''' \subsetneqq \Delta$ et posons $\Omega'' = g_\alpha^{-1}(\Delta'')$, $\Omega''' = g_\alpha^{-1}(\Delta''')$.

La suite de fonctions $(P_{\alpha_n}^k)_{n \in \mathbb{N}}$ converge vers g_α uniformément sur tout compact de Ω. On en déduit d'après le lemme 4.1.2 qu'il existe un rang $n_0 \in \mathbb{N}$ tel que pour $n \geq n_0$, on a :

$$P_{\alpha_n}^k \text{ est injective sur } \Omega'''; \qquad (11)$$

$$\Delta' \subset P_{\alpha_n}^k(\Omega'') \subset \Delta . \qquad (12)$$

D'après (11), on peut appliquer le lemme 4.1.1 à $P_{\alpha_n}^k$ sur Ω''', par suite il existe une constante $\mu \geq 1$ (indépendante de n) tel que :

$$\frac{1}{\mu}|(P_{\alpha_n}^k)'(\omega_0)| \leq |(P_{\alpha_n}^k)'(\omega)| \leq \mu|(P_{\alpha_n}^k)'(\omega_0)| \text{ pour } \omega \in \overline{\Omega}'' \qquad (13)$$

Or $\lim\limits_{n \to +\infty} (P_{\alpha_n}^k)'(\omega_0) = (P_\alpha^k)'(\omega_0)$, par suite il existe un rang $n_1 \geq n_0$ tel que pour $n \geq n_1$, on a :

$$\tfrac{1}{2}|(P_\alpha^k)'(\omega_0)| \leq |(P_{\alpha_n}^k)'(\omega_0)| \leq 2|(P_\alpha^k)'(\omega_0)| \qquad (14)$$

En posant $\mu_1 = 2\mu|(P_\alpha^k)'(\omega_0)|$, $\mu_2 = \frac{1}{2\mu}|(P_\alpha^k)'(\omega_0)|$, on obtient d'après (13) et (14) que pour $n \geq n_1$, on a :

$$\mu_2 \leq |(P_{\alpha_n}^k)'(\omega)| \leq \mu_1 \quad \text{pour } \omega \in \Omega'' \tag{15}$$

et on a :

$$\text{mes}\left(P_{\alpha_n}^k((\mathbb{C} - K(P_{\alpha_n})) \cap \Omega'')\right) = \iint_{(\mathbb{C}-K(P_{\alpha_n}))\cap\Omega''} |(P_{\alpha_n}^k)'(\omega)|^2 du dv$$

$$\text{où } \omega = u + iv$$

Ainsi d'après (15), on aura pour $n \geq n_1$:

$$\mu_2^2 \text{mes}(((\mathbb{C} - K(P_{\alpha_n})) \cap \Omega'') \leq \text{mes}(P_{\alpha_n}^k((\mathbb{C} - K(P_{\alpha_n})) \cap \Omega'')$$

$$\leq \mu_1^2 \text{mes}((\mathbb{C} - K(P_{\alpha_n})) \cap \Omega''), \tag{16}$$

d'autre part $P_{\alpha_n}^k((\mathbb{C} - K(P_{\alpha_n})) \cap \Omega'') = (\mathbb{C} - K(P_{\alpha_n})) \cap P_{\alpha_n}^k(\Omega'')$ car

$$P_{\alpha_n}^{-1}(\mathbb{C} - K(P_{\alpha_n})) = \mathbb{C} - K(P_{\alpha_n}) = P_{\alpha_n}(\mathbb{C} - K(P_{\alpha_n})).$$

On en déduit d'après (12) et (16) que pour $n \geq n_1$, on a:

$$\text{mes}((\mathbb{C} - K(P_{\alpha_n})) \cap \Delta') \leq \mu_1^2 \text{mes}((\mathbb{C} - K(P_{\alpha_n})) \cap \Omega) \tag{17}$$

$$\text{mes}((\mathbb{C} - K(P_{\alpha_n})) \cap \Omega') \leq \mu_2^{-2} \text{mes}((\mathbb{C} - K(P_{\alpha_n})) \cap \Delta) \tag{18}$$

d'où de (17) et (18) on tire que pour qu'une composante connexe Ω de $\overset{\circ}{K}(P_\alpha)$ vérifie $\lim\limits_{n \to +\infty} \text{mes}(K(P_{\alpha_n}) \cap \Omega) = \text{mes}(\Omega)$ il faut et il suffit que Δ vérifie $\lim\limits_{n \to +\infty} \text{mes}(K(P_{\alpha_n}) \cap \Delta) = \text{mes}(\Delta)$.

Ainsi ii) implique qu'on a $\lim\limits_{n \to +\infty} \text{mes}(K(P_{\alpha_n}) \cap \Delta) = \text{mes}(\Delta)$ et par suite pour toute composante connexe Ω_j de $\overset{\circ}{K}(P_\alpha)$, on a : $\lim\limits_{n \to +\infty} \text{mes}(K(P_{\alpha_n}) \cap \Omega_j) = \text{mes}(\Omega_j)$.

Ecrivons $\overset{\circ}{K}(P_\alpha) = \bigcup\limits_{j \in \mathbb{N}} \Omega_j$, on a :

$$\lim_{n \to +\infty} \text{mes}(K(P_{\alpha_n}) \cap \overset{\circ}{K}(P_\alpha)) = \lim_{n \to +\infty} \sum_{j \in \mathbb{N}} \text{mes}(K(P_{\alpha_n}) \cap \Omega_j). \tag{19}$$

Remarquons que $\text{mes}(K(P_{\alpha_n}) \cap \Omega_j) \leq \text{mes}(\Omega_j)$ pour tout $j, n \in \mathbb{N}$ avec de plus la série $\sum\limits_{j \in \mathbb{N}} \text{mes}(\Omega_j)$ converge (vers $\text{mes}(\overset{\circ}{K}(P_\alpha))$).

Par conséquent on peut permuter les symboles " $\lim\limits_{n\to+\infty}$ " et " $\sum\limits_{j\in\mathbb{N}}$ " dans (19) et on obtient :

$$\lim_{n\to+\infty} \mathrm{mes}\,(K(P_{\alpha_n})\cap \overset{\circ}{K}(P_\alpha)) = \sum_{j\in\mathbb{N}} \mathrm{mes}\,(\Omega_j) = \mathrm{mes}\,(\overset{\circ}{K}(P_\alpha))$$

d'où ii) ⇒ i) et le théorème se trouve démontré.

4.2. Motivation de la conjecture réduite.

Comme on l'a déjà mentionné dans l'introduction, le travail qui va suivre est une tentative pour aborder la conjecture réduite (équivalente à la conjecture de Douady d'après le théorème 3), c'est-à-dire : $\lim\limits_{n\to+\infty} \mathrm{mes}\,(K(P_{\theta_n})\cap \Delta) = \mathrm{mes}\,(\Delta)$?

Le modèle obtenu pour réaliser cette étude est une fonction holomorphe, injective $G_n : \mathbb{H} \to \mathbb{C}$ (où $\mathbb{H} = \{\omega \in \mathbb{C}/\mathrm{R\acute{e}}\ \omega < 0\}$), de la forme $G_n(\omega) = \omega + \lambda_n e^\omega + h_n(\omega)$ (qui en un sens on prétend qu'elle possède un comportement dynamique comparable à celui de la fonction $\omega \mapsto \omega + \lambda_n e^\omega$).

Commençons par démontrer deux lemmes, pour se faire la main :

4.2.1. Lemme. *Soit* $P(z) = e^{2\pi i \frac{p}{q}} z + a_2 z^2$ *avec* $0 \neq a_2 \in \mathbb{C}$, p *et* q *étant deux entiers premiers entre eux. Alors l'itérée* $q^{\mathrm{ième}}$ *de* P *est de la forme* $P^q(z) = z + a_{q+1} z^{q+1} + \mathcal{O}(z^{q+2})$ *avec* $a_{q+1} \neq 0$.

Démonstration : Voir [J1], §2, Lemme 2.

4.2.2. Lemme. *Soit* $f(z) = z + a_{q+1} z^{q+1} + \sum\limits_{k=q+2}^{+\infty} a_k z^k$ *avec* $a_{q+1} \neq 0$, *et*

$$\varphi(z) = \mu_1 z + \sum_{k=2}^{+\infty} \mu_k z^k$$ *avec* $\mu_1 \neq 0$, *deux fonctions analytiques au voisinage de 0. Alors dans un petit voisinage de 0 la fonction* $F = \varphi \circ f \circ \varphi^{-1}$ *est de la forme :*

$$F(z) = z + \frac{1}{\mu_1^q} a_{q+1} z^{q+1} + \mathcal{O}(z^{q+2}).$$

Démonstration : Il s'agit d'un calcul facile en partant de l'identité $F \circ \varphi = \varphi \circ f$.

4.2.3. Notations et changement de coordonnées.

Soient α un nombre diophantien, $(\theta_n = p_n/q_n)_{n\in\mathbb{N}}$ la suite des réduites de α dans le développement en fraction continue, $P_\theta(z) = e^{2\pi i \theta} z + z^2$, $k \geq 2$ un entier naturel donné et $(\varphi_{k,\theta})_{\theta\in J}$ qu'on notera tout simplement $(\varphi_\theta)_{\theta\in J}$ la famille $\mathbb{R}$-analytique donnée par le théorème 2 (avec $\alpha \in J$) et associée à la famille $(P_\theta)_{\theta \in \mathbb{R}}$. Δ désigne toujours le disque de Siegel maximal de P_α en 0, $\varphi : \Delta \longrightarrow \mathbb{D} = \{z \in \mathbb{C}/|z| < 1\}$ une application linéarisante de P_α en 0.

Fixons dans tout ce qui suit $r \in]0, 1[$ et considérons deux réels auxiliaires r_1, r_2 vérifiant $0 < r < r_1 < r_2 < 1$.

On pose $\Delta_j = \varphi^{-1}(\mathbb{D}_{r_j}) \subset \Delta$ avec $\mathbb{D}_{r_j} = \{z \in \mathbb{C}/|z| < r_j\}$ pour $j \in \{1, 2\}$. Comme $P_\alpha : \Delta \longrightarrow \Delta$ et $\varphi : \Delta \longrightarrow \mathbb{D}$ sont deux isomorphismes avec $\lim_{n \to +\infty} P_{\theta_n} = P_\alpha$ et (d'après le théorème 2) $\lim_{n \to +\infty} \varphi_{\theta_n} = \varphi$ uniformément sur tout compact de Δ, alors il en résulte d'après les théorèmes 1,2 et le lemme 4.1.2, qu'il existe un rang $n_0 \in \mathbb{N}$ tel que les propriétés suivantes sont vraies pour $n \geq n_0$:

$$\overline{\Delta}_1 \subset \Delta_2(\theta_n, q_n); \tag{20}$$

$$\Delta_2 \subset \mathrm{Dom}(\varphi_{\theta_n}); \tag{21}$$

$$P_{\theta_n} \text{ et } \varphi_{\theta_n} \text{ sont injectives sur } \Delta_2; \tag{22}$$

$$\overline{\mathbb{D}}_r \subset \varphi_{\theta_n}(\Delta_1). \tag{23}$$

Ainsi pour $|z| \leq r$, on a $\varphi_{\theta_n}^{-1}(z) \in \Delta_1$ est bien défini, $P_{\theta_n}^j(\varphi_{\theta_n}^{-1}(z)) \in \Delta_2$ pour $0 \leq j \leq q_n$ et les propriétés (21) et (22) montrent que les applications $z \mapsto (\varphi_{\theta_n} \circ P_{\theta_n}^j \circ \varphi_{\theta_n}^{-1})(z)$ sont holomorphes et injectives sur $\mathbb{D}_r$ (pour $0 \leq j \leq q_n$).

On en déduit d'après la remarque 3.6, qu'il existe une constante $M > 0$ et un entier $n_1 \in \mathbb{N}$ tel que pour $|z| \leq r$ et $n \geq n_1$, on a :

$$|(\varphi_{\theta_n} \circ P_{\theta_n}^j \circ \varphi_{\theta_n}^{-1})(z) - e^{2\pi i (j\theta_n)}z| \leq M |\theta_n - \alpha|^{k - \frac{1}{2}}. \tag{24}$$

On note $F_{\theta_n}(z) = (\varphi_{\theta_n} \circ P_{\theta_n}^{q_n} \circ \varphi_{\theta_n}^{-1})(z)$ pour $|z| \leq r$ et $n \geq n_0$. Remarquons que d'après le lemme 4.2.1, on peut écrire :

$$P_{\theta_n}^{q_n}(z) = z + a_{q_n+1}z^{q_n+1} + \mathcal{O}(z^{q_n+2}) \qquad \text{avec } a_{q_n+1} \neq 0.$$

Et en utilisant le lemme 4.2.2, on obtient pour $|z| < r$:

$$F_{\theta_n}(z) = z + \frac{a_{q_n+1}}{(\varphi'(0))^{q_n}}z^{q_n+1} + \mathcal{O}(z^{q_n+2}). \tag{25}$$

Soit $N_0 \geq \max(n_0, n_1)$ tel que $|\theta_n - \alpha|^{k - \frac{1}{2}} < \frac{r}{2M}$ pour $n \geq N_0$.

Introduisons une nouvelle coordonnée $\omega \in \mathbb{H} = \{\omega \in \mathbb{C}/\mathrm{Ré}\ \omega < 0\}$ par $z = \Pi_n(\omega) = re^{\frac{\omega}{q_n}}$ et définissons la fonction $G_n(\omega)$ pour $\omega \in \mathbb{H}$ et $n \geq N_0$ par :

$$G_n(\omega) = \omega + q_n \log\left(\frac{F_{\theta_n}(z)}{z}\right) \qquad \text{avec } z = \Pi_n(\omega). \tag{26}$$

Dans cette formule log désigne la branche principale du logarithme (i.e.
$-\pi < \text{Im } \log(.) \le \pi$).

4.2.4. Propriétés.

Pour $n \ge N_0$, on a les propriétés suivantes :
(P_1) $G_n : \mathbb{H} \to \mathbb{C}$ est définie, holomorphe sur $\mathbb{H}$ et vérifiant :

$$\Pi_n \circ G_n = F_{\theta_n} \circ \Pi_n$$

(P_2) G_n est injective sur $\mathbb{H}$ et on a :

$$G_n(\omega + 2\pi i q_n) = G_n(\omega) + 2\pi i q_n \qquad \text{pour tout } \omega \in \mathbb{H}$$

(P_3) $G_n(\omega) = \omega + \mathcal{O}((\theta_n - \alpha)^{k-1})$ uniformément sur $\mathbb{H}$.
(P_4) $G_n(\omega) - \omega \sim \lambda_n e^{\omega}$ pour Ré $\omega \to -\infty$; où $\lambda_n = q_n a_{q_n+1}(\frac{r}{\varphi'(0)})^{q_n}$. De
plus on a : $0 < |\lambda_n| \le 2$.
(P_5) $G_n(\omega + 2\pi i) = G_n(\omega) + 2\pi i + \mathcal{O}((\theta_n - \alpha)^{2k-2})$ uniformément sur tout
demi-plan $\mathbb{H}_\ell = \{\omega \in \mathbb{C}/\text{Ré } \omega < -\ell\}$ avec $\ell > 0$.

Démonstration :

(P_1): D'après l'inégalité (24), en prenant $j = q_n$, on obtient :

$$|F_{\theta_n}(z) - z| \le M|\theta_n - \alpha|^{k-\frac{1}{2}} \qquad \text{pour } |z| \le r \quad \text{et } n \ge N_0.$$

Par suite d'après le principe du maximum, on aura :

$$|\frac{F_{\theta_n}(z)}{z} - 1| \le \frac{M}{r}|\theta_n - \alpha|^{k-\frac{1}{2}} < \frac{1}{2} \quad \text{pour } |z| \le r \text{ et } n \ge N_0. \qquad (27)$$

Ainsi

$$G_n(\omega) = \omega + q_n \log\left(\frac{F_{\theta_n}(z)}{z}\right) = \omega + q_n \log\left(1 + \left(\frac{F_{\theta_n}(z)}{z} - 1\right)\right)$$

avec $z = \Pi_n(\omega) \in \mathbb{D}_r$, donc G_n est définie et holomorphe sur $\mathbb{H}$. De plus,

$$(\Pi_n \circ G_n)(\omega) = re^{\frac{G_n(\omega)}{q_n}} = re^{\frac{\omega}{q_n}} \cdot \frac{F_{\theta_n}(z)}{z} = F_{\theta_n}(z) = (F_{\theta_n} \circ \Pi_n)(\omega).$$

(P_2): du fait que $\Pi_n(\omega + 2\pi i q_n) = \Pi_n(\omega)$, on en déduit immédiatement que
$G_n(\omega + 2\pi i q_n) = G_n(\omega) + 2\pi i q_n$ pour tout $\omega \in \mathbb{H}$. Montrons que G_n est
injective sur $\mathbb{H}$: Soit $\omega_1, \omega_2 \in \mathbb{H}$ tel que $G_n(\omega_1) = G_n(\omega_2)$.
Il en résulte de la propriété (P_1) que $F_{\theta_n}(\Pi_n(\omega_1)) = F_{\theta_n}(\Pi_n(\omega_2))$,
mais $\Pi_n(\omega_1), \Pi_n(\omega_2) \in \mathbb{D}_r$ et F_{θ_n} est injective sur $\mathbb{D}_r$, donc on doit avoir
$\Pi_n(\omega_1) = \Pi_n(\omega_2)$ et il existe $m \in \mathbb{Z}$ tel que $\omega_1 = \omega_2 + 2\pi i m q_n$. Ceci donne

$G_n(\omega_1) = G_n(\omega_2 + 2\pi i m q_n)$ ou encore $G_n(\omega_1) = G_n(\omega_2) + 2\pi i m q_n$, d'où $m = 0$ et $\omega_1 = \omega_2$.

(P_3): Tout d'abord on a : $|\log(1+z)| \leq 2|z|$ pour $|z| \leq \frac{1}{2}$, en effet

$$\log(1+z) = \sum_{m=1}^{+\infty} (-1)^{m-1} \frac{z^m}{m} \qquad \text{pour } |z| < 1;$$

donc $|\log(1+z)| \leq \sum_{m=1}^{+\infty} |z|^m \leq |z| \sum_{m=0}^{+\infty} (\frac{1}{2})^m = 2|z|$ pour $|z| \leq \frac{1}{2}$.

D'autre part, $G_n(\omega) = \omega + q_n \log\left(1 + \left(\frac{F_{\theta_n}(z)}{z} - 1\right)\right)$ avec $z = \Pi_n(\omega)$ et d'après (27), on a : $|\frac{F_{\theta_n}(z)}{z} - 1| < \frac{1}{2}$ pour $|z| < r$. Ceci entraîne que :

$$|G_n(\omega) - \omega| \leq 2q_n |\frac{F_{\theta_n}(z)}{z} - 1| \leq \frac{2M}{r} q_n |\theta_n - \alpha|^{k-\frac{1}{2}} \qquad \text{(d'après (27))}$$

or $q_n < |\theta_n - \alpha|^{-\frac{1}{2}}$ et par suite on obtient :

$$|G_n(\omega) - \omega| \leq \frac{2M}{r} |\theta_n - \alpha|^{k-1} \qquad \text{pour } \omega \in \mathbb{H} \quad \text{et } n \geq N_0,$$

c'est-à-dire $G_n(\omega) = \omega + \mathcal{O}((\theta_n - \alpha)^{k-1})$ uniformément sur $\mathbb{H}$.

(P_4): D'après (25), on a : $\frac{F_{\theta_n}(z)}{z} - 1 \sim \frac{a_{q_n+1}}{(\varphi'(0))^{q_n}} z^{q_n}$ pour $z \to 0$.

Comme $G_n(\omega) - \omega = q_n \log\left(1 + \left(\frac{F_{\theta_n}(z)}{z} - 1\right)\right)$ avec $z = \Pi_n(\omega)$, alors on aura pour Ré $\omega \to -\infty$:

$$G_n(\omega) - \omega \sim q_n \left(\frac{F_{\theta_n}(z)}{z} - 1\right) \sim q_n a_{q_n+1} \left(\frac{r}{\varphi'(0)}\right)^{q_n} e^\omega = \lambda_n e^\omega.$$

Il est clair que $|\lambda_n| > 0$ car $a_{q_n+1} \neq 0$. Montrons que $|\lambda_n| \leq 2$:

On pose $S = \{h : \mathbb{D} \to \mathbb{C}, \text{ holomorphe, injective avec } h'(0) - 1 = h(0) = 0\}$. Un résultat sur la théorie des fonctions univalentes (voir [Go] page 494) permet de dire que si $h(z) = z + s_{m+1}z^{m+1} + s_{m+2}z^{m+2} + \dots$ est un élément de S, alors on a $|s_{m+1}| \leq \frac{2}{m}$ et l'inégalité est optimale. Maintenant d'après (25) l'application :

$$z \mapsto \frac{1}{r} F_{\theta_n}(rz) = z + a_{q_n+1}\left(\frac{r}{\varphi'(0)}\right)^{q_n} z^{q_n+1} + \dots$$

est un élément de la classe S et par suite on obtient :

$$|a_{q_n+1}\left(\frac{r}{\varphi'(0)}\right)^{q_n}| \le \frac{2}{q_n}$$

ou encore $|\lambda_n| \le 2$.

(P_5) : * **Etape 1.** Considérons $\widetilde{F}_{\theta_n}(z) = (\varphi_{\theta_n} \circ P_{\theta_n}^{q_{n-1}} \circ \varphi_{\theta_n}^{-1})(z)$ pour $|z| \le r$
et $n \ge N_0$. D'après l'inégalité (24), en prenant $j = q_{n-1}$, on obtient :

$$|\widetilde{F}_{\theta_n}(z) - e^{2\pi i \frac{q_{n-1}p_n}{q_n}} z| \le M|\theta_n - \alpha|^{k-\frac{1}{2}}.$$

Compte tenu de l'égalité $q_n p_{n-1} - p_n q_{n-1} = (-1)^n$ (voir préliminaires 1.1),
alors on a :

$$\frac{q_{n-1}p_n}{q_n} = p_{n-1} + \frac{(-1)^{n+1}}{q_n} \ , \ \text{et } e^{2\pi i \frac{q_{n-1}p_n}{q_n}} = e^{2\pi i \frac{(-1)^{n+1}}{q_n}}$$

et on aura $|\widetilde{F}_{\theta_n}(z) - e^{2\pi i \frac{(-1)^{n+1}}{q_n}} z| \le M|\theta_n - \alpha|^{k-\frac{1}{2}}$ pour $|z| \le r$ et $n \ge N_0$.
Par conséquent, d'après le principe du maximum, on a :

$$\left|\frac{\widetilde{F}_{\theta_n}(z)}{e^{2\pi i \frac{(-1)^{n+1}}{q_n}} z} - 1\right| \le \frac{M}{r}|\theta_n - \alpha|^{k-\frac{1}{2}} < \frac{1}{2} \quad \text{pour } |z| \le r \quad \text{et } n \ge N_0.$$

Ainsi la fonction $\widetilde{G}_n : \mathbb{H} \to \mathbb{C}$ donnée par :

$$\widetilde{G}_n(\omega) = \omega + (-1)^{n+1} 2\pi i + q_n \log\left(\frac{\widetilde{F}_{\theta_n}(z)}{e^{2\pi i \frac{(-1)^{n+1}}{q_n}} z}\right) \quad \text{avec } z = \Pi_n(\omega)$$

est définie et holomorphe sur $\mathbb{H}$. De plus on voit facilement qu'elle vérifie
l'égalité $\Pi_n \circ \widetilde{G}_n = \widetilde{F}_{\theta_n} \circ \Pi_n$.

* **Etape 2.** Soit $\ell > 0$ et $\mathbb{H}_\ell = \{\omega \in \mathbb{C}/\text{Ré } \omega < -\ell\}$.
Posons $\varepsilon_n = \frac{2M}{r}|\theta_n - \alpha|^{k-1}$ et soit $N_1 \in \mathbb{N}$ tel que $N_1 \ge N_0$ et $\varepsilon_n \le \frac{\ell}{2}$
pour tout $n \ge N_1$.
Ceci entraîne que si $n \ge N_1$, on a : $G_n(\mathbb{H}_\ell) \subset \mathbb{H}$ et $\widetilde{G}_n(\mathbb{H}_\ell) \subset \mathbb{H}$.
Montrons que pour $n \ge N_1$, on a : $G_n \circ \widetilde{G}_n = \widetilde{G}_n \circ G_n$ sur $\mathbb{H}_\ell$. On a :

$$\Pi_n \circ (G_n \circ \widetilde{G}_n) = (\Pi_n \circ G_n) \circ \widetilde{G}_n = F_{\theta_n} \circ (\Pi_n \circ \widetilde{G}_n) = (F_{\theta_n} \circ \widetilde{F}_{\theta_n}) \circ \Pi_n$$

de même $\Pi_n \circ (\widetilde{G}_n \circ G_n) = (\widetilde{F}_{\theta_n} \circ F_{\theta_n}) \circ \Pi_n$, or $F_{\theta_n} \circ \widetilde{F}_{\theta_n} = \widetilde{F}_{\theta_n} \circ F_{\theta_n}$ car
$P_{\theta_n}^{q_n} \circ P_{\theta_n}^{q_{n-1}} = P_{\theta_n}^{q_{n-1}} \circ P_{\theta_n}^{q_n}$. Donc $\Pi_n \circ (G_n \circ \widetilde{G}_n) = \Pi_n \circ (\widetilde{G}_n \circ G_n)$ et on en

déduit qu'il existe $m \in \mathbb{Z}$ tel que $(G_n \circ \widetilde{G}_n)(\omega) = (\widetilde{G}_n \circ G_n)(\omega) + 2\pi i m q_n$ pour tout $\omega \in \mathbb{H}_\ell$. On peut écrire pour tout $\omega \in \mathbb{H}_\ell$:

$$2\pi i m q_n = [G_n(\widetilde{G}_n(\omega)) - \widetilde{G}_n(\omega)] + [\widetilde{G}_n(\omega) - G_n(\omega)] + [G_n(\omega) - \widetilde{G}_n(G_n(\omega))].$$

(28)

D'autre part on a :

$$\lim_{\substack{\text{Ré } \omega \to -\infty}} G_n(\omega) - \omega = 0 \quad \text{et} \quad \lim_{\substack{\text{Ré } \omega \to -\infty}} \widetilde{G}_n(\omega) - \omega = (-1)^{n+1} 2\pi i ,$$

maintenant par passage à la limite quand Ré$\omega \to -\infty$ dans (28), on obtient $m = 0$ et par suite $G_n \circ \widetilde{G}_n = \widetilde{G}_n \circ G_n$ sur $\mathbb{H}_\ell$.

* Etape 3. Posons

$$H_n(\omega) = q_n \log\left(\frac{F_{\theta_n}(z)}{z}\right) \quad \text{et} \quad \widetilde{H}_n(\omega) = q_n \log\left(\frac{\widetilde{F}_{\theta_n}(z)}{e^{2\pi i \frac{(-1)^{n+1}}{q_n}} z}\right)$$

où $z = \Pi_n(\omega) = r e^{\frac{\omega}{q_n}}$ pour $\omega \in \mathbb{H}$ et $n \geq N_1$.

En d'autres termes : $G_n(\omega) = \omega + H_n(\omega)$ et $\widetilde{G}_n(\omega) = \omega + (-1)^{n+1} 2\pi i + \widetilde{H}_n(\omega)$. D'après la démonstration de la propriété (P_3), on a :

$$|H_n(\omega)| \leq \varepsilon_n \quad \text{pour } \omega \in \mathbb{H} \quad \text{et } n \geq N_1 \tag{29}$$

un même raisonnement donne :

$$|\widetilde{H}_n(\omega)| \leq \varepsilon_n \quad \text{pour } \omega \in \mathbb{H} \quad \text{et } n \geq N_1. \tag{30}$$

On en déduit d'après la formule de Cauchy que pour $\omega \in \mathbb{H}$,

$$|H_n'(\omega)| \leq \frac{\varepsilon_n}{|\text{Ré } \omega|} ; \tag{31}$$

$$|\widetilde{H}_n'(\omega)| \leq \frac{\varepsilon_n}{|\text{Ré } \omega|} . \tag{32}$$

* Etape 4. Pour $\omega \in \mathbb{H}_\ell$ et $n \geq N_1$, on a:

$$\widetilde{G}_n(G_n(\omega)) = G_n(\omega) + (-1)^{n+1} 2\pi i + \widetilde{H}_n(G_n(\omega))$$

$$= \omega + H_n(\omega) + (-1)^{n+1} 2\pi i + \widetilde{H}_n(\omega) + \widetilde{\psi}_n(\omega)$$

où on a posé $\widetilde{\psi}_n(\omega) = \widetilde{H}_n(\omega + H_n(\omega)) - \widetilde{H}_n(\omega)$.

En utilisant la formule suivante

$$\widetilde{H}_n(\omega + a) - \widetilde{H}_n(\omega) = a \int_0^1 \widetilde{H}_n'(\omega + at) dt$$

on obtient :
$$|\widetilde{\psi}_n(\omega)| \le |H_n(\omega)| \cdot \sup_{|\xi-\omega|\le\varepsilon_n} |\widetilde{H}_n'(\xi)|.$$

On en déduit d'après (29), (32) et le fait que $\varepsilon_n \le \frac{\ell}{2}$, qu'on a :

$$|\widetilde{\psi}_n(\omega)| \le \frac{2\varepsilon_n^2}{\ell} \qquad \text{pour } \omega \in \mathbb{H}_\ell \quad \text{et } n \ge N_1. \qquad (33)$$

D'autre part, on a:

$$G_n(\widetilde{G}_n(\omega)) = \widetilde{G}_n(\omega) + H_n(\widetilde{G}_n(\omega))$$
$$= \omega + (-1)^{n+1}2\pi i + \widetilde{H}_n(\omega) + H_n(\omega + (-1)^{n+1}2\pi i) + \psi_n(\omega)$$

où on a posé

$$\psi_n(\omega) = H_n(\omega + (-1)^{n+1}2\pi i + \widetilde{H}_n(\omega)) - H_n(\omega + (-1)^{n+1}2\pi i).$$

Un raisonnement analogue utilisant (30) et (31), montre :

$$|\psi_n(\omega)| \le \frac{2\varepsilon_n^2}{\ell} \qquad \text{pour } \omega \in \mathbb{H}_\ell \quad \text{et } n \ge N_1. \qquad (34)$$

Maintenant en vertu de l'identité $G_n \circ \widetilde{G}_n = \widetilde{G}_n \circ G_n$, on obtient :

$$H_n(\omega+(-1)^{n+1}2\pi i)-H_n(\omega) = \widetilde{\psi}_n(\omega) - \psi_n(\omega) \qquad \text{pour } \omega \in \mathbb{H}_\ell \quad \text{et } n \ge N_1.$$

Par suite d'après (33) et (34), on a $|H_n(\omega + (-1)^{n+1}2\pi i) - H_n(\omega)| \le \frac{4\varepsilon_n^2}{\ell}$ pour $\omega \in \mathbb{H}_\ell$ et $n \ge N_1$. On en déduit que $|H_n(\omega + 2\pi i) - H_n(\omega)| \le \frac{4\varepsilon_n^2}{\ell}$ pour $\omega \in \mathbb{H}_\ell$ et $n \ge N_1$. Ceci exprime qu'on a $G_n(\omega + 2\pi i) = G_n(\omega) + 2\pi i + \mathcal{O}\big((\theta_n - \alpha)^{2k-2}\big)$ uniformément sur $\mathbb{H}_\ell$.

4.2.5. Quelques questions

Soit $\Delta' \subsetneq \Delta$ un sous-disque de Siegel de P_α en 0 et notons par $\widehat{\Delta}'$ son symétrique par rapport au point critique $\omega = -\frac{1}{2}e^{2\pi i\alpha}$ de P_α. Désignons par $\Omega(P_{\theta_n})$ la composante connexe de $K(P_{\theta_n}) - \{0\}$, contenant le point critique $\omega_n = -\frac{1}{2}e^{2\pi i\theta_n}$ de P_{θ_n}.

Question 1. Existe-t-il un rang $n_0 \in \mathbb{N}$ tel que :

$$(K(P_{\theta_n}) - \Omega(P_{\theta_n})) \cap \widehat{\Delta}' = \phi \qquad \text{pour tout } n \ge n_0?$$

Au vu de la définition de la fonction G_n, alors on peut montrer qu'une réponse positive à la question 1, entraîne une réponse positive à la question suivante :

Question 2. Existe-t-il un rang $N_0 \in \mathbb{N}$ tel que pour tout $n \geq N_0$, la fonction G_n n'a pas de points fixes dans $\mathbb{H}$?

Remarque. On prétend que les cinq propriétés de 4.2.4 et une réponse positive à la question 2 ci-dessus sont suffisantes pour poursuivre cette étude et aboutir à une réponse positive à la conjecture de Douady énoncée dans l'introduction.

Remerciements.
Je remercie le Professeur A. Douady qui m'a proposé ce sujet, ainsi que le Professeur J.H. Hubbard pour ses remarques très utiles.

Bibliographie
[Go] G.M. GOLUZIN, *Geometric theory of functions of a complex variable*, vol. 26, Translations of Mathematical monographs, A.M.S. 1969.
[J1] H. Jellouli, Indice holomorphe et multiplicateur, dans ce volume.
[Po] C. POMMERENKE, *Univalent functions*, Vandenhoeck and Ruprecht, Göttingen, 1975.

Institut préparatoire aux études d'ingénieur de Nabeul, Département de Mathématiques et d'Informatique, Merazka, 8000 Nabeul, Tunisie.

INDICE HOLOMORPHE ET MULTIPLICATEUR

HABIB JELLOULI

Résumé : Soit $P_{\lambda_0}(z) = \lambda_0 z + z^2$ où $\lambda_0 = e^{2\pi i p/q}$, $(p,q) \in \mathbb{Z} \times \mathbb{N}^*$, p et q étant premiers entre eux et $z \in \mathbb{C}$. Il est bien connu qu'au voisinage du point fixe parabolique 0, l'itérée q-ème de P_{λ_0} est analytiquement conjuguée à une fonction de la forme $z \mapsto z + z^{q+1} + A z^{2q+1} + \mathcal{O}(z^{2q+2})$. Le nombre complexe A est un invariant de conjugaison analytique.

On considère des perturbations de P_{λ_0} dans la famille des polynômes quadratiques $P_\lambda(z) = \lambda z + z^2$ où λ est dans un petit disque $D(\lambda_0, r)$ centré en λ_0 et de rayon r.

Pour r assez petit, le multiplicateur $\rho(\lambda)$ du cycle de longueur q le plus proche de 0, définit une fonction holomorphe en $\lambda \in D(\lambda_0, r)$. On utilise la notion d'indice holomorphe pour donner les valeurs des dérivées première et seconde du multiplicateur en λ_0, et par des propriétés dynamiques des P_λ, on donne un encadrement de la partie réelle du nombre complexe A. Enfin on utilise ces résultats dans l'étude d'explosion des cylindres de Fatou-Ecalle.

Abstract : Let $P_{\lambda_0}(z) = \lambda_0 z + z^2$ where $\lambda_0 = e^{2\pi i p/q}$, $(p,q) \in \mathbb{Z} \times \mathbb{N}^*$, $(p,q) = 1$ and $z \in \mathbb{C}$. It is well known that the q-th iterate $P_{\lambda_0}^q$ of P_{λ_0} is analytically conjugate to a holomorphic function of the form $z \mapsto z + z^{q+1} + A z^{2q+1} + \mathcal{O}(z^{2q+2})$ in a neighbourhood of 0, where A is a complex number invariant under analytic conjugation.

We consider perturbations of P_{λ_0} in the quadratic polynomials $P_\lambda(z) = \lambda z + z^2$ with $\lambda \in \mathbb{D}(\lambda_0, r)$, r small.

For r sufficiently small, the multiplier $\rho(\lambda)$ of the q-periodic cycle close to 0, is holomorphic as a function of $\lambda \in D(\lambda_0, r)$. We use the holomorphic index to calculate the values of the first and second derivatives of $\rho(\lambda)$ at λ_0. Furthermore we bound the real part of A from above and below, using the dynamical properties of P_λ. Finally we use these results in the study of the explosion of Fatou-Ecalle cylinders.

1. Notion d'indice holomorphe

Soit $\mathcal{U}$ un domaine de $\mathbb{C}$, $z_0 \in \mathcal{U}$ et $f : \mathcal{U} \to \mathbb{C}$ une fonction holomorphe pour laquelle z_0 est un point fixe $(f(z_0) = z_0)$. On définit l'indice holomorphe (voir [M]) de f au point z_0 et on note $i(f, z_0)$ comme étant le résidu

$$i(f, z_0) = \frac{1}{2\pi i} \oint_{|z - z_0| = \varepsilon} \frac{1}{z - f(z)} dz \quad .$$

où le cercle d'intégration $|z - z_0| = \varepsilon$ est orienté positivement et ε choisi assez petit vérifiant $f(z) - z \neq 0$ pour $0 < |z - z_0| \leq \varepsilon$.

Le lemme suivant dont on aura besoin exprime essentiellement que l'indice holomorphe est un invariant de conjugaison analytique

Lemme 1.

1) *Si z_0 est un point fixe simple de f, alors on a*

$$i(f, z_0) = \frac{1}{1 - f'(z_0)} \quad .$$

2) *Si z_0 est un point fixe de f et $F = h \circ f \circ h^{-1}$ où h est holomorphe, injective au voisinage de z_0, alors on a*

$$i(f, z_0) = i(F, h(z_0)) \quad .$$

Preuve : 1) Par hypothèse, $f(z) = z_0 + a_1(z - z_0) + a_2(z - z_0)^2 + \ldots\ldots$ avec $a_1 \neq 1$. Ceci donne le développement en série de Laurent

$$\frac{1}{z - f(z)} = \frac{1}{(1 - a_1)(z - z_0)} + \sum_{k=0}^{+\infty} b_k(z - z_0)^k \quad ;$$

par suite

$$i(f, z_0) = \frac{1}{1 - a_1} = \frac{1}{1 - f'(z_0)} \quad .$$

2) 1er cas : $f'(z_0) \neq 1$. On a: $F(h(z_0)) = h(z_0)$, $F'(h(z_0)) = f'(z_0) \neq 1$. D'après 1), on obtient

$$i(f, z_0) = \frac{1}{1 - f'(z_0)} = \frac{1}{1 - F'(h(z_0))} = i(F, h(z_0)) \quad .$$

2e cas : $f'(z_0) = 1$. Désignons par $m \geq 2$ l'ordre de multiplicité de z_0 comme point fixe de f. Soit $\varepsilon > 0$ tel que $f(z) - z \neq 0$ et $f'(z) \neq 1$ pour $0 < |z - z_0| \leq \varepsilon$.

Posons

$$c = \inf_{|z-z_0|=\varepsilon} |f(z) - z| > 0 \quad ,$$

$$f_t(z) = f(z) + t \quad , \qquad F_t = h \circ f_t \circ h^{-1} \quad \text{pour} \quad |t| < \mu \le \frac{c}{2} \quad ;$$

ceci donne $|f_t(z) - z| \ge \frac{c}{2}$ pour $|z - z_0| = \varepsilon$.

D'après le principe de l'argument, le nombre $N(t)$ des points fixes de f_t à l'intérieur du cercle $\gamma_\varepsilon = \{z/|z - z_0| = \varepsilon\}$ est donné par

$$N(t) = \frac{1}{2\pi i} \oint_{\gamma_\varepsilon} \frac{f_t'(z) - 1}{f_t(z) - z} dz \quad .$$

D'après cette formule, $N(t)$ est une fonction continue et est à valeurs dans $\mathbb{N}$, donc constante. Compte tenu du fait que $N(0) = m$, alors $N(t) = m$ pour $|t| < \mu$. Désignons par $\beta_1(t), \beta_2(t), \ldots, \beta_m(t)$ les points fixes de f_t avec $\beta_1(0) = \beta_2(0) = \ldots = \beta_m(0) = z_0$, et on a $\beta_j(t) \ne z_0$, $f_t'(\beta_j(t)) = f'(\beta_j(t)) \ne 1$ pour $0 < |t| < \mu$ et $1 \le j \le m$.

Par suite, d'après le 1er cas, on a $i(f_t, \beta_j(t)) = i(F_t, h(\beta_j(t)))$, ce qui donne

$$\sum_{j=1}^m i(f_t, \beta_j(t)) = \sum_{j=1}^m i(F_t, h(\beta_j(t))) \quad ;$$

soit

$$\frac{1}{2\pi i} \oint_{\gamma_\varepsilon} \frac{1}{z - f_t(z)} dz = \frac{1}{2\pi i} \oint_{h(\gamma_\varepsilon)} \frac{1}{z - F_t(z)} dz \quad . \tag{1}$$

Comme les contours d'intégration γ_ε et $h(\gamma_\varepsilon)$ sont fixes, il est clair que les deux membres de (1) sont continus en t pour $|t| < \mu$ et par suite, par passage à la limite, quand t tend vers 0 dans (1), on aura $i(f, z_0) = i(F, h(z_0))$.

2. Polynômes quadratiques et multiplicateur

On considère la famille $(P_\lambda)_{\lambda \in \mathbb{C}}$ de polynômes quadratiques définis sur $\mathbb{C}$ par $P_\lambda(z) = \lambda z + z^2$.

On note P_λ^n la n-ème itérée de P_λ. On appelle cycle périodique, de longueur $m \in \mathbb{N}^*$ pour P_λ, une suite de points distincts $\{z_0, z_1, \ldots, z_{m-1}\}$ telle que $z_j = P_\lambda(z_{j-1})$ et $z_0 = P_\lambda(z_{m-1})$.

Ainsi, $P_\lambda^m(z_j) = z_j$ pour $0 \le j \le m-1$, et la même valeur est obtenue en évaluant la dérivée $(P_\lambda^m)'$ en chaque point z_j du cycle (en effet $(P_\lambda^m)'(z_0) = \prod_{i=0}^{m-1} P_\lambda'(z_i) = (P_\lambda^m)'(z_j)$ pour $0 \le j \le m-1$).

Cette valeur commune ρ est appelée le multiplicateur (ou encore la valeur propre) de ce cycle.

Un cycle de multiplicateur ρ est dit attractif (resp. répulsif; resp. indifférent; resp. indifférent parabolique; resp. superattractif) si $|\rho| < 1$ (resp. $|\rho| > 1$; $|\rho| = 1$; $\rho = e^{2\pi i \alpha}$ avec $\alpha \in \mathbb{Q}$; $\rho = 0$).

Dans toute la suite, on fixe $\lambda_0 = e^{2\pi i p/q}$ où $(p, q) \in \mathbb{Z} \times \mathbb{N}^*$ avec p et q premiers entre eux.

Lemme 2. *L'ordre de multiplicité de 0 comme point fixe du polynôme $P_{\lambda_0}^q$ est $(q + 1)$.*

Preuve : $P_{\lambda_0}(z) = e^{2\pi i p/q} z + z^2$. Soit m l'ordre de multiplicité du point fixe 0 de $P_{\lambda_0}^q$ ($2 \leq m \leq 2^q$); donc on a : $P_{\lambda_0}^q(z) = z + a_m z^m + a_{m+1} z^{m+1} + \ldots$ avec $a_m \neq 0$.

Il s'agit de montrer que $m = q + 1$. En vertu de l'identité $P_{\lambda_0}^q \circ P_{\lambda_0} = P_{\lambda_0} \circ P_{\lambda_0}^q$, on obtient

$$(e^{2\pi i p/q} z + z^2) + a_m (e^{2\pi i p/q} z + z^2)^m + \mathcal{O}(z^{m+1})$$
$$= e^{2\pi i p/q}(z + a_m z^m + \ldots) + (z + a_m z^m + \ldots)^2 \ .$$

En égalant les termes en z^m, on aura

$$a_m e^{2\pi i m p/q} = e^{2\pi i p/q} a_m \quad , \quad \text{ce qui donne} \quad e^{2\pi i (m-1) p/q} = 1 \quad ;$$

par suite, il existe $q_1 \in \mathbb{N}^*$ tel que $m = q_1 q + 1$.

Montrons que $q_1 = 1$. On a: $P_{\lambda_0}^q(z) = z(1 + a_{q_1 q+1} z^{q_1 q} + \ldots)$ avec $a_{q_1 q+1} \neq 0$. $\{z \in \mathbb{C} / a_{q_1 q+1} z^{q_1 q} \in \mathbb{R}_-\}$ définit $q_1 q$ demi-droites appelées axes d'attraction de $P_{\lambda_0}^q$.

Pour chaque axe d'attraction L, l'ensemble $B_L = \{z \in \mathbb{C}$ tel que la suite $P_{\lambda_0}^{nq}(z)$ tend vers 0 suivant la direction $L\}$ est un ouvert, et il existe une composante connexe A_L de B_L qui contient, pour chaque $z \in B_L$, les termes $P_{\lambda_0}^{nq}(z)$ pour n assez grand. A_L est appelé un bassin parabolique attaché à 0 (car $0 \in \overline{A_L}$) et ces bassins sont en nombre $q_1 q$. De plus $P_{\lambda_0}(A_L) = A_{L'}$, où $L' = \lambda_0 L = \{\lambda_0 z, z \in L\}$; d'où $P_{\lambda_0}^q(A_L) = A_L$ et les $q_1 q$ bassins se répartissent en q_1 cycles. On sait que chaque cycle doit contenir au moins un point critique de P_{λ_0}. Or P_{λ_0} admet un seul point critique et par suite on a un seul cycle de bassins paraboliques attachés à 0, autrement dit $q_1 = 1$. Soit donc $m = q + 1$, et ceci achève la preuve du lemme 2.

Lemme 3. *Il existe $\varepsilon > 0$ et $r > 0$ tels que, pour tout $\lambda \in \mathbb{C}$ avec $0 < |\lambda - \lambda_0| < r$, le polynôme P_λ^q admet exactement q points fixes dans le disque pointé $0 < |z| < \varepsilon$; de plus ces q points forment un cycle de longueur q pour P_λ.*

Preuve : $P_{\lambda_0}(z) = \lambda_0 z + z^2$ où $\lambda_0 = e^{2\pi i p/q}$. Remarquons que 0 est un point fixe simple de $P_{\lambda_0}^k$ pour $1 \leq k \leq (q-1)$, tandis que, d'après le lemme 2,

0 est un point fixe de $P_{\lambda_0}^q$ de multiplicité $(q+1)$. Comme les zéros d'une fonction holomorphe (non identiquement nulle) sont isolés, alors il existe $\varepsilon \in]0,1[$ tel que $P_{\lambda_0}^k(z) - z \neq 0$ pour $0 < |z| \leq 2\varepsilon$ et $1 \leq k \leq q$. On pose

$$\gamma_\varepsilon = \{z \in \mathbb{C}/|z| = \varepsilon\} \quad , \quad \gamma_{2\varepsilon} = \{z \in \mathbb{C}/|z| = 2\varepsilon\} \quad ,$$

$$c = \inf_{1 \leq k \leq q} \{|P_{\lambda_0}^k(z) - z| \quad \text{avec} \quad z \in \gamma_\varepsilon \cup \gamma_{2\varepsilon}\} > 0 \quad .$$

Il existe $r \in]0, 1-\varepsilon[$ tel que $|P_\lambda^k(z) - z| \geq \frac{c}{2}$ pour $|\lambda - \lambda_0| < r$, $z \in \gamma_\varepsilon \cup \gamma_{2\varepsilon}$ et $1 \leq k \leq q$. D'après le principe de l'argument, le nombre $N_{k,\varepsilon}(\lambda)$ (resp. $N_{k,2\varepsilon}(\lambda)$) des points fixes de P_λ^k dans le disque $|z| < \varepsilon$ (resp. $|z| < 2\varepsilon$) pour $|\lambda - \lambda_0| < r$ et $1 \leq k \leq q$ est donné par

$$N_{k,\varepsilon}(\lambda) = \frac{1}{2\pi i} \oint_{\gamma_\varepsilon} \frac{(P_\lambda^k)'(z) - 1}{P_\lambda^k(z) - z} dz$$

$$\left(\text{resp.} \quad N_{k,2\varepsilon}(\lambda) = \frac{1}{2\pi i} \oint_{\gamma_{2\varepsilon}} \frac{(P_\lambda^k)'(z) - 1}{P_\lambda^k(z) - z} dz\right).$$

D'après ces formules, il est clair que les fonctions $N_{k,\varepsilon}(\lambda)$ et $N_{k,2\varepsilon}(\lambda)$ sont continues en λ et à valeurs dans $\mathbb{N}$ par suite constantes. Ainsi $N_{k,\varepsilon}(\lambda) = N_{k,\varepsilon}(\lambda_0)$ et $N_{k,2\varepsilon}(\lambda) = N_{k,2\varepsilon}(\lambda_0)$ pour $|\lambda - \lambda_0| < r$ et $1 \leq k \leq q$. Soit donc

$$\begin{cases} N_{k,\varepsilon}(\lambda) = N_{k,2\varepsilon}(\lambda) = 1 & \text{si} \quad 1 \leq k \leq q-1 \\ N_{k,\varepsilon}(\lambda) = N_{k,2\varepsilon}(\lambda) = q+1 & \text{si} \quad k = q \end{cases} .$$

On en déduit que 0 est le seul point fixe de P_λ^k $(1 \leq k \leq (q-1))$ dans le disque $|z| < 2\varepsilon$, tandis que P_λ^q, pour $0 < |\lambda - \lambda_0| < r$, admet 0 comme point fixe simple et q points fixes dans le disque pointé $0 < |z| < \varepsilon$. Remarquons que P_λ^q, pour $|\lambda - \lambda_0| < r$ n'admet pas de points fixes dans la couronne $\varepsilon \leq |z| < 2\varepsilon$ (car $N_{q,\varepsilon}(\lambda) = N_{q,2\varepsilon}(\lambda) = q+1$). D'autre part, on a $|z| < \varepsilon \Rightarrow |P_\lambda(z)| < 2\varepsilon$; en effet

$$|P_\lambda(z)| = |\lambda z + z^2| \leq |\lambda| \, |z| + |z|^2 \leq (1 + |\lambda - \lambda_0|)|z| + |z|^2$$

$$\leq (1 + r)\varepsilon + \varepsilon^2 < (2 - \varepsilon)\varepsilon + \varepsilon^2 = 2\varepsilon \quad (\text{car } r < 1 - \varepsilon) \quad .$$

Ainsi, si z_0 désigne l'un des points fixes de P_λ^q avec $0 < |z_0| < \varepsilon$, on aura $z_j = P_\lambda^j(z_0)$ pour $0 \leq j \leq q-1$, tous des points fixes de P_λ^q distincts deux à deux dans le disque pointé $0 < |z| < \varepsilon$; par suite les z_j, $0 \leq j \leq q-1$, représentent les q points fixes de P_λ^q pour $|\lambda - \lambda_0| < r$ et forment bien un cycle de longueur q.

Notation : Pour $0 < |\lambda - \lambda_0| < r$, on note par $\rho(\lambda)$ le multiplicateur du cycle périodique de longueur q pour P_λ dans le disque pointé $0 < |z| < \varepsilon$ (donné par le lemme 3).

Lemme 4. $\rho : \lambda \mapsto \rho(\lambda)$ *est une fonction holomorphe sur le disque* $|\lambda - \lambda_0| < r$; *de plus* $\rho(\lambda_0) = 1$.

Preuve : Pour $0 < |\lambda - \lambda_0| < r$, on désigne par $\beta_1(\lambda), \beta_2(\lambda), \ldots, \beta_q(\lambda)$ les q points distincts du cycle de longueur q dans le disque pointé $0 < |z| < \varepsilon$ donné par le lemme 3. On suppose $\beta_j(\lambda) = P_\lambda(\beta_{j-1}(\lambda))$ pour $2 \leq j \leq q$. On construit le polynôme

$$Q_\lambda(z) = (z - \beta_1(\lambda))(z - \beta_2(\lambda)) \ldots \ldots (z - \beta_q(\lambda))$$

$$= z^q + b_{q-1}(\lambda)z^{q-1} + \ldots \ldots + b_1(\lambda)z + b_0(\lambda)$$

où, pour $1 \leq k \leq q$, $b_{q-k}(\lambda) = (-1)^k \sum_{1 \leq j_1 < j_2 < \ldots < j_k \leq q} \beta_{j_1}(\lambda)\beta_{j_2}(\lambda) \ldots \beta_{j_k}(\lambda)$ sont les fonctions symétriques élémentaires des points $\beta_j(\lambda)(1 \leq j \leq q)$. On considère les q polynômes symétriques de Newton en $\beta_j(\lambda)$ $(1 \leq j \leq q)$ définis par

$$\sigma_k(\lambda) = \sum_{j=1}^{q} (\beta_j(\lambda))^k \quad , \quad 1 \leq k \leq q \quad .$$

On sait que chaque fonction symétrique élémentaire s'exprime comme un polynôme en les polynômes symétriques de Newton

$$b_{q-1}(\lambda) = -\sigma_1(\lambda)$$

$$b_{q-2}(\lambda) = \tfrac{1}{2} \left[(\sigma_1(\lambda))^2 - \sigma_2(\lambda) \right]$$

$$\ldots \ldots$$

D'autre part, d'après le théorème du résidu, on a pour $0 < |\lambda - \lambda_0| < r$ et $1 \leq k \leq q$

$$\sigma_k(\lambda) = \frac{1}{2\pi i} \oint_{|z|=\varepsilon} z^k \frac{(P_\lambda^q)'(z) - 1}{P_\lambda^q(z) - z} dz \quad .$$

Par suite, $\lambda \mapsto \sigma_k(\lambda)$ est holomorphe sur le disque $|\lambda - \lambda_0| < r$ avec $\sigma_k(\lambda_0) = 0$. On en déduit que $\lambda \mapsto b_{q-k}(\lambda)$ est holomorphe sur le disque $|\lambda - \lambda_0| < r$ avec $b_{q-k}(\lambda_0) = 0$ et $Q_{\lambda_0}(z) = z^q$. On a

$$\rho(\lambda) = (P_\lambda^q)'(\beta_1(\lambda)) = \prod_{j=1}^{q} P_\lambda'(\beta_j(\lambda)) = \prod_{j=1}^{q} (\lambda + 2\beta_j(\lambda)) = (-2)^q Q_\lambda(-\frac{\lambda}{2}).$$

Il en résulte que $\lambda \mapsto \rho(\lambda)$ est définie et holomorphe sur le disque $|\lambda - \lambda_0| < r$ avec $\rho(\lambda_0) = (-2)^q Q_{\lambda_0}(-\frac{\lambda_0}{2}) = \lambda_0^q = 1$.

Remarque : Il est naturel de penser à $\rho(\lambda_0) = 1$ comme étant $(P_{\lambda_0}^q)'(0)$ où dans ce cas le cycle de longueur q est formé de points tous égaux à 0.

3. Dérivées du multiplicateur et l'indice holomorphe

Dans tout ce paragraphe, on conserve les notations utilisées en 2.

Proposition 1. $\rho'(\lambda_0) = -\dfrac{q^2}{\lambda_0}$ *et* $\rho''(\lambda_0) = \dfrac{q^2 + q^3(2A-1)}{\lambda_0^2}$ *où* $A = i(P_{\lambda_0}^q, 0)$.

Preuve : $\lambda \mapsto \rho(\lambda)$ étant holomorphe sur $|\lambda - \lambda_0| < r$ avec $\rho(\lambda_0) = 1$, donc il existe $0 < r' < r$ vérifiant $\rho(\lambda) \neq 1$ pour $0 < |\lambda - \lambda_0| < r'$. Pour ε donné par le lemme 3 et $|\lambda - \lambda_0| < r'$, on pose

$$\tau(\lambda) = \frac{1}{2\pi i} \oint_{|z|=\varepsilon} \frac{1}{z - P_\lambda^q(z)} dz \quad .$$

Il est clair que $\lambda \mapsto \tau(\lambda)$ est holomorphe dans le disque $|\lambda - \lambda_0| < r'$ avec $\tau(\lambda_0) = i(P_{\lambda_0}^q, 0) = A$; par suite $\tau(\lambda) = A + O(\lambda - \lambda_0)$. D'autre part, d'après le théorème du résidu, on a pour $0 < |\lambda - \lambda_0| < r'$

$$\tau(\lambda) = i(P_\lambda^q, 0) + \sum_{j=1}^{q} i(P_\lambda^q, \beta_j(\lambda))$$

$$= \frac{1}{1 - \lambda^q} + \frac{q}{1 - \rho(\lambda)} \quad \text{(d'après le lemme 1)} \ .$$

Ceci donne

$$\rho(\lambda) = 1 - \frac{q(\lambda^q - 1)}{1 + (\lambda^q - 1)\tau(\lambda)}$$

$$= 1 - q \left[\frac{q}{\lambda_0}(\lambda - \lambda_0) + \frac{q(q-1)}{2\lambda_0^2}(\lambda - \lambda_0)^2 + \mathcal{O}((\lambda - \lambda_0)^3) \right]$$
$$\left[1 - \left(\frac{q}{\lambda_0}(\lambda - \lambda_0) + \mathcal{O}((\lambda - \lambda_0)^2) \right)\left(A + \mathcal{O}(\lambda - \lambda_0) \right) \right]$$

$$= 1 - \left[\frac{q^2}{\lambda_0}(\lambda - \lambda_0) + \frac{q^2(q-1)}{2\lambda_0^2}(\lambda - \lambda_0)^2 + \mathcal{O}((\lambda - \lambda_0)^3) \right]$$
$$\left[1 - \frac{qA}{\lambda_0}(\lambda - \lambda_0) + \mathcal{O}((\lambda - \lambda_0)^2) \right]$$

$$= 1 - \frac{q^2}{\lambda_0}(\lambda - \lambda_0) + \left(\frac{q^3 A}{\lambda_0^2} - \frac{q^2(q-1)}{2\lambda_0^2} \right)(\lambda - \lambda_0)^2 + \mathcal{O}((\lambda - \lambda_0)^3)$$

$$= 1 - \frac{q^2}{\lambda_0}(\lambda - \lambda_0) + \frac{q^2 + q^3(2A-1)}{2\lambda_0^2}(\lambda - \lambda_0)^2 + \mathcal{O}((\lambda - \lambda_0)^3) \ .$$

D'où la proposition.

Proposition 2. *Etant donné* $P_{\lambda_0}(z) = e^{2\pi i p/q} z + z^2$, *alors il existe une fonction analytique* φ *injective au voisinage de 0 avec* $\varphi(0) = 0$, *tel que*

$F_{\lambda_0} = \varphi \circ P_{\lambda_0}^q \circ \varphi^{-1}$ est de la forme $F_{\lambda_0}(z) = z + z^{q+1} + A z^{2q+1} + \mathcal{O}(z^{2q+2})$. Le nombre complexe A est un invariant de conjugaison analytique vérifiant

$$\frac{q+1}{2} - 2^{q-1} \leq R\acute{e}(A) \leq \frac{q+1}{2} \; .$$

Preuve : D'après le lemme 2, on peut écrire $P_{\lambda_0}^q(z) = z + a z^{q+1} + \dots$ avec $a \neq 0$. Dans ces conditions, il est bien connu ([Be]) qu'il existe une fonction analytique φ injective au voisinage de 0, $\varphi(0) = 0$, tel que $F_{\lambda_0} = \varphi \circ P_{\lambda_0}^q \circ \varphi^{-1}$ prend la forme $F_{\lambda_0}(z) = z + z^{q+1} + A z^{2q+1} + \mathcal{O}(z^{2q+2})$.

Un calcul simple donne $A = i(F_{\lambda_0}, 0)$; donc, d'après le lemme 1, on a aussi $A = i(P_{\lambda_0}^q, 0)$ qui est un invariant de conjugaison analytique.

La formule de l'indice holomorphe (voir [M]) pour une fraction rationnelle $f : \overline{\mathbb{C}} \to \overline{\mathbb{C}}$, non identiquement égale à l'identité, exprime que

$$\sum_{f(z)=z, \; z \in \overline{\mathbb{C}}} i(f, z) = 1 \; .$$

Le cas d'un polynôme Q, ∞ est un point fixe superattractif, donc d'après le lemme 1, on a a $i(Q, \infty) = 1$ et on aura

$$\sum_{Q(z)=z, \; z \in \mathbb{C}} i(Q, z) = 0 \; .$$

Dans notre cas, $P_\lambda(z) = \lambda z + z^2$ et l'itérée q-ème P_λ^q admet 2^q points fixes (non nécessairement distincts).

Conservons les notations du 2) et désignons, pour $0 < |\lambda - \lambda_0| < r$, $\beta_1(\lambda), \beta_2(\lambda), \dots, \beta_q(\lambda)$ les q points distincts du cycle de longueur q dans le disque pointé $0 < |z| < \varepsilon$ donné par le lemme 3, et posons $\beta_0(\lambda) = 0$, $\beta_{q+1}(\lambda), \dots, \beta_{2^q-1}(\lambda)$ les autres points fixes de P_λ^q dans $\mathbb{C}$. Ainsi on a

$$\sum_{j=0}^{2^q-1} i(P_\lambda^q, \beta_j(\lambda)) = 0 \; .$$

Soit

$$\tau(\lambda) = \frac{1}{2\pi i} \oint_{|z|=\varepsilon} \frac{1}{z - P_\lambda^q(z)} dz = i(P_\lambda^q, 0) + \sum_{j=1}^{q} i(P_\lambda^q, \beta_j(\lambda))$$

$$= \frac{1}{1-\lambda^q} + \frac{q}{1-\rho(\lambda)} \quad \text{(d'après la preuve de la proposition 1)} \; .$$

Rappelons que $\lim_{\lambda \to \lambda_0} \tau(\lambda) = A$. D'autre part, on sait que le paramètre $\lambda_0 = e^{2\pi i p/q}$ se trouve sur le bord de l'ensemble de Mandelbrot M; par suite on peut choisir une suite $(\lambda_n)_{n \in \mathbb{N}^*}$ de nombres complexes avec $\lambda_n \notin M \ \forall n \in \mathbb{N}^*$ et $\lim_{n \to +\infty} \lambda_n = \lambda_0$. Il en résulte que tout point fixe du polynôme $P^q_{\lambda_n}$ est répulsif; autrement dit $|(P^q_{\lambda_n})'(\beta_j(\lambda_n))| > 1$ pour $0 \le j \le 2^q - 1$. Or la fonction $\omega \mapsto \frac{1}{1-\omega}$ transforme $\{\omega \in \mathbb{C}/|\omega| > 1\}$ sur $\{\omega \in \mathbb{C}/\text{Ré } \omega < \frac{1}{2}\}$; il en résulte, d'après le lemme 1, que

$$\text{Ré}(i(P^q_{\lambda_n}, \beta_j(\lambda_n))) < \tfrac{1}{2} \quad, \forall n \in \mathbb{N}^* \quad \text{et} \quad \forall \ 0 \le j \le 2^q - 1 \ ;$$

et on a

$$\tau(\lambda_n) = i(P^q_{\lambda_n}, 0) + \sum_{j=1}^{q} i(P^q_{\lambda_n}, \beta_j(\lambda_n)) = - \sum_{j=q+1}^{2^q-1} i(P^q_{\lambda_n}, \beta_j(\lambda_n)) \ .$$

On en déduit que

$$\frac{q+1}{2} - 2^{q-1} < \text{Ré } (\tau(\lambda_n)) < \frac{q+1}{2} \quad,$$

et, par passage à la limite quand $n \to +\infty$, on obtient

$$\frac{q+1}{2} - 2^{q-1} \le \text{Ré}(A) \le \frac{q+1}{2} \ .$$

Ceci achève la preuve de la Proposition 2.

4. Application à l'étude d'explosion des cylindres

Pour des propriétés générales sur les phénomènes de bifurcation parabolique et d'explosion des cylindres de Fatou-Ecalle, on pourra consulter les références suivantes [Br], [DH], [L], [S1] et [S2].

Rappels et Notations.

Considérons le polynôme quadratique complexe $P_{\lambda_0}(z) = \lambda_0 z + z^2$ où $\lambda_0 = e^{2\pi i p/q}$. L'origine $z = 0$ est un point fixe parabolique et on a exactement q axes d'attraction et q axes de répulsion dans un petit voisinage de 0 où $P^q_{\lambda_0}$ et $P^{-q}_{\lambda_0}$ sont bien définis. On peut trouver q domaines $\Omega^+_{\lambda_0,k}$ ($k \in \mathbb{Z}/q\mathbb{Z}$) contenant chacun un axe attractif et stable par $P^q_{\lambda_0}$ et q domaines $\Omega^-_{\lambda_0,k}$ ($k \in \mathbb{Z}/q\mathbb{Z}$) contenant chacun un axe répulsif et stable par $P^{-q}_{\lambda_0}$ (le choix de $\Omega^+_{\lambda_0,1}$ restant bien sûr arbitraire avec $\Omega^-_{\lambda_0,1}$ suivant $\Omega^+_{\lambda_0,1}$).

L'espace quotient $\mathcal{C}^+_{\lambda_0,k}$ (resp. $\mathcal{C}^-_{\lambda_0,k}$) de $\Omega^+_{\lambda_0,k}$ (resp. $\Omega^-_{\lambda_0,k}$) sous l'action de $P^q_{\lambda_0}$ (resp. $P^{-q}_{\lambda_0}$) est un cylindre topologique conformément isomorphe à $\mathbb{C}^*$ comme surface de Riemann et appelé cylindre de Fatou-Ecalle.

On notera $S_{\lambda_0,k}^+$ (resp. $S_{\lambda_0,k}^-$) la zone limitée par $\partial\Omega_{\lambda_0,k}^+$ (resp. $\partial\Omega_{\lambda_0,k}^-$) inclus et son image par $P_{\lambda_0}^q$ (resp. $P_{\lambda_0}^{-q}$) non incluse. Ces zones sont appelées des régions fondamentales.

Considérons maintenant des perturbations de P_{λ_0} dans la famille des polynômes quadratiques $P_\lambda(z) = \lambda z + z^2$ où $\lambda = \lambda_0 e^{2\pi i\alpha}$ assujetti à la condition $\alpha \neq 0$ et $|\arg\alpha| < \frac{\pi}{4}$.

Sous ces hypothèses et pour λ assez voisin de λ_0, on a un cycle de longueur q pour P_λ noté $(\beta_1(\lambda), \beta_2(\lambda), \ldots, \beta_q(\lambda))$ donné par le lemme 3 et on peut trouver $2q$ régions fondamentales $S_{\lambda,k}^+$ et $S_{\lambda,k}^-$ ($k \in \mathbb{Z}/q\mathbb{Z}$) dont le bord de chacune est formé de deux courbes (une image de l'autre par P_λ^q) joignant 0 et le point $\beta_k(\lambda)$ du cycle de longueur q.

Les cylindres quotients $C_{\lambda,k}^+ = S_{\lambda,k}^+/P_\lambda^q$ et $C_{\lambda,k}^- = S_{\lambda,k}^-/P_\lambda^{-q}$ sont encore conformément isomorphes à $\mathbb{C}^*$ comme surface de Riemann (en convenant d'envoyer $\beta_k(\lambda)$ en 0 et 0 en ∞).

Partons d'un point de $S_{\lambda,k}^+$, en itérant P_λ^q un grand nombre de fois, on arrive à $S_{\lambda,k}^-$ et par suite les cylindres $C_{\lambda,k}^+$ et $C_{\lambda,k}^-$ sont désormais identifiés par l'action de P_λ^q. Cette identification induit un isomorphisme G_λ de $\mathbb{C}^*$ sur $\mathbb{C}^*$ de la forme $G_\lambda(z) = G_\lambda(1) \cdot z$. Par construction des cylindres, un voisinage de $\beta_k(\lambda)$ dans $S_{\lambda,k}^-$ est envoyé par itérations de P_λ^q sur un voisinage de $\beta_k(\lambda)$ dans $S_{\lambda,k}^+$ (ce phénomène reste vrai pour $\lambda = \lambda_0$), ceci induit une application holomorphe bijective F_λ d'un voisinage de zéro dans $\mathbb{C}^*$ vers un voisinage de zéro dans $\mathbb{C}^*$ prolongeable par continuité par $F_\lambda(0) = 0$ et vérifiant $F_\lambda'(0) \neq 0$.

Dans ([DH],II) page 75, on a le développement limité suivant :

$$(F_\lambda'(0)) \cdot G_\lambda(1) = \exp\left(\frac{4\pi^2}{\log(\rho(\lambda))}\right) = \exp\left(\frac{4\pi^2}{\rho'(\lambda_0)(\lambda - \lambda_0)} + K + o(1)\right)$$

où log désigne la branche principale du logarithme, $\rho(\lambda)$ désigne le multiplicateur du cycle $(\beta_1(\lambda), \ldots, \beta_q(\lambda))$ de longueur q donné par le lemme 3 et K est une constante (dépendant de λ_0).

Notre résultat dans cette section montre que la constante K s'exprime en fonction du nombre complexe $A = i(P_{\lambda_0}^q, 0)$ (indice holomorphe de $P_{\lambda_0}^q$ en zéro).

Proposition 3. *Pour* $\lambda = \lambda_0 e^{2\pi i\alpha}$ *assez voisin de* λ_0, *assujetti à la condition* $\alpha \neq 0$ *et* $|\arg\alpha| < \frac{\pi}{4}$, *on a le développement suivant :*

$$(F_\lambda'(0)) \cdot G_\lambda(1) = \exp\left[\frac{2\pi i}{q^2\alpha} + \frac{4\pi^2}{q}\left(\frac{q+1}{2} - A\right) + o(1)\right]$$

où $A = i(P_{\lambda_0}^q, 0)$.

Preuve : D'après ([DH], II) page 75, on a :

$$(F'_\lambda(0)) \cdot G_\lambda(1) = \exp\left(\frac{4\pi^2}{\log(\rho(\lambda))}\right)$$

on peut écrire :

$$\log(\rho(\lambda)) = \log(\rho(\lambda_0)) + \frac{\rho'(\lambda_0)}{\rho(\lambda_0)}(\lambda - \lambda_0)$$
$$+ \frac{\rho''(\lambda_0)\rho(\lambda_0) - (\rho'(\lambda_0))^2}{2(\rho(\lambda_0))^2}(\lambda - \lambda_0)^2 + o((\lambda - \lambda_0)^2)$$

comme $\rho(\lambda_0) = 1$, on aura :

$$\log(\rho(\lambda)) = \rho'(\lambda_0)(\lambda - \lambda_0)\left[1 + \frac{\rho''(\lambda_0) - (\rho'(\lambda_0))^2}{2\rho'(\lambda_0)}(\lambda - \lambda_0) + o(\lambda - \lambda_0)\right]$$

d'où :

$$\frac{1}{\log(\rho(\lambda))} = \frac{1}{\rho'(\lambda_0)(\lambda - \lambda_0)}\left[1 - \frac{\rho''(\lambda_0) - (\rho'(\lambda_0))^2}{2\rho'(\lambda_0)}(\lambda - \lambda_0) + o(\lambda - \lambda_0)\right]$$
$$= \frac{1}{\rho'(\lambda_0)(\lambda - \lambda_0)} - \tfrac{1}{2}\left(\frac{\rho''(\lambda_0)}{(\rho'(\lambda_0))^2} - 1\right) + o(1)$$
$$= \frac{1}{\lambda_0\rho'(\lambda_0)(e^{2\pi i\alpha} - 1)} - \tfrac{1}{2}\left(\frac{\rho''(\lambda_0)}{(\rho'(\lambda_0))^2} - 1\right) + o(1)$$
$$= \frac{1}{\lambda_0\rho'(\lambda_0)2\pi i\alpha}\left(1 - \frac{2\pi i\alpha}{2} + o(\alpha)\right) - \tfrac{1}{2}\left(\frac{\rho''(\lambda_0)}{(\rho'(\lambda_0))^2} - 1\right) + o(1)$$
$$= \frac{1}{\lambda_0\rho'(\lambda_0)2\pi i\alpha} - \frac{1}{2\lambda_0\rho'(\lambda_0)} - \tfrac{1}{2}\left(\frac{\rho''(\lambda_0)}{(\rho'(\lambda_0))^2} - 1\right) + o(1)\ .$$

Tenant compte de la proposition 1, on a :

$$(F'_\lambda(0)) \cdot G_\lambda(1) = \exp\left(\frac{4\pi^2}{\log(\rho(\lambda))}\right)$$
$$= \exp\left[\frac{2\pi i}{q^2\alpha} + \frac{4\pi^2}{q}\left(\frac{q+1}{2} - A\right) + o(1)\right]\ .$$

Ceci achève la preuve de la proposition 3.

Remarque. : Le développement limité donné par la proposition 3 nous permet de retrouver l'inégalité Ré$(A) \leq \frac{q+1}{2}$ déjà démontrée dans la proposition 2, en effet : prenons $\lambda = \lambda_0 e^{2\pi i\alpha}$ avec $\alpha > 0$ assez petit, on aura:

$$|(F'_\lambda(0)) \cdot G_\lambda(1)| = \exp\left[\frac{4\pi^2}{q}\left(\frac{q+1}{2} - \text{Ré}(A)\right) + o(1)\right] \tag{2}$$

d'autre part on a:

$$\left| (F'_\lambda(0)) \cdot G_\lambda(1) \right| = \left| \exp\left(\frac{4\pi^2}{\log(\rho(\lambda))} \right) \right| = \exp\left(\frac{4\pi^2 \log|\rho(\lambda)|}{|\log(\rho(\lambda))|^2} \right) \qquad (3)$$

or l'origine $z = 0$ est un point fixe indifférent pour $P_\lambda(z) = \lambda z + z^2$ car $|\lambda| = 1$, par suite le multiplicateur $\rho(\lambda)$ du cycle de longueur q donné par le lemme 3 doit vérifier $|\rho(\lambda)| > 1$ (voir [Bl]).

Ainsi en utilisant (2) et (3) et par passage à la limite quand α tend vers 0^+, on obtient

$$\exp\left[\frac{4\pi^2}{q} \left(\frac{q+1}{2} - \mathrm{R\acute{e}}(A) \right) \right] \geq 1$$

d'où

$$\mathrm{R\acute{e}}(A) \leq \frac{q+1}{2}.$$

Remerciements :

Je tiens à remercier les Professeurs A. Douady, J.H. Hubbard, J. Milnor et M. Shishikura pour leurs aides et encouragements.

Bibliographie

[Be] A.F. BEARDON, Iteration of rational maps, Springer-Verlag, New-York, 1991.

[Bl] P. BLANCHARD, *Complex analytic dynamics on the Riemann sphere*, Bull. Amer. Math. Soc. 11 (1984), 85-141.

[Br] B. BRANNER, *Fatou coordinates and Ecalle cylinders*, manuscrit 1993.

[DH] A. DOUADY et J.H. HUBBARD, Etudes dynamiques des polynômes complexes I, II, Publications Mathématiques d'Orsay, 1984.

[L] P. LAVAURS, *Systèmes dynamiques holomorphes : explosion de points périodiques paraboliques*, thèse de doctorat de l'Université de Paris-Sud, Orsay, France, 1989.

[M] J. MILNOR, *Dynamics in one complex variable, Introductory lectures*, Preprint of IMS, SUNY at Stony Brook, 1990/5.

[S1] M. SHISHIKURA, *On the parabolic bifurcation of holomorphic maps*, Proc. of Nogoya Conf., Sept. 1990.

[S2] —, *The Hausdorff dimension of the boundary of the Mandelbrot set and Julia sets*, Annals of Math. **147**(1998), 225-267.

Institut préparatoire aux études d'ingénieur de Nabeul, Département de Mathématiques et d'Informatique, Merazka, 8000 Nabeul, Tunisie.

An alternative proof of Mañé's theorem on non-expanding Julia sets

Mitsuhiro Shishikura and Tan Lei

Abstract

We give a proof of the following theorem of Mañé: A forward invariant compact set in the Julia set of a rational map is either expanding, contains parabolic points or critical points, or intersects the ω-limit set of a recurrent critical point. Moreover the boundary of a Siegel disc is contained in the ω-limit set of some recurrent critical point. We establish also a semi-local version of the result.

1 Statements

A classical theorem of Fatou says that for a given rational map f, if no critical points are in its Julia set $J(f)$, then f is uniformly expanding on a neighborhood of $J(f)$. Such a map is called *hyperbolic*. A local version of this says that if $x \in J(f)$ is not in the closure of the critical orbits, there is a neighborhood U of x such that the diameters of the components of $f^{-n}U$ shrink to zero as $n \to \infty$.

The next best category of rational maps are those with no parabolic orbits but with each critical point in the Julia set having a finite orbit. They are called *sub-hyperbolic* maps. Such maps expand also uniformly on a neighborhood of the Julia set, but with respect to a Riemannian metric with certain singularities.

If we now allow parabolic orbits, but still require critical points in the Julia set to have a finite orbit, we are into *geometrically finite* maps. These maps have also some sort of expansion on the Julia set, but only on sector neighborhoods of the parabolic points. See for example [TY].

A critical point in the Julia set having a finite orbit is a particular case of being *non-recurrent*, in the following sense: Denote by $\omega(c)$, the ω-limit set of c, to be $\{z \in \overline{\mathbb{C}} \mid \text{there exists } n_k \to \infty \text{ such that } z = \lim f^{n_k}(c)\}$. We say that c is *recurrent* if $c \in \omega(c)$. Mañé has a result that establishes expansion properties (or contraction properties of f^{-1}), at points or on compact invariant subsets of the Julia set away from the parabolic points and the ω-limit sets of recurrent critical points. This is an important result with many applications, for example combined with conditions on the critical orbits (such as Collet-Eckmann conditions), one may obtain metrical or geometrical information about the Julia set of more general type of rational maps.

In this paper we present an alternative proof of Mañé's result, along with the first application by Mañé, which is about the boundary of Siegel discs.

265

We then apply it to a semi-local setting, following ideas of H. Kriete (see [Kr]). There are many further applications in the literature, see for example [Yin], in this volume, [CJY] and [PR].

Mañé's result can be stated in the form of the following three theorems:

Denote by f a given rational map of degree at least two, by $J(f)$ its Julia set (which may or may not be $\overline{\mathbb{C}}$), and by N an integer depending only on f (to be made precise in §2).

Theorem 1.1. *(point version) If a point $x \in J(f)$ is not a parabolic periodic point and is not contained in the ω-limit set of a recurrent critical point, then for all $\varepsilon > 0$ there exists a neighborhood U of x such that, for each $n \geq 0$ and each connected component V' of $f^{-n}(U)$,*

(a) *the spherical diameter of V' is $\leq \varepsilon$ and $\deg(f^n : V' \to U) \leq N$.*

(b) *For all $\varepsilon_1 > 0$ there exists $n_0 > 0$ such that if $n \geq n_0$, the spherical diameter of V' is $\leq \varepsilon_1$.*

Theorem 1.2. *(compact set version) Let $\Lambda \subset J(f)$ be a compact invariant set (i.e $f(\Lambda) \subset \Lambda$) not containing critical points or parabolic periodic points. If Λ is disjoint from the ω-limit set of every recurrent critical point, then it is expanding (see §3 for the precise definition). In particular a Cremer periodic point must be contained in $\omega(c)$ for some recurrent critical point c.*

Theorem 1.3. *(an application) Let Γ be either the orbit of a Cremer periodic point, the boundary of a Siegel disc, or a connected component of the boundary of a Herman ring. There exists a recurrent critical point c such that $\omega(c) \supset \Gamma$.*

A recent application of these results by Yin ([Yi]) shows that if $J(f)$ contains no recurrent critical points, then it is *shallow* (a notion introduced by C. McMullen), and therefore has Hausdorff dimension < 2 (a result first obtained by Urbanski).

Section 2 presents a proof of Theorem 1.1 due to M. Shishikura. It is edited by Tan Lei based on a hand-written manuscript of M. Shishikura, and is presented in a form that is also valid for the case $J(f) = \overline{\mathbb{C}}$, and for the generalization in Section 4. Appendix E is written by M. Shishikura. The rest is written by Tan Lei. Section 3 contains proofs of Theorems 1.2 and 1.3 following Mañé. Section 4 translates the above three theorems into a semi-local version, and gives a brief description of an interesting application of H. Kriete. The appendices contain some classical estimates on the Poincaré metric and alternative proofs of some of the lemmas.

Our proof of Theorem 1.1 is in spirit the same as Mañé's original one, but differs in presentation. It emphasizes on the use of Poincaré metric, and gives a direct argument rather than by contradiction. It also provides a bit of more quantitative information.

The authors would like to thank C. McMullen and A. Manning for helpful discussions and K. Astala for allowing us to include his elementary proof of one of our lemmas.

2   Proof of Theorem 1.1

For $z \in \mathbb{C}$, denote by $D_r(z)$ the open disc centred at z of radius r, and by $\mathbb{D}$ the unit disc. We state first a general result:

Lemma 2.1. *For any $0 < r < 1$ and any positive integer p, there exists a constant $C(p, r) > 0$ such that for any holomorphic proper map $g : V \to \mathbb{D}$ of degree $\leq p$, with V simply connected, each connected component of $g^{-1}(\overline{D_r}(0))$ has diameter $\leq C(p, r)$ with respect to the Poincaré metric on V. Moreover $\lim_{r \to 0} C(p, r) = 0$.*

Proof. (For a more elementary proof due to K. Astala, see Appendix D.) Let A_i be the (finitely many) maximal concentric open annuli in $\mathbb{D}$ surrounding $\overline{D_r}(0)$ such that A_i does not contain critical values of g. For $E \subset V$ a component of $g^{-1}(\overline{D_r}(0))$ and each i, there is $A_i' \subset V$ such that A_i' surrounds E and $g : A_i' \to A_i$ is a covering. So

$$\text{mod } A_i' = \frac{1}{\deg(g|_{A_i'})} \text{ mod } A_i \geq \frac{1}{p} \text{ mod } A_i \quad \text{and}$$

$$\text{mod } (V \smallsetminus E) \geq \sum_i \text{mod } A_i' \geq \frac{1}{p} \sum_i \text{mod } A_i = \frac{\log(1/r)}{2\pi p} .$$

The rest is classical (cf. Appendix B). ∎

Next we define a universal constant.

Definition of N_0: There exist $z_1, \cdots, z_{N_0-1} \in \mathbb{D}$ such that $\{\frac{2}{3} \leq |z| \leq 1\} \subset \bigcup_{i=1}^{N_0-1} D_{\frac{1}{3}}(z_i)$.

Let f be the given rational map of degree at least two.

Definition of Ω_0 and J': Set $\Omega_0 = \overline{\mathbb{C}}$ if $J(f) = \overline{\mathbb{C}}$. Otherwise, by Sullivan's classification of the dynamics in the Fatou set we can find a non-empty compact set L as a disjoint union of finitely many closed Jordan domains and closed sub-annuli in the Herman rings and their preimages such that $f(L) \subset L$, L contains all critical points in the Fatou set and $L \cap J(f) = \{$parabolic periodic points$\}$. Set $\Omega_0 = \overline{\mathbb{C}} \smallsetminus L$. In both cases we have $f^{-1}(\Omega_0) \subset \Omega_0$ and $\Omega_0 \cap J(f) = J(f) \smallsetminus \{$parabolic periodic points$\}$.

In case $J(f) = \overline{\mathbb{C}}$ set $J' = \overline{\mathbb{C}}$. In case $J(f) \neq \overline{\mathbb{C}}$, i.e. $\overline{\Omega_0} \neq \overline{\mathbb{C}}$, we may then assume $\overline{\Omega_0} \subset \mathbb{C}$ (so it is bounded). We set $K' = \{z, f^n(z) \in \overline{\Omega_0} \text{ for all } n \geq 0\}$ and $J' = \partial K'$. Then $f^{-1}(J') \subset J'$ and the Hausdorff distance between J'

and $\partial f^{-n}(\Omega_0)$ tends to 0 as $n \to \infty$, with respect to the Euclidean metric (in fact both K' and J' are totally invariant but we will not need these properties). Notice that in this case $J(f) \subset J'$, moreover L can be chosen so that $J(f) \neq J'$ if and only if some Siegel disc or Herman ring of f contains postcritical points.

In both cases $\Omega_0 \cap J'$ contains no parabolic periodic points. The critical points in Ω_0 are all contained in J' and are of the following three types:
I. c is a recurrent critical point, i.e. $c \in \omega(c)$.
II. c is a non-recurrent critical point with $c \in \omega(c')$ for some recurrent critical point c'.
III. c is a non-recurrent critical point not contained in the ω-limit set of any recurrent critical point.

Definition of c_i, N and C_0: Denote by $c_1, \cdots, c_\nu$ the critical points of type III, $\deg(f, c)$ the local degree of f at c, and by N a positive integer greater than or equal to $\prod_{i=1}^{\nu} \deg(f, c_i)$ (this makes sense even when there are no critical points of type III)[1]. Set $C_0 = N \cdot N_0 \cdot C(N, 2/3)$, where $C(\cdot, \cdot)$ is the constant given by Lemma 2.1.

Denote by $d(z, E)$ the Euclidean distance between a point $z \in \mathbb{C}$ and a closed subset E of $\mathbb{C}$, and by $\mathrm{diam}_W(W')$ the diameter of W' with respect to the Poincaré metric of W (assuming that W is a hyperbolic Riemann surface and $W' \subset W$).

Lemma 2.2. *(1) Let $x \in \Omega_0 \cap J'$. There exists Ω_1 with $x \in \Omega_1$ and satisfying:*

$$(*) \begin{cases} \Omega_1 \subset \Omega_0 \text{ is open and hyperbolic, not necessarily connected,} \\ f^{-1}(\Omega_1) \subset \Omega_1, \ d_{\Omega_1}(f(c_i), f(c_j)) > C_0, \text{ if } f(c_i) \neq f(c_j) \\ d_{\Omega_1}(c_i, f^n(c_i)) > C_0, \text{ for } i = 1, \cdots, \nu \text{ and } n \geq 1, \end{cases}$$

where d_{Ω_1} is the Poincaré metric on Ω_1.

(2). Given Ω_1 satisfying $(*)$ above, for all $C > 0$ and $\varepsilon_1 > 0$, there exists n_0 such that if $n \geq n_0$ then for all $W' \subset W \subset f^{-n}(\Omega_1)$ such that W is simply connected, $W' \cap J' \neq \emptyset$ and $\mathrm{diam}_W(W') \leq C$, the set W' has spherical diameter $< \varepsilon_1$.

Proof. (For a different proof relating directly the Poincaré metric to the spherical metric, see Appendix E.)

(1). For each $i = 1, \cdots, \nu$, there exists a repelling periodic point t_i arbitrarily close to c_i (and therefore with $f(t_i)$ close to $f(c_i)$) such that the orbit of t_i does not contain x. We choose these points so that for $Z_0 = \bigcup_i \mathrm{Orbit}(t_i)$, the set $\overline{\mathbb{C}} \setminus Z_0$ is a hyperbolic surface, $d_{\overline{\mathbb{C}} \setminus Z_0}(f(c_i), f(c_j)) > C_0$

---

[1]Note that N can be chosen depending only on the degree d of f, for example $= d^{2d-2}$.

if $f(c_i) \neq f(c_j)$ and $d_{\overline{\mathbb{C}} \smallsetminus Z_0}(c_i, f^n(c_i)) > C_0$ for $i = 1, \cdots, \nu$ and $n \geq 1$. Now set $\Omega_1 = \Omega_0 \smallsetminus Z_0 \subset \overline{\mathbb{C}} \smallsetminus Z_0$. It satisfies all the conditions.

(2) With the help of a Möbius transformation we may assume $0, \infty \notin \Omega_1$. We will need the following inequality. Let $a \in W' \subset W \subset \Omega \subset \overline{\mathbb{C}} \smallsetminus \{0, \infty\}$ with W a simply connected domain and $\operatorname{diam}_W(W') \leq C$; we have

$$(**) \quad \operatorname{diam}_{\text{spherical}} W' \leq C' \cdot \inf \left\{ d(a, \partial\Omega), \frac{1}{|a|} \right\}$$

with C' a constant depending only on C. See Appendix A for a proof.

Case $J' \neq \overline{\mathbb{C}}$. We may then assume $J' \subset \overline{D_R}(0)$ for some $R > 0$. For W', W as in (2), choose $a \in W' \cap J'$ we get from $(**)$,

$$\operatorname{diam}_{\text{spherical}} W' \leq C' \cdot d(a, \partial f^{-n}(\Omega_1)) \ .$$

Since the Hausdorff distance (with respect to the Euclidean metric) between J' and $\partial f^{-n}(\Omega_1)$ tends to 0 as $n \to \infty$, there is n_0 such that for $n \geq n_0$ and every $a \in J'$ we have $d(a, \partial f^{-n}(\Omega_1)) \leq \varepsilon_1/C'$.

Case $J' = J(f) = \overline{\mathbb{C}}$. Let $S = C'/\varepsilon_1$. For a pair (W', W) as in (2), if there is $a \in W'$ with $|a| > S$, then by the inequality $(**)$ the spherical diameter of W' is less than ε_1. It remains to consider a pair W', W with $W' \subset \overline{D_S}(0)$. Choose $Z \subset \overline{\mathbb{C}} \smallsetminus \Omega_1$ a periodic orbit (which is surely non-exceptional). Now $J(f) \cap \overline{D_S}(0)$ is covered by finitely many discs $D_r(a_j)$ with $r = \varepsilon_1/(2C')$ and $a_j \in J(f) \cap \overline{D_S}(0)$. By properties of the Julia set, for each j, there is $N(j)$ such that $f^n(D_r(a_j)) \supset Z$ for $n \geq N(j)$. Hence $D_r(a_j) \cap f^{-n}(Z) \neq \emptyset$ for $n \geq N(j)$. Now let $n_0 = \max_j N(j)$. Fix $n \geq n_0$. Then for any $a \in J(f) \cap \overline{D_S}(0)$, there is j such that $a \in D_r(a_j)$. So $d(a, \partial f^{-n}(\Omega_1)) \leq d(a, f^{-n}(Z)) \leq 2r = \varepsilon_1/C'$. ∎

Lemma 2.3. *For any Ω_1 satisfying $(*)$ above, let $U_0 \subset \mathbb{C}$ be a round disc such that $U_0 \subset \Omega_1 \smallsetminus \bigcup_c \text{recurrent critical point} \omega(c)$ and $\operatorname{diam}_{\Omega_1} U_0 \leq C_0$. Then, for every $n \geq 0$,*

deg(n). *for every $D_s(z) \subset U_0$ with $0 < s < d(z, \partial U_0)/2$, and every connected component V' of $f^{-n}(D_s(z))$, V' is simply connected and $\deg(f^n : V' \to D_s(z)) \leq N$;*

diam(n). *for every $D_r(w) \subset U_0$ with $0 < r < d(w, \partial U_0)/2$, and every connected component V of $f^{-n}(D_r(w))$, $\operatorname{diam}_{\Omega_1} V \leq C_0$.*

Proof. We claim at first that $f^{-n}(U_0)$ for any $n \geq 0$ contains no critical points other than $c_1, \cdots, c_\nu$. First note that $f(\omega(c)) = \omega(c)$ for a recurrent critical point c. So for $\Omega_2 = \Omega_1 \smallsetminus \bigcup_c \text{recurrent critical point } \omega(c)$, we have $f^{-1}(\Omega_2) \subset \Omega_2$. In particular $f^{-n}(U_0) \subset \Omega_2$ for all $n \geq 0$. Secondly recall that the critical points in Ω_1 are in three different types, I, II and III, and those of types I and II are contained in $\bigcup_c$ recurrent critical point $\omega(c)$ (notice that for c a recurrent critical point, $\omega(c)$ coincides with the closure of the orbit $\{f^n(c), \; n > 0\}$). Therefore Ω_2, hence $f^{-n}(U_0)$, contains only critical points of type III, i.e. $c_1, \cdots, c_\nu$.

Now let us prove the assertion by induction on n. For $n = 0$, it is obvious. Suppose it is true up to $n - 1$. We will prove **deg(n)** and **diam(n)** in two distinct steps.

Step 1. deg(n) *follows essentially from* **diam(0)**, $\cdots$, **diam(n-1)**.

Let $D_s(z)$ be as in the Lemma. Let V' be a component of $f^{-n}(D_s(z))$. Then $V_j = f^j(V')$ are components of $f^{-(n-j)}(D_s(z))$ $(j = 0, 1, \cdots, n)$. So, by the hypothesis of induction, for $j = 1, \cdots, n$, the set V_j is simply connected and $\mathrm{diam}_{\Omega_1}(V_j) \leq C_0$.

Now V' is also a component of $f^{-1}(V_1)$. Since $\mathrm{diam}_{\Omega_1}(V_1) \leq C_0$, V_1 contains at most one critical value in $\{f(c_i), i = 1, \cdots, \nu\}$ (by Condition $(*)$), and contains no critical values of other types of critical points. Therefore V' is simply connected and $\deg(f : V_j \to V_{j+1}) = \deg(f, c_i)$ if $c_i \in V_j$ and $= 1$ otherwise.

Fix $i \in \{1, \cdots, \nu\}$. If $c_i \in V_j$ for some maximal j between 1 and n, then $f^m(c_i) \notin V_j$ for $m \geq 1$, by the induction assumption $\mathrm{diam}_{\Omega_1}(V_j) \leq C_0$ and Condition $(*)$. Hence $c_i \notin V_l$ for $0 \leq l < j$. This means that c_i appears at most once in $V_0 = V', V_1, \cdots, V_n = D_s(z)$. Since no critical points other than $c_1, \cdots, c_\nu$ occur in these components, we have

$$\deg(f^n : V' \to D_s(z)) = \prod_{j=0}^{n-1} \deg(f : V_j \to V_{j+1}) \leq \prod_i \deg(f, c_i) \leq N \; .$$

Thus **deg(n)** is proved for any $D_s(z)$ as in the Lemma.

Step 2. *Given* $D_r(w)$ *as in the Lemma,* **diam(n)** *for* $D_r(w)$ *follows from* **deg(n)** *applied to several* $D_s(z)$*'s.*

Let V be a component of $f^{-n}(D_r(w))$. Apply **deg(n)** to $D_r(w)$ we know that V is simply connected and $\deg(f^n : V \to D_r(w)) \leq N$.

Now choose $z_1, \cdots, z_{N_0-1} \in D_r(w)$ such that

$$D_r(w) \smallsetminus D_{\frac{2}{3}r}(w) \subset \bigcup_{i=1}^{N_0-1} D_{\frac{1}{3}r}(z_i) \; .$$

We have, for $i = 1, \cdots, N_0 - 1$,

$$d(z_i, \partial U_0) \geq d(w, \partial U_0) - d(w, z_i) \geq d(w, \partial U_0) - r > 2r - r = r \; .$$

So we can apply **deg(n)** to $D_{\frac{r}{2}}(z_i)$ and V' a component of $f^{-n}(D_{\frac{r}{2}}(z_i))$ to conclude that V' is simply connected and $\deg(f^n : V' \to D_{\frac{r}{2}}(z_i)) \leq N$.

Since $\frac{1}{3}r = \frac{2}{3}\cdot\frac{1}{2}r$, by Lemma 2.1 each connected component of $f^{-n}(D_{\frac{1}{3}r}(z_i))$ in V' has diameter $\leq C(N, \frac{2}{3})$ with respect to $d_{V'}$, and hence with respect to d_{Ω_1} since $V' \subset f^{-n}(\Omega_1) \subset \Omega_1$.

Similarly each connected component of $f^{-n}(D_{\frac{2}{3}r}(w))$ has diameter at most $C(N, \frac{2}{3})$ with respect to d_{Ω_1}.

Now $D_r(w)$ is the union of the following N_0 open connected sets: $D_{\frac{2}{3}r}(w)$ and $D_{\frac{1}{3}r}(z_i) \cap D_r(w)$ for $i = 1, \cdots, N_0 - 1$. Since $f^n : V \to D_r(w)$ is a proper map of degree $\leq N$, the preimage by f^{-n} of each of the above N_0 sets has at most N connected components in V, and V is covered by these components. Therefore V is covered by at most $N \cdot N_0$ sets of diameter $\leq C(N, \frac{2}{3})$. Hence $\operatorname{diam}_{\Omega_1} V \leq N \cdot N_0 \cdot C(N, \frac{2}{3}) = C_0$ since V is connected.

This completes the induction. ∎

Proof of Theorem 1.1. Let $x \in \Omega_0 \cap J'$ not contained in the ω-limit set of a recurrent critical point. By Lemma 2.2(1), there is Ω_1 satisfying $(*)$ and

$$x \in (\Omega_1 \cap J') \setminus \bigcup_{c \text{ recurrent critical point}} \omega(c) .$$

We may assume $0, \infty \notin \Omega_1$. Fix $\varepsilon > 0$. We will prove Theorem 1.1 for all $x' \in (\Omega_1 \cap J') \setminus \bigcup_{c \text{ recurrent critical point}} \omega(c)$ with bounds depending only on Ω_1.

First note that there exists $\delta > 0$, if $V' \subset V \subset \Omega_1$ with V simply connected and $\operatorname{diam}_V(V') \leq \delta$ then the spherical diameter of V' is less than ε (see Appendix A). Take $0 < \rho < 1$ such that $C(N, \rho) \leq \delta$ (where $C(N, \rho)$ is given by Lemma 2.1).

Let $x' \in \Omega_1 \cap J' \setminus \bigcup_c \text{ recurrent critical point } \omega(c)$. Let U_0 be a round disc such that $x' \in U_0 \subset \Omega_1 \setminus \bigcup_c \text{ recurrent critical point } \omega(c)$ and $\operatorname{diam}_{\Omega_1} U_0 \leq C_0$ (as in Lemma 2.3).

Let $r > 0$ be such that $r < \frac{1}{2}d(x', \partial U_0)$. Define $U = D_{\rho r}(x') \subset D_r(x')$.

Fix $n \geq 0$. Let V be a connected component of $f^{-n}(D_r(x'))$ and V' be a connected component of $f^{-n}(U)$ in V. Then by Lemma 2.3 both V and V' are simply connected and $\deg(f^n : V \to D_r(x')) \leq N$. So $\operatorname{diam}_V V' \leq C(N, \rho) \leq \delta$ (Lemma 2.1). Hence V' has spherical diameter $< \varepsilon$. This proves (a). For (b), let $\varepsilon_1 > 0$ and $C = C(N, \rho)$. Let n_0 be the integer given in Lemma 2.2(2) (which depends only on Ω_1) and assume $n \geq n_0$. For V and V' as above (relative to f^{-n}) we have $V' \subset V \subset f^{-n}(\Omega_1)$, V is simply connected, $V' \cap J' \neq \emptyset$ (since $x' \in J' \cap U$ and $f^{-1}(J') \subset J'$) and $\operatorname{diam}_V(V') \leq C(N, \rho) = C$. So we can apply Lemma 2.2(2) to $W = V$ and $W' = V'$ to conclude that the spherical diameter of V' is less than ε_1.

This ends the proof of Theorem 1.1. ∎

3 Proofs of Theorems 1.2 and 1.3

Proof of Theorem 1.2. We say that a compact set Λ is *expanding* if there is N such that for any $n \geq N$, $\min_{z \in \Lambda} ||(f^n)'(z)|| > 1$, where $|| \cdot ||$ is with respect to the spherical metric.

Assume that Λ is a forward invariant compact set not containing critical points and parabolic points. We may assume $0, \infty \notin \Lambda$ and thus work with Euclidean metric rather than spherical metric.

Assume by contradiction that Λ is not expanding. In other words, there are $n_k \to \infty$, $z_k \in \Lambda$, such that $|(f^{n_k})'(z_k)| \leq 1$. We will show that any accumulation point of $\{f^{n_k}(z_k)\}_{k \in \mathbb{N}}$ is in the ω-limit set of some recurrent critical point.

Assume not. We may assume the entire sequence $f^{n_k}(z_k)$ converges to a point $x \in \Lambda$. Then x satisfies the conditions of Theorem 1.1. Take $\varepsilon > 0$ such that $d(\Lambda, c) > 2\varepsilon$ for every critical point c. Let U be a round disc centred at x associated to ε given by Theorem 1.1. For k large, $f^{n_k}(z_k)$ is in U. Let V_k be the component of $f^{-n_k}(U)$ containing z_k. By Theorem 1.1, $\text{diam}(f^j(V_k)) \leq \varepsilon$ for $0 \leq j \leq n_k$. Since $f^j(z_k) \in \Lambda$ for all j, we have $f^j(V_k)$ does not contain critical points of f. Therefore $f^{n_k} : V_k \to U$ is a bijection. Let $\varphi_k : U \to V_k$ be the inverse. Then the family $\{\varphi_k, \ k \in \mathbb{N}\}$ is normal and any limit function φ must be constant, since $\text{diam}V_k \to 0$ as $k \to \infty$ (Theorem 1.1). This contradicts the fact that

$$|\varphi'(x)| = \lim_{k \to \infty} |\varphi_k'(f^{n_k}(z_k))| = \lim_{k \to \infty} \frac{1}{|(f^{n_k})'(z_k)|} \geq 1 \ .$$

∎

Proof of Theorem 1.3. The orbit of a Cremer periodic point is in J', invariant and non-expanding. So by Theorem 1.2 it intersects $\omega(c)$ for some recurrent critical point c. It is therefore contained in $\omega(c)$.

Now let Γ denotes either the boundary of a Siegel disc or a component of the boundary of a Herman ring. We want to show that there is a recurrent critical point c such that $\Gamma \subset \omega(c)$. The proof consists of two steps.

1. $\Gamma \subset \bigcup_c$ recurrent critical point $\omega(c)$.

Since there are only finitely many (recurrent) critical points, the right hand set is closed. Note also that there are only finitely many critical points and parabolic periodic points.

Assume the assertion is not true. Then there is $x \in \Gamma$ satisfying the conditions of Theorem 1.1. There is therefore a connected open neighborhood V of x such that components of $f^{-nq}(V)$ intersecting Γ have diameter tending to 0. On the other hand, consider the "harmonic measure" μ of Γ, namely that induced by the boundary map of the conformal linearization map $\varphi : A \to B$, with B the Siegel disc with boundary Γ or respectively the Herman ring with one boundary component Γ, and with $A = \mathbb{D}$, or respectively an annulus. It is known that μ is non-atomic with support the whole Γ, f^q preserves μ and is ergodic with respect to it. From the properties of V and $f^{-nq}(V)$ we know that $V \cap \Gamma$ has positive harmonic measure and $f^{-nq}(V) \cap \Gamma$ has harmonic measure tending to 0 as n tends to ∞. This is a contradiction since f^q preserves the harmonic measure.

2. *There is one recurrent critical point c such that $\Gamma \subset \omega(c)$.*

For any recurrent critical point c, we have $f(\Gamma \cap \omega(c)) \subset (\Gamma \cap \omega(c))$. As $f^q : \Gamma \to \Gamma$ is ergodic with respect to μ, the set $\Gamma \cap \omega(c)$ has either 0 or full harmonic measure. But there are only finitely many such critical points. By 1 there is c recurrent such that $\Gamma \cap \omega(c)$ has full harmonic measure. Now since $\Gamma \cap \omega(c)$ is closed and the support of the harmonic measure is the whole set Γ, we must have $\Gamma \cap \omega(c) = \Gamma$. ∎

4 A semi-local version of the above results

Although we stated and proved Theorems 1.1, 1.2 and 1.3 for a rational map as a global dynamical systems of $\overline{\mathbb{C}}$, our proof actually works, with only minor modification, for restrictions of a rational map in a sub-domain of $\overline{\mathbb{C}}$, considered as a semi-local dynamical system, together with semi-local recurrent critical points and semi-local invariant compact sets. We may then conclude that only semi-local recurrent critical points are relevant to semi-local invariant subsets.

More precisely, let $F : \overline{\mathbb{C}} \to \overline{\mathbb{C}}$ be a rational map. Let $Q' \subset Q$ be two open subsets of $\overline{\mathbb{C}}$ with $Q' \neq Q$ and $\overline{Q} \neq \overline{\mathbb{C}}$, such that $F|_{Q'} : Q' \to Q$ is a holomorphic proper map. Consider $f = F|_{\overline{Q'}} : \overline{Q'} \to \overline{Q}$ as a dynamical system that is only defined on $\overline{Q'}$. (This is a bit like polynomial-like mappings, although it is not required that Q' is relatively compact in Q, and no quasi-conformal technique is needed).

Now critical points of F in Q' are also critical points of f. We say that a critical point c *escapes* if there is $k \geq 0$ such that $c, f(c), \cdots, f^k(c) \in Q$ but $f^k(c) \notin Q'$. We call $\{c, f(c), \cdots, f^k(c)\}$ the f-orbit of c. Although such critical points may very well be recurrent under F, they will not play relevant roles to the f-invariant sets in Q'.

We modify the definition of Ω_0 as follows: Define L for the rational map F as before. Let L' be the union of L and the f-orbits of escaping critical points. Set $\Omega_0 = Q \setminus L'$. Since $f(L \cap Q') \subset F(L) \subset L$, we have $f(L' \cap Q') \subset L'$. Therefore $f^{-1}(\Omega_0) \subset \Omega_0$.

Now we can define $K' = \{z \in \overline{Q'}, f^n(z) \in \overline{f^{-1}(\Omega_0)} \text{ for all } n \geq 0\}$ and $J' = \partial K'$. Then again

(A) $f^{-1}(J') \subset J'$ and

(B) the Hausdorff distance between J' and $\partial f^{-n}(\Omega_0)$ tends to 0, as $n \to \infty$, with respect to the Euclidean metric.

We can now restate Theorems 1.1, 1.2 and 1.3 for f provided we make the following changes: In the statements of the theorems replace "$x \in J(f)$" by "$x \in \Omega_0 \cap J'''$", "recurrent critical point" by "f-recurrent critical point", "$\Lambda \subset J(f)$" by "$\Lambda \subset \Omega_0 \cap J'''$", and finally "$\Gamma$" by "$\Gamma \subset \Omega_0 \cap J'$ with $f^n(\Gamma) \subset \Omega_0$ for all n".

For the proof of these theorems, we make the following changes: In the proof of Lemma 2.2(1) choose the points t_i whose f-orbits escape, and set Z_0 to be the union of these f-orbits. Since there are such kind of points arbitrarily close to each c_i, Condition $(*)$ is easily achieved. In the proof of Lemma 2.2(2), the study of the case $J' = \overline{\mathbb{C}}$ can be deleted. Notice that the only properties of J' needed is (B) (in the proof of Lemma 2.2(2)) and (A) (at the very end of §2). This completes the proof of Theorems 1.1, 1.2 and 1.3 in this generalized setting.

We now describe briefly an application by H. Kriete (see [Kr]),·of this generalized version of Mañé's result: Let F be a polynomial. Then Goldberg-Milnor's Fixed Point Portrait (cf. [GM]) provides regions in $\mathbb{C}$ bounded by periodic external rays that separate Cremer points and Siegel discs. Apply Theorem 1.3 to F restricted to each of these regions we obtain a critical point which is recurrent within the region and whose orbit accumulates to the Cremer point or Siegel boundary of the region.

A   The inequality

Recall that $d(a, E)$ denotes the Euclidean distance between a point $a \in \mathbb{C}$ and a closed subset E of $\mathbb{C}$.

Lemma A.1. *Let $W \neq \mathbb{C}$ be an open and simply connected proper subset of $\mathbb{C}$. Let $a, b \in W$. Then*

$$|b - a| \leq (e^{2d_W(a,b)} - 1) \cdot d(a, \partial W) .$$

Proof. Let $F : \mathbb{D} \to W$ be a conformal map with $F(0) = a$. Set $r = |F^{-1}(b)|$. Then

$$d_W(a, b) = d_{\mathbb{D}}(0, F^{-1}(b)) = \log \frac{1 + r}{1 - r} .$$

Denote by $h(z) = (F(z) - a)/F'(0)$ the normalized univalent map. By Koebe 1/4 and distortion theorems we have

$$\frac{d(a, \partial W)}{|F'(0)|} = d(0, \partial h(W)) \geq \frac{1}{4} \quad \text{and} \quad \frac{|b - a|}{|F'(0)|} = |h(F^{-1}(b))| \leq \frac{r}{(1 - r)^2} .$$

Combining these we get

$$|b - a| \leq \frac{r}{(1 - r)^2}|F'(0)| \leq \frac{4r}{(1 - r)^2}d(a, \partial W) = \left(\frac{(1 + r)^2}{(1 - r)^2} - 1\right)d(a, \partial W) .$$

The right hand side is equal to $(e^{2d_W(a,b)} - 1) \cdot d(a, \partial W)$. ∎

Lemma A.2. *Let $a \in W' \subset W \subset \Omega \subset \mathbb{C} \setminus \{0, \infty\}$ with W open and simply connected. Assume $\mathrm{diam}_W(W') \leq C$. Then*

$$\mathrm{diam}_{\mathrm{spherical}}W' \leq C' \inf\left\{d(a, \partial\Omega), \frac{1}{|a|}\right\}$$

where $C' = 2C'''(e^{2C} - 1)$ with C''' a universal constant.

Proof. Let C''' be the Lipschitz constant between the spherical metric and the Euclidean metric. We have

$$
\begin{aligned}
\mathrm{diam}_{\mathrm{spherical}}W' &\leq C'''\mathrm{diam}_{\mathrm{Euclidean}}W' \leq 2C''' \sup_{b\in W'} |b - a| \\
&\leq 2C''' \sup_{b\in W'} (e^{2d\,w(a,b)} - 1)d(a, \partial W) \\
&\leq 2C'''(e^{2C} - 1)d(a, \partial W) = C'd(a, \partial\Omega) .
\end{aligned}
$$

For a suitable choice of the spherical metric, the map $H(z) = 1/z$ is an isometry. Apply the above inequality to $H(a) \subset H(W') \subset H(W) \subset H(\Omega)$ we get

$$\mathrm{diam}_{\mathrm{spherical}}W' = \mathrm{diam}_{\mathrm{spherical}}H(W') \leq C'd(H(a), \partial H(\Omega)) \leq \frac{C'}{|a|} ,$$

where the last inequality is due to the facts that $H(a) = \frac{1}{a}$ and $0 \notin H(\Omega)$. ∎

From this one deduces easily that for any $\varepsilon > 0$, there is δ such that if $C \leq \delta$ then $\mathrm{diam}_{\mathrm{spherical}}W' \leq \varepsilon$. This fact was needed at the end of §2 for $\Omega = \Omega_1$. ∎

B   Control between modulus and diameter

Lemma B.1. *For $\delta \in (0, \infty)$, there are two strictly decreasing continuous functions $u(\delta)$ and $v(\delta)$ with*

$$\lim_{\delta\to 0^+} u(\delta) = \lim_{\delta\to 0^+} v(\delta) = +\infty; \quad \lim_{\delta\to +\infty} u(\delta) = \lim_{\delta\to +\infty} v(\delta) = 0 ,$$

satisfying the following property: Given any pair (V, E) with $V \subset \overline{\mathbb{C}}$ a hyperbolic open simply connected subset (i.e. $\#(\overline{\mathbb{C}} \setminus V) > 2$) and $E \subset V$ compact so that $V \setminus E$ is an open annulus, then for m the modulus of $V \setminus E$ and δ the diameter of E with respect to the Poincaré metric on V, we have

$$u(\delta) \leq m \leq v(\delta) .$$

Proof. We may assume $V = \mathbb{D}$, and E contains 0 and s with $s \in (0,1)$ and $d_{\mathbb{D}}(0,s) = \delta$. We have

$$\log \frac{1+s}{1-s} = \delta \text{ therefore } s = s(\delta) = \frac{e^\delta - 1}{e^\delta + 1} .$$

Note that $E \subset \overline{D}_s(0)$ so $A \supset \mathbb{D} \smallsetminus \overline{D}_s(0)$. As a consequence

$$m \geq \text{ mod } (\mathbb{D} \smallsetminus \overline{D}_s(0)) = \frac{1}{2\pi} \log \frac{1}{s} = \frac{1}{2\pi} \log \frac{e^\delta + 1}{e^\delta - 1} =: u(\delta) .$$

To get $v(\delta)$, we use the solution of Grötzsch extremal problem (cf. [Al1], Chapter III) that $m \leq \text{ mod } (\mathbb{D} \smallsetminus [0, s(\delta)]) =: v(\delta)$. ∎

In the case of Lemma 2.1, we have $v(\delta) \geq m \geq (\log(1/r))/(2\pi p)$ so

$$\delta \leq v^{-1}\left(\frac{\log(1/r)}{2\pi p}\right) =: C(p,r) .$$

The function $C(p,r)$ as a function of r is strictly increasing and continuous, with $\lim_{r \to 0^+} C(p,r) = 0$ and $\lim_{r \to 1^-} C(p,r) = +\infty$.

C   Bounded distortion

We use Lemma 2.1 to recover a distortion estimate in [CJY] (although we don't need it in our paper). In the same setting as Lemma 2.1, denote by $B_V(z,s)$ the Poincaré disc in V centred at z with Poincaré radius s. Define $B_{\mathbb{D}}(w,t)$ similarly. Then there is a constant C_1 depending only on s and p such that, for any $z \in V$,

$$B_{\mathbb{D}}(g(z), C_1) \subset g(B_V(z,s)) \subset B_{\mathbb{D}}(g(z), s) .$$

Proof. The right hand inclusion is due to Schwarz Lemma. As for the left hand side, we may assume $g(z) = 0$. Let $B_{\mathbb{D}}(0,t)$ be the largest disc contained in $g(B_V(z,s))$. Let E be the connected component of $g^{-1}(B_V(0,t))$ containing z. Obviously $\overline{E} \cap \partial B_V(z,s) \neq \emptyset$. Therefore the Poincaré diameter δ of E satisfies $\delta \geq s$. By Lemma 2.1 and Lemma B.1, $s \leq \delta \leq C(p, r(t)) =: C_p(t)$, where

$$r(t) = \frac{e^t - 1}{e^t + 1} .$$

Clearly $C_p(t)$ is a continuous strictly increasing function of t. As a consequence $t \geq (C_p)^{-1}(s) =: C_1(p,s)$.

D   An elementary proof of Lemma 2.1

This proof is due to K. Astala, who kindly allowed us to include it here.

In the setting of Lemma 2.1 we may assume $V = \mathbb{D}$. Therefore $g : \mathbb{D} \to \mathbb{D}$ is a Blaschke product of degree at most p, more precisely

$$g(z) = \prod_{i=1}^{l} \frac{z - a_i}{1 - \bar{a}_i z}$$

with $a_i \in \mathbb{D}$ and $l \leq p$.

Let $z' \in \mathbb{D}$ such that $|g(z')| \leq r$. Then there is at least one i such that

$$\left| \frac{z' - a_i}{1 - \bar{a}_i z'} \right| \leq r^{1/p} .$$

As $p_i(z) := \frac{z - a_i}{1 - \bar{a}_i z}$ is an isometry for the Poincaré metric, we have $d_{\mathbb{D}}(z', a_i) = d_{\mathbb{D}}(p_i(z'), 0) \leq \log \frac{1 + r^{1/p}}{1 - r^{1/p}} =: M$. As a consequence $z' \in \bigcup_{i=1}^{l} B_{\mathbb{D}}(a_i, M)$.

Therefore $g^{-1}(\overline{D}_r) \subset \bigcup_{i=1}^{l} B_{\mathbb{D}}(a_i, M)$. Now any connected component of $g^{-1}(\overline{D}_r)$ would have diameter with respect to $d_{\mathbb{D}}$ at most $p \times 2M =: C'(p, r)$. From the definition of M one can see clearly that $C'(p, r) \to 0$ and $r \to 0$.

E   Alternative proof of Lemma 2.2

The Poincaré metric $\lambda(z)|dz|$ of $\mathbb{C} \setminus \{0, 1\}$ has an estimate

$$\log \lambda(z) = \log \frac{1}{|z| \log \frac{1}{|z|}} + O(1) \text{ as } z \to 0 .$$

See [Al2, p.18, (1-24)] (or also [McM, p13, Theorem 2.3]). An easy calculation shows that for any $C > 0$ and $\epsilon > 0$, there exists $\delta > 0$ such that if $y \in \mathbb{C}$ and $0 < |y| < \delta$ then

$$\{z | \, d_{\mathbb{C} \setminus \{0,1\}}(z, y) \leq C\} \subset D_\epsilon(0) .$$

Using affine transformations sending 0, 1, ∞ to 1, w, ∞, one can show that for any $0 < r < R < \infty$, $C > 0$ and $\epsilon > 0$, there exists $\delta > 0$ such that if $r < |w| < R$, $d_{\text{spherical}}(w, y) < \delta$ and $y \in \mathbb{C} \setminus \{0, 1, w\}$, then

$$\text{diam}_{\text{spherical}}\{z | \, d_{\mathbb{C} \setminus \{0,w\}}(z, y) \leq C\} < \epsilon.$$

The same argument can be used with $\{0, \infty\}$ replaced by $\{0, 1\}$ or $\{1, \infty\}$. Therefore we conclude that for any $C > 0$ and $\epsilon > 0$, there exists $\delta > 0$ such that if $d_{\text{spherical}}(w, y) < \delta$ and $y \in \mathbb{C} \setminus \{0, 1, w\}$ then

$$\text{diam}_{\text{spherical}}\{z | \, d_{\mathbb{C} \setminus \{0,1,w\}}(z, y) \leq C\} < \epsilon .$$

Now let us define Ω_1. Let t be a repelling periodic point which is not in the orbits of x and c_i's. Let $Z_0 = \cup_{n=0}^M f^{-n}(t)$, where M is to be determined later and $\Omega_1 = \Omega_0 - Z_0$. By the choice of t, we know that x and the orbits of c_i's belong to Ω_1. We may suppose that $0, 1, \infty \in Z_0$.

To prove (1), set $C = C_0$ and

$$\epsilon = \min\{d_{\text{spherical}}(f(c_i), f(c_j)) | f(c_i) \neq f(c_j)\} \cup$$

$$\{d_{\text{spherical}}(c_i, f^n(c_j)) | i = 1, \ldots, \nu, \ n \geq 1\},$$

and we obtain δ satisfying the above. Since the inverse orbit of t is dense in the Julia set, we can choose large M so that for any $z \in J(f)$ there exists an element $w \in Z_0 = \cup_{n=0}^M f^{-n}(t)$ with $d_{\text{spherical}}(z, w) < \delta$. In particular, if $y = f(c_i)$, $w \in Z_0$ and $d_{\text{spherical}}(w, y) < \delta$ then

$$\text{diam}_{\text{spherical}}\{z| \ d_{\Omega_1}(z, y) \leq C_0\} \leq \text{diam}_{\text{spherical}}\{z| \ d_{\mathbb{C} \setminus \{0,1,w\}}(z, y) \leq C_0\} < \epsilon.$$

Therefore $d_{\text{spherical}}(f(c_i), f(c_j)) > C_0$ if $f(c_i) \neq f(c_j)$. The same argument applies to $d_{\text{spherical}}(c_i, f^n(c_j))$.

To prove (2), set $C = C$ and $\epsilon = \epsilon_1$ and we obtain δ satisfying the above property. Choose n_0 so that every point of the Julia set is within spherical distance δ from $f^{-n_0}(Z_0)$. Then for any $y \in J' \cap \Omega_0 \subset J(f)$, there exists $w \in Z_0$ such that $d_{\text{spherical}}(w, y) < \delta$ and hence

$$\text{diam}_{\text{spherical}}\{z| \ d_{f^{-n}(\Omega_1)}(z, y) \leq C\}$$

$$\leq \text{diam}_{\text{spherical}}\{z| \ d_{\mathbb{C} \setminus \{0,1,w\}}(z, y) \leq C\} < \epsilon_1.$$

It follows that if $n \geq n_0$, $W' \subset f^{-n}(\Omega_1)$ and $\text{diam}_{f^{-n}(\Omega_1)}(W') \leq C$ then W' has a spherical diameter less tahn ϵ_1. ∎

Remark. [McM, p.39, Theorem 3.6] uses a similar argument.

References

[Al1] L.V. Ahlfors, Lectures on Quasiconformal Mappings, Wadsworth, 1987.

[Al2] L.V. Ahlfors, Conformal Invariants, McGraw-Hill, 1973.

[CJY] L. C. Carleson, P. W. Jones and J.-C. Yoccoz, Julia and John, *Bol. Soc. Bras. Mat.* vol. 25, p. 1-30 (1994).

[GM] L Goldberg and J. Milnor, Fixed points of polynomial maps, part II, *Ann. Sci. Éc. Norm. Sup.* vol. 26 (1993), p. 51-98.

[Ma] R. Mañé, On a theorem of Fatou, *Bol. Soc. Bras. Mat.* vol. 24, p. 1-11 (1993).

[Kr] H. Kriete, Recurrence and periodic points, manuscript, February 1998.

[McM] C. T. McMullen, Complex Dynamics and Renormalization, Princeton University Press, 1994.

[PR] F. Przytycki and S. Rohde, Porosity of Collet-Eckmann Julia sets, *Fund Math.* 155 (1998), pp. 189-199.

[TY] Tan Lei and Yin Yongcheng, Local connectivity of the Julia set for geometrically finite rational maps, *Science in China (Series A)*, vol. 39, number 1 (1996), pp. 39-47.

[Yi] Yin Yongcheng, Geometry and dimension of Julia sets, in this volume.

Mitsuhiro Shishikura, Hiroshima University, Department of Mathematics, Kagamiyama, Higashi-Hiroshima 739-8526, Japan.
email : mitsu@math.sci.hiroshima-u.ac.jp

Tan Lei, Department of Mathematics, University of Warwick, Coventry CV4 7AL, United Kingdom.
e-mail: tanlei@maths.warwick.ac.uk .

GEOMETRY AND DIMENSION OF JULIA SETS

YIN YONGCHENG

ABSTRACT. We show that the Julia set of a critically non-recurrent rational map is shallow, in the sense of McMullen (except the case that the Julia set is the whole sphere). This recovers, as a corollary, a theorem of M. Urbanski with a different approach.

1991 Mathematical Subject Classification: 30D05, 58F08

1. INTRODUCTION

Let $f : \overline{\mathbb{C}} \to \overline{\mathbb{C}}$ be a rational map with degree $\deg f \geq 2$. A point z is said to be a *periodic point* if $f^k(z) = z$ for some $k \geq 1$. The minimal of such k is called the *period*. For a periodic point z_0, denote the *multiplier* of z_0 by $\lambda = (f^k)'(z_0)$. The periodic point z_0 is either *attracting*, *indifferent* or *repelling* according to $|\lambda| < 1$, $|\lambda| = 1$ or $|\lambda| > 1$. In the indifferent case, we say z_0 is *parabolic* if λ is the root of unity.

The *Julia set*, denote by $J(f)$, is the closure of all repelling periodic points. Its complement is called *Fatou set*, denote by $F(f)$. A connected component of $F(f)$ is called a *Fatou component*. D. Sullivan proved that each Fatou component U is preperiodic, this means there exist integers $m \geq 1$, $n \geq 0$ so that

$$f^{m+n}(U) = f^n(U),$$

and every periodic Fatou component is either an attracting basin, a parabolic basin, a Siegel disk or a Herman ring. For the classical results of complex dynamics, see [Be], [Mi] and [CG].

Definition 1. A rational map $f : \overline{\mathbb{C}} \to \overline{\mathbb{C}}$ is called *critically non-recurrent* if each critical point $c \in J(f)$ is non-recurrent, i.e. $c \notin \omega(c)$.

Denote $B(x, \varepsilon)$ a ball of venter x and radius ε in the spherical metric on $\overline{\mathbb{C}}$, $D(x, \epsilon)$ a disk of venter x and radius ϵ in the Euclidean metric in $\mathbb{C}$.

The following definition is given by C.McMullen, [Mc].

The research is partially supported by NSF of China (No.19531060) and Morningside Centre of Mathematics, Chinese Academy of Sciences.

Definition 2. A compact subset $X \subset \overline{\mathbb{C}}$ is *shallow* if there exists $0 < k < 1$ such that any ball $B(z, r)$, $z \in X$, $r > 0$, contains a smaller ball $B(y, kr) \subset \overline{\mathbb{C}} \setminus X$.

It's easy to prove that if X is shallow, then its box dimension $BD(X) < 2$ and its Hausdorff dimension $HD(X) \leq BD(X) < 2$.

The main result of this note is the following statement.

Main Theorem. *For a critically non-recurrent rational map* $f : \overline{\mathbb{C}} \to \overline{\mathbb{C}}$, *the Julia set* $J(f)$ *is shallow or* $J(f) = \overline{\mathbb{C}}$.

A directly consequence is

Corollary. *(Urbanski, [Ur]) For a critically non-recurrent rational map* $f : \overline{\mathbb{C}} \to \overline{\mathbb{C}}, HD(J(f)) < 2$ *or* $J(f) = \overline{\mathbb{C}}$.

Remark. C. McMullen, F. Przytycki and S. Rohde proved the same theorems for quadratic polynomials with Siegel disk of bounded type rotation numbers and Collet-Eckmann rational maps without parabolic points, [Mc], [PR].

Notations. 1. d is the spherical metric on $\overline{\mathbb{C}}$.
2.   The distortion $\mathrm{Dist}(X, z)$ of a compact set X about $z \notin X$ is $\max_{y \in X} d(y, z) / \min_{y \in X} d(y, z)$.
3.   Two positive numbers A and B are K-commensurable or simply commensurable if $K^{-1} \leq A/B \leq K$ for some $K \geq 1$ independent of A and B. $A \asymp B$ means A and B are commensurable.

Acknowledgement. The work was done during the program on complex dynamics hold at Morningside Centre of Mathematics in 1998. Thanks to Prof. Yang Lo and Prof. Li Zhong organised this program and to all members in the seminar on complex dynamics at Beijing.

2. BACKGROUND MATERIALS

2.1 The distortion theorem. The distortion theorem of univalent maps is a powerful tool in the study of complex dynamics. The following distortion theorem of the version for p-valent maps is well-known (see [CJY] or [ST] appendix C).

Lemma 2.1. *Let U_1 and U_2 be simply connected domains in $\overline{\mathbb{C}}$ and $g : U_1 \to U_2$ be a proper holomorphic map of $\deg g \leq p$. then*

$$\{w | \rho_{U_2}(w, g(z_0)) \leq C^{-1}\} \subset g(\{z | \rho_{U_1}(z, z_0) \leq r\}) \subset \{w | \rho_{U_2}(w, g(z_0)) \leq r\}$$

for any $z_0 \in U_1$, where ρ_{U_1} and ρ_{U_2} are hyperbolic metrics of U_1 and U_2, C is a constant depending only on p and r.

Denote ρ the hyperbolic metric in $D(0, 1)$ and $B_\rho(0, r)$ the hyperbolic ball of centre at 0 and radius r.

Lemma 2.2. *Let* $h : D(0,1) \to D(0,1)$ *be a proper holomorphic map of degree at most* p *and* $h(0) = 0$. *Then there exists a constant* R^* *depending only on* p *and* R *such that* $D(0, R^*) \subset h(B_\rho(0,R))$ *and the component of* $h^{-1}(D(0, R^*))$ *containing* 0 *is a subset of* $B_\rho(0, R)$.

Proof. Assume $\deg h = p' \le p$. Let $0 = z_0, z_1, \cdots, z_{p'-1}$ be the preimages of $h(z_0) = 0$ and $h_j : D(0,1) \to D(0,1)$ be the univalent map so that $h_j(z_j) = 0$, $0 \le j \le p' - 1$. Then

$$h(z) = e^{i\theta} \prod_{j=0}^{p'-1} h_j(z)$$

for some $\theta \in \mathbb{R}$.

There are at most p' many points of $\{z_j \,|\, 0 \le j \le p' - 1\}$ in the closed hyperbolic ball $\overline{B_\rho(0,R)}$. We can find a hyperbolic circle γ of centre at 0 in $B_\rho(0, R)$ such that

$$\rho(z, z_j) \ge \frac{R}{2(p' - 1)}$$

for $z \in \gamma$ and $0 \le j \le p' - 1$. Then

$$\rho(h_j(z), 0) = \rho(h_j(z), h_j(z_j)) = \rho(z, z_j) \ge \frac{R}{2(p' - 1)}.$$

Change it to the Euclidean metric,

$$d(h_j(z), 0) \ge \frac{e^{\frac{R}{2(p'-1)}} - 1}{e^{\frac{R}{2(p'-1)}} + 1} = C_1.$$

Then

$$d(h(z), 0) = \prod_{j=0}^{p'-1} d(h_j(z), 0) \ge C_1^{p'} = R^*(p', R).$$

and $R^* = \min\{R^*(p', R) : 1 \le p' \le p\}$ is the constant which we need.

QED

From lemma 2.2, we have a useful corollary in spherical metric.

Corollary 2.3. *Let* $U \ni z_0$ *be a simply connected domain and* $g : U \to B(w_0, 2\delta)$ *be a proper holomorphic map of* $\deg g \le p$, $w_0 = g(z_0)$. *If* g *maps* $B(z_0, r) \subset U$ *into* $B(w_0, \delta)$, *then there exists a constant* K *depending only on* p *so that* $B(w_0, r^*) \subset g(B(z_0, r))$ *and the component of* $g^{-1}(B(w_0, r^*))$ *containing* z_0 *is a subset of* $B(z_0, r)$, *where* $r^* \ge \operatorname{diam} g(B(z_0, r))/K$.

Proof. Assume $(\partial g(B(z_0, r))) \cap (\partial B(w_0, \delta)) \ne \emptyset$ at first. Let U' be the component of $g^{-1}(B(w_0, \delta))$ containing z_0. Then $\mod (U \setminus U') \ge \frac{\log 2}{2\pi p}$.

284   Yin Yongcheng

For any two points $x, y \in \partial B(z_0, r)$, $\rho_U(x, z_0)$ and $\rho_U(y, z_0)$ are $C(p)$-commensurable with 1 for a constant $C(p)$ depending only on p. Take $R = \min_{x \in \partial B(z_0, r)} \rho_U(x, z_0)$. By lemma 2.2, there exists $r^* = 2\delta R^*$ such that $B(w_0, r^*) \subset g(B(z_0, r))$ and the component of $g^{-1}(B(w_0, r^*))$ containing z_0 is a subset of $B(z_0, r)$. Denote $K = 1/R^*$. Then $r^* \geq \operatorname{diam} g(B(z_0, r))/K$.

In the case $(\partial g(B(z_0, r))) \cap (\partial B(w_0, \delta)) = \emptyset$, take $\delta' = d(w_0, \partial g(B(z_0, r)))$ $\leq \delta$. We replace U by the component of $g^{-1}(B(w_0, 2\delta'))$ containing z_0. Use the same method as above, we get the constant K.

QED

2.2 Mañé's theorem In [Ma](see also [ST]), R. Mañé proved a beautiful theorem. It is a key lemma in the proof of our main result.

Lemma 2.4. *(Mañé's theorem) Let $f : \overline{\mathbb{C}} \to \overline{\mathbb{C}}$ be a rational map. If a point $x \in J(f)$ is not a parabolic periodic point and is not contained in the ω-limit set of a recurrent critical point, then for all $\varepsilon > 0$ there exists a neighbourhood U of x such that*

(a) *For all $n > 0$, every component of $f^{-n}(U)$ has diameter $< \varepsilon$;*

(b) *There exists $p > 0$ such that for all $n \geq 0$ and every component V of $f^{-n}(U)$,*

$$\operatorname{degree}(f^n : V \to U) \leq p.$$

2.3 Dynamics near the parabolic point. Let $f_0 : \overline{\mathbb{C}} \to \overline{\mathbb{C}}$ be a rational map with a non-degenerate parabolic fixed point 0, i.e. $f_0'(0) = 1$ and $f_0''(0) \neq 0$. Choose a neighbourhood $U_0 \ni 0$ so that f_0 maps U_0 homeomorphically onto a neighbourhood $U_0' \ni 0$.

Lemma 2.5. *(Fatou coordinates) Take f_0, U_0 and U_0' as above. Then there exist simply connected domains $D_\pm$ compactly contained in $U_0 \cap U_0'$, whose union forms a punctured neighbourhood of 0, satisfying*

$$f_0^{\pm}(\overline{D}_\pm) \subset D_\pm \cup \{0\} \quad and \quad \cap_{n \geq 0} f_0^{\pm n}(\overline{D}_\pm) = \{0\}.$$

Moreover, there exist univalent maps $\Phi_\pm : D_\pm \to \mathbb{C}$ such that

(1) $\Phi_\pm(f_0(z)) = \Phi_\pm(z) + 1$,

(2) *Range $(\Phi_+) \supset \{\zeta \mid -\frac{3}{4}\pi < \arg(\zeta - R_0) < \frac{3}{4}\pi\}$ and Range $(\Phi_-) \supset$ $\{\zeta \mid \frac{1}{4}\pi < \arg(\zeta + R_0) < \frac{7}{4}\pi\}$ for some $R_0 > 0$.*

The domains D_+ and D_- in lemma 2.5 are called attracting petal and repelling petal respectively.

Set $\tilde{V}_0^* = \{\zeta \,|\, \mathrm{Re}\,\zeta < -R_0\}$, $\tilde{V}_0 = \{\zeta \,|\, \mathrm{Re}\,\zeta < -R_0,\ |\mathrm{Im}\,\zeta| > R_0\}$, $\tilde{\Omega}_0 = \{\zeta \,|\, -R_0 - 1 < \mathrm{Re}\,\zeta < -R_0,\ |\mathrm{Im}\,\zeta| < R_0\}$, $\tilde{\Omega}_j = \tilde{\Omega}_0 - j$, $j \geq 1$ and $V_0^* = \Phi_-^{-1}(\tilde{V}_0^*)$, $V_0 = \Phi_-^{-1}(\tilde{V}_0)$, $\Omega_j = \Phi_-^{-1}(\tilde{\Omega}_j)$, $j \geq 0$. From lemma 2.5, V_0 is contained in the Fatou set $F(f_0) = \overline{\mathbb{C}} \smallsetminus J(f_0)$ and $f_0 : \Omega_{j+1} \to \Omega_j$ is homeomorphic for $j \geq 0$.

3. Proof of the main theorem

Let $f : \overline{\mathbb{C}} \to \overline{\mathbb{C}}$ be a rational map. In this section, we always assume $J(f) \neq \overline{\mathbb{C}}$.

For $x \in J(f)$, $r > 0$, define

$$h(x,r) = \sup\{\xi \,|\, \text{there is } B(y,\xi) \subset F(f) \cap B(x,r)\}.$$

Then $h : J(f) \times \mathbb{R}_+ \to \mathbb{R}_+$ is continuous.

Since $\mathrm{int}(J(f)) = \phi$, hence

$$h(r) := \inf\{h(x,r) \,|\, x \in J(f)\} > 0$$

for any $r > 0$.

Proof of the main theorem. First of all, we suppose f is a critically non-recurrent rational map having no parabolic periodic points. From lemma 2.4, there are $\delta_0 > 0$ and $p < \infty$ so that for any $x \in J(f)$, $n > 0$ and any component V of $f^{-n}(B(x, 2\delta_0))$, V is simply connected and

$$\deg(f^n : V \to B(x, 2\delta_0)) \leq p.$$

For any $z_0 \in J(f)$ and $0 < r \leq \delta_0$, look at the forward images $B_m = f^m(B(z_0, r))$, $z_m = f^m(z_0)$, $m \geq 0$.

Let n_0 be the smallest integer so that $\mathrm{diam} B_{n_0+1} > \delta_0$. Then $l_0 \delta_0 \leq \mathrm{diam} B_{n_0} \leq \delta_0$ for some $0 < l_0 < 1$. By corollary 2.3, there are a constant K depending only on p and a ball $B(z_{n_0}, r_0) \subset B_{n_0}$ with $r_0 \geq \mathrm{diam} B_{n_0}/K$ such that the component of $f^{-n_0}(B(z_{n_0}, r_0))$ containing z_0 is a subset of $B(z_0, r)$. There exists $B(y_0, \frac{1}{2}h(r_0)) \subset B(z_{n_0}, r_0) \cap F(f)$. Let D_0 be a component of $f^{-n_0}(B(y_0, \frac{1}{2}h(r_0)))$ contained in $B(z_0, r)$, $y \in D_0 \cap f^{-n_0}(y_0)$. Then $\mathrm{Dist}(\partial D_0, y) \leq M$ for some M depending only on p and $\mathrm{diam} D_0 \asymp r$. Therefore, there exists $0 < k < 1$ which does not depend on z_0 and r so that $B(y, kr) \subset D_0 \subset B(z_0, r) \cap F(f)$. The Julia set $J(f)$ is shallow.

Now let f be a non-recurrent rational map having parabolic periodic points. For simplicity, we suppose f has only one non-degenerate parabolic fixed point 0.

Take $\Omega_j (j \geq 0)$ as in section 2. The set $X_0 = J(f) \setminus (\cup_{j \geq 1} \Omega_j \cup \{0\})$ is a compact subset of $J(f)$. By lemma 2.4, there exist $\delta_1 > 0$ and $p_1 < \infty$ so that for all $x \in X_0$, $n > 0$, $0 < \delta \leq \delta_1$ and any component V of $f^{-n}(B(x, 2\delta))$, V is simply connected and

$$\deg(f^n : V \to B(x, 2\delta)) \leq p_1.$$

Let K_1 be the constant depending only on p_1 in corollary 2.3. Suppose D contains a ball $B(z, \epsilon)$ with $\epsilon \geq \mathrm{diam} D / K_1$. Then $f(D)$ contains a ball $B(f(z), \epsilon^*)$ such that the component of $f^{-1}(B(f(z), \epsilon^*))$ containing z is a subset of $B(z, \epsilon)$, where $\epsilon^* \geq \mathrm{diam} f(D) / K_0$ for a constant K_0 depending only on K_1.

Choose N large enough so that $N \geq 128 K_0 R_0$ and $d(0, \partial \Omega_n) > 4 K_0 \mathrm{diam} \Omega_n$ for $n \geq N$. Take $\delta_2 = \eta \mathrm{diam} \Omega_N$ for a suitable η so that

$$B(x, 2\delta_2) \subset V_0^* = V_0 \cup (\cup_{j \geq 0} \Omega_j)$$

for any $x \in \cup_{j=1}^N \Omega_j$.

Now we prove that the Julia set $J(f)$ is shallow.

For any $z_0 \in J(f)$ and $0 < r \leq \delta_0 = \min(\delta_1, \delta_2)$, denote $B_m = f^m(B(z_0, r))$, $z_m = f^m(z_0)$, $m \geq 0$. Let n_0 be the smallest integer such that $\mathrm{diam} B_{n_0+1} > \delta_0$. Then $\mathrm{diam} B_m \leq \delta_0$ for $m \leq n_0$ and $\mathrm{diam} B_{n_0} \asymp \delta_0$.

If $z_{n_0} \in X_0$, then the same argument as in no parabolic periodic points case shows there exists $B(y, k_1 r) \subset B(z_0, r) \cap F(f)$ for some $0 < k_1 < 1$.

Suppose $z_{n_0} \in (\cup_{j \geq 1} \Omega_j) \cup \{0\}$ and $m_0 \geq 0$ is the smallest integer such that $z_j \in (\cup_{j \geq 1} \Omega_j) \cup \{0\}$ for all $m_0 \leq j \leq n_0$. Then B_{m_0} contains a ball $W = B(z_{m_0}, \epsilon^*)$ with $\epsilon^* \geq \mathrm{diam} B_{m_0} / K_0$ such that the component of $f^{-m_0}(W)$ containing z_0 is a subset of $B(z_0, r)$.

We claim that there exists a ball $B(y_0, r_0) \subset B_{m_0} \cap F(f)$ such that $f^{-m_0}(B(y_0, r_0))$ has a component contained in $B(z_0, r)$ for some $r_0 \asymp \mathrm{diam} B_{m_0}$.

(1) If $z_{m_0} = 0$, take $r_0 = \frac{1}{2} \epsilon^*$.

(2) If $z_{n_0} \in \Omega_{j_0}$ for $1 \leq j_0 \leq N$, then $B_{n_0} \subset B(z_{n_0}, \delta_0)$ and $B(z_{n_0}, 2\delta_0) \subset V_0^*$. Let B' be a component of $f^{-(n_0-m_0)}(B(z_{n_0}, 2\delta_0))$ containing B_{m_0}. Then $f^{n_0-m_0} : B' \to B(z_{n_0}, 2\delta_0)$ is univalent and $\mathrm{mod}\,(B' \setminus B_{m_0}) \geq \frac{1}{2\pi} \log 2$. By distortion theorem of univalent map, $\mathrm{Dist}(\partial W_1, z_{n_0}) \asymp 1$ and there is $B(y_0^*, r_0^*) \subset W_1 \cap F(f)$ for some $r_0^* \asymp \delta_0$, where $W_1 = f^{n_0-m_0}(W)$. Hence there is $B(y_0, r_0) \subset W \cap F(f)$ for some $r_0 \asymp \mathrm{diam} B_{m_0}$.

(3) If $z_{n_0} \in \Omega_{j_0}$ for $j_0 > N$, $z_{m_0} \in \Omega_{j_0} + (n_0 - m_0) = \Omega_{i_0}$.

(3a) When $\mathrm{diam} B_{m_0} > 2 K_0 \mathrm{diam} \Omega_{i_0}$, $d(z_{m_0}, \partial W) > 2 \mathrm{diam} \Omega_{i_0}$, take $r_0 = \frac{1}{4} d(z_{m_0}, \partial W)$.

(3b) When $\mathrm{diam}B_{m_0} \leq 2K_0\mathrm{diam}\Omega_{i_0}$, $D = B(z_{m_0}, 2\mathrm{diam}B_{m_0}) \subset V^*$. Let $\tilde{D} = \Phi_-(D)$, $\tilde{W} = \Phi_-(W)$, $\tilde{W}_1 = \Phi_-(W_1)$ and $\tilde{z}_{m_0} = \Phi_-(z_{m_0})$. Then $\mathrm{Dist}(\partial\tilde{W}, \tilde{z}_{m_0}) \leq 16\,\mathrm{Dist}(\partial W, z_{m_0}) \leq 16K_0$.

If $d(\tilde{z}_{m_0}, \partial\tilde{W}) > 4R_0$, then there is a ball $\tilde{B} = B(\tilde{y}_0, \frac{1}{2}d(\tilde{z}_{m_0}, \partial\tilde{W})) \subset \tilde{V}_0$ such that $\Phi_-^{-1}(\tilde{B}) = B$ contains a ball $B(y_0, r_0) \subset W \cap V_0 \subset W \cap F(f)$ for $r_0 \asymp \mathrm{diam}B_{m_0}$.

If $d(\tilde{z}_{m_0}, \partial\tilde{W}) \leq 4R_0$, then $\tilde{W} \subset B(\tilde{z}_{m_0}, 64K_0R_0)$ and $\tilde{W}_1 = \tilde{W} + (n_0 - m_0) \subset B(\tilde{z}_{n_0}, 64K_0R_0)$, $\tilde{z}_{n_0} = \tilde{z}_{m_0} + (n_0 - m_0) = \Phi_-(z_{n_0})$. Since $\tilde{z}_{n_0} \in \tilde{\Omega}_{j_0}$, $j_0 > N \geq 128K_0R_0$, we have $\tilde{D}' = B(\tilde{z}_{n_0}, 128K_0R_0) \subset \tilde{V}_0^*$, $D' = \Phi_-^{-1}(\tilde{D}') \subset V^*$, $B_{n_0} \subset D'$ and $\mathrm{mod}\,(D' \smallsetminus B_{n_0}) \geq \frac{1}{2\pi}\log 2$. Hence $\mathrm{Dist}\,(\partial W_1, z_{n_0}) \leq 16\mathrm{Dist}\,(\partial\tilde{W}_1, \tilde{z}_{n_0}) = 16\mathrm{Dist}\,(\partial\tilde{W}, \tilde{z}_{m_0}) \leq 16^2K_0$.

Let D'' be the component of $f^{-(n_0-m_0)}(D')$ containing B_{m_0}. Then $f^{n_0-m_0} : D'' \to D'$ is univalent. By distortion theorem, there is a ball $B(y_0, r_0) \subset W \cap F(f)$ for some $r_0 \asymp \mathrm{diam}B_{m_0}$. The proof of our claim is completed.

Now we come back to the proof of the main theorem.

If $m_0 = 0$, we are done. If $m_0 \geq 1$, then $B_{m_0-1} \cap f^{-1}(B(y_0, r_0))$ contains a ball with radius $r_0' \asymp \mathrm{diam}B_{m_0-1}$. The same argument as in no parabolic points case shows that there is $B(y, k_2r) \subset B(z_0, r) \cap F(f)$ for some $0 < k_2 < 1$. Take $k = \min(k_1, k_2)$. Then for any $z_0 \in J(f)$, $r > 0$, $B(z_0, r)$ contains a ball $B(y, kr) \subset F(f)$. This means $J(f)$ is shallow.

QED

REFERENCES

[Be] A.Beardon, *Iteration of Rational Functions*, Springer-Verlag, 1991.

[CG] L.Carleson and T.Gamelin, *Complex Dynamics*, Spring-Verlag, 1993.

[CJY] L. Carleson, P.W. Jones and J.-C. Yoccoz, *Julia and John*, Bol. Soc. Bras. Mat. **25, N.1** (1994), 1-30.

[Ma] R.Mañé, *On a theorem of Fatou*, Bol. Soc. Bras. Mat. **24 N.1** (1993), 1-11.

[Mc] C.McMullen, *Self-similarity of Siegel disks and the Hausdorff dimension of Julia sets*, Acta Math. **180** (1998), 247-292.

[Mi] J.Milnor, *Dynamics in one complex variable*, Stony Brook Preprint **5** (1990).

[PR] F.Przytycki and S.Rohde, *Porosity of Collet-Eckmann Julia sets*, Fund Math. **155** (1998), 189-199.

[ST] M.Shishikura and Tan Lei, *An alternative proof of Mañé's theorem on non-expanding Julia sets*, in this volume.

[Ur] M.Urbanski, *Rational functions with no recurrent critical points*, Ergod. Th. and Dynam. Sys. **14** (1994), 391-414.

INSTITUTE OF MATHEMATICS, ZHEJIANG UNIVERSITY, HANGZHOU, 310027, P.R.CHINA

E-mail address: yin@math.zju.edu.cn

On a theorem of M. Rees for matings of polynomials

Mitsuhiro Shishikura

Abstract

We show the following theorem of M. Rees: if the formal mating of
two postcritically finite polynomials is (weakly) equivalent to a rational
map, then the topological mating is conjugate to the rational map.

The dynamics of complex polynomials has been extensively studied and many
results are obtained. For example, a polynomial of Thurston's type admits
a description of the dynamics in terms of external angles, etc (see [DH1]).
However, much less is known for rational maps in general (see [R] for degree
two case). The *mating* is a way to construct a rational map or a branched
covering of S^2 from a pair of polynomials with connected Julia sets. When
this is possible, we can understand the dynamics of the rational map in terms
of that of the polynomials. The notion of the mating was also introduced by
Douady and Hubbard [D].

There are two notions of the mating —the formal mating and the topological mating. In fact there is another, the degenerate mating, which is a
slightly modified version of the formal mating. To define the formal mating,
we take two copies of $\mathbb{C}$ each with the circle at infinity and make a topological
sphere by gluing the copies together along the circles at infinity. Then define
a mapping on the sphere using one polynomial on one copy and the other
polynomial on the other copy. To define the topological mating, take the
disjoint union of the filled-in Julia sets of the polynomials and identify the
point with external angle θ for one polynomial with the point with external
angle $-\theta$ for the other polynomial. Then there is a naturally induced map
on the quotient.

Now the question is when they are "equivalent" to a rational map (matability). For the case of degree two, there are several works [L], [W], [R] and
the final result on the combinatorial matability is obtained by Tan Lei [T].

As for the relationship between the two matings, M. Rees [R] proved:

*If the formal mating of two hyperbolic polynomials of Thurston's type is
Thurston equivalent to a rational map, then the topological mating is conjugate
to the rational map. ([R] §0.15).*

The main theorem (§1.7) of this paper claims that this is also the case even
if the polynomials are merely of Thurston's type and the formal mating can
be replaced by the degenerate mating. This generalization has an interesting
consequence. If we start with two polynomials of Thurston's type which have
empty filled-in Julia sets (consequently with Lebesgue measure zero), when

they are "matable", by gluing these Julia sets (which are "dendrites"), we obtain not only a topological 2-sphere but also a complex structure on it such that the induced dynamics is analytic with respect to this structure. For example, this is the case when $z^2 + i$ is mated with itself.

The main idea of the proof is similar to [R]. However, the critical points in the Julia set cause a lot of complications. The reader should notice that many of our arguments are trivial in hyperbolic case. M. Rees says she has also a proof for this generalization, so the author does not insist on the priority for it. This paper is aimed simply to provide a proof available for the sake of future development of the theory of matings.

In §1, we define the matings and state the main theorem. In §2, when two polynomials are matable, we construct a semi-conjugacy h from the formal mating to the rational map. In §3, we study the properties of h. In §4, we conclude that two points are mapped by h to the same point if and only if they are ray-equivalent. Here we use a generalization of Douady's lemma (Remark 4.3(3)) instead of Lemma 3.7 in [R]. Together with this, the use of the map G (Lemma 4.5) is the new idea, which made it possible to prove the theorem in non-hyperbolic case.

I would like to thank A. Douady, J. H. Hubbard, Mary Rees and Tan Lei for helpful discussions, especially Tan Lei introduced me the subject of the mating, told me the importance of the theorem, finally encouraged me to include this paper in this volume and helped the arrangement for retyping the paper. I would like to thank also IHES and Akizuki project (founded by Toyosaburo Taniguchi) for supporting my visit to IHES in the year 1989/90, during which this paper was written. The original version of this paper had been circulated as IHES preprint M/90/48 (1990).

1 Matings of Polynomials and Main Theorem

Let f_1, f_2 be monic complex polynomials of degree d (≥ 2). We consider only the case where f_1, f_2 are of *Thurston's type*, i.e., all the critical points are periodic.

We are concerned with two notions of the matings of f_1 and f_2 —the formal mating and the topological mating. The degenerate mating is a slightly modified version of the formal one.

1.1 Formal mating. We add to the complex plane $\mathbb{C}$ "the circle at infinity" which is symbolically denoted by $\{\infty \cdot e^{2\pi i\theta} \mid \theta \in \mathbb{R}/\mathbb{Z}\}$. Define

$$\tilde{\mathbb{C}} = \mathbb{C} \cup \{\infty \cdot e^{2\pi i\theta} \mid \theta \in \mathbb{R}/\mathbb{Z}\},$$

which is homeomorphic to a closed disc with the natural topology. Extend the polynomials $f_j\colon \mathbb{C} \to \mathbb{C}$ to $\tilde{\mathbb{C}}$ by setting

$$f_j(\infty \cdot e^{2\pi i\theta}) = \infty \cdot e^{2\pi id\theta}.$$

Let $\tilde{\mathbb{C}}_1$, $\tilde{\mathbb{C}}_2$ be two copies of $\tilde{\mathbb{C}}$, and from now on we consider that f_j acts on $\tilde{\mathbb{C}}_j$, i.e. $f_j\colon \tilde{\mathbb{C}}_j \to \tilde{\mathbb{C}}_j$ ($j = 1,2$). Let $\sim$ be the equivalence relation in $\tilde{\mathbb{C}}_1 \sqcup \tilde{\mathbb{C}}_2$ (disjoint union) which identifies $\infty \cdot e^{2\pi i\theta}$ in $\tilde{\mathbb{C}}_1$ and $\infty \cdot e^{-2\pi i\theta}$ in $\tilde{\mathbb{C}}_2$ for $\theta \in \mathbb{R}/\mathbb{Z}$. Then the quotient by $\sim$ is topologically a two dimensional sphere. So denote

$$S^2 = \tilde{\mathbb{C}}_1 \sqcup \tilde{\mathbb{C}}_2/\sim .$$

In this paper, S^2 is a notation specially reserved for this quotient. Finally, the *formal mating* $f_1 \perp\!\!\!\perp f_2\colon S^2 \to S^2$ is defined by

$$F = f_1 \perp\!\!\!\perp f_2 = \begin{cases} f_1 & \text{on } \tilde{\mathbb{C}}_1 \\ f_2 & \text{on } \tilde{\mathbb{C}}_2 \end{cases},$$

which is well defined. This defines an orientation preserving branched covering of S^2 onto itself. The *equator* S^1 of S^2 is the quotient image of the circles at infinity of $\tilde{\mathbb{C}}_1$ and $\tilde{\mathbb{C}}_2$. Here, S^1 is also a notation reserved for the equator. However, there is a natural correspondence between the equator and the standard unit circle by $\infty \cdot e^{2\pi i\theta} \in \tilde{\mathbb{C}}_1 \leftrightarrow e^{2\pi i\theta}$, and the restriction of the formal mating, $F\colon S^1 \to S^1$ corresponds to the map $z \mapsto z^d$ on the unit circle.

Remark. Since we added the circle to $\mathbb{C}$, we have to destroy the conformal structure in order to introduce a differentiable structure on S^2. However, we can keep the conformal structure on a neighbourhood of the filled-in Julia sets $K_{f_j}(\subset \tilde{\mathbb{C}}_j)$. Thus we can talk about conformal mappings in $\operatorname{int} K_{f_1} \sqcup \operatorname{int} K_{f_2}$.

We are interested in the relationship between $F = f_1 \perp\!\!\!\perp f_2$ and rational maps. Since we have lost the conformality at least near the equator, it is too optimistic to expect that F is conjugate to a rational map. So we introduce a weaker notion of equivalence than the conjugacy.

1.2   Thurston equivalence. Let $f\colon S^2 \to S^2$ be an orientation preserving branched covering. Define

$$\Omega_f = \{\text{critical points of } f\} \quad \text{and}$$
$$P_f = \bigcup_{n>0} f^n(\Omega_f).$$

We say f is of *Thurston's type*, if $\#P_f < \infty$. Two such maps f, g are defined to be *Thurston equivalent* if and only if there exist orientation preserving homeomorphisms $\theta, \theta' : S^2 \to S^2$ such that $\theta = \theta'$ on P_f, $\theta(P_f) = P_g$, θ and θ' are isotopic relative to P_f, and the following diagram commutes:

$$
\begin{array}{ccc}
S^2 & \xrightarrow{\theta'} & S^2 \\
{\scriptstyle f}\downarrow & & \downarrow{\scriptstyle g} \\
S^2 & \xrightarrow{\theta} & S^2.
\end{array}
$$

Definition. f_1 and f_2 are *strongly matable* if $f_1 \perp\!\!\!\perp f_2$ is Thurston equivalent to a rational map.

Remark. Thurston gave a topological criterion for a branched covering to be equivalent to a rational map (see [DH2]). Namely, if a branched covering f is not equivalent to a rational map, then there exists a so-called *Thurston obstruction*, which consists of disjoint simple closed curves in $S^2 \setminus P_f$ with certain homotopical growth property under f^{-n}. This criterion is in general difficult to check. However, in case of degree two, there are extensive works [R], [T] using Levy cycle criterion which is proved to be equivalent to the Thurston's. For higher degree, it is known [ST] that Levy cycle criterion is not equivalent to Thurston's.

In 1.4, we will see how to reduce F when a "removable obstruction" exists.

1.3   Ray-equivalence. Let K_j be the filled-in Julia set of f_j $(j = 1, 2)$. By Douady and Hubbard [DH1], there exists a conformal mapping $\varphi_{K_j} : \mathbb{C} \setminus K_j \to \mathbb{C} \setminus \bar{\mathbb{D}}$, where $\mathbb{D}$ is the open unit disc, such that:

$$\varphi_{K_j} \circ f_j(z) = (\varphi_{K_j}(z))^d;$$

$$\varphi_{K_j}(z)/z \to 1 \text{ as } z \to \infty;$$

$\varphi_{K_j}^{-1}$ extends continuously to the boundary $\partial\mathbb{D}$ (and gives the parameterisation of ∂K_j).

Moreover $\varphi_{K_j}^{-1}$ also extends continuously to the circle at infinity by

$$\varphi_{K_j}^{-1} \mid S^1 = \mathrm{id}_{S^1}.$$

The *external ray* for f_j of angle $\theta \in \mathbb{R}/\mathbb{Z}$ is

$$R_\theta^{(j)} = \varphi_{K_k}^{-1}(\{\, re^{2\pi i\theta} \mid 1 < r < \infty \,\})(\subset \tilde{\mathbb{C}}_j),$$

and the *closed external ray* of angle θ is its closure in S^2, i.e.,

$$\overline{R_\theta^{(j)}} = R_\theta^{(j)} \cup \{\varphi_{K_j}^{-1}(e^{2\pi i\theta})\} \cup \{\infty \cdot e^{2\pi i\theta} \in \tilde{\mathbb{C}}_j\}.$$

We define the *ray equivalence* $\underset{\text{ray}}{\sim}$ on S^2 as the equivalence relation generated by

$$x \underset{\text{ray}}{\sim} y \text{ if } x \text{ and } y \text{ are in the same closed external ray.}$$

Hence if x and y are on two external rays which touch each other either at the equator S^1 or at the Julia sets $\partial K_1 \sqcup \partial K_2$, then x and y are ray-equivalent. in other words, two point are ray-equivalent if and only if they are connected by a path consisting of closed external rays.

Remark. If an external ray in a ray-equivalence class is periodic (as a set), then all other rays in this class are also periodic with the same period. It follows that if a ray-equivalence class contains a preperiodic point, then this class contains only finitely many external rays.

1.4 Degenerate mating. $F': S^{2'} \to S^{2'}$. Let

$$Y' = \{\text{ray equivalence class containing at least two points of } \Omega_F \cup P_F\},$$
$$Y = \{\xi \mid \xi \text{ is a ray equivalence class intersecting } \Omega_F \cup P_F \text{ and}$$
$$\text{there exist } m, n \geq 0 \ \eta \in Y' \text{ such that } f^m(\xi) = f^n(\eta)\}.$$

If $Y = \emptyset$, we define $S^{2'} = S^2$ and $F' = F$. If $Y \neq \emptyset$ and all $\xi \in Y$ are trees (i.e., contain no non-trivial loop), then we define the degenerate mating $F' = S^{2'} \to S^{2'}$ of f_1 and f_2 as follows: In S^2, shrink each $\xi \in Y$ to a point (see the remark above), then this makes a new topological sphere, which is denoted by $S^{2'}$. Let $\pi: S^{2'} \to S^{2'}$ be the natural projection.

There exists a continuous map $\bar{F}: S^{2'} \to S^{2'}$ such that $\bar{F} \circ \pi = \pi \circ F$ and it is not a branched covering only in a neighbourhood of $\pi(F^{-1}(Y) \smallsetminus Y)$. Each component of $\pi(F^{-1}(Y) \smallsetminus Y)$ is a tree and its image by $\bar{F}$ is a point. So we can modify $\bar{F}$ in a small neighbourhood of $\pi(F^{-1}(Y) \smallsetminus Y)$ to obtain a branched covering $F': S^{2'} \to S^{2'}$. This is well-defined up to Thurston equivalence.

If $Y \neq \emptyset$ and Y contains ξ which is not a tree, then we do not define the degenerate mating.

When $Y \neq \emptyset$ and all $\xi \in Y$ are trees, take disjoint closed disc neighbourhoods U_ξ for $\xi \in Y$. Then $\{\partial U_\xi \mid \xi \in Y\}$ gives a Thurston's obstruction for F. The construction of F' means that this obstruction is removable in the sense that we can reduce it so that the new branched covering is defined.

1.5 Matability.

Definition. Two monic polynomials are matable if and only if the degenerate mating exists and is Thurston equivalent to a rational map.

1.6 Topological mating. Using the ray-equivalence, we can define the quotient space $S^2/\underset{\text{ray}}{\sim}$ and the *topological mating* $F^* = [f_1 \underline{\perp\!\!\!\perp} f_2]^* : S^2/\underset{\text{ray}}{\sim} \to$ $S^2/\underset{\text{ray}}{\sim}$ as the induced map from F. Since every ray lands on a point in ∂K_{f_1} or ∂K_{f_2}, we have

$$S^2/\underset{\text{ray}}{\sim} = K_{f_1} \sqcup K_{f_2}/\underset{\text{ray}}{\sim},$$

where the point in ∂K_{f_1} with external angle θ is identified with the point in ∂K_{f_2} with external angle $-\theta$ ($\theta \in \mathbb{R}/\mathbb{Z}$).

In general, $S^2/\underset{\text{ray}}{\sim}$ is not always homeomorphic to the sphere. It is not known if the quotient is always a Hausdorff space. In [ST], we gave an example for which $S^2/\underset{\text{ray}}{\sim}$ is a sphere but f_1 and f_2 are not matable.

1.7 Main theorem. Our main result in this paper is:

Main theorem. Let f_1, f_2 be monic polynomials of Thurston's type of degree d (≥ 2). If the degenerate mating $F' = (f_1 \underline{\perp\!\!\!\perp} f_2)'$ is Thurston equivalent to a rational map $R \colon \bar{\mathbb{C}} \to \bar{\mathbb{C}}$, then there exists a continuous semi-conjugacy $h \colon S^2 \to \bar{\mathbb{C}}$ such that:

(i) the following diagram commutes:

$$
\begin{array}{ccc}
S^2 & \xrightarrow{\ F\ } & S^2 \\
{\scriptstyle h}\downarrow & & \downarrow{\scriptstyle h} \\
\bar{\mathbb{C}} & \xrightarrow{\ R\ } & \bar{\mathbb{C}},
\end{array}
$$

where $F = f_1 \underline{\perp\!\!\!\perp} f_2$ is the formal mating;

(ii) h is a uniform limit of orientation preserving homeomorphisms;

(iii) h is conformal in $\operatorname{int} K_{f_1} \sqcup \operatorname{int} K_{f_2}$ onto $\bar{\mathbb{C}} \smallsetminus J_R$ and $h^{-1}(\bar{\mathbb{C}} \smallsetminus J_R) = \operatorname{int} K_{f_1} \sqcup \operatorname{int} K_{f_2}$;

(iv) for $x, y \in S^2$, $h(x) = h(y)$ if and only if x and y are ray-equivalent.

In other words, the topological mating $F^* \colon S^2/\underset{\text{ray}}{\sim} \to S^2/\underset{\text{ray}}{\sim}$ is conjugate to R by $\bar{h} \colon S^2/\underset{\text{ray}}{\sim} \to \bar{\mathbb{C}}$ induced by h.

Remark. If both f_1 and f_2 are hyperbolic, this result was obtained by M. Rees (Lamination Map Conjugacy Theorem [R] §0.15). However, she stated the theorem in terms of laminations.

Corollary. Under the hypothesis of the theorem, the ray-equivalence is closed. Moreover, $h|_E \colon E \to J_R$ induces a semi-conjugacy between $z^d|_{S^1}$ and $R|_{J_R}$. This semi-conjugacy is a uniform limit of topological embeddings.

For further properties of h, see Proposition 3.1.

2 Construction of the semi-conjugacy

In this section, we show the existence of semi-conjugacy from the mating to a rational map, when the polynomials are matable.

2.1. <u>Theorem</u>. Suppose that the degenerate mating $F' = (f_1 \perp\!\!\!\perp f_2)'$ of polynomials f_1 and f_2 is Thurston equivalent to a rational map $R \colon \bar{\mathbb{C}} \to \bar{\mathbb{C}}$. Then there exists a continuous mapping $h \colon S^2 \to \bar{\mathbb{C}}$, satisfying (i), (ii) and (iii) of Main Theorem 1.7.

<u>Proof</u>. The main idea of the proof is to use the lifts of the equivalence maps θ, θ' and to show their convergence using the expanding metric for R.

To make the idea clearer, we first deal with the case $F' = F$, i.e. the case where there is no degenerate Levy cycle for F. We shall see later the case $F' \neq F$, i.e. the case where F' is really degenerate.

Assuming that $F' = F$ is Thurston equivalent to R, we shall construct sequences θ_n, H_n ($n = 0, 1, \dots$) and a neighbourhood V in S^2 of super-attracting periodic orbits of F with $F(\bar{V}) \subset V$ ($V = \emptyset$ if there is no super-attracting orbit), satisfying the following conditions:

(a) $\theta_n \colon S^2 \to \bar{\mathbb{C}}$ is an orientation preserving homeomorphism;

(b) $\theta_n \circ F = R \circ \theta_{n+1}$;

(c) $\theta_n = \theta_{n+1}$ on $P_F \cup F^{-n}(V)$, $\theta_n(P_F) = P_R$ and θ_n is conformal in $F^{-n}(V)$;

(d1) $H_n \colon S^2 \times [0, 1] \to \bar{\mathbb{C}}$ is continuous;

(d2) $H_n(\cdot, t)$ is a homeomorphism for any $t \in [0, 1]$;

(d3) $H_n(\cdot, 0) = \theta_n$, $H_n(\cdot, 1) = \theta_{n+1}$;

(d4) $H_n(x, t) = \theta_n(x)$ for $x \in P_F \cup F^{-n}(V)$, $t \in [0, 1]$.

First of all, by the definition of Thurston equivalence, we have $\theta_0 = \theta$, $\theta_1 = \theta'$ and H_0 satisfying (a)-(d4) except the conditions on $F^{-n}(V)$ in (c) and (d4). When F has a super-attracting orbit, we need to modify θ_0, θ_1 and H_0 as follows so that these conditions are satisfied.

Let $\{\zeta_j \mid j = 0, 1, \dots, p - 1\}$ be a super-attracting orbit of F ($\zeta_j = F^j(\zeta_0)$ and $\zeta_0 = F^p(\zeta_0)$), and denote $\zeta_j' = \theta_0(\zeta_j)$, $k_j = \deg_{\zeta_j} F$ and $k = k_0 k_1 \dots k_{p-1}$. For each ζ_j' ($j = 0, 1, \dots, p - 1$), there exists a local coordinate function w_j from a neighbourhood W_j of ζ_j' into $\mathbb{C}$ such that $w_j(\zeta_j') = 0$ and $R^p \colon w_j \to (w_j)^k$ in this coordinate. A *Dehn twist* $D_t^{(j)}$ around ζ_j' of angle t is a homeomorphism of $\bar{\mathbb{C}}$ defined by

$$D_t^{(j)} \colon w_j \to e^{itv(|w_j|)} \cdot w_j \quad \text{in } W_j$$

and

$$D_t^{(j)} = \text{id} \qquad \text{in } \bar{\mathbb{C}} \smallsetminus W_j,$$

where v is a continuous function such that $v(r) = 1$ $(0 \le r \le r_1)$, $v(r) = 0$ $(r \ge r_2)$ with $0 < r_1 < r_2$ and $\{|w_j| \le r_2\} \subset W_j$.

For two homeomorphisms $\theta, \theta' \colon S^2 \to \bar{\mathbb{C}}$, we write $\theta \sim \theta'$ is there is an isotopy $H \colon S^2 \times [0,1] \to \bar{\mathbb{C}}$ such that $H(\cdot, 0) = \theta$, $H(\cdot, 1) = \theta'$ and $H(x,t) = \theta(x)$ for $x \in P_F$ and $t \in [0,1]$; we write $\theta \approx \theta'$ if in addition there is a neighbourhood V of $\{\zeta_j\}$ such that $H(x,t) = \theta(x)$ for $x \in V$ and $t \in [0,1]$.

Since $\{\zeta_j\}$ and $\{\zeta_j'\}$ are super-attracting and $\deg_{\zeta_j} F = \deg_{\zeta_j'} R$, there exists a local conformal conjugacy between F and R near these orbits. Hence there exists a homeomorphism $\theta_0' \colon S^2 \to \bar{\mathbb{C}}$ such that $\theta_0' \sim \theta_0$, $\theta_0' \circ F = R \circ \theta_0'$ near ζ_j and θ_0' is conformal near ζ_j. Since F and R are branched coverings branching only over P_F and P_R respectively, we can lift the isotopy between θ_0 and θ_0' to an isotopy starting from θ_1. So we obtain a homeomorphism $\theta_1' \colon S^2 \to \bar{\mathbb{C}}$ such that $\theta_0' \circ F = R \circ \theta_1'$ and $\theta_1' \sim \theta_1 \sim \theta_0 \sim \theta_0'$. Hence $R \circ \theta_0' = R \circ \theta_1'$ near ζ_j. It follows that

$$\theta_1' \approx D_{t_0}^{(0)} \circ D_{t_1}^{(1)} \circ \cdots \circ D_{t_{p-1}}^{(p-1)} \circ \theta_0'$$

for some $t_j \in \frac{1}{k_j}\mathbb{Z}$. On the other hand, for each j and $s \in \mathbb{R}$, there is a homeomorphism $D \colon \bar{\mathbb{C}} \to \bar{\mathbb{C}}$ such that $D_s^{(j+1)} \circ R = R \circ D$ and $D \approx D_{s/k_j}^{(j)}$, where $D_s^{(p)} = D_s^{(0)}$ and $\approx$ is defined similarly as above.

Let $\theta_0^* = D_{s_0}^{(0)} \circ D_{s_1}^{(1)} \circ \cdots \circ D_{s_{p-1}}^{(p-1)} \circ \theta_0'$, where s_j are determined below. Then there exists a homeomorphism $\theta_1^* \colon S^2 \to \bar{\mathbb{C}}$ such that $\theta_1^* \sim \theta_1'$ and $\theta_0^* \circ F = R \circ \theta_1^*$. By the above remark, we have

$$\theta_1^* \approx D_{s_1/k_0}^{(0)} \circ D_{s_2/k_0}^{(1)} \circ \cdots \circ D_{s_p/k_{p-1}}^{(p-1)} \circ \theta_1'$$
$$\approx D_{t_0+s_1/k_0}^{(0)} \circ \cdots \circ D_{t_{p-1}+s_p/k_{p-1}}^{(p-1)} \circ \theta_0',$$

where $s_p = s_0$. Hence $\theta_0^* \approx \theta_1^*$ if $s_j = t_j + s_{j+1}/k_j$ $(j = 0,1,\ldots,p-1)$. And this equation has a solution (in $\frac{1}{(k-1)}\mathbb{Z}$) since $k = k_0 k_1 \ldots k_{p-1} > 1$. Using this solution (s_j), we redefine new θ_0 (resp. θ_1) to be θ_0^* (resp. θ_1^*) and new H_0 to be the isotopy fixing P_F and a neighbourhood V of $\{\zeta_j\}$. If there are several super-attracting orbits, this procedure can be done for those orbits. Finally we can take V so that $F(\bar{V}) \subset V$. Thus for new θ_0, θ_1 and H_0, the conditions (a)–(d4) are all satisfied.

Now we define θ_n, H_n inductively. Suppose we have θ_{n-1}, θ_n and H_{n-1} satisfying the conditions. Considering F and R as branched coverings, we can lift H_{n-1} by F and R to obtain $H_n \colon S^2 \times [0,1] \to \bar{\mathbb{C}}$ such that: H_n is continuous; $H_n(\cdot, 0) = \theta_n$; and

(e) $$H_{n-1}(F(x), t) = R(H_n(x, t)).$$

Then define $\theta_{n+1} = H_n(\cdot, 1)$. It is easy to check the conditions for θ_{n+1}, H_n. Thus we have constructed the sequences θ_n, H_n.

To show the convergence of θ_n, we need the following:

2.2. Theorem (Douady-Hubbard [DH1]). Let $R: \bar{\mathbb{C}} \to \bar{\mathbb{C}}$ be a rational map of Thurston's type and $U(\subset \bar{\mathbb{C}})$ an open set whose closure does not contain super-attracting orbits. There exist $\lambda > 1$ and a Riemannian metric $\|\cdot\|$ on U with a finite number of singularities such that

if z, $R(z) \in U$ and $v \in T_z \bar{\mathbb{C}}$, then $\|T_z R(v)\| \geq \lambda \|v\|$;

and at the singularity, $\|\cdot\|$ has the form $u(z)|dz|/|z|^\beta$ with $0 < \beta < 1$ and $u(z) > 0$ continuous. This metric defines the same topology on U as the standard one. Moreover, the metric d defined by $\|\cdot\|$ satisfies

(spherical distance of x and y) $\leq K \cdot d(x, y)$ if $x, y \in U$.

Even though this theorem says that R is infinitesimally expanding on J_R, R is not locally expanding, even not locally injective whenever $\Omega_R \cap J_R \neq \emptyset$. So we use the following notion.

2.3. Definition. Let $U: [0, 1] \to U \smallsetminus P_R$ be an arc. The *homotopic length* of α is

h-length$(\alpha) = \inf\{$length$(\alpha') \mid \alpha'$ is a rectifiable arc in $U \smallsetminus P_R$,

homotopic (in $U \smallsetminus P_R$) to α fixing end points$\}$,

where length(α') is with respect to $\|\cdot\|$.

2.4. Lemma. (i) $d(\alpha(0), \alpha(1)) \leq h$-length$(\alpha)$.

(ii) Let U' be a set such that $\overline{U'} \subset U$. For any $\epsilon > 0$, there exists $\delta > 0$ such that if α is an arc in $U' \smallsetminus P_R$ with diam$(\alpha) < \delta$ then h-length$(\alpha) < \epsilon$, where diam$(\cdot)$ is the diameter with respect to the spherical metric.

(iii) If α is an arc such that $\alpha, R(\alpha) \subset U \smallsetminus P_R$, then h-length$(R(\alpha)) \geq \lambda \cdot h$-length$(\alpha)$.

The proof is left to the reader.

In Theorem 2.2, we may take U so that $U' = \bar{\mathbb{C}} \smallsetminus \theta_0(V) \subset U$. For $x \in S^2 \smallsetminus (V \cap P_F)$, let us consider the arc $H_n(x, \cdot): [0, 1] \to \bar{\mathbb{C}}$. If $x \in S^2 \smallsetminus (F^{-n}(V) \cup P_F)$, then the image of this arc is contained in $\bar{\mathbb{C}} \smallsetminus (\theta_n(F^{-n}(V)) \cup P_R) \subset \bar{\mathbb{C}} \smallsetminus (\theta_n(V) \cup P_R) = U' \smallsetminus P_R$. So we can define the homotopic length for H_n. By (e) and

Lemma 2.4(ii), if $x \in S^2 \setminus (F^{-n}(V) \cup P_F)$, then $F(x) \in S^2 \setminus (F^{-(n-1)}(V) \cup P_F)$ and

$$h\text{-length}(H_n(x, \cdot)) \leq \frac{1}{\lambda} \left(h\text{-length}(H_{n-1}(F(x), \cdot)) \right).$$

On the other hand,

$$C = \sup_{x \in S^2 \setminus (V \cup P_F)} h\text{-length}(H_0(x, \cdot)) < \infty,$$

because of the uniform continuity of H_0 and Lemma 2.4(ii). Using the condition (c), we have

$$\sup_{x \in S^2} d(\theta_n(x), \theta_{n+1}(x)) = \sup_{x \in S^2 \setminus (F^{-n}(V) \cup P_F)} d(H_n(x, 0), H_n(x, 1))$$

$$\leq \sup_{x \in S^2 \setminus (F^{-n}(V) \cup P_F)} h\text{-length}(H_n(x, \cdot))$$

$$\leq C \left(\frac{1}{\lambda} \right)^n.$$

This shows the uniform convergence of θ_n with respect to the spherical metric of $\bar{\mathbb{C}}$. Let $h = \lim_{n \to \infty} \theta_n$. Then h must satisfy $h \circ F = R \circ h$ by (b). Other properties of h are also easy to check, using the fact that

$$\text{int} \, K_{f_1} \sqcup \text{int} \, K_{f_2} = \bigcup_{n \geq 0} F^{-n}(V),$$

$$\bar{\mathbb{C}} \setminus J_R = \bigcup_{n \geq 0} R^{-n}(\theta_0(V)) = \bigcup_{n \geq 0} \theta_{n+1}(F^{-n}(V)).$$

Thus the theorem is proved in the case $F' = F$.

Let us consider the case $F' \neq F$. Let $\pi \colon S^2 \to S^{2'}$ be the collapsing map, and $\bar{F} \colon S^{2'} \to S^{2'}$ such that $\pi \circ F = \bar{F} \circ \pi$ (see Definition 1.4). By the assumption, there exist the equivalence maps $\theta, \theta' \colon S^{2'} \to \bar{\mathbb{C}}$ such that $\theta \circ F' = R \circ \theta'$ etc. Define

$$Q_{F'} = \{ x \in S^{2'} \mid \pi^{-1}(x) \neq 1 \, pt \},$$
$$Q_F = \pi^{-1}(Q_{F'}), \quad Q_R = \{ \theta(x) \mid x \in Q_{F'} \}.$$

Note that Q_F consists of trees and $Q_{F'} \subset \Omega_{F'} \cup P_{F'}$, $Q_R \subset \Omega_R \cup P_R$.

We construct the sequences θ_n, H_n and V as before with the modified conditions (a')–(d4'):

(a') $\theta_n \colon S^2 \to \bar{\mathbb{C}}$ is continuous, $\theta_n | S^2 \setminus F^{-n}(Q_F) \colon S^2 \setminus F^{-n}(Q_F) \to \bar{\mathbb{C}} \setminus R^{-n}(Q_R)$ is orientation preserving homeomorphism, and for every $z \in R^{-n}(Q_R)$, $\theta_n^{-1}(z)$ is a tree;

(b') = (b); (c') = (c) with P_F, P_R replaced by $P_F \cup Q_F$, $P_R \cup Q_R$, respectively;

(d1') = (d1);

(d2') $H_n(x,t) \notin P_R$ for $x \notin P_F \cup Q_F$, $t \in [0,1]$;

(d3') = (d3); (d4') = (d4) with P_F replaced by $P_F \cup Q_F$.

First, we may assume that $\theta = \theta'$ on $P_{F'} \cup Q_{F'}$ and the isotopy between θ and θ' fixes $Q_{F'}$ as well as $P_{F'}$, by lifting θ, θ' and the isotopy by F' and R and replacing them by the lifts. We can modify them again, as before, so that θ and θ' coincide and are conformal in a neighbourhood $V' = \pi(V)$ of the super-attracting orbits of F', and the isotopy fixes V'.

By the definition of F', there exists a continuous map $\eta = S^{2'} \to S^{2'}$ such that $\bar{F} = F' \circ \eta$; η is homotopic to the identity by a homotopy which differs from the identity only in a small neighbourhood of $\bar{F}^{-1}(Q_{F'}) \smallsetminus Q_{F'}$. Now define $\theta_0 = \theta \circ \pi$ and $\theta_1 = \theta' \circ \eta \circ \pi$, and construct the homotopy H_0 from (isotopy between θ and θ') $\circ \pi$ and $\theta' \circ$ (homotopy between η and id) $\circ \pi$. Then it is easy to see that θ_0, θ_1, H_0 satisfy the conditions (a')–(d4'). By the lifting argument, we can construct sequences θ_n, H_n, since the condition (d2') is enough to guarantee the possibility of lifting. Then using (b') and the homotopy H_n, one can check (a') for θ_{n+1}.

Note that we can take H_0 with the additional condition:

$$H_0(x,t) \notin \theta_0(V) \text{ for } x \notin V.$$

Hence for all $n \geq 0$,

$$H_n(x,t) \notin R^{-n}(\theta_0(V)) \text{ for } x \notin F^{-n}(V).$$

Now the same argument as before applies to show that θ_n converges.

Finally it is easy to see that θ_n's can be approximated by orientation preserving homeomorphisms (by (a')). $\square$

3   Properties of the semi-conjugacy

Here are some properties of h.

3.1. Proposition. Let F, R and h be as in Theorem 2.1. Then we have:

(i) h is surjective; for any $z \in \bar{\mathbb{C}}$, $h^{-1}(z)$ is closed, connected and non-separating (i.e., its complement is connected).

(ii) If x and y in S^2 are ray-equivalent, then $h(x) = h(y)$.

(iii) $F\colon h^{-1}(z) \to h^{-1}(R(z))$ is a map of degree $\deg_z R$. More precisely, for $z \in \bar{\mathbb{C}}$, $x \in h^{-1}(R(z))$,

$$\sum_{y \in F^{-1}(x) \cap h^{-1}(z)} \deg_y F = \deg_z R,$$

where $\deg_y F$, $\deg_z R$ are the local degrees of F, R at y, z, respectively. In particular, if z is non critical for R, then $F|_{h^{-1}(z)}$ is injective.

Proof. (i) These properties follow from the fact that h is a uniform limit of homeomorphisms. The surjectivity and the closeness of $h^{-1}(z)$ are easy to see.

Suppose that $h^{-1}(z)$ is not connected, i.e., there are open sets U_1 U_2 in S^2 such that $h^{-1}(z) \subset U_1 \cup U_2$, $U_1 \cap U_2 = \emptyset$ and $U_1 \cap h^{-1}(z) \neq \emptyset$, $U_2 \cap h^{-1}(z) \neq \emptyset$. Then $h(S^2 \smallsetminus U_1 \cup U_2)$ is a compact set not containing z. Take a connected neighbourhood of z such that $\bar{V} \cap h(S^2 \smallsetminus U_1 \cup U_2) = \emptyset$. Since h is a uniform limit of homeomorphisms, there exists a homeomorphism $h'\colon S^2 \to \bar{\mathbb{C}}$ such that

$$d(h, h') < \min\{\text{distance } (z, \partial V), \text{distance } (\bar{V}, h(S^2 \smallsetminus U_1 \cup U_2))\}.$$

Then it follows that $h'(S^2 \smallsetminus U_1 \cup U_2) \cap \bar{V} = \emptyset$, hence $h'^{-1}(V) \subset U_1 \cup U_2$, and $h'(h^{-1}(z)) \subset V$ hence $U_1 \cap h'^{-1}(V) \neq \emptyset$, $U_2 \cap h'^{-1}(V) \neq \emptyset$. This contradicts with the fact that $h'^{-1}(V)$ is connected.

Suppose that $h^{-1}(z)$ is separating, i.e., there exist non empty sets V_1 and V_2 in S^2 such that $V_1 \cap V_2 = \emptyset$ and $S^2 \smallsetminus h^{-1}(z) = V_1 \cup V_2$. For any $w \in \bar{\mathbb{C}} \smallsetminus \{z\}$, either $h^{-1}(w) \subset V_1$ or $h^{-1}(w) \subset V_2$, since $h^{-1}(w)$ is connected. Define $W_i = \{w \in \bar{\mathbb{C}} \smallsetminus \{z\} \mid h^{-1}(w) \subset V_i\}$ ($i = 1, 2$). Then $W_1 \cap W_2 = \emptyset$ and $W_1 \cup W_2 = \bar{\mathbb{C}} \smallsetminus \{z\}$. By the continuity of h, the set function $w \to h^{-1}(w)$ is "upper semi-continuous", hence W_i are open. Since V_i were non-empty, so are W_i. This contradicts the fact that $\bar{\mathbb{C}} \smallsetminus \{z\}$ is connected.

(ii) Fix $r_0 > 1$, $\theta \in \mathbb{R}/\mathbb{Z}$, $j = 1$ or 2, and consider the arc

$$\alpha = \varphi_{K_j}^{-1}(\{re^{2\pi i\theta} \mid r_0 \leq r \leq \infty\}).$$

Then $F^n(\alpha) = \varphi_{K_j}^{-1}(\{re^{2\pi i d^n \theta} \mid r_0^{d^n} \leq r \leq \infty\})$ and

$$\text{diam } F^n(\alpha) \to 0 \text{ (in } S^2) \text{ as } n \to \infty.$$

By the uniform continuity of h, we have

$$\text{diam } R^n(h(\alpha)) = \text{diam } h(F^n(\alpha)) \to 0.$$

Note that $R^n(h(\alpha)) \subset J_R$.

If $(\bigcup_n R^n(h(\alpha))) \cap P_R \neq \emptyset$, there is an m such that $R^m(h(\alpha))$ contains a repelling periodic point. Since $R^m(h(\alpha))$ is connected and diam $R^n(h(\alpha)) \to 0$ $(n \to \infty)$, $R^m(h(\alpha))$ must be one point. Then so is $h(\alpha)$, since $h(\alpha)$ is connected and $h(\alpha) \subset R^{-m}(R^m(h(\alpha)))$.

If $(\bigcup_n R^n(h(\alpha))) \cap P_R = \emptyset$, then we can use the homotopic length (Definition 2.3). By Lemma 2.4, we have

$$h\text{-length}(h(\alpha)) \leq (\frac{1}{\lambda})^n (h\text{-length}(R^n(h(\alpha)))) \to 0.$$

Hence in any case we have $h(\varphi_{K_j}^{-1}(r_0 e^{2\pi i\theta})) = h(\varphi_{K_j}^{-1}(\infty \cdot e^{2\pi i\theta}))$. By the continuity of $\varphi_{K_j}^{-1}$ and h, we conclude that $h(\{\varphi_{K_j}^{-1}(re^{2\pi i\theta}) \mid 1 \leq r \leq \infty\})$ is one point. Then the definition of the ray-equivalence implies that each ray-equivalence class is mapped to one point by h.

(iii) Define the function

$$N(z, x) = \sum_{y \in F^{-1}(x) \cap h^{-1}(z)} \deg_y F$$

for $z \in \bar{\mathbb{C}}$ and $x \in S^2$. It is easy to see that this function is lower semi-continuous. On the other hand, for fixed z, $N(z, x)$ is upper semi-continuous as a function of $x \in h^{-1}(R(z))$, since

$$N(z, x) = d - \sum_{z' \in R^{-1}(R(z)) \smallsetminus \{z\}} N(z', x).$$

Since $h^{-1}(R(z))$ is connected and N is integer valued, $N(z, x)$ is constant on this set and it is denoted by $N(z)$. For any $z \in \bar{\mathbb{C}}$, $N(z) \geq 1$, by the surjectivity of h. If $R(z)$ is not a critical value, we have $N(z) = 1$, since $\sum_{z' \in R^{-1}(R(z))} N(z') = d$ and $\#R^{-1}(R(z)) = d$. When w is a critical value of R, by approximating w by non critical values, one can show that

$$N(z) \geq \deg_z R \text{ for } z \in R^{-1}(w).$$

Again using $\sum_{z \in R^{-1}(w)} N(z) = d$, we obtain $N(z) = \deg_z R$ for any $z \in \bar{\mathbb{C}}$. $\square$

4   Proof of (iv) of Main Theorem

In this section, we prove the property (iv) of the semi-conjugacy h. The key of the proof is:

4.1. <u>Proposition</u>. Let f_1, f_2, R and h be as in Theorem 2.1. Then $\#h^{-1}(z) \cap S^1$ is bounded for $z \in \bar{\mathbb{C}}$, where S^1 is the equator of S^2.

Assuming this result, we can now complete:

Proof of Main Theorem. By Proposition 3.1(ii), $h^{-1}(z)$ consists of ray-equivalence classes. By the above proposition, there are only finite number of ray-equivalence classes in $h^{-1}(z)$ and each of them contains a finite number of external rays. Then it follows that each ray-equivalence class is closed. So $h^{-1}(z)$ can contain only one ray equivalence class, otherwise $h^{-1}(z)$ would be disconnected. In other words $h(x) = h(y)$ if and only if x and y are ray equivalent. $\square$

Proof of Proposition 4.1.

4.2. Lemma. Let X_n be compact metric spaces and $g_n \colon X_n \to X_{n+1}$ maps $(n = 0, 1, 2, \ldots)$. Suppose that there exist $0 < \mu < 1$, $\delta > 0$ and an integer $N > 0$ such that

(i) for $n \geq 0$, $x, y \in X_n$,
    if $d(g_n(x), g_n(y)) \leq \delta$, then $d(x, y) \leq \mu d(g_n(x), g_n(y))$;

(ii) each X_n can be covered by N sets of diameter $\leq \delta$.

Then $\#X_n \leq N$ $(n = 0, 1, 2, \ldots)$.

4.3. Remark. (1) The condition (i) implies that g_n are injective.
(2) In our application, X_n are always subsets of S^1 and the metric is the one which is induced from $\mathbb{R}/\mathbb{Z}$ (hence the total length of S^1 is 1). Then the condition (ii) is trivial, in fact one can take an integer N greater than $1/\delta$.
(3) Lemma 4.2 is a generalisation of Douady's lemma which says:
Let X be a compact metric space and $g \colon X \to X$ a continuous locally expanding map. If g is injective on X, then X is finite.
To prove this from Lemma 4.2, put $X_n = X$ and $g_n = g$, and check the condition (i) from the local expanding property and injectivity of g. (See the proof of Lemma 4.4).

Proof of Lemma 4.2. Note that, by condition (i), if $U \subset X_{n+1}$ is a set with $\text{diam}(U) \leq \delta$ then $U' = g_n^{-1}(U)$ has:

$$\text{diam}\, U' \leq \mu \, \text{diam}\, U \leq \mu \delta.$$

By (ii), X_n is covered by N sets $U_1, \ldots, U_N$ with $\text{diam}\, U_i \leq \delta$. Then X_{n-1} will be covered by N sets $U'_1, \ldots U'_N$ with $\text{diam}\, U'_i \leq \mu \, \text{diam}\, U_i \leq \mu \delta$, where $U'_i = g_n^{-1}(U_i)$. Repeating this procedure, one obtains N sets $U^*_1, \ldots, U^*_N$ with $\text{diam}\, U^*_i \leq \mu^n \delta$ which cover X_0. Since $\mu^n \delta \to 0$ as $n \to \infty$, X_0 cannot contain more than N points. A similar argument applies to X_n. $\square$

4.4. Lemma. Let V be any open neighbourhood in $\bar{\mathbb{C}}$ of $\Omega_R \cap J_R$. Then there exists $\delta > 0$ such that
(i') For $z \in \bar{\mathbb{C}} \smallsetminus V$, $x, y \in h^{-1}(z) \cap S^1$, if $d(F(x), F(y)) \leq \delta$, then $d(x, y) = \frac{1}{d} d(F(x), F(y))$.

Proof. If not, one can choose $x_n, y_n \in S^1$ such that

$$h(x_n) = h(y_n) \in \bar{\mathbb{C}} \smallsetminus V, \ d(F(x_n), F(y_n)) \to 0 \text{ and } d(x_n, y_n) \geq \frac{1}{2d}.$$

(Note that if $d(x, y) \leq \frac{1}{2d}$ then $d(F(x), F(y)) = d \cdot d(x, y)$.)
Extracting a convergent subsequence, we may assume that $x_n \to x$, $y_n \to y$.
Then we have $h(x) = h(y) \notin \Omega_R \cap J_R$, $d(F(x), F(y)) = 0$ and $d(x, y) \geq \frac{1}{2d} > 0$. This contradicts with the fact that $F|_{h^{-1}(h(x))}$ is injective. (Proposition 3.1(iii)).   □

If R is hyperbolic, we may take $V = \emptyset$. In this case, applying Lemma 4.2 to $X_n = h^{-1}(R^n(z)) \cap S^1$ and $g_n = F|_{X_n}$, one obtains Proposition 4.1.

Now we consider non hyperbolic case. The problem is the behaviour of $h^{-1}(z)$ in the neighbourhood of $\Omega_R \cap J_R$. For $z \in \Omega_R \cup P_R$, there are only finite number of external rays in $h^{-1}(z)$, by Remark 4.3(3).

4.5. Lemma. Let $\omega \in \Omega_R \cap J_R$. Then there exists a neighbourhood U of $h^{-1}(\omega)$ in S^2, a homeomorphism $G: U \to G(U) \subset S^2$, and a neighbourhood V of ω in $\bar{\mathbb{C}}$ satisfying:
If $z \in V$, then $h^{-1}(z) \subset U$ and there is $w = \bar{G}(z) \in \bar{\mathbb{C}}$ such that $G: h^{-1}(z) \to h^{-1}(w)$ and it is injective; and
(i″) There exists $\delta > 0$ such that for $z \in V$, $x, y \in h^{-1}(z) \cap S^1$ with $d(G(x), G(y)) \leq \delta$, we have $d(x, y) \leq \frac{1}{d}d(G(x), G(y))$.

Proof. Let $l > 0$ be the integer such that $\alpha = R^l(\omega)$ is the first periodic point in the orbit of ω. Then $h^{-1}(\alpha)$ is periodic as a set under F. Let $p > 0$ be the period of external rays in $h^{-1}(\alpha)$, so that all the external rays in $h^{-1}(\alpha)$ are fixed under F^p. We construct G so that it satisfies the commutative diagram

$$
\begin{array}{ccc}
(U, h^{-1}(\omega)) & \xrightarrow{\ G\ } & (S^2, h^{-1}(\omega)) \\
{\scriptstyle F^l}\downarrow & & \downarrow{\scriptstyle F^l} \\
(S^2, h^{-1}(\alpha)) & \xrightarrow{\ F^p\ } & (S^2, h^{-1}(\alpha)).
\end{array}
$$

In the above diagram we consider F^l as a branched covering from a simply connected neighbourhood of $h^{-1}(\omega)$ to a simply connected neighbourhood of $h^{-1}(\alpha)$. Then we can lift F^p by F^l to obtain G, since F^p is homotopic to identity in this neighbourhood by a homotopy fixing the points in $h^{-1}(\alpha) \cap P_R$.

Now we verify the properties of G. By continuity of h there exists a neighbourhood V of ω such that if $z \in V$ then $h^{-1}(z) \subset U$. By the above diagram, we have $G(h^{-1}(z)) \subset F^{-l}(F^{l+p}(h^{-1}(\omega)) = h^{-1}(R^{-l}(R^{l+p}(\omega)))$ which is a disjoint union of $h^{-1}(z_i)$'s ($z_i \in R^{-l}(R^{l+p}(\omega))$). Since $h^{-1}(z)$ is connected, $h^{-1}(z)$ must be contained in one of $h^{-1}(z_i)$'s. Let $w = \bar{G}(z) = z_i$. Note that $F^p|_{S^1}$ in a neighbourhood of $h^{-1}(\alpha) \cap S^1$ is locally an expanding map with factor d^p, then so is $G|_{S^1}$ in a neighbourhood of $h^{-1}(\omega) \cap S^1$. So one obtains (ii'), restricting the neighbourhood if necessary.   □

Let $\Omega_R \cap J_R = \{\omega_1, \ldots, \omega_k\}$. Denote the $U, G, V, \bar{G}, \delta$ in Lemma 4.5 for $\omega = \omega_i$ by $U_i, G_i, V_i, \bar{G}_i, \delta_i$. We may take V_i's disjoint. Let $V = \bigcup_i V_i$ and denote by δ_0 the δ in Lemma 4.4. Finally let $\mu = 1/d$, $\delta = \min_{0 \le i \le K} \delta_i$ and an integer $N \ge 1/\delta$.

Now, take any $z \in J_R$ and set $z_0 = z$. Define X_n, g_n, z_{n+1} inductively as follows:

$$X_n = h^{-1}(z_n) \cap S^1;$$

if $z_n \notin V$, then $g_n = F|_{X_n}$ and $z_{n+1} = R(z_n)$;

if $z_n \in V_i$, then $g_n = G_i|_{X_n}$ and $z_{n+1} = \bar{G}(z_n)$.

Then $\{X_n, g_n\}$, together with μ, δ, N as above, satisfies the assumptions (i), (ii) of Lemma 4.2. Hence

$$\#h^{-1}(z) \cap S^1 = \#X_0 \le N$$

and this bound does not depend on z. Thus Proposition 4.1 is proved.   □

References

[DH1] A. Douady and J. H. Hubbard, Etudes dynamiques des polynômes complexes, I and II, avec la collaboration de P. Lavaurs, Tan Lei et P. Sentenac, *Publication Mathématique d'Orsay*, 84-02 (1984), 85-04 (1985).

[DH2] A. Douady and J. H. Hubbard, A proof of Thurston's topological characterization of rational functions. *Acta Math.*, 171 (1993), 263–297.

[D]   A. Douady, Systèmes dynamiques holomorphes, (Bourbaki seminar, Vol. 1982/83) Astérisque, 105-106 (1983), 39–63.

[L]   S. Levy, Critically finite rational maps, Ph. D. Thesis, Princeton University, 1985.

[R]   M. Rees, A partial description of parameter space of rational maps of degree two: Part I. *Acta Math.*, 168 (1992), 11–87.

[ST]   M. Shishikura and Tan Lei, A family of cubic rational maps and mating of cubic polynomials, preprint, to appear.

[T]   Tan Lei, Matings of quadratic polynomials, *Erg. Th. & Dyn. Sys.*, 12 (1992), 589–620.

[W]   B. Wittner. On the bifurcation loci of rational maps of degree two. Ph. D. Thesis, Cornell University, 1986.

Hiroshima University, Department of Mathematics, Kagamiyama, Higashi-Hiroshima 739-8526, JAPAN
email : mitsu@math.sci.hiroshima-u.ac.jp

Le théorème d'intégrabilité des structures presque complexes

Adrien Douady
d'après des notes de X. Buff

Abstract

We give a proof of the theorem of Morrey-Bojarski-Ahlfors-Bers in the L^2 framework, using Fourier transform. We prove successively the following results:

Th. 1 : For $\mu \in L^\infty_{comp}(\mathbf{C})$ with $\|\mu\|_{L^\infty} < 1$, there is a unique $\Phi = \Phi_\mu$ in $H^1_{loc}(\mathbf{C})$ which is a solution of the Beltrami equation defined by μ , such that $\Phi(z) - z$ tends to 0 at infinity. Moreover, if μ depends analytically on a parameter λ , then Φ_μ depends analytically on λ .

Th. 2 : If μ is C^∞ on a neighborhood of z_0 with $|\mu(z_0)| < 1$, the Beltrami equation has local solutions which provide local C^∞ charts.

Th. 3 : If $\mu \in C^\infty_{comp}(\mathbf{C})$ with $\|\mu\|_{L^\infty} < 1$, the map Φ_μ is a C^∞ diffeomorphism of $\mathbf{C}$ onto itself.

Th. 4 : For μ in $L^\infty_{comp}(\mathbf{C})$ with $\|\mu\|_{L^\infty} < 1$, the map Φ_μ is a homeomorphism of $\mathbf{C}$ onto itself.

Th. 5 : For μ in $L^\infty(\mathbf{C})$ with $\|\mu\|_{L^\infty} < 1$, there exists a unique $\Phi = \Phi^\natural_\mu$ in $H^1_{loc}(\mathbf{C})$ which is a solution of the Beltrami equation defined by μ, and a homeomorphism of $\mathbf{C}$ onto itself fixing 0 and 1. Moreover, if μ depends analytically on a parameter λ, then $\Phi^\natural_\mu$ depends analytically on λ.

Nous donnons une démonstration du Théorème de Morrey-Bojarski-Ahlfors-Bers dans le cadre L^2, en utilisant la transformée de Fourier. Pour l'enchaînement des résultats, voir le paragraphe 3 de l'article.

1 Motivation

Soient U et V deux ouverts de $\mathbf{C}$ et $\phi : U \to V$ un difféomorphisme C^∞ (au sens réel) préservant l'orientation, ce que nous écrivons $\phi \in Diff^+(U; V)$. Pour $x \in U$, notons $\mathcal{E}_\phi(x)$ l'ellipse $T_x\phi^{-1}(S^1)$, considérée à homothétie réelle près, et posons $\mu_\phi(x) = \frac{\partial\phi/\partial\bar{z}}{\partial\phi/\partial z}(x)$. On a $|\mu_\phi(x)| < 1$. Les conditions suivantes sont équivalentes :

(1) L'application ϕ est conforme, i.e. holomorphe ;
(2) $\mathcal{E}_\phi$ est le champ de cercles ;
(3) $\mu_\phi = 0$.

La donnée du champ d'ellipses $\mathcal{E}_\phi$ sur U (toujours à homothétie réelle près en chaque point) et celle de la fonction $\mu_\phi : U \to \mathbf{D}$ sont équivalentes :

l'ellipticité (rapport des longueurs des axes) K de $\mathcal{E}_\phi(x)$ et $k = |\mu_\phi(x)|$ sont liés par $K = \frac{1+k}{1-k}$, i.e. $k = \frac{K-1}{K+1}$, et $Arg(\mu) = -2\theta$ où θ est l'argument du petit axe de $\mathcal{E}_\phi(x)$.

On dit que ϕ est un difféomorphisme K-quasi-conforme si le champ $\mathcal{E}_\phi$ est à ellipticité bornée par K, i.e. si $||\mu_\phi||_{L^\infty} \leq \frac{K-1}{K+1}$. Nous verrons comment définir la quasi-conformité pour des applications qui ne sont pas des difféomorphismes.

La question suivante est naturelle : *Etant donné un ouvert U de $\mathbf{C}$ et un champ d'ellipses $\mathcal{E}$ sur U, peut-on trouver un difféomorphisme $\phi : U \to V$ tel que $\mathcal{E} = \mathcal{E}_\phi$?*

Question équivalente : *Etant donné une fonction $\mu : U \to \mathbf{D}$, peut on trouver un difféomorphisme $\phi : U \to V$ tel que $\mu = \mu_\phi$?*

Si on veut pour ϕ un difféomorphisme C^∞, une condition nécessaire est bien sûr que μ soit C^∞. En termes de $\mathcal{E}$, cette condition s'exprime moins bien. Nous verrons (Th. 2 et Th.3) que cette condition est suffisante localement, et aussi globalement si $U = \mathbf{C}$ et μ à support compact.

Mais pour les besoins de la chirurgie holomorphe comme pour ceux de l'étude des groupes kleiniens, on est amené à considérer l'*équation de Beltrami*

$$\frac{\partial \phi}{\partial \overline{z}} = \mu \cdot \frac{\partial \phi}{\partial z}$$

pour des fonctions μ qui ne sont pas continues. On cherche alors des solutions ϕ qui ne sont pas C^1, et les dérivées partielles $\frac{\partial \phi}{\partial z}$ et $\frac{\partial \phi}{\partial \overline{z}}$ doivent être prises au sens des distributions.

Une condition naturelle à imposer à ϕ est alors d'être dans l'espace de Sobolev H^1_{loc}. C'est en tous cas une bonne présentation de la théorie. Nous traitons le cas $\mu \in L^\infty_{comp}$ avec $||\mu||_{L^\infty} < 1$, nous réservant de nous débarrasser de l'hypothèse μ à support compact à la fin. Les résultats que nous présentons sont dûs à Morrey, Bojarski et Ahlfors-Bers. La démonstration que nous donnons utilise seulement des techniques L^2 et la transformée de Fourier.

2   Quelques espaces fonctionnels

Soit U un ouvert de C. On note $C(U)$ (resp. $C^1(U)$, $C^\infty(U)$) l'espace vectoriel des fonctions continues sur U à valeurs dans $\mathbf{C}$. On note $C_{comp}(U)$ l'espace des fonctions à support compact dans U, et on définit de même C^1_{comp}, $C^\infty_{comp}(U)$.

L'espace $L^2(U)$ est le complété de $C_{comp}(U)$ pour la norme L^2 définie par

$$(||f||_{L^2})^2 = \iint |f|^2 = \iint_{z \in U} |f(z)|^2 \, dx dy$$

où $z = x + iy$. C'est un espace de Hilbert complexe.

Nous préférons cette définition à celle qui consiste à voir les éléments de $L^2(U)$ comme des fonctions de carré intégrable définies à une fonction négligeable près. En effet, suivant une morale héritée de Laurent Schwartz, nous rechignons à évaluer une fonction en un point lorsqu'elle n'est pas continue. Il nous arrivera cependant de parler des éléments de $L^2(U)$ comme de "fonctions".

On désigne par $L^2_{comp}(U)$ l'espace des fonctions de $L^2(U)$ à support compact dans U, et par $L^2_{loc}(U)$ l'espace des fonctions sur U qui sont localement dans L^2. On définit de même $L^1(U)$, $L^1_{comp}(U)$, $L^1_{loc}(U)$. L'epace $L^1_{loc}(U)$ est le plus grand des espaces de fonctions sur U que nous considérons.

Pour $L^\infty(U)$ il faut être un peu plus soigneux : la boule fermée de rayon r dans $L^\infty(U)$ est la fermeture dans $L^1_{loc}(U)$ de l'ensemble des $f \in C_{comp}(U)$ tels que $||f||_{L^\infty} \leq r$ (la topologie sur $L^1_{loc}(U)$ étant celle induite par le produit des $L^1(U')$ pour U' relativement compact dans U).

On écrit L^2 pour $L^2(\mathbf{C})$, etc.

Pour $f \in L^1_{loc}(U)$ et (f_n) une suite d'éléments de $L^1_{loc}(U)$, on dit que f_n tend vers f *au sens des distributions* sur U si

$$(\forall h \in C^\infty_{comp}(U)) \quad \iint f_n h \to \iint f h .$$

Pour f et g dans $L^1_{loc}(U)$, on dit que $g = \frac{\partial f}{\partial x}$ au sens des distributions si

$$(\forall h \in C^\infty_{comp}(U)) \quad \iint g h = -\iint f \frac{\partial h}{\partial x} .$$

La fonction g est uniquement déterminée (comme élement de $L^1_{loc}(U)$) par cette condition.

On définit l'espace de Sobolev $H^1_{loc}(U)$ comme l'espace vectoriel des fonctions $f \in L^1_{loc}(U)$ admettant des dérivées partielles $\frac{\partial f}{\partial x}$ et $\frac{\partial f}{\partial y}$ au sens des distributions dans $L^2_{loc}(U)$. Pour $f \in H^1_{loc}(U)$, on définit $\frac{\partial f}{\partial z}$ et $\frac{\partial f}{\partial \bar z}$ par les formules habituelles $\frac{\partial f}{\partial z} = \frac{1}{2}(\frac{\partial f}{\partial x} - i\frac{\partial f}{\partial y})$, $\frac{\partial f}{\partial \bar z} = \frac{1}{2}(\frac{\partial f}{\partial x} + i\frac{\partial f}{\partial y})$. Si $\frac{\partial f}{\partial \bar z} = 0$, f est holomorphe (*lemme de Weyl*).

Une fonction de $H^1_{loc}(U)$ n'est pas nécessairement continue, et c'est bien dommage car cela faciliterait la démonstration du Th. 4. Par exemple la fonction $z \mapsto \log|\log|z||$ appartient à $H^1_{loc}(\mathbf{C})$.

3 Enoncés

Théoreme 1.- *Soit* $\mu \in L^\infty_{comp}(\mathbf{C})$ *avec* $||\mu||_{L^\infty} < 1$. *Alors il existe une fonction* $\Phi = \Phi_\mu \in H^1_{loc}(\mathbf{C})$ *et une seule qui soit solution de l'équation de Beltrami*

$$\frac{\partial \Phi}{\partial \bar z} = \mu \frac{\partial \Phi}{\partial z}$$

définie par μ, et telle que $\Phi(z) - z$ tende vers 0 quand $z \to \infty$.

De plus, si μ dépend analytiquement (resp. continûment) d'un paramètre λ, la fonction Φ_μ dépend analytiquement (resp. continûment) de λ.

Théorème 2.- *Si μ est C^∞ au voisinage de z_0 avec $|\mu(z_0)| < 1$, l'équation de Beltrami définie par μ admet une solution locale qui est un difféomorphisme C^∞ d'un voisinage de z_0 sur un ouvert de $\mathbf{C}$.*

Théorème 3.- *Si $\mu \in C^\infty_{comp}(\mathbf{C})$ avec $\|\mu\|_{L^\infty} < 1$, l'application Φ_μ est un difféomorphisme C^∞ de $\mathbf{C}$ sur lui-même.*

Théorème 4.- *Pour $\mu \in L^\infty_{comp}(\mathbf{C})$ avec $\|\mu\|_{L^\infty} < 1$, l'application Φ_μ est un homéomorphisme de $\mathbf{C}$ sur lui-même.*

Théoreme 5.- *Pour $\mu \in L^\infty(\mathbf{C})$ avec $\|\mu\|_{L^\infty} < 1$, il existe un $\Phi = \Phi^\natural_\mu \in H^1_{loc}(\mathbf{C})$ et un seul qui soit solution de l'équation de Beltrami*

$$\frac{\partial \Phi}{\partial \overline{z}} = \mu \frac{\partial \Phi}{\partial z}$$

et qui soit un homéomorphisme $\mathbf{C} \to \mathbf{C}$ avec $\Phi(0) = 0$ et $\Phi(1) = 1$.

De plus, si μ dépend analytiquement (resp. continûment) d'un paramètre λ, la fonction $\Phi^\natural_\mu$ dépend analytiquement (resp. continûment) de λ.

4   Rappels sur la transformation de Fourier

On écrit L^2 pour $L^2(\mathbf{C})$, etc.

Pour $f \in L^1$, on définit la transformée de Fourier $\hat{f}$ de f par la formule:

$$\hat{f}(\varsigma) = \hat{f}(\xi + i\eta) = \iint f(x + iy)e^{-2i\pi(x\xi + y\eta)}\, dx dy.$$

On a alors $\hat{f} \in C_0$, i.e. $\hat{f}$ est continue et tend vers 0 à l'infini,

$$\widehat{f * g} = \hat{f} \cdot \hat{g} \quad , \quad \|\hat{f}\|_{L^\infty} \leq \|f\|_{L^1} \quad \text{et} \quad \|\hat{f}\|_{L^2} = \|f\|_{L^2}\, .$$

La dernière égalité permet de définir la transformation de Fourier $f \mapsto \hat{f}$ comme isométrie de L^2. Pour f et g dans L^2, on a

$$\widehat{f \cdot g} = \hat{f} * \hat{g} \quad , \quad \widehat{\hat{f}}(x) = f(-x) \quad \text{et} \quad \|f\|_{L^\infty} \leq \|\hat{f}\|_{L^1}\, .$$

Pour $f \in L^1_{loc}$ telle que $\hat{f}$ soit définie, et que $\frac{\partial f}{\partial x} \in L^2$ et $\frac{\partial f}{\partial y} \in L^2$, on a

$$\widehat{\frac{\partial f}{\partial x}} = 2i\pi\xi\hat{f} \quad \text{et} \quad \widehat{\frac{\partial f}{\partial y}} = 2i\pi\eta\hat{f}$$

d'où

$$\frac{\widehat{\partial f}}{\partial \bar{z}} = i\pi\zeta\hat{f} \quad \text{et} \quad \frac{\widehat{\partial f}}{\partial z} = i\pi\bar{\zeta}\hat{f}.$$

On dit qu'une fonction $f \in L^\infty$ est à décroissance rapide si $f \in \mathcal{O}(1/|\zeta|^k)$ pour tout $k \in \mathbf{N}$. On appelle *espace de Schwartz* et on note $\mathcal{S}$ l'espace vectoriel des fonctions C^∞ à décroissance rapide et dont toutes les dérivées sont à décroissance rapide.

L'espace de Schwartz est stable par multiplication, par convolution et par transformée de Fourier: on a $f \in \mathcal{S}$ si et seulement si $\hat{f} \in \mathcal{S}$. Par ailleurs, si f est C^∞ à support compact, alors $\hat{f}$ est à décroissance rapide, et si $\hat{f} \in L^1_{loc}$ est à décroissance rapide, alors f est C^∞.

Définition : *On appelle transformation de Hilbert-Beurling l'isométrie $\mathcal{L} : L^2 \to L^2$ définie par*

$$\widehat{\mathcal{L}(f)} = \frac{\bar{\zeta}}{\zeta}\hat{f}.$$

Si $f \in L^2_{comp}$, la fonction $F = \frac{1}{\pi z} * f$ appartient à L^2_{loc} et vérifie $\frac{\partial F}{\partial \bar{z}} = f$ au sens des distributions, et $F(z) \in \mathcal{O}(\frac{1}{z})$ à l'infini. C'est la seule solution de $\frac{\partial F}{\partial \bar{z}} = f$ qui tende vers 0 à l'infini. On a alors $\frac{\partial F}{\partial z} = \mathcal{L}(f)$, la fonction $\mathcal{L}(f)$ est holomorphe en dehors du support de f, et dans $\mathcal{O}(\frac{1}{z^2})$ à l'infini.

Remarque : Il n'y a pas de façon naturelle de choisir pour tout $f \in L^2$ une solution F de $\frac{\partial F}{\partial \bar{z}} = f$.

5 Dans le cadre L^2

Démonstration du Théorème 1 : L'équation de Beltrami s'écrit, en posant $\Phi(z) = z + \varphi(z)$ et $f = \partial\varphi/\partial\bar{z}$,

$$f = \mu \cdot (1 + \mathcal{L}(f)).$$

L'opérateur $\mu\cdot : h \mapsto \mu \cdot h$ de L^2 dans lui même est de norme $k < 1$ et $\mathcal{L}$ est une isométrie, donc $1 - \mu\cdot\mathcal{L} : L^2 \to L^2$ est inversible et si on note $\bar{\mu}$ l'élément μ de L^2, on obtient

$$f = (1 - \mu\cdot\mathcal{L})^{-1}\bar{\mu} \quad,$$

soit

$$f = \bar{\mu} + \mu\cdot\mathcal{L}(\bar{\mu}) + \mu\cdot\mathcal{L}(\mu\cdot\mathcal{L}(\bar{\mu})) + \ldots.$$

Ceci démontre l'existence et l'unicité de la solution $f \in L^2$. A partir de f, on obtient φ par $\varphi = \frac{1}{\pi z} * f$ et Φ par $\Phi(z) = z + \varphi(z)$. Alors φ est holomorphe au voisinage de l'infini et tend vers 0 à l'infini.

Si μ_λ dépend de façon **C**-analytique (resp. continue) d'un paramètre $\lambda \in \Lambda$ (variété **C**-analytique) (resp. espace topologique), l'opérateur $\mu_\lambda \cdot$ dépend analytiquement (resp continûment) de λ, donc aussi $(1 - \mu_\lambda \cdot \mathcal{L})^{-1}$. Il en est de même de $\vec{\mu_\lambda}$, donc aussi de f_λ, φ_λ et finalement Φ_λ.   cqfd.

6   Un cas particulier

Une partie du Théorème 3 est que Φ, ou ce qui revient au même φ, est C^∞. Ceci pourrait se déduire du théorème d'hypo-ellipticité des opérateurs différentiels elliptiques, dont un cas particulier est le suivant :

Théorème. *Si sur un ouvert U de* **C** *on a* $\frac{\partial \phi}{\partial \bar{z}} + u \frac{\partial \phi}{\partial z} = v$ *avec u et v dans* C^∞ *et* $\|u\|_{L^\infty} < 1$, *alors* $\phi \in C^\infty$.

Mais il semble difficile de déduire le Th. 3 de ce théorème. C'est pourquoi nous choisissons de redémontrer le théorème d'hypo-ellipticité dans la situation qui nous intéresse, avec une démonstration qui donne pour le même prix le Th. 2, à partir duquel nous obtiendrons le Th. 3.

Pour le moment on va démontrer la proposition suivante.

Proposition 1.- *Supposons* $\mu \in C^\infty_{comp}$.
  a) si $\|\hat{\mu}\|_{L^1} = \ell < 1$, *alors* Φ_μ *est* C^∞ ;
  b) si $\|\hat{\mu}\|_{L^1} = \ell < 1/3$, *alors* Φ_μ *est un difféomorphisme.*

Démonstration : On a $f = \mu \cdot \mathcal{L}(f) + \mu$. Par conséquent on peut écrire

$$\hat{f} = \hat{\mu} * \left(\frac{\bar{\zeta}}{\zeta} \hat{f} \right) + \hat{\mu}.$$

On sait que $\hat{\mu}$ est à décroissance rapide et on va montrer que $\hat{f}$ est à décroissance rapide.

Lemme. *Soit $h : \mathbf{R}^+ \to \mathbf{R}^+$ une fonction continue, tendant vers 0 mais qui ne soit pas à décroissance rapide. Alors il existe un entier $k \in \mathbf{N}$ et une suite (x_n) avec $x_n \to \infty$, telle que $x_n^k h(x_n) \to \infty$, et*

$$\sup_{|x - x_n| \leq \sqrt{x_n}} \frac{h(x)}{h(x_n)} \to 1.$$

Démonstration : Soit k un entier tel que $h(x) \notin \mathcal{O}(1/x^k)$. Pour tout $n \in \mathbf{N}$,

$$A_n = \left\{ x \mid h(x) \geq \frac{n}{x^k} \right\}$$

est un fermé non vide de **R** ; notons x_n un point où la fonction h atteint son maximum sur A_n. Pour tout $x \in \mathbf{R}$, on a alors

$$h(x) \leq \frac{n}{x^k} \leq \frac{x_n^k}{x^k} h(x_n) \text{ si } x \notin A_n, \quad \text{et} \quad h(x) \leq h(x_n) \text{ si } x \in A_n.$$

Par conséquent

$$1 \leq \sup_{|x - x_n| \leq \sqrt{x_n}} \frac{h(x)}{h(x_n)} \leq \left(\frac{x_n}{x_n - \sqrt{x_n}} \right)^k .$$

cqfd.

Reprenons la démonstration de la Partie $a)$ de la Prop. 1. Supposons que $\hat{f}$ n'est pas à décroissance rapide. On va appliquer le lemme à

$$h(x) = \sup_{|\zeta| = x} |\hat{f}(\zeta)|.$$

On trouve donc une suite (ζ_n) telle que $x_n = |\zeta_n|$ satisfasse aux conditions du lemme avec $h(x_n) = |f(\zeta_n)|$. Écrivons

$$\hat{f} = \hat{\mu}_n * \left(\frac{\overline{\zeta}}{\zeta} \hat{f} \right) + (\hat{\mu} - \hat{\mu}_n) * \left(\frac{\overline{\zeta}}{\zeta} \hat{f} \right) + \hat{\mu},$$

où $\hat{\mu}_n = \hat{\mu} \cdot \chi_{D(0, \sqrt{x_n})}$. Alors, $\|\hat{\mu}_n\|_{L^1} \leq \ell$, de sorte que

$$\left| \hat{\mu}_n * \left(\frac{\overline{\zeta}}{\zeta} \hat{f} \right) (\zeta_n) \right| \leq \ell \sup_{D(\zeta_n, \sqrt{x_n})} |\hat{f}| = \ell |\hat{f}(\zeta_n)| (1 + o(1)).$$

Par ailleurs, pout tout $k' \in N$ on a

$$\|\hat{\mu} - \hat{\mu}_n\|_{L^1} \in \mathcal{O}\left(\frac{1}{x_n^{k'}} \right) \quad \text{et} \quad \hat{\mu}(\zeta_n) \in \mathcal{O}\left(\frac{1}{x_n^{k'}} \right).$$

Donc

$$|\hat{f}(\zeta_n)| \leq \ell |\hat{f}(\zeta_n)| (1 + o(1)) + \mathcal{O}\left(\frac{1}{x_n^{k'}} \right),$$

avec $\ell < 1$ et $\hat{f}(\zeta_n)$ qui n'est pas dans $\mathcal{O}(1/x_n^k)$, d'où une contradiction.

Ceci montre que $\hat{f}$ est á décroissance rapide, et $\hat{g}$ aussi puisque ces deux fonctions ont même module. Par suite f et g sont C^∞, donc aussi φ et Φ.

Passons maintenant à la démonstration de $b)$. On suppose que $\|\hat{\mu}\|_{L^1} \leq 1/3$, et de

$$\hat{f} = \hat{\mu} * \left(\frac{\overline{\zeta}}{\zeta} \hat{f} \right) + \hat{\mu},$$

on déduit $\|\hat{f}\|_{L^1} \leq \|\hat{\mu}\|_{L^1} \cdot \|\hat{f}\|_{L^1} + \|\hat{\mu}\|_{L^1}$. Donc

$$\|f\|_{L^\infty} \leq \|\hat{f}\|_{L^1} \leq \frac{\|\hat{\mu}\|_{L^1}}{1 - \|\hat{\mu}\|_{L^1}} < \frac{1}{2}.$$

En posant $g = \partial\phi/\partial z$, et donc $\hat{g} = \overline{\zeta}/\zeta \cdot \hat{f}$, on a

$$\|\hat{g}\|_{L^1} = \|\hat{f}\|_{L^1},$$

d'où

$$\|g\|_{L^\infty} \leq \|\hat{g}\|_{L^1} < \frac{1}{2}.$$

En tout point $z \in \mathbf{C}$, on a donc

$$\left|\frac{\partial\Phi}{\partial z}\right| > 1 - \frac{1}{2} = \frac{1}{2} \quad \text{et} \quad \left|\frac{\partial\Phi}{\partial \overline{z}}\right| < \frac{1}{2}.$$

Il en résulte que Φ est un difféomorphisme local. Il est propre car $\Phi(z) = z + o(1)$. Donc c'est un difféomorphisme.   cqfd.

7   Existence de solutions locales, cas C^∞

Le Théorème 2 est une consquence immédiate de la proposition suivante :

Proposition 2. *Soit $\mu \in C^\infty(\mathbf{C})$ et $z_0 \in \mathbf{C}$. On suppose que $|\mu(z_0)| < 1$. Alors il existe un difféomorphisme $\Psi : \mathbf{C} \to \mathbf{C}$ tel que*

$$\frac{\partial\Psi}{\partial\overline{z}} = \mu \cdot \frac{\partial\Psi}{\partial z}$$

au voisinage de z_0.

Démonstration : Quitte à faire une translation, on peut supposer que $z_0 = 0$.

1) On va d'abord traiter le cas où $\mu(z_0) = 0$. On choisit une fonction $h \in C^\infty_{comp}$ telle que $h = 1$ au voisinage de 0. On définit alors

$$h_\varepsilon(z) = h\left(\frac{z}{\varepsilon}\right)$$

de sorte que

$$\hat{h}_\varepsilon(\zeta) = \varepsilon^2\hat{h}(\varepsilon\zeta), \quad \frac{\partial\hat{h}_\varepsilon}{\partial\xi}(\zeta) = \varepsilon^3\frac{\partial\hat{h}}{\partial\xi}(\varepsilon\zeta) \quad \text{et} \quad \frac{\partial\hat{h}_\varepsilon}{\partial\eta}(\zeta) = \varepsilon^3\frac{\partial\hat{h}}{\partial\eta}(\varepsilon\zeta).$$

On vérifie alors que

$$\|\hat{h}_\varepsilon\|_{L^1} = \|\hat{\eta}\|_{L^1}, \quad \left\|\frac{\partial\hat{h}_\varepsilon}{\partial\xi}\right\|_{L^1} = \varepsilon\left\|\frac{\partial\hat{h}}{\partial\xi}\right\|_{L^1} \quad \text{et} \quad \left\|\frac{\partial\hat{h}_\varepsilon}{\partial\eta}\right\|_{L^1} = \varepsilon\left\|\frac{\partial\hat{h}}{\partial\eta}\right\|_{L^1}.$$

Posons $\mu_\varepsilon = h_\varepsilon \cdot \mu$.

Lemme.- *On a $\|\hat{\mu}_\varepsilon\|_{L^1} \to 0$ quand $\varepsilon \to 0$.*

Démonstration : On peut supposer $\mu \in C^\infty_{comp}$. Alors, $\hat{\mu} \in \mathcal{S}$ et de $\mu(0) = 0$, on déduit

$$\iint \hat{\mu}(\zeta)d\xi d\eta = 0.$$

Sous-lemme.- *Soit $w \in \mathcal{S}$ telle que $\iint w(x,y)\,dxdy = 0$. Alors on peut trouver u et v dans $\mathcal{S}$ telles que*

$$w = \frac{\partial u}{\partial x} + \frac{\partial v}{\partial y}\,.$$

Démonstration : Soit $h \in C^\infty_{comp}(\mathbf{R}; \mathbf{R})$ telle que $\int_{\mathbf{R}} h = 1$. Posons

$$p(x) = \int_{y \in \mathbf{R}} w(x,y)dy$$

et écrivons $w = w_1 + w_2$ avec

$$w_1(x,y) = p(x)h(y) \quad \text{et} \quad w_2(x,y) = w(x,y) - p(x)h(y)\,.$$

On a

$$\int_{x \in \mathbf{R}} p(x)\,dx = 0,$$

donc

$$(\forall y \in \mathbf{R}) \quad \int_{x \in \mathbf{R}} w_1(x,y)\,dx = 0, \quad \text{et} \quad (\forall x \in \mathbf{R}) \quad \int_{y \in \mathbf{R}} w_2(x,y)dy = 0.$$

Posons

$$u(x,y) = \int_{t=-\infty}^{x} w_1(t,y)dt = -\int_{x}^{+\infty} w_1(t,y)$$

et

$$v(x,y) = \int_{t=-\infty}^{y} w_2(x,t)dt = -\int_{y}^{+\infty} w_2(x,t)dt.$$

Alors on a $\partial u/\partial x = w_1$, $\partial v/\partial y = w_2$ et u et v sont dans $\mathcal{S}$.   cqfd.

Reprenons la démonstration du lemme. Appliquant le sous-lemme à $\hat{\mu}$, on peut écrire

$$\hat{\mu}_\varepsilon = \hat{h}_\varepsilon * \hat{\mu} = \hat{h}_\varepsilon * \left(\frac{\partial u}{\partial \xi} + \frac{\partial v}{\partial \eta}\right) = \frac{\partial \hat{h}_\varepsilon}{\partial \xi} * u + \frac{\partial \hat{h}_\varepsilon}{\partial \eta} * v.$$

Par conséquent

$$\|\hat{\mu}_\varepsilon\|_{L^1} \le \left\|\frac{\partial \hat{h}_\varepsilon}{\partial \xi}\right\|_{L^1} \|u\|_{L^1} + \left\|\frac{\partial \hat{h}_\varepsilon}{\partial \eta}\right\|_{L^1} \|v\|_{L^1}$$

$$\leq \varepsilon \left(\left\| \frac{\partial \hat{h}}{\partial \xi} \right\|_{L^1} \|u\|_{L^1} + \left\| \frac{\partial \hat{h}}{\partial \eta} \right\|_{L^1} \|v\|_{L^1} \right).$$

cqfd.

Nous reprenons maintenant la démonstration de la proposition 2.

Pour ε assez petit, $\|\hat{\mu}_\varepsilon\|_{L^1} < 1/3$. D'après la proposition 1.b), il existe un difféomorphisme $\Psi : \mathbf{C} \to \mathbf{C}$ tel que

$$\frac{\partial \Psi}{\partial \bar{z}} = \mu_\varepsilon \frac{\partial \Psi}{\partial z} = \mu \frac{\partial \Psi}{\partial z}$$

au voisinage de z_0.

2) On s'intéresse maintenant au cas où $\mu(z_0) = \mu_0$ avec $|\mu_0| < 1$.

Soit $A : \mathbf{C} \to \mathbf{C}$ telle que

$$A(z) = z + \mu_0 \bar{z}.$$

En posant $\Psi = \widetilde{\Psi} \circ A$, l'équation pour $\widetilde{\Psi}$ est une équation de Beltrami

$$\frac{\partial \widetilde{\Psi}}{\partial \bar{z}} = \tilde{\mu} \frac{\partial \widetilde{\Psi}}{\partial z}$$

avec $\tilde{\mu} \in C^\infty$ et $\tilde{\mu}(0) = 0$.   cqfd.

8   Propriétés topologiques des solutions locales

Le Théorème 2 admet comme conséquence :

Proposition 3.- *Soit $\mu \in C^\infty$ et z_0 tel que $|\mu(z_0)| < 1$, et soit $\phi \in H^1_{loc}(U)$ une solution de l'équation de Beltrami*

$$\frac{\partial \phi}{\partial \bar{z}} = \mu \cdot \frac{\partial \phi}{\partial z}$$

sur un voisinage U de z_0. Alors

a) ϕ est C^∞ au voisinage de z_0,

b) ou bien ϕ est une constante au voisinage de z_0, ou bien il existe $\nu > 0$ tel que ϕ soit en z_0 un revêtement ramifié de degré ν,

c) si $\nu = 1$, ϕ est un difféomorphisme au voisinage de z_0.

Démonstration : D'après la proposition 2, on peut trouver un difféomorphisme $\Psi : \mathbf{C} \to \mathbf{C}$ tel que

$$\frac{\partial \Psi}{\partial \bar{z}} = \mu \cdot \frac{\partial \Psi}{\partial z}$$

au voisinage de z_0. Posons $V = \Psi(U)$, $z_1 = \Psi(z_0)$ et $\tilde{\phi} = \phi \circ \Psi^{-1}$, de sorte que $\phi = \tilde{\phi} \circ \Psi$. Alors $\tilde{\phi} \in H^1_{loc}$, et $\partial\tilde{\phi}/\partial\bar{z} = 0$ au voisinage de z_1. Par suite, $\tilde{\phi}$ est holomorphe au voisinage de z_1 d'après le lemme de Weyl, donc vérifie les propriétés a), b) et c) (au voisinage de z_1) et il en est de même de ϕ (au voisinage de z_0). cqfd.

9 Démonstration du théorème 3

Sous les hypothèses du théorème 3, l'application $\Phi : \mathbf{C} \to \mathbf{C}$ est C^∞ et propre. Son degré topologique est 1 car $\Phi(z) = z + o(1)$. Si Φ était constante sur un ouvert non vide U, on aurait $U \neq \mathbf{C}$, donc U aurait un point frontière z_0 et on aurait une contradiction avec la proposition 3.b. Donc Φ n'est constante sur aucun ouvert non vide et en chaque point $z \in \mathbf{C}$ elle a un degré local $\nu_z > 0$.

Pour tout $w \in \mathbf{C}$, le degré total au dessus de w est égal au degré topologique, i.e.

$$\sum_{z \in \Phi^{-1}(w)} \nu_z = 1.$$

Par suite $\Phi^{-1}(w)$ ne contient qu'un seul point et le degré local de Φ en ce point est 1. Autrement dit, Φ est bijective et est un difféomorphisme local. C'est donc un difféomorphisme. Ceci démontre le Théorème 3.

10 Inégalités de Grötzsch

Proposition 4.- *Soient A_1, A_2 et A_3 des anneaux complexes, $\varphi_1 : A_1 \to A_3$ et $\varphi_2 : A_2 \to A_3$ des plongements C^∞ non homotopiquement triviaux et disjoints. On suppose que φ_1 est K_1-quasi-conforme et que φ_2 est K_2-quasi-conforme. Alors*

$$\mathrm{mod}(A_3) \geq \frac{1}{K_1}\mathrm{mod}(A_1) + \frac{1}{K_2}\mathrm{mod}(A_2) .$$

Démonstration : Voir [Ah, Theorems 3 et 4]. cqfd.

Corollaire 1.- *Soient R_1 et R_2 tels que $R_2 > R_1 > 0$, et soit ϕ un plongement C^∞ de D_{R_2} dans $\mathbf{C}$. On suppose que ϕ est holomorphe sur D_{R_1} avec $\phi(0) = 0$ et $\phi'(0) = 1$, et K-quasiconforme sur $D_{R_2} \setminus \overline{D_{R_1}}$. Alors $\phi(D_{R_2})$ contient D_{R_3} avec*

$$R_3 = \frac{1}{4}R_2^{1/K}R_1^{1-1/K} .$$

Démonstration : Posons $U = \phi(D_{R_2})$ et soit $\phi_U : U \to D_{R_4}$ un isomorphisme holomorphe tel que $\phi_U(0) = 0$ et $\phi'_U(0) = 1$; posons $\phi_4 = \phi_U \circ \phi$. Pour r petit, soit $h(r) = r(1 + o(1))$ tel que $\phi(D_r) \supset D_{h(r)}$. Alors ϕ_4 induit un plongement de $D_{R_2} \setminus \overline{D_r}$ dans $D_{R_4} \setminus \overline{D_{h(r)}}$, holomorphe sur $D_{R_1} \setminus \overline{D_r}$ et K-quasi-conforme sur $D_{R_2} \setminus \overline{D_{R_1}}$. D'après la proposition 4, on a

$$\frac{1}{2\pi} \log \frac{R_4}{h(r)} = \mathrm{mod}(D_{R_4} \setminus \overline{D_{h(r)}}) \geq \mathrm{mod}(D_{R_1} \setminus \overline{D_r}) + \frac{1}{K} \mathrm{mod}(D_{R_2} \setminus \overline{D_{R_1}})$$

$$= \frac{1}{2\pi} \log \frac{R_1}{r} + \frac{1}{K} \log \frac{R_2}{R_1}.$$

D'où

$$\log(R_4) + \varepsilon \geq \left(1 - \frac{1}{K}\right) \log R_1 + \frac{1}{K} \log R_2,$$

avec un ε arbitrairement petit, donc qu'on peut supprimer. Autrement dit,

$$R_4 \geq R_2^{1/K} R_1^{1-1/K}.$$

D'après le théorème du 1/4 de Koebe, U contient un disque de rayon R_3 avec $R_3 = R_4/4$. \hfill cqfd.

Corollaire 2.- *Plaçons nous sous les hypothèses du théorème 3 avec $\|\mu\|_{L^\infty} = k < 1$, et supp$(\mu) \subset D_R$, et posons $K = (1 + k)/(1 - k)$. Alors*
a) Pour z_1 et z_2 dans $\mathbf{C}$ avec $|z_2 - z_1| \leq r$, on a

$$|\Phi(z_2) - \Phi(z_1)| \leq 4r^{1/K}(2R + r)^{1-1/K},$$

b) on a $\Phi(D_R) \subset D_{2R}$, et pour tout $z \in \mathbf{C}$, on a $|\Phi(z) - z| \leq 3R$.

Démonstration : *a)* : Pour $|z_1| \leq R + r$, on applique le corollaire 1 à $\varphi : D_{1/r} \to \mathbf{C}$ défini par

$$\frac{1}{\Phi(z) - \Phi(z_1)} = \varphi\left(\frac{1}{z - z_1}\right).$$

Pour $|z_1| \geq R + r$, on remarque que pour t fixé avec $|t| \leq r$, la fonction $z \mapsto \Phi(z + t) - \Phi(z)$ est holomorphe sur $\mathbf{C} \setminus D_{R+r}$, bornée par $4r^{1/K}(2R+r)^{1-1/K}$ sur le cercle de rayon $R + r$, et tend vers 0 à l'infini. L'inégalité *a)* découle alors du principe du maximum.

b) : La première assertion est une variante du Théorème du $\frac{1}{4}$ de Koebe, en utilisant le fait que Φ est univalente sur $\mathbf{C} - \overline{\mathbf{D}}_R$ et est sous la forme $\Phi(z) = z + b_1 z^{-1} + \cdots$ à l'infini (voir [Po, page 8, formule (8)]). Il en résulte que $|\Phi(z) - z| \leq 3R$ sur $\overline{\mathbf{D}}_R$. La fonction $z \mapsto \Phi(z) - z$ est holomorphe sur $\mathbf{C} - \mathbf{D}_R$ et tend vers 0 à l'infini ; comme elle est majorée par $3R$ sur le cercle de rayon R, elle l'est aussi sur $\mathbf{C} - \mathbf{D}_R$. \hfill cqfd.

11 Démonstration du théorème 4

Soit $\mu \in L^\infty_{comp}$ à support dans D_R et vérifiant $\|\mu\|_{L^\infty} = k < 1$. Choisissons une fonction $\eta \geq 0$ dans C^∞_{comp} telle que $\iint \eta = 1$, et pour $\varepsilon > 0$ posons $\eta_\varepsilon(z) = \frac{1}{\varepsilon^2}\eta(\frac{z}{\varepsilon})$, de sorte que $\iint \eta_\varepsilon = 1$. Soit (ε_n) une suite tendant vers 0, et posons $\mu_n = \eta_{\varepsilon_n} * \mu$. On a alors

- $\mu_n \in C^\infty_{comp}$;
- $\mathrm{supp}(\mu_n) \in \overline{D_R}$ pour n assez grand;
- $\|\mu_n\|_{L^\infty} \leq k$;
- $\mu_n \to \mu$ dans L^2.

Pour chaque n, en appliquant le théorème 3, on trouve un difféomorphisme Φ_n de $\mathbf{C}$ sur lui-même tel que

$$\frac{\partial \Phi_n}{\partial \bar{z}} = \mu_n \cdot \frac{\partial \Phi_n}{\partial z}.$$

Les Φ_n forment une famille équi-continue d'après le corollaire 2 de la proposition 4, et d'après le théorème d'Ascoli, on peut extraire un suite qui converge uniformément sur tout compact vers une application continue $\tilde{\Phi} : \mathbf{C} \to \mathbf{C}$. Les Φ_n^{-1} satisfont aux mêmes conditions, avec le même k, puisqu'ils sont K-quasi-conformes avec le même K que les Φ_n, mais avec $R' = 3R$. Ils forment donc également une famille équicontinue, et il en résulte que $\tilde{\Phi}$ est un homéomorphisme.

Pour démontrer le Théorème 4, il suffit de montrer que $\tilde{\Phi} = \Phi_\mu$. Posons

$$\Phi_n(z) = z + \varphi_n(z) \quad , \quad f_n = \frac{\partial \varphi_n}{\partial \bar{z}} \quad \text{et} \quad g_n = \frac{\partial \varphi_n}{\partial z}.$$

On a

$$\|g_n\|_{L^2} = \|f_n\|_{L^2} \leq \frac{1}{1-k}\|\mu_n\|_{L^2}.$$

Donc quitte à extraire encore une sous-suite on peut supposer que (f_n) et (g_n) convergent faiblement vers des fonctions $\tilde{f}$ et $\tilde{g}$ dans L^2. Posons $\tilde{\varphi} = \lim \tilde{\varphi}_n$, de sorte que $\tilde{\phi}(z) = z + \tilde{\varphi}(z)$. Il reste à démontrer le lemme suivant.

Lemme.

  a) $\tilde{f} = \partial\tilde{\varphi}/\partial\bar{z}$ *(au sens des distributions)* ;
  b) $\tilde{g} = \partial\tilde{\varphi}/\partial z$ *(au sens des distributions)* ;
  c) $\tilde{f} = \mu \cdot \tilde{g} + \mu$.

Démonstration :
  a) : Soit $h \in C^\infty_{comp}$. On a

$$\iint \tilde{f} \cdot h = \lim \iint f_n \cdot h = -\lim \iint \varphi_n \cdot \frac{\partial h}{\partial \bar{z}} = -\iint \tilde{\varphi} \cdot \frac{\partial h}{\partial \bar{z}}.$$

$b)$: idem avec $\partial/\partial z$ au lieu de $\partial/\partial \bar{z}$.

$c)$: On a

$$\mu\tilde{g} - \mu_n g_n = \mu(\tilde{g} - g_n) + (\mu - \mu_n)g_n.$$

Dans cette somme, $\mu(g - g_n)$ tend vers 0 faiblement dans L^2 car pour tout $h \in L^2$,

$$\int h\mu(g - g_n) = \int (h\mu)(g - g_n)$$

tend vers 0. De plus, on a $\|\mu - \mu_n\|_{L^2} \to 0$ et $\|g_n\|_{L^2}$ est borné, donc

$$\|(\mu - \mu_n)g_n\|_{L^1} \leq \|\mu - \mu_n\|_{L^2}\|g_n\|_{L^2} \to 0.$$

Par suite $\mu_n g_n$ tend vers $\mu\tilde{g}$ et $f_n = \mu_n g_n + \mu_n$ tend vers $\mu\tilde{g} + \mu$ au sens des distributions. D'autre part, f_n tend vers $\tilde{f}$ faiblement dans L^2, donc aussi au sens des distributions. Par suite $\tilde{f} = \mu \cdot \tilde{g} + \mu$. cqfd.

Ceci achève la démonstration du Théorème 4.

12   Un scrupule

Si μ dépend continûment (resp. analytiquement) d'un paramètre $\lambda \in \Lambda$, nous avons vu que Φ_λ est un homéomorphisme $\mathbf{C} \to \mathbf{C}$ et que ϕ_λ dépend continûment (resp. analytiquement) de λ pour la topologie L^2. Mais la dépendance est-elle continue (resp. analytique) pour la topologie de la convergence uniforme sur tout compact? La réponse est **OUI** dans les deux cas:

$a)$: *Cas continu.* Pour tout λ_0 il existe un voisinage Λ' de λ_0, un $k < 1$ et un R tels que, pour $\lambda \in \Lambda'$, μ_λ soit de norme L^∞ bornée par k et à support dans $\mathbf{D}_R$ (c'est du moins l'hypothèse qu'il faut faire pour que ça marche...).

Alors les Φ_λ forment une famille équicontinue, donc relativement compacte pour la topologie de la convergence uniforme sur tout compact. Et sur une telle famille, toutes les topologies raisonnables coïncident, en tous cas la topologie L^2_{loc} et la topologie de la convergence uniforme sur les compacts.

$b)$ *Cas analytique.* On peut utiliser le lemme suivant :

Lemme.- *Soient E_1 et E_2 deux espaces de Fréchet, $\iota : E_1 \to E_2$ une application linéaire continue injective. Soient Λ une variété $\mathbf{C}$-analytique et $F_1 : \Lambda \to E_1$ une application. Posons $F_2 = \iota \circ F_1$. On suppose F_1 continue et F_2 $\mathbf{C}$-analytique. Alors F_1 est $\mathbf{C}$-analytique.*

Démonstration : On peut supposer que Λ est un voisinage de 0 dans $\mathbf{C}^n$ contenant $(\overline{\mathbf{D}}_1)^n$. Pour $k = (k_1, ..., k_n)$ posons

$$c_k = \int ... \int_{[0,1]^n} e^{-2i\pi<k,t>} F_1\left(e^{2i\pi(t_1+...+t_n)}\right)dt_1...dt_n .$$

Comme F_2 est analytique, on a $F_2(z) = \sum_{k \in \mathbb{N}^n} \iota(c_k) z^k$ pour $|z| < 1$, et comme ι est injective, on a $F_1(z) = \sum_{k \in \mathbb{N}^n} c_k z^k$.   cqfd.

13   Solutions locales, cas L^∞

Proposition 5.- *Soit* $\mu \in L^\infty_{comp}$ *avec* $\|\mu_{L^\infty}\| < 1$, *considérons* $\Phi = \Phi_\mu$ *et soit* U *un ouvert de* $\mathbf{C}$.

   a) Si f *est une fonction holomorphe sur* $\Phi(U)$, *on a* :

$$\frac{\partial(f \circ \Phi)}{\partial z} = (f' \circ \Phi) \cdot \frac{\partial \Phi}{\partial z} \ ;$$

$$\frac{\partial(f \circ \Phi)}{\partial \bar{z}} = (f' \circ \Phi) \cdot \frac{\partial \Phi}{\partial \bar{z}} \ .$$

En particulier, $f \circ \Phi \in H^1_{loc}(U)$, *et est solution sur* U *de l'équation de Beltrami définie par* μ.

   b) Réciproquement, si $\phi \in H^1_{loc}(U)$ *est solution sur* U *de l'équation de Beltrami définie par* μ, *alors* ϕ *est de la forme* $f \circ \phi_\mu$, *où* f *est holomorphe sur* $\Phi(U)$.

Démonstration : *a)* : C'est immédiat si $\mu \in C^\infty_{comp}$, puisqu'alors Φ est C^∞. Sinon, considérons les Φ_n définis dans le paragraphe 11. Pour tout ouvert V relativement compact dans U, on a $\Phi_n(V) \subset \Phi(U)$ pour n assez grand, $\frac{\partial \Phi_n}{\partial z} \to \frac{\partial \Phi}{\partial z}$ dans $L^2(V)$, $f' \circ \Phi_n \to f' \circ \Phi$ uniformément sur V. Donc

$$(f' \circ \Phi_n) \cdot \frac{\partial \Phi_n}{\partial z} \to (f' \circ \Phi) \cdot \frac{\partial \Phi}{\partial z} \ .$$

Or $(f' \circ \Phi_n) \cdot \frac{\partial \Phi_n}{\partial z} = \frac{\partial(f \circ \Phi_n)}{\partial z}$.

Démonstration analogue pour la deuxième formule. Les autres assertions de (a) en découlent immédiatement.

   b) : Ici encore le résultat est évident pour $\mu \in C^\infty_{comp}$, et nous allons approcher Φ par une suite (Φ_n) comme au paragraphe 11, mais la démonstration est plus délicate. Il convient de considérer, plutôt que des dérivées partielles $\frac{\partial \phi}{\partial z}$ et $\frac{\partial \phi}{\partial \bar{z}}$ les formes différentielles $\partial \phi = \frac{\partial \phi}{\partial z} dz$ et $\bar{\partial} \phi = \frac{\partial \phi}{\partial \bar{z}} d\bar{z}$.

La norme L^2 des formes différentielles de degré 1 est invariante par les applications conformes et quasi-invariante par les applications quasi-conformes: Si $\phi : W_1 \to W_2$ est un difféomorphisme conforme (resp. K-quasi-conforme), pour toute forme différentielle α de degré 1 sur W_2, on a $\|\phi^* \alpha\|_{L^2} = \|\alpha\|_{L^2}$ (resp. $\frac{1}{K} \|\phi^* \alpha\|_{L^2} \leq \|\alpha\|_{L^2} \leq K \|\phi^* \alpha\|_{L^2}$).

Soit $\phi \in H^1_{loc}(U)$ une solution sur U de l'équation de Beltrami définie par μ, et notons f la fonction $\phi \circ \Phi^{-1}$ sur $\Phi(U)$. Soit V un ouvert relativement compact dans U. On veut montrer que f est holomorphe sur $\Phi(V)$. Pour n assez grand, on a $\Phi(V) \subset \Phi_n(U)$, posons alors $f_n = \phi \circ \Phi_n^{-1}$ sur $\Phi(V)$.

Un calcul donne

$$\Phi_n^* \overline{\partial} f_n = (\frac{\partial \phi}{\partial \overline{z}} - \mu_n \frac{\partial \phi}{\partial z}) \frac{d\overline{z} + \mu_n dz}{1 - |\mu_n|^2}$$

soit, en tenant compte de $\frac{\partial \phi}{\partial \overline{z}} = \mu \frac{\partial \phi}{\partial z}$,

$$\Phi_n^* \overline{\partial} f_n = (\mu - \mu_n) \frac{\partial \phi}{\partial z} \cdot \frac{d\overline{z} + \mu_n dz}{1 - |\mu_n|^2} .$$

Lemme.- *Soit (u_n) une suite dans L^{∞}_{comp} avec $\|u_n\|_{L^{\infty}} \leq m$ indépendant de n et $\|u_n\|_{L^2} \to 0$, et soit $g \in L^2$. Alors $\|u_n g\|_{L^2} \to 0$.*

Démonstration : Soit $\varepsilon > 0$, et soit $g_1 \in C_{comp}$ telle que $\|g - g_1\|_{L^2} < \frac{\varepsilon}{2m}$. Pour tout n, on a $\|u_n(g - g_1)\|_{L^2} < \frac{\varepsilon}{2}$, et pour n assez grand, on a $\|u_n\|_{L^2} < \frac{\varepsilon}{2\|g_1\|_{L^{\infty}}}$, d'où $\|u_n g_1\|_{L^2} < \frac{\varepsilon}{2}$ et $\|u_n g\|_{L^2} < \varepsilon$.   cqfd.

Fin de la démonstration de la Prop. 5 : On a $\|\mu - \mu_n\|_{L^2} \to 0$, $\frac{\partial \phi}{\partial z} \in L^2$ indépendant de n, et $\|\frac{d\overline{z} + \mu_n dz}{1 - |\mu_n|^2}\|_{L^{\infty}} \leq \frac{1+k}{1-k^2} = \frac{1}{1-k}$. Il en résulte en vertu du lemme que $\|\Phi^* \overline{\partial} f_n\|_{L^2} \to 0$, donc $\|\overline{\partial} f_n\|_{L^2(V)} \to 0$.

D'autre part $f_n = \phi \circ \Phi_n^{-1} \to f$ uniformément sur V, donc $\overline{\partial} f = 0$ sur V au sens des distributions, et f est holomorphe d'après le lemme de Weyl.   cqfd.

Corollaire.- *Soient U un ouvert de $\mathbf{C}$ et $\mu \in L^{\infty}(U)$ avec $\|\mu\|_{L^{\infty}} < 1$. Il existe alors une solution $\Phi \in H^1_{loc}(U)$ de l'équation de Beltrami définie par μ sur U qui est un homéomorphisme de U sur un ouvert V de $\mathbf{C}$.*

Si Φ satisfait à ces conditions, toute $\varphi \in H^1_{loc}(U)$ qui est solution de la même équation de Beltrami est de la forme $F \circ \phi$, où F est une fonction holomorphe sur V.

Démonstration : Prolongeons μ par 0 sur $\mathbf{C} - U$. Alors la restriction de Φ_{μ} à U répond à la question. Une fonction Φ solution de l'équation de Beltrami qui est injective sur U est nécessairement de la forme $G \circ \Phi_{\mu}$ où G est un isomorphisme de $\Phi_{\mu}(U)$ sur un ouvert V.   cqfd.

14   Démonstration du Théorème 5

Soit $\mu \in L^{\infty}(\mathbf{C})$ avec $\|\mu\|_{L^{\infty}} \leq k < 1$. Pour $R > 1$, définissons μ_R par $\mu_R = \mu$ sur $\mathbf{D}_R$ et $\mu_R = 0$ sur $\mathbf{C} - \mathbf{D}_R$. Posons $\Phi_R = \Phi_{\mu_R}$, et soit $\Phi_R^{\natural}$

l'homéomorphisme de $\mathbf{C}$ de la forme $A \circ \Phi_R$, où $A : \mathbf{C} \to \mathbf{C}$ est une application affine ajustée de façon que $\Phi_R^\natural$ laisse fixes 0 et 1.

Pour $R' > R$, posons $\psi_{R,R'} = \Phi_{R'}^\natural \circ (\Phi_R^\natural)^{-1}$, de sorte que $\Phi_{R'}^\natural = \psi_{R,R'} \circ \Phi_R^\natural$. D'après la Prop. 5, $\psi_{R,R'}$ est holomorphe, et évidemment univalente, sur $U_R = \Phi_R^\natural(\mathbf{D}_R)$.

Proposition 6.- *a) : L'application $\psi_{R,R'}$ a une limite ψ_R quand $R' \to \infty$; c'est une fonction holomorphe et univalente sur U_R.*

b) $\psi_{R,R'}$ tend vers l'identité quand R et R' tendent vers l'infini. .

Lemme.- *Il existe une constante $\kappa^\natural$ telle que, pour tout $R \geq 2$, pour toute fonction ϕ holomorphe et univalente sur $\mathbf{D}_R$, laissant fixes 0 et 1, et tout z tel que $|z| \leq \frac{R}{2}$, on ait*

$$|\phi(z) - z| \leq \frac{\kappa^\natural |z|^2}{R} .$$

Démonstration : D'après un théorème de Koebe, l'ensemble S des $\varphi : \mathbf{D} \to \mathbf{C}$ holomorphes univalentes telles que $\varphi(0) = 0$ et $\varphi'(0) = 1$ est compact (pour la convergence uniforme sur les compacts de $\mathbf{D}$). Par suite, il existe une constante κ telle que, pour toute $\varphi \in S$ et tout z tel que $|z| < \frac{1}{2}$, on ait $|\varphi(z) - z| \leq \kappa |z|^2$.

Pour $R > 1$, notons S_R (resp. $S_R^\natural$) l'ensemble des $\varphi : \mathbf{D}_R \to \mathbf{C}$, holomorphes univalentes et telles que $\varphi(0) = 0$ et $\varphi'(0) = 1$ (resp. fixant 0 et 1). Pour $\phi \in S_R$ et $|z| \leq \frac{R}{2}$, on a $|\phi(z) - z| \leq \frac{\kappa |z|^2}{R}$. En posant $\phi^\natural = \frac{\phi}{\phi(1)}$, de sorte que $\phi^\natural \in S^\natural$, on obtient

$$|\phi^\natural(z) - z| \leq \frac{\kappa |z|^2}{R} + \left| \frac{1}{\phi(1)} - 1 \right| \cdot |\phi(z)| \leq \frac{\kappa |z|^2}{R} + \left| \frac{1}{\phi(1)} - 1 \right| \left(|z| + \frac{\kappa |z|^2}{R} \right)$$

Or $|\phi(1) - 1| \leq \frac{\kappa}{R}$, et $|\frac{1}{\phi(1)} - 1| \leq \frac{M}{R}$ pour une certaine constante M. On obtient alors

$$|\phi^\natural(z) - z| \leq \frac{\kappa |z|^2}{R} + \frac{M}{R} \left(|z| + \frac{\kappa |z|^2}{R} \right) \leq \frac{\kappa^\natural |z|^2}{R} ,$$

avec $\kappa^\natural = 2\kappa + M$. cqfd.

Démonstration de la Prop. 6 : Démontrons d'abord l'assertion (*b*). On a

$$\mathrm{mod}(U_R - [0,1]) \geq \frac{1}{K} \mathrm{mod}(\mathbf{D}_R - [0,1]) \geq \frac{1}{K} .$$

Par suite, par le Théorème du $\frac{1}{4}$ de Koebe, $U_R \supset \mathbf{D}_{\tilde{R}}$ avec $\tilde{R} = \frac{1}{4} R^{\frac{1}{K}}$. La fonction $\psi_{R,R'}$ est holomorphe sur ce disque et on a $|\psi_{R,R'}(z) - z| \leq \frac{\kappa^\natural |z|^2}{\tilde{R}}$ dès que $\tilde{R} > 2|z|$, d'où (*b*).

Démontrons maintenant (a). Pour $R < R' < R''$, on a $\psi_{R,R''} = \psi_{R',R''} \circ \psi_{R,R'}$. Fixons R ; si on prend une suite de valeurs de R' tendant vers ∞, les $\psi_{R,R'}$ forment une suite de Cauchy pour la norme de la convergence uniforme de tout compact de U_R. Cette suite a une limite ψ_R, qui ne dépend pas de la suite de valeurs de R' choisie. D'où (a).   cqfd.

Démonstration du Théorème 5 : Pour $R < R'$, on a $\psi_R = \psi_{R'} \circ \psi_{R,R'}$ sur U_R, d'où $\psi_R \circ \Phi_R^\natural = \psi_{R'} \circ \Phi_{R'}^\natural$ sur $\mathbf{D}_R$. Il existe donc une fonction $\Phi = \Phi_\mu^\natural \in H^1_{loc}(\mathbf{C})$ induisant $\psi_R \circ \Phi_R^\natural$ sur $\mathbf{D}_R$ pour tout $R > 2$. Cette fonction est un homéomorphisme $\mathbf{C} \to \mathbf{C}$, solutioin de l'équation de Beltrami définie par μ laissant fixe 0 et 1.

Montrons l'unicité. Si $\Psi \in H^1_{loc}(\mathbf{C})$ est une autre solution de l'équation de Beltrami définie par μ, la fonction Ψ est de la forme $F \circ \Phi_\mu^\natural$ où F est holomorphe sur $\mathbf{C}$ d'après le Cor. de la Prop. 5. Si de plus Ψ est injective et laisse fixes 0 et 1, il en est de même de F, donc F est l'identité et $\Psi = \Phi_\mu^\natural$.

Enfin si μ dépend d'un paramètre $\lambda \in \Lambda$, pour tout sous-ensemble Λ' de Λ tel que $(\exists k < 1)(\forall \lambda \in \Lambda')\|\mu_\lambda\|_{L^\infty} \le k$, $\Phi_{\mu_\lambda,R}^\natural$ tend vers $\Phi_{\mu_\lambda}^\natural$ uniformément sur $\Lambda' \times A$ pour tout compact A de $\mathbf{C}$. Si μ_λ dépend continûment (resp. $\mathbf{C}$-analytiquement) de λ, $\Phi_{\mu_\lambda,R}^\natural$ dépend continûment (resp. $\mathbf{C}$-analytiquement) de λ pour tout R, et il en est de même de $\Phi_{\mu_\lambda}^\natural$.   cqfd.

References

[Ah] L. V. Ahlfors, Lectures on Quasiconformal mappings, Wadsworth & Brooks/Cole, 1987.

[Po] Ch. Pommerenke, Boundary Behaviour of Conformal Maps, Springer-Verlag, 1992.

Adrien Douady, Département de Mathématique, Université de Paris-Sud, Bât. 425, 91405 Orsay, France.
    e-mail : Adrien.Douady@ens.fr

Xavier Buff, Universite Paul Sabatier, Laboratoire Emile Picard, UFR MIG, 118 route de Narbonne, 31062 Toulouse Cedex, France.
    e-mail : buff@topo.math.u-psud.fr

BIFURCATION OF PARABOLIC FIXED POINTS

MITSUHIRO SHISHIKURA

Abstract : We present here the foundation of parabolic implosion theory. We show that, in the bifurcation of parabolic fixed points with one attracting petal, Fatou coordinates exist and depend continuously on the parameter. We study also properties of Ecalle cylinders, transition maps and return maps.

§1. INTRODUCTION

This paper is devoted to the study of the bifurcation of parabolic periodic points for holomorphic maps of one complex variable. Local behavior of parabolic periodic points was already studied by Fatou ([Fa]) and Leau ([Le]) (see also Camacho, [Ca]). Fatou constructed a partial conjugacy between parabolic dynamics and a translation, which we call a "Fatou coordinate" (see §2.2).

However once we perturb maps which have parabolic periodic points, very complicated but interesting bifurcation phenomena can occur via an interaction between local and global dynamics.

For example, the Julia set can vary discontinuously (Douady, [Do]). The aim of this paper is to present a tool to study such a bifurcation, both qualitatively and quantitatively.

The study of the bifurcation of parabolic periodic points was begun by Douady and Hubbard in their Orsay lecture note [DH]. They introduced the "Ecalle cylinders" for parabolic periodic points and for their perturbation, and applied it to the study of the external rays of the Mandelbrot set. The theory of Ecalle cylinders was then developed by Lavaurs in his thesis [La], and he applied this technique to prove many results–such as the "limit shape" of the Mandelbrot set at $c = \frac{1}{4}$, the limits $\liminf_{n \to \infty} J_{c_n}$ and $\limsup_{n \to \infty} K_{c_n}$ for $c_n \to \frac{1}{4}$, the non-local-connectivity of the cubic Mandelbrot set (the connectedness locus for cubic polynomials), etc. Recently the author proved that the boundary of the Mandelbrot set has Hausdorff dimension two, and in the proof, the Ecalle cylinders were also an essential tool (see [Sh]).

Typeset by $\mathcal{A}_{\mathcal{M}}\mathcal{S}$-TEX

The difficulty in the study of the bifurcation of parabolic periodic points comes from the fact that after a small perturbation, the parabolic periodic point bifurcates into two or more periodic orbits and the behavior of the orbits near the parabolic periodic point changes drastically according to the parameter. The incoming and outgoing Ecalle cylinders are defined to be the quotients of fundamental regions for the dynamics near the parabolic periodic point. (See §2 for precise definition.) The fundamental regions can be taken so that they depend continuously on the parameter, so the correspondence between the original dynamical plane and the cylinders is "continuous" in a suitable sense. The orbits near the parabolic periodic point are traced by the transition map from the incoming cylinder to the outgoing cylinder, and this is the part which represents the drastic change according to the parameter. Through the incoming (resp. outgoing) cylinder, orbits come in to (resp. go out from) the neighborhood of the parabolic periodic point, and the transition map connects the incoming orbits to the outgoing orbits.

If we consider orbits which return many times to either of the fundamental regions, it is natural to study the first return map, i.e., the map which assigns to a point in the fundamental region its first orbit point which comes back to that region. By the quotient, there is an induced first return map on the Ecalle cylinder. So one can study iteration of the first return map instead of the original map.

Moreover it turns out that if the perturbed map has a new fixed point which is parabolic or irrationally indifferent, then the corresponding return map also has parabolic or irrationally indifferent fixed point at the end of the cylinder. So we have a chance to iterate this procedure. In fact, iterating this procedure means that we are tracing parameters such that the angle of multiplier of the fixed points correspond to successive continued fraction expansion of the limit. It is also experimentally known that such a limit has quite "thick" Julia set and the Mandelbrot set also look "thick" at this parameter see for example, [Ma], [Mi]). This observation can be justified to some extent, for example, we can use this technique to prove that the boundary of the Mandelbrot set has Hausdorff dimension two.

This note is organized as follows: In section 2, we study parabolic fixed points locally and construct Fatou coordinates, Ecalle cylinders and the transition map. Bifurcation of such fixed points are studied locally in section 3, and the return map is defined. The extension and global properties these objects are discussed in section 4.

This paper is a revision of Chapters 2-4 of: The parabolic bifurcation of rational maps, 19° Colóquio Brasileiro de Matemática, IMPA, 1993. The author would like to thank the editor, Tan Lei, for her help and encouragement in writing this paper, and R. Oudkerk, for various comments and for producing the drawings from our sketches.

FIGURE 1. The Julia sets of $z^2 + c$ for $c = \frac{1}{4}$ and $c = 0.25393 + 0.00046i$

§2. LOCAL THEORY ON PARABOLIC FIXED POINTS

§2.1. Preliminary coordinate change and petals.

2.1.1. In this note, we are concerned with parabolic periodic points of analytic maps. For simplicity, we only treat the case where the period is one (i.e. fixed point) and the multiplier is one. Moreover by a coordinate change we may assume that the fixed point is at $z = 0$. Therefore we are interested in the class of maps satisfying

$$f_0(0) = 0 \text{ and } f_0'(0) = 1.$$

We need a further assumption on the *non-degeneracy* of the fixed point, that is

$$f_0''(0) \neq 0.$$

If this is the case, by another coordinate change, we may assume that $f_0''(0) = 1$, hence

$$f_0(z) = z + z^2 + \dots.$$

So define

$$\mathcal{F}_0 = \left\{ f_0 \;\middle|\; \begin{array}{c} f_0 \text{ is defined and analytic in a neighborhood of } 0, \\ f_0(0) = 0, \ f_0'(0) = 1, \text{ and } f_0''(0) = 1 \end{array} \right\}.$$

328   *Mitsuhiro Shishikura*

2.1.2. For an $f_0 \in \mathcal{F}_0$, define f_0^* by conjugating by $w = I(z) = -\dfrac{1}{z}$:

$$f_0^* = I \circ f_0 \circ I^{-1}.$$

Then f_0^* is defined in a neighborhood of $w = \infty$, $f_0^*(\infty) = \infty$, and it is easy to see that

$$f_0^*(w) = w + 1 + \frac{a}{w} + O(\frac{1}{w^2}) \text{ as } w \to \infty,$$

where a is a constant. So f_0^* is, as the first approximation, close to the translation $T(w) = w + 1$. Since each orbit of T lies on a horizontal line, and the images of horizontal lines by $z = I^{-1}(w)$ are circles tangent to the real axis at $z = 0$, each orbit of f_0 also lies on a curve which "looks like" a circle tangent to the real axis at $z = 0$.

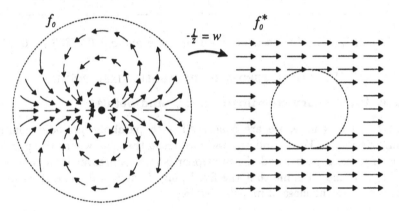

FIGURE 2. Orbits of f_0 and f_0^*

2.1.3. Let, for a large $b > 0$, $\Omega^- = \{z \in \mathbb{C}|\ \text{Re}(z) > b - |\text{Im}(z)|\}$ and $\Omega^+ = \{z \in \mathbb{C}|\ \text{Re}(z) < -b + |\text{Im}(z)|\}$. It is easy to see from the form of f_0^* that if b is sufficiently large, then on $\overline{\Omega^-} \cup \overline{\Omega^+}$, f_0^* is defined and injective, and satisfies

$$|f_0^*(z) - (z+1)| < \frac{1}{4}.$$

Then it follows that

$$f_0^*(\overline{\Omega^-}) \subset \Omega^- \cup \{\infty\} \quad \text{and} \quad f_0^*(\Omega^+ \cup \{\infty\}) \supset \overline{\Omega^+}.$$

If $z \in \Omega^-$, then $(f_0^*)^n(z)$ are defined for all $n \geq 0$ and converge to ∞ as $n \to \infty$. Conversely if there is a neighborhood of z_0 on which the iterates $(f_0^*)^n(z)$ are defined for all $n \geq 0$ and converge to ∞ as $n \to \infty$, then $(f_0^*)^n(z_0) \in \Omega^-$ for some n.

2.1.4. Now define $\Omega_- = I^{-1}(\Omega^-)$, $\Omega_+ = I^{-1}(\Omega^+)$. We call Ω_- an attracting petal, and Ω_+ a repelling petal.

Remark. In this section, we use the subscript $\pm$ ($\Omega_\pm$, $\Phi_\pm$, $S_\pm$, etc.) to denote objects in the original coordinate (in which f_0 is defined) and the superscript $\pm$ ($\Omega^\pm$, $\Phi^\pm$, $S^\pm$, etc.) to denote corresponding objects in the coordinate $w = -1/z$ (in which f_0^* is defined). The difference (for example, between $S_\pm$ and $S^\pm$) seems to be subtle, however this allows us to avoid more complicated notations and to save symbols. Anyway they represent the "same" objects in different coordinates.

They satisfy:

Ω_- and Ω_+ are simply connected domains bounded by Jordan curves passing through 0;

$\Omega_- \cup \Omega_+ \cup \{0\}$ is a neighborhood of 0, on which f_0 is defined and injective; $f_0(\overline{\Omega_-}) \subset \Omega_- \cup \{0\}$ and $f_0(\Omega_+ \cup \{0\}) \supset \overline{\Omega_+}$.

2.1.5. For an analytic function f which has a parabolic fixed point ζ, the *parabolic basin* of is

$$B = \left\{ z \;\middle|\; \begin{array}{l} z \text{ has a neighborhood on which } f^n \ (n = 1, 2, \dots) \\ \text{are defined and } f^n \to \zeta \text{ uniformly as } n \to \infty \end{array} \right\}.$$

Note that ζ itself is not in the parabolic basin.

Lemma 2.1.6. *For $f_0 \in \mathcal{F}_0$, the parabolic basin of $z = 0$ can be expressed by*

$$B = \bigcup_{n \geq 0} f_0^{-n}(\Omega_-).$$

Proof. It immediately follows from the fact on Ω^- in 2.1.3. $\square$

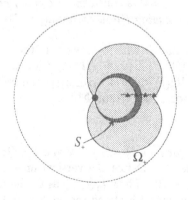

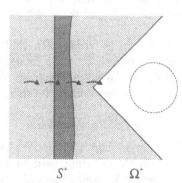

FIGURE 3. Ω_+, S_+ $\qquad\qquad\qquad$ Ω^+, S^+

§2.2. Fatou coordinates.

As we have seen in 2.1, the conjugate f_0^* of $f_0 \in \mathcal{F}_0$ looks like the translation $T(w) = w + 1$. In fact, we see here that there are partial conjugacies between f_0 and T.

Proposition 2.2.1. *Let $f_0 \in \mathcal{F}_0$ and f_0^*, $\Omega^{\pm}$ etc. be as in 2.1. There exist conformal mappings (analytic diffeomorphisms) $\Phi^- : \Omega^- \to \mathbb{C}$ and $\Phi^+ : \Omega^+ \to \mathbb{C}$ satisfying*

$$\Phi^-(f_0^*(w)) = \Phi^-(w) + 1 \text{ and } \Phi^+(f_0^*(w)) = \Phi^+(w) + 1.$$

If

$$f_0^*(w) = w + 1 + \frac{a}{w} + O(\frac{1}{w^2}),$$

then

$$\Phi^{\pm}(w) = w - a \log w + c_{\pm,0} + o(1)$$

as w tend to ∞ within sectors $\{z| \ \mathrm{Re}(z) > b - k|\mathrm{Im}(z)|\}$ ($\subset \Omega^-$) for Φ^- and $\{z| \ \mathrm{Re}(z) < -b + k|\mathrm{Im}(z)|\}$ ($\subset \Omega^+$) for Φ^+, where $c_{\pm,0}$ are constants and k is any constant between 0 and 1. These $\Phi^{\pm}$ are unique up to additive constant.

The proof will be given in §2.5 and §2.6.

2.2.2. *Remark.* As a convention, the branches of logarithm in the above expression are chosen so that they coincide in the upper component of $\Omega^- \cap \Omega^+$ and differ by $2\pi i$ in the lower. Note here that if a is not zero, then Φ^- and Φ^+ cannot coincide on the both components of $\Omega^- \cap \Omega^+$, because of the difference of the branches of the logarithm. In this case, f_0 cannot be conjugated to T in any neighborhood of 0.

Definition. The map $\Phi_- = \Phi^- \circ I : \Omega_- \to \mathbb{C}$ is called the *incoming Fatou coordinate* of f_0 and $\Phi_+ = \Phi^+ \circ I$ the *outgoing Fatou coordinate* of f_0.

2.2.4. It can be easily seen that for any k with $0 < k < 1$, there is a large $b' > 0$ such that the image $\Phi^-(\Omega^-)$ (resp. $\Phi^+(\Omega^+)$) contains a set $\{z| \ \mathrm{Re}(z) > b' - k|\mathrm{Im}(z)|\}$ (resp. $\{z| \ \mathrm{Re}(z) < -b' + k|\mathrm{Im}(z)|\}$) and on this set there is a well-defined inverse branch of Φ^- (resp. Φ^+).

§2.3. The Ecalle cylinders.

2.3.1. Let z_0^- and z_0^+ be points satisfying $\mathrm{Re}\, z_0^+ < -b - 2$ and $b < \mathrm{Re}\, z_0^-$, where b is the constant used to define $\Omega^{\pm}$. Define the vertical lines $\ell^- = \{z_0^- + iy|y \in \mathbb{R}\}$ and $\ell^+ = \{z_0^+ + iy|y \in \mathbb{R}\}$. Then $f_0^*(\ell^-)$ lies to the right of ℓ^-, both contained in Ω^- and they bound a closed region S^- ($\subset \Omega^-$), which looks like a vertical strip. Similarly $f_0^*(\ell^+)$ and ℓ^+ bound a closed region S^+ ($\subset \Omega^+$). It is easy to see that every orbit of f_0^* in Ω^- (resp. Ω^+) lands on S^- (resp. S^+) exactly once, except when the orbit lands on one of the boundary lines for S^- (resp. S^-), in which case it lands on the other boundary line as well.

Let $S_- = I^{-1}(S^-)$ and $S_+ = I^{-1}(S^+)$. Their shape look like a croissant with the tips at the fixed point $z = 0$. They are called *fundamental regions* for the dynamics of $f_0|_{S_\pm}$.

2.3.2. Definition. (Ecalle cylinders) Construct the quotient of S_- by identifying $z \in I^{-1}(\ell^-)$ and $f_0(z) \in I^{-1}(f_0^*(\ell^-))$:

$$C_- = S_-/\sim, \quad \text{where } z \sim f_0(z).$$

Obviously $C_- = S_-/\sim$ is topologically a cylinder and can be naturally identified with the quotient $\Omega_-/\sim$, where $z \sim f_0(z)$ if both z and $f_0(z)$ belong to Ω_-. Therefore C_- has a structure of a Riemann surface, since f_0 is conformal. Similarly $C_+ = S_+/\sim= \Omega_+/\sim$ is defined. We call C_- the *incoming Ecalle cylinder* and C_+ the *outgoing Ecalle cylinder*.

Let $\pi : \mathbb{C} \to \mathbb{C}^* = \mathbb{C} \setminus \{0\}$ be the universal covering map $\pi(z) = e^{2\pi i z}$.

Lemma 2.3.3. *The maps $\pi \circ \Phi_-$ and $\pi \circ \Phi_+$ induces isomorphisms*

$$\overline{\Phi}_- : C_- = S_-/\sim \to \mathbb{C}^* \text{ and } \overline{\Phi}_+ : C_- = S_+/\sim \to \mathbb{C}^*.$$

Proof. It follows from Lemma 2.5.4 in §2.5. $\square$

§2.4. Transition map.

It follows from 2.2.4 that the map $\tilde{\mathcal{E}}_{f_0} = \Phi_- \circ (\Phi_+)^{-1} = \Phi^- \circ (\Phi^+)^{-1}$ can be defined in a region $\{w \mid |\mathrm{Im}(w)| > b'' + 2|\mathrm{Re}\,w|\}$ for large $b'' > 0$. By its definition it satisfies

$$\tilde{\mathcal{E}}_{f_0}(w + 1) = \tilde{\mathcal{E}}_{f_0}(w) + 1$$

whenever the both sides are defined. Using the relation, one can extend $\tilde{\mathcal{E}}_{f_0}$ to the region $\{w \mid |\mathrm{Im}\,w| > \eta_0\}$ for some $\eta_0 > 0$. The relation implies that

$$\mathcal{E}_{f_0} = \pi \circ \tilde{\mathcal{E}}_{f_0} \circ \pi^{-1} : \{0 < |w| < e^{-2\pi \eta_0}\} \cup \{e^{2\pi \eta_0} < |w| < \infty\} \to \mathbb{C}^*$$

is well-defined, where $\pi(w) = e^{2\pi i w}$. The asymptotic behavior of $\Phi^\pm$ implies that $\tilde{\mathcal{E}}_{f_0}(w) - w$ tend to a limit as $\mathrm{Im}\,w \to \infty$ or $-\infty$. (The limits are different for $\mathrm{Im}\,w \to \infty$ and $-\infty$.) Hence $\mathcal{E}_{f_0}$ extends analytically to 0 and ∞, so that $\mathcal{E}_{f_0}(0) = 0$, $\mathcal{E}_{f_0}(\infty) = \infty$, $\mathcal{E}'_{f_0}(0) \neq 0$, and $\mathcal{E}'_{f_0}(\infty) \neq 0$.

Definition. We call $\mathcal{E}_{f_0}$ the *transition map* or *Ecalle transformation*. Since the Fatou coordinates $\Phi_\pm$ are determined up to additive constants, the transition map is defined up to pre- and post-composition of linear maps of the form $z \mapsto \alpha z$.

Normalization. We normalize Φ^- and Φ^+ by adding constants so that

$$\Phi^-(w) - \Phi^+(w) \to 0 \quad \text{when } w \in \Omega^- \cap \Omega^+ \text{ and } \operatorname{Im} w \to \infty.$$

Then automatically

$$\Phi^-(w) - \Phi^+(w) \to 2\pi i a \quad \text{when } w \in \Omega^- \cap \Omega^+ \text{ and } \operatorname{Im} w \to -\infty.$$

Under this normalization, it is easy to see that the transition map satisfies

$$\mathcal{E}'_{f_0}(0) = 1.$$

Note that we can still add the same constant to Φ^- and Φ^+, so there is one more degree of freedom to normalize.

Comparing with the definitions in 2.3, we can interpret the transition map as a map from the outgoing Ecalle cylinder $\mathcal{C}_+$ to the incoming Ecalle cylinder, defined only in a neighborhood of the ends of $\mathcal{C}_+$.

§2.5. The existence of the Fatou coordinates.

2.5.1. In this section, we prove the existence of the Fatou coordinates in a more generalized setting. This form is necessary for the application to the perturbed maps. To state the continuity of the construction, it is convenient to define the following topology.

Definition. (Compact-open topology together with the domain of definition) In what follows, a map f is always associated with its domain of definition $\mathcal{D}(f)$ (an open set of $\overline{\mathbb{C}}$ or $\mathbb{C}/\mathbb{Z}$ etc.); that is, two maps are considered as distinct, if they have distinct domains, even if one is an extension of the other. A neighborhood of an analytic map $f : \mathcal{D}(f) \to \overline{\mathbb{C}}$ is a set containing

$$\{g : \mathcal{D}(g) \to \overline{\mathbb{C}} \mid g \text{ is analytic}, \mathcal{D}(g) \supset K \text{ and } \sup_{z \in K} d(g(z), f(z)) < \varepsilon\},$$

where K is a compact set in $\mathcal{D}(f)$, $\varepsilon > 0$ and $d(\cdot, \cdot)$ is the spherical metric. (If the map is to some other space, then d should be replaced by an appropriate metric.) A sequence $\{f_n\}$ converges to f in this topology if and only if for any compact set $K \subset \mathcal{D}(f)$ there exists an n_0 such that

$K \subset \mathcal{D}(f_n)$ $(n \geq n_0)$ and $f_n|_K \to f_K$ uniformly on K as $n \to \infty$. The system of these neighborhoods defines "the compact-open topology together with the domain of definition", which is unfortunately not Hausdorff, since an extension of f is contained in any neighborhood of f.

For $b_1, b_2 \in \mathbb{C}$ with $\operatorname{Re} b_1 < \operatorname{Re} b_2$, define

$$\mathcal{Q}(b_1, b_2) = \{z \in \mathbb{C}|\ \operatorname{Re}(z-b_1) > -|\operatorname{Im}(z-b_1)|,\ \operatorname{Re}(z-b_2) < |\operatorname{Im}(z-b_2)|\ \}.$$

If $b_1 = -\infty$ (resp. $b_2 = \infty$), the condition involving b_1 (resp. b_2) should be removed.

Proposition 2.5.2. *Let F be a holomorphic function defined in $\mathcal{Q}$ with $\mathcal{Q} = \mathcal{Q}(b_1, b_2)$ and $\operatorname{Re} b_2 > \operatorname{Re} b_1 + 2$ (here b_1 or b_2 may be $-\infty$ or ∞). Suppose*

$$|F(z) - (z+1)| < \frac{1}{4}, \quad \text{and } |F'(z) - 1| < \frac{1}{4} \text{ for } z \in \mathcal{Q}.$$

Then
(0) F is univalent on $\mathcal{Q}$.
(i) Let $z_0 \in \mathcal{Q}$ be a point such that $\operatorname{Re} b_1 < \operatorname{Re} z_0 < \operatorname{Re} b_2 - 5/4$. Denote by S the closed region (a strip) bounded by the two curves $\ell = \{z_0 + iy | y \in \mathbb{R}\}$ and $F(\ell)$. Then for any $z \in \mathcal{Q}$, there exists a unique $n \in \mathbb{Z}$ such that $F^n(z)$ is defined and belongs to $S - F(\ell)$.
(ii) There exists a univalent function $\Phi : \mathcal{Q} \to \mathbb{C}$ satisfying

$$\Phi(F(z)) = \Phi(z) + 1$$

whenever both sides are defined. Moreover Φ is unique up to addition of a constant.
(iii) Fix a point $z_0 \in \mathcal{Q}$. If we normalize Φ by $\Phi(z_0) = 0$, then the correspondence $F \mapsto \Phi$ is continuous with respect to the compact-open topology.

Proof. (0) and (i) are easy and left to the reader.
(ii) Let $z_0 \in \mathcal{Q}$ be as in (i). Define $h_1 : \{z|\ 0 \leq \operatorname{Re} z \leq 1\} \to \mathcal{Q}$ by

$$h_1(x + iy) = (1 - x)(z_0 + iy) + xF(z_0 + iy), \quad \text{for } 0 \leq x \leq 1, \ y \in \mathbb{R}.$$

Then we have

$$\frac{\partial h_1}{\partial x} = F(z_0 + iy) - (z_0 + iy), \quad \frac{\partial h_1}{\partial y} = ixF'(z_0 + iy) + i(1 - x).$$

Hence

$$\left|\frac{\partial h_1}{\partial z} - 1\right| = \frac{1}{2}|\{F(z_0) - (z_0 + iy + 1)\} + x(F'(z_0 + iy) - 1)| \leq \frac{1}{4},$$

$$\left|\frac{\partial h_1}{\partial \bar{z}}\right| = \frac{1}{2}\left|\{F(z_0) - (z_0 + iy + 1)\} - x(F'(z_0 + iy) - 1)\right| \le \frac{1}{4}.$$

Therefore $\left|\frac{\partial h_1}{\partial \bar{z}} / \frac{\partial h_1}{\partial z}\right| < 1/3$ and h_1 is a quasi-conformal mapping onto the strip S, and satisfies $h_1^{-1}(F(z)) = h_1^{-1}(z) + 1$ for $z \in \ell$. Let σ_0 be the standard conformal structure of $\mathbb{C}$, and take the pull-back $\sigma = h_1^* \sigma_0$ on $\{z| \ 0 \le \mathrm{Re}\, z \le 1\}$. Then extend σ to $\mathbb{C}$ by $\sigma = (T^n)^* \sigma$ on $\{z| -n \le \mathrm{Re}\, z \le -n+1\}$, where $T(z) = z+1$. By Ahlfors-Bers measurable mapping theorem [A], there exists a unique quasi-conformal mapping $h_2 : \mathbb{C} \to \mathbb{C}$ such that $h_2^* \sigma_0 = \sigma$ and $h_2(0) = 0$, $h_2(1) = 1$. By the definition of σ, T preserves σ. Hence $h_2 \circ T \circ h_2^{-1}$ preserves the standard conformal structure σ_0, i.e. conformal, therefore it must be an affine function. Since T has no fixed point in $\mathbb{C}$, so does $h_2 \circ T \circ h_2^{-1}$, hence it is a translation. Using $h_2 \circ T \circ h_2^{-1}(0) = 1$, we have $h_2 \circ T \circ h_2^{-1} = T$.

Now define Φ by $\Phi = h_2 \circ h_1^{-1}$ on S , and extend to the whole $\mathcal{Q}$ using the relation $\Phi(F(z)) = \Phi(z) + 1$. Then Φ is well-defined by (i), continuous and homeomorphic by the above relations on h_1^{-1} and h_2. Moreover Φ is analytic outside the orbit of ℓ, then analytic in the whole $\mathcal{Q}$ by Morera's theorem. Thus we have obtained the desired univalent function Φ.

If Φ' is another such function, then $\Phi''(z) = \Phi' \circ \Phi^{-1}$ commutes with T at least in $\Phi(\mathcal{Q})$. Hence $\Phi''(z) - z$ extends to $\mathbb{C}$ as a periodic function, then Φ'' extends to $\mathbb{C}$ as a holomorphic function commuting with T. Similarly Φ''^{-1} also has this property, therefore φ must be an affine function. However an affine function commuting with the translation T is also a translation by a constant. Hence the assertion follows.

(iii) Let us consider F and F_0 defined in the same $\mathcal{Q}$ and satisfying the condition of the Proposition. As in (ii), we can construct h_1, h_2, Φ for F and $h_{1,0}$, $h_{2,0}$, Φ_0 for F_0. It is easy to see that on any compact set of $\{z|0 \le \mathrm{Re}\, z \le 1\}$, $\frac{\partial h_1}{\partial \bar{z}} / \frac{\partial h_1}{\partial z} \to \frac{\partial h_{1,0}}{\partial \bar{z}} / \frac{\partial h_{1,0}}{\partial z}$ as $F \to F_0$. Hence $\sigma_F = h_1^* \sigma_0 \to \sigma_{F_0} = h_{1,0}^* \sigma_0$ and $h_2 \to h_{2,0}$ on any compact set as $F \to F_0$. It follows from the definition of the extension of Φ that $\Phi \to \Phi_0$ as $F \to F_0$. $\square$

2.5.3. A strip S as in Proposition 2.5.2 (ii), is called a *fundamental region* for the map $F|_{\mathcal{Q}}$. The quotient space

$$\mathcal{C} = S/\sim, \quad \text{where } \ell \ni z \sim F(z) \in F(\ell)$$
$$= \mathcal{Q}/\sim, \quad \text{where } z \sim F(z) \text{ if } z \in \mathcal{Q} \cap F^{-1}(\mathcal{Q})$$

is topologically a cylinder which is called the *Ecalle cylinder*. Moreover, $\mathcal{C}$ has a natural structure of a Riemann surface, using F near ℓ as a coordinate patching.

Lemma 2.5.4. *Let F, $\mathcal{Q}$, S be as above. Then $\pi \circ \Phi$ induces an isomorphism*

$$\bar{\Phi} : \mathcal{C} = S/\sim \to \mathbb{C}^*.$$

Proof. It is easy to see from the construction that $\overline{\Phi}$ is a covering map. Moreover it induces an isomorphism from the fundamental groups, since the generator of the fundamental group of C corresponds to F and that of $\mathbb{C}^*$ is T. □

2.5.6. It is easy to see that Proposition 2.5.2 and Lemma 2.5.4 imply Proposition 2.2.1 (the existence part) and Lemma 2.3.3 respectively.

§2.6. An estimate for the Fatou coordinate.

2.6.1. Let $F(z)$ be an analytic function such that $v(z) = F(z) - z$ is close to 1 (for example, as in Proposition 2.5.2). Note that if $v \equiv 1$ then $F(z) = z + v(z) = z + 1$ is the time one map of the constant flow $\frac{dz}{dt} = 1$. So it is natural idea to compare the orbit of F with that of the flow $\frac{dz}{dt} = v(z)$. This flow can be solved by the integration

$$t = \int^z \frac{d\zeta}{v(\zeta)} + C.$$

If we regard t in this formula as a complex function, the correspondence $z \mapsto t$ gives a new coordinate in which the flow is *actually* the constant flow. So according to the above idea, the Fatou coordinate must have a similar behavior as $t(z)$. The following statement justifies this observation.

Proposition 2.6.2. *Suppose that Φ and v are holomorphic functions in a region $\mathcal{U}$ satisfying:*

$$\Phi \quad \text{is univalent in } \mathcal{U}, \quad |v(z) - 1| < 1/4 \quad \text{for } z \in \mathcal{U} \quad \text{and}$$

$$\Phi(z + v(z)) = \Phi(z) + 1, \quad \text{if} \quad z, z + v(z) \in \mathcal{U}.$$

(i) *There exist universal constants $R_1, C_1, C_2 > 0$ such that if $\mathcal{U} = \{z \mid |z - z_0| < R\}$ for $R \geq R_1$, then*

$$\left| \Phi'(z_0) - \frac{1}{v(z_0)} \right| \leq C_1 \left(\frac{1}{R^2} + |v'(z_0)| \right) \leq \frac{C_2}{R}.$$

(ii) *Suppose $\mathcal{U} = \{z \in \mathbb{C}^* \mid \theta_1 < \arg z < \theta_2\}$ $(\theta_2 < \theta_1 + 2\pi)$ and $|v'(z)| \leq C/|z|^{1+\nu}$ $(z \in \mathcal{U})$ for some $C, \nu > 0$. For $z_0 \in \mathcal{U}$ and θ_1', θ_2' with $\theta_1 < \theta_1' < \theta_2' < \theta_2$, there exists $R_2, C_3 > 0$ and $\xi \in \mathbb{C}$ such that*

$$\left| \Phi(z) - \int_{z_0}^z \frac{d\zeta}{v(\zeta)} - \xi \right| \leq C_3 \left(\frac{1}{|z|} + \frac{C}{|z|^\nu} \right),$$

for z satisfying $\theta_1' < \arg z < \theta_2'$, $dist(z, \mathbb{C} \setminus \mathcal{U}) > R_1$. Moreover C_3 depend only on θ_i, θ_i'.

See [Y] for a similar estimate.

Proof. (i) We may suppose $z_0 = 0$. We take R so that $R \gg 1$. It follows from Koebe's distortion theorem [P] that if $|z| < R - 2$, then

$$\frac{|v(z)|}{(1 + |v(z)|/2)^2} \le \left| \frac{\Phi(z + v(z)) - \Phi(z)}{\Phi'(z)} \right| \le \frac{|v(z)|}{(1 - |v(z)|/2)^2},$$

since Φ is univalent in $\{\zeta | \ |\zeta - z| < 2\ \}$. Hence $C \le |\Phi'(z)| \le C'$ if $|z| < R - 2$. We have $|\Phi''(z)| \le C/R$ if $|z| < R/2$. (In fact, by Cauchy's formula, $\Phi''(z)$ can be expressed in terms of an integral of $\Phi'(\zeta)/(\zeta - z)^2$ over the contour $\{\zeta | \ |\zeta - z| = R/3\}$, then use the above estimate.) Using Taylor's formula

$$\Phi(z + a) = \Phi(z) + a\Phi'(z) + a^2 \int_0^1 (1 - t)\Phi''(z + at)dt,$$

we obtain $|1 - \Phi'(z)v(z)| \le C/R$ if $|z| < R/2 - 5/4$ (this implies $|z| < R/2$ and $|z + v(z)| < R/2$). Again by Cauchy's formula, $|(1 - \Phi'(z)v(z))'| = |\Phi''(z)v(z) + \Phi'(z)v'(z)| \le C/R^2$ if $|z| < R/4$. It also follows from Cauchy's formula that $|v'(z)| \le C/R$ and $|v''(z)| \le C/R^2$ for $|z| < R/2$. Therefore $|v'(z)| \le |v'(0)| + C/R^2$ for $|z| < 5/4$ Hence we have $|\Phi''(z)| \le C(1/R^2 + |v'(0)|)$ for $|z| < 5/4$. By the above Taylor's formula again, we have $|1 - \Phi'(0)v(0)| \le C(1/R^2 + |v'(0)|) \le C/R$. So we obtain the desired inequality.

(ii) Note that in the sector $\theta_1' < \arg z < \theta_2'$ we have $dist(z, \mathbb{C} \setminus \mathcal{U}) > C_4|z|$ for some constant $C_4 > 0$. So the result in (i) applies to the region $\{w | \ |w - z| < C_4|z|\}$, and gives

$$\left| \Phi'(z) - \frac{1}{v(z)} \right| \le C_1 \left(\frac{1}{(C_4|z|)^2} + |v'(z_0)| \right) \le C_1' \left(\frac{1}{|z|^2} + \frac{1}{|z|^{1+\nu}} \right).$$

Integrating this formula along a path in the smaller sector, we obtain the inequality. $\square$

2.6.3. *Proof of the estimate in Proposition 2.2.1.* Let $v(z) = f_0^*(z) - z = 1 + a/z + O(1/z^2)$, $\Phi = \Phi^\pm$ and $\mathcal{U} = \{z \in \mathbb{C}^* | \ \frac{\pi}{4} < \arg z < \frac{3\pi}{4}\}$. Since $|v'(z)| \le C/|z|^2$, we can apply Proposition 2.6.2 and obtain an estimate in a smaller sector:

$$\left| \Phi^\pm(z) - \int_{z_0}^z \frac{d\zeta}{v(\zeta)} - \xi \right| \le C_3' \frac{1}{|z|}.$$

Since $1/v(z) = 1 - a/z + O(1/z^2)$, we have

$$\Phi^-(z) = z - a \log z + const + O(1/z).$$

$\square$

§3. LOCAL THEORY OF THE PARABOLIC BIFURCATION

§3.1. The phenomenon. Let $f_0 \in \mathcal{F}_0$ be as in §2, i.e.

$$f_0(z) = z + z^2 + \ldots.$$

As we have seen in §2.1.2, the orbits of f_0 are contained in curves which "look like" circles tangent to the real axis at $z = 0$. In this section, we study how such orbits bifurcate after perturbation.

First the parabolic fixed point $z = 0$ has multiplicity two as a fixed point, i.e., as the solution of the fixed point equation $f_0(z) - z = 0$. Therefore in general, it bifurcates into two fixed points after perturbation. Their multipliers are close to one, and depending on the perturbation, they can be attracting, repelling, or indifferent.

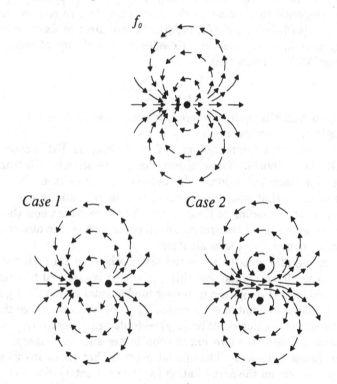

FIGURE 4. Orbits for f_0 and its perturbation

Two examples of perturbed orbits are depicted in the Figure. One notices a big difference for orbits starting from a point on the left of the origin, for example, on the negative real axis. For f_0, such an orbit simply converges to the origin; in Case 1, the orbit converges to an attracting fixed point which bifurcated from the origin; however in Case 2, the orbit can go through an "opening" between the two fixed points and leave the neighborhood. Such a new type of orbits can give rise to a drastic change of the global dynamics (such as the inflation of the Julia set). However it takes an extremely long time for these orbits to go through the opening, so we need to consider a large number of iterates of the map in order to see this phenomenon. And the effect of the perturbation might accumulate during the iteration, so it becomes difficult to control.

For the perturbed system as Case 1, there is not a drastic change in the global dynamics, so we concentrate on Case 2. The distinction between Case 1 and Case 2 can be given by the multiplier λ of a bifurcated fixed point. Roughly speaking, Case 1 corresponds to λ which is closer to the real axis than to the unit circle (or $\arg(\lambda - 1) \in [-\frac{\pi}{4}, \frac{\pi}{4}] \cup [\frac{3\pi}{4}, \frac{5\pi}{4}]$) and Case 2 corresponds to λ closer to the unit circle than to the real axis (or $\arg(\lambda - 1) \in [\pi/4, 3\pi/4] \cup [5\pi/4, 7\pi/4]$). (There are two fixed points and two multipliers λ, λ' to consider. However by the theory of holomorphic indices (see [Mi]) we know that

$$\frac{1}{\lambda - 1} + \frac{1}{\lambda' - 1}$$

is continuous hence bounded, whereas λ's are close to 1. So we can take either one in the above discussion.)

We analyse such a bifurcation as in Case 2 using the Fatou coordinates and the Ecalle cylinders. These objects for f_0 are already constructed in §2, using "fundamental regions" S_- and S_+. An important fact (which we prove later) is that such fundamental regions continue to exist after a perturbation corresponding to Case 2. The difference is that now the tips of the croissant shape are at two distinct fixed points. So we can also construct Ecalle cylinders and Fatou coordinates.

This setting allows us to trace the orbits which go through the opening between the fixed points. For this purpose, it is essential to know how an orbit coming in through the incoming fundamental region $S_{-,f}$ goes out through the outgoing fundamental region $S_{+,f}$. In fact, we will see that this correspondence (incoming point in $S_{-,f}$) $\mapsto$ (outgoing point in $S_{+,f}$) induces a translation between the two Fatou coordinates and an isomorphism between the Ecalle cylinders. The crucial point is that this correspondence depends sensitively on the perturbation (or the parameter), however the dependence can be controlled in terms of λ (or more precisely $1/\log \lambda$ where $\log 1 = 0$). In fact, in a suitable coordinate with a normalization, the above correspondence can be written as $z \mapsto z - 2\pi i/\log \lambda$. (See Proposition 3.2.2 (iii).)

In Case 2, points near the tips of the croissants $S_{+,f}$ have orbits which return many times to $S_{\pm,f}$. So the (first) return map can be defined on

$S_{+,f}$, and its iteration can be used to analyse the long term behavior of those orbits. Such a return map induces a (well-defined, continuous and analytic) map $\mathcal{R}_f$ on the Ecalle cylinder $\mathcal{C}_{+,f}$. Studying the dynamics of $\mathcal{R}_f$, we are able to obtain more information on the Julia sets.

In this section, we state these facts more precisely and give the proof.

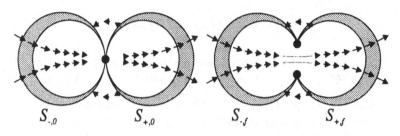

FIGURE 5. Fundamental domains $S_{\pm,f}$

§3.2. Statements.

3.2.1. Let $f_0(z) = z + z^2 + \cdots \in \mathcal{F}_0$ be as in §2. If f is an analytic mapping close to f_0, then it has two fixed points near 0 counted with the multiplicity. By changing the coordinate by a translation near the identity, we can shift one of them to the origin, so f reduces to the form

$$f(z) = \lambda z + O(z^2).$$

Note here that there are in general two choices– which fixed point to put at the origin. In fact, for a parametrized family of maps containing f_0, it is in general impossible to reduce them to the above form by coordinate changes depending continuously on the parameter. However this problem can be avoided either by taking a branched double cover of the parameter space, or restricting the family to the part Case 2 in the discussion in 3.1 (for example maps belonging to $\mathcal{F}_1$ defined below).

Anyway we may suppose that the map under consideration belongs to

$$\mathcal{F} = \left\{ f \;\middle|\; \begin{array}{c} f \text{ is defined and analytic in a neighborhood of } 0, \\ f(0) = 0, \; f'(0) \neq 0 \end{array} \right\}.$$

For $f \in \mathcal{F}$, we express the derivative $f'(0)$ as

$$f'(0) = \exp(2\pi i \alpha(f))$$

where $\alpha(f) \in \mathbb{C}$ and $-\frac{1}{2} < \operatorname{Re} \alpha(f) \leq \frac{1}{2}$. Our study is focused on the maps in the following class:

$$\mathcal{F}_1 = \{ f \in \mathcal{F} \mid |\arg \alpha(f)| < \pi/4 \}.$$

Let $\sigma(f)$ be the other fixed point of f near the origin ($\sigma(f) = 0$ if the fixed point is degenerate). Since $\sigma(f)$ is a solution of $f(z) - z = 0$, with $f'(0) = \exp(2\pi i \alpha(f))$ and $f''(0) \sim f_0''(0) = 1$, it is a solution of $(e^{2\pi i \alpha} - 1)z + z(1 + O(\alpha) + O(z)) = 0$. So $\sigma(f) = (-2\pi i \alpha + O(\alpha^2))(1 + O(\alpha) + O(z))$, and the $O(z)$ term tends of 0 as $\alpha \to 0$. Therefore $\sigma(f) = O(\alpha)$ and

$$\sigma(f) = -2\pi i \alpha(f)(1 + o(1)) \text{ as } f \to f_0.$$

Let us denote the objects defined for $f_0 \in \mathcal{F}_0$ in §2 with an additional subscript "0", i.e., $\Phi_{\pm}$, $\Phi^{\pm}$, $S_{\pm}$, etc are denoted by $\Phi_{\pm,0}$, $\Phi_0^{\pm}$, $S_{\pm,0}$, etc.

Proposition 3.2.2. *Let $f_0 \in \mathcal{F}_0$. There exists a neighborhood $\mathcal{N}_0$ of f_0 (in the sense of the topology defined in 2.5.1) such that if $f \in \mathcal{N}_0 \cap \mathcal{F}_1$, then there exist closed Jordan domains $S_{-,f}$, $S_{+,f}$ and analytic functions $\Phi_{-,f}$, $\Phi_{+,f}$ satisfying the following:*
(i) $S_{\pm,f}$ is bounded by an arc $\ell_{\pm}$ and its image $f(\ell_{\pm})$ such that $\ell_{\pm}$ joins two fixed points $\{0, \sigma(f)\}$, $\ell_{\pm} \cap f(\ell_{\pm}) = \{0, \sigma(f)\}$ and $S_{-,f} \cap S_{+,f} = \{0, \sigma(f)\}$;
(ii) $\Phi_{\pm,f}$ is defined, analytic and injective in a neighborhood of $S_{\pm,f}' = S_{\pm,f} \smallsetminus \{0, \sigma(f)\}$;
(iii) if $z \in S_{-,f}'$ then there is an $n \geq 1$ such that $f^n(z) \in S_{+,f}'$ and for the smallest such n,

$$\Phi_{+,f}(f^n(z)) = \Phi_{-,f}(z) - \frac{1}{\alpha(f)} + n \; ;$$

(iv) when $f \in \mathcal{F}_1$ tends to f_0, $S_{\pm,f} \to S_{\pm,0} \cup \{0\}$ in the Hausdorff metric, and $\Phi_{\pm,f} \to \Phi_{\pm,0}$ in the topology defined in 2.5.1.

The proof will be given later in §3.4.

It is often easier to work with $\varphi_f = (\Phi_{+,f})^{-1}$ and for applications, the following is more powerful.

Proposition 3.2.3. *Let φ_0 be the inverse function $(\Phi_{+,0})^{-1}$ defined in*

$$\mathcal{Q}_0 = \{w : |\arg(-w - b')| < 2\pi/3\}$$

for large $\xi_0 > 0$ (see 2.2.4). There exists a neighborhood $\mathcal{N}_0$ of f_0 such that if $f \in \mathcal{N}_0 \cap \mathcal{F}_1$, then there exist large $\xi_0, \eta_0 > 0$, and analytic maps $\varphi_f : \mathcal{Q}_f \to \overline{\mathbb{C}}$ and $\hat{\mathcal{R}}_f : \{w \in \mathbb{C} : |\operatorname{Im} w| > \eta_0\} \to \mathbb{C}$, where

$$\mathcal{Q}_f = \{w \in \mathbb{C} : |\arg(-w - \xi_0)| < 2\pi/3 \text{ and } |\arg(w + \frac{1}{\alpha(f)} - \xi_0)| < 2\pi/3 \},$$

satisfying the following:
(i) $\varphi_f(\mathcal{Q}_f) \subset \mathcal{D}(f)$; *If* $w, w+1 \in \mathcal{Q}_f$, *then*

$$\varphi_f(w+1) = f \circ \varphi_f(w);$$

$\varphi_f(w) \to 0$ *when* $w \in \mathcal{Q}_f$ *and* $\operatorname{Im} w \to +\infty$; $\varphi_f(w) \to \sigma(f)$ *when* $w \in \mathcal{Q}_f$ *and* $\operatorname{Im} w \to -\infty$;
(ii) If $|w| > \eta_0$, *then*

$$\tilde{\mathcal{R}}_f(w+1) = \tilde{\mathcal{R}}_f(w) + 1;$$

$\tilde{\mathcal{R}}_f(w) - w$ *tends to* $-1/\alpha(f)$ *when* $\operatorname{Im} w \to \infty$, *to a constant when* $\operatorname{Im} w \to -\infty$;
(iii) If $w \in \mathcal{Q}_f$ *satisfies* $|w| > \eta_0$ *and* $w' = \tilde{\mathcal{R}}_f(w) + n \in \mathcal{Q}_f$ *for some* $n \in \mathbb{Z}$, *then*

$$f^n(\varphi_f(w)) = \varphi_f(w') \text{ if } n \geq 0, \quad f^{-n}(\varphi_f(w')) = \varphi_f(w) \text{ if } n < 0;$$

(iv) when $f \in \mathcal{F}_1$ *and* $f \to f_0$,

$$\varphi_f \to \varphi_0 \quad \text{and} \quad \tilde{\mathcal{R}}_f + \frac{1}{\alpha(f)} \to \tilde{\mathcal{E}}_{f_0}.$$

Moreover the convergence of $\tilde{\mathcal{R}}_f + \frac{1}{\alpha(f)}$ *is uniform with respect to the Euclidean metric.*

The proof will be given in §3.5.

3.2.4. The induced return map $\mathcal{R}_f$ on the Ecalle cylinder.
By Proposition 3.2.3 (ii), $\tilde{\mathcal{R}}_f$ induces a well-defined mapping $\mathcal{R}_f$ on $\mathbb{C}^*$ via $\pi(w) = e^{2\pi i w}$:

$$\mathcal{R}_f = \pi \circ \tilde{\mathcal{R}}_f \circ \pi^{-1} : \{0 < |z| < e^{-2\pi\eta_0}\} \cup \{e^{2\pi\eta_0} < |z| < \infty\} \to \mathbb{C}^*.$$

Moreover it extends to 0 and ∞ analytically, and satisfies: $\mathcal{R}_f(0) = 0$, $\mathcal{R}_f(\infty) = \infty$, the fixed point $z = 0$ has multiplier

$$\mathcal{R}'_f(0) = \exp(-2\pi i \frac{1}{\alpha(f)})$$

and ∞ has also a non-zero multiplier.

Suppose that $w, w' \in \mathcal{Q}_f$, $|w| > \eta_0$ and $\mathcal{R}_f(\pi(w)) = \pi(w')$. Then by definition, $\tilde{\mathcal{R}}_f(w) + n = w'$ for an integer n, and therefore $f^n(\varphi_f(w)) = \varphi_f(w')$ $(n \geq 0)$, or $f^{-n}(\varphi_f(w')) = \varphi_f(w)$ $(n < 0)$. Note that

$$\tilde{\mathcal{R}}_f(w) - (w - \frac{1}{\alpha(f)}) \to \tilde{\mathcal{E}}_{f_0}(w) - w,$$

hence this expression is bounded in $\{|\operatorname{Im} w| > y_0\}$ (taking y_0 larger if necessary). On the other hand, $\operatorname{Re} 1/\alpha(f) \to +\infty$ when $\mathcal{F}_1 \ni f \to f_0$. Therefore for $f \in \mathcal{F}_1$ sufficiently close to f_0, we have $\operatorname{Re} \tilde{\mathcal{R}}_f(w) \ll \operatorname{Re} w$ and $2\pi/3 < \arg(\tilde{\mathcal{R}}_f(w) - w) < 4\pi/3$. So in this case, the above integer n is positive (or even large) if $\operatorname{Re} w$ and $\operatorname{Re} w'$ stay bounded. This is also the case if

$$|\arg(w' + \frac{1}{2\alpha(f)} - w)| < 2\pi/3.$$

If n is large, we can interpret that one iteration of $\mathcal{R}_f$ corresponds to a large number of forward iteration of f via φ_f. So we call $\mathcal{R}_f$ the *induced return map* and this map is defined on (a subset of) $\mathbb{C}^*$, which is identified with the "outgoing Ecalle cylinder" $\mathcal{C}_{+,f} = S_{+,f}/ \sim$. Together with these facts, Proposition 3.2.3 (iii) implies

(iii') If U, U' are domains contained in $\mathcal{Q}_f$ such that
   $\mathcal{R}_f^m$ is defined on $\pi(U)$ and $\mathcal{R}_f^m(\pi(U)) \subset \pi(U')$ for some $m \geq 1$,
   $\varphi_f|_U, \pi|_{U'}$ are injective, and
   $|\arg(w' + \frac{1}{2\alpha(f)} - w)| < 2\pi/3$ for $w \in U$, $w' \in U'$,
then there exists an $n > m$ such that

$$f^n = \varphi_f \circ (\pi|_{U'})^{-1} \circ \mathcal{R}_f^m \circ \pi \circ (\varphi_f|_U)^{-1} \quad \text{on } \varphi_f(U).$$

From Proposition 3.2.3 (iv), we have:
(iv') $e^{2\pi i/\alpha(f)}\mathcal{R}_f \to \mathcal{E}_{f_0}$ when $f \in \mathcal{N}_0 \cap \mathcal{F}_1$ and $f \to f_0$.

It follows that if $\{f_n\}$ is a sequence in $\mathcal{N}_0 \cap \mathcal{F}_1$ such that $f_n \to f_0$ and $1/\alpha(f_n) - k_n \to -\beta$ (constant) as $n \to \infty$, where k_n are integers, then there exists a limit

$$\lim_{n \to \infty} \mathcal{R}_{f_n} = e^{2\pi i \beta}\mathcal{E}_{f_0}.$$

Here the condition on α can be rewritten in the form:

$$\alpha(f) = \frac{1}{k_n - \beta_n} \quad \text{and} \quad \beta_n \to \beta.$$

It is sometimes convenient to consider the induced map on $\mathbb{C}/\mathbb{Z}$ instead of $\mathbb{C}^*$. Define the quotient map $\pi_1 : \mathbb{C} \to \mathbb{C}/\mathbb{Z}$ by $\pi_1(w) = w \mod \mathbb{Z}$, and the identifying map $\pi_2 : \mathbb{C}/\mathbb{Z} \to \mathbb{C}^*$ by $\pi_2(w \bmod \mathbb{Z}) = \exp(2\pi i w)$. Then π is decomposed as $\pi_2 \circ \pi_1$. Note that π_2 sends the upper (resp. lower) end of $\mathbb{C}/\mathbb{Z}$ to $z = 0$ (resp. ∞). Let $\hat{\mathcal{R}}_f = \pi_1 \circ \tilde{\mathcal{R}}_f \circ \pi_1^{-1} = \pi_2^{-1} \circ \mathcal{R}_f \circ \pi_2$. The dynamics of $\hat{\mathcal{R}}_f$ near the upper end of $\mathbb{C}/\mathbb{Z}$ is close to the translation by $-1/\alpha \bmod \mathbb{Z}$, and near the lower end it is close to a translation by a constant.

3.2.5. To study the quantitative aspect, it is often more convenient to work with $\varphi_f^* = I \circ \varphi_f$. Let us define $D(\eta) = \{w \in \mathbb{C} |\ |w - i\eta| \leq |\eta|/4\}$ and $D'(\eta) = \{w \in \mathbb{C} |\ |w - i\eta| \leq |\eta|/2\}$, for $\eta \in \mathbb{R}$.

As a corollary of Proposition 2.2.1 and Proposition 3.2.3 (iv), we have the following:

For large $\eta > 0$, there exists a neighborhood $\mathcal{N}_1(\eta)(\subset \mathcal{N}_0)$ of f_0, depending on η, such that if $f \in \mathcal{N}_1(\eta) \cap \mathcal{F}_1$, then φ_f is defined and injective on $D(\pm\eta)$, the image $\varphi_f^*(D(\pm\eta))$ is contained in $D'(\pm\eta)$ and $\frac{1}{2} < |(\varphi_f^*)'| < 2$ on $D(\pm\eta)$.

3.2.6. Remark. For $f \in \mathcal{F}$ with $3\pi/4 < \arg \alpha(f) < 5\pi/4$, instead of being in $\mathcal{F}_1$, we can obtain a similar result as Propositions 3.2.2 and 3.2.3 with the following changes:

the lower end of $\mathbb{C}/\mathbb{Z}$ corresponds to the fixed point 0;

in the formulae and definitions, $\frac{1}{\alpha(f)}$ should be replaced by $-\frac{1}{\alpha(f)}$, except for the multiplier in 3.2.4, $\mathcal{R}'_f(\infty) = \exp(-2\pi i \frac{1}{\alpha(f)})$;

in Proposition 3.2.3 (i), $\varphi_f(w) \to \sigma(f)$ ($w \in \mathcal{Q}_f$ and $\operatorname{Im} w \to +\infty$), $\varphi_f(w) \to 0$ ($w \in \mathcal{Q}_f$ and $\operatorname{Im} w \to -\infty$).

§3.3. A preliminary coordinate change– unwrapping coordinate.

We would like to prove the statements in §3.2 using Propositions 2.5.2 and 2.6.2, in the same frame work as the Fatou coordinates for the unperturbed system f_0 (i.e., Proposition 2.2.1 and Lemma 2.3.3). For this purpose, we introduce a new coordinate w (unwrapping coordinate) in which the map looks like a translation. For this purpose, we do not need the assumption on the argument of $\alpha(f)$.

3.3.1. Recall that 0 and $\sigma = \sigma(f)$ are the only fixed points of f in a small neighborhood of 0, if $f \in \mathcal{F}$ is sufficiently close to f_0. So

$$u(z) = u_f(z) = \frac{f(z) - z}{z(z - \sigma)}$$

is analytic and non-zero in the neighborhood. In particular, for $f = f_0$, $\sigma(f_0) = 0$, $u_{f_0}(z) = (f_0(z) - z)/z^2 = 1 + O(z)$. It is easy to see, for example by using an integral representation, that the correspondences $f \mapsto \sigma(f)$ and $f \mapsto u_f(z)$ are continuous (where we suppose that u_f is defined is a fixed small neighborhood $\mathcal{V}$ of *zero*). Moreover we may suppose that $\sigma(f) \in \mathcal{V}$ and $1/2 < |u_f(z)| < 2$ on $\mathcal{V}$.

Recall that

$$\sigma(f) = -2\pi i \alpha(f)(1 + o(1)) \quad \text{as } f \to f_0.$$

3.3.2. Suppose that $f \in \mathcal{F}$ is as in 3.3.1 and $\alpha(f) \neq 0$ (i.e. $\sigma(f) \neq 0$).

Let us introduce a new coordinate $w \in \mathbb{C}$ by

$$z = \tau_f(w) \equiv \frac{\sigma}{1 - e^{-2\pi i \alpha w}},$$

where $\sigma = \sigma(f)$ and $\alpha = \alpha(f)$. Also define the translation $T_f(w)$ by

$$T_f(w) = w - \frac{1}{\alpha(f)}.$$

Here are the important properties of τ_f for $f \in \mathcal{F}$ sufficiently close to f_0 with $\alpha(f) \neq 0$:

(i) The map $\tau_f : \mathbb{C} \to \overline{\mathbb{C}} \smallsetminus \{0, \sigma(f)\}$, $w \mapsto z = \tau_f(w)$ is a universal covering, whose covering transformation group is generated by the translation T_f; $\tau_f(w) \to 0$ as $\operatorname{Im} \alpha w \to +\infty$, and $\tau_f(w) \to \sigma$ as $\operatorname{Im} \alpha w \to -\infty$.

(ii) Let $\tau_0(w) = I^{-1}(w) = -1/w$. Then

$$\sup_{w:\, |\operatorname{Re}\alpha w| < 3/4} d_{\overline{\mathbb{C}}}\left(\tau_f(w), \tau_0(w)\right) \to 0 \quad \text{as } f \to f_0,$$

where $d_{\overline{\mathbb{C}}}(\cdot, \cdot)$ is the spherical metric on $\overline{\mathbb{C}}$. Hence on any compact set, $\tau_f(w)$ converges to $\tau_0(w)$ in the spherical metric when $f \to f_0$.

(iii) For any $\varepsilon > 0$, there exists $R_0 > 0$ independent of f such that if

$$w \in \mathbb{C} \smallsetminus \bigcup_{n \in \mathbb{Z}} T_f^n D_{R_0}, \quad \text{where } D_{R_0} = \{w' \mid |w'| < R_0\},$$

then $|\tau_f(w)| < \varepsilon$.

Proof. (i) is obvious. For (ii), consider the function

$$\frac{1}{1 - e^{-2\pi i \zeta}} - \frac{1}{2\pi i \zeta}.$$

It is easy to see that this function is analytic in $\{\zeta : |\operatorname{Re}\zeta| < k\}$ for any $0 < k < 1$ (for example $k = 3/4$), and bounded on the boundary of this region (including the limit $\operatorname{Im}\zeta \to \pm\infty$). Hence by the maximum value principle, it is bounded on the region as well. This implies that $|\tau_f - \frac{-\sigma}{2\pi i \alpha}\tau_0| \leq C|\sigma|$

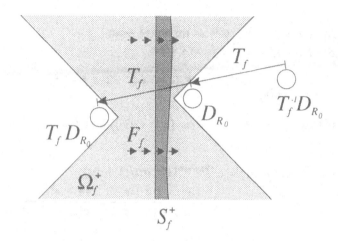

FIGURE 6. $F_f(w)$ and T_f in the unwrapping coordinate

therefore the assertion holds, since $\frac{-\sigma}{2\pi i \alpha} \to 1$.

(iii) It follows from (ii) that if R_0 is large, $\tau_f(\partial D_{R_0})$ is contained in a small disk around 0. Since τ_f is a covering map, the image of D_{R_0}, $\tau_f(D_{R_0})$ must contain the exterior of the small disk. Then the image of $\mathbb{C} \smallsetminus \bigcup_{n \in \mathbb{Z}} T_f^n D_{R_0}$ cannot go to the exterior. So the assertions are proved. $\square$

3.3.3. Now we define the lift of f to the w-coordinate. First note that if $f \in \mathcal{F}$ sufficiently close to f_0 and $R_0 > 0$ is large,

$$|\sigma(f)u_f(z)/(1 + zu_f(z))| \le C|\sigma| \le 1/2$$

for $w \in \mathbb{C} \smallsetminus \bigcup_{n \in \mathbb{Z}} T_f^n D_{R_0}$ and $z = \tau_f(w)$ (by 3.3.1 and 3.3.2 (iii)).

Define $F_f : \mathcal{D}(F_f) \to \mathbb{C}$, where $\mathcal{D}(F_f) \equiv \mathbb{C} \smallsetminus \bigcup_{n \in \mathbb{Z}} T_f^n D_{R_0}$, by

$$F_f(w) = w + \frac{1}{2\pi i \alpha(f)} \log\left(1 - \frac{\sigma(f)u_f(z)}{1 + zu_f(z)}\right) \quad \text{with } z = \tau_f(w)$$

where we use the branch of logarithm with $-\pi < \operatorname{Im} \log(\cdot) \le \pi$.

If $f \in \mathcal{F}$ sufficiently close to f_0, $\alpha(f) \ne 0$ and $R_0 > 0$ is large, then F_f has the following properties:

(i) $f \circ \tau_f = \tau_f \circ F_f$ and $T_f \circ F_f = F_f \circ T_f$;

(ii) For $w \in \mathcal{D}(F_f)$,

$$|F_f(w) - (w+1)| < \frac{1}{4}, \quad |F_f'(w) - 1| < \frac{1}{4}$$

(iii) $F_f(w) = w + 1 + O(1/w^2)$ as $\operatorname{Im} \alpha w \to +\infty$;

(iv) Denote $F_0 = f_0^* = \tau_0^{-1} \circ f_0 \circ \tau_0$, then

$$\sup_{w \in \mathcal{D}(F_f)} |F_f(w) - F_0(w)| \to 0 \quad \text{when } f \to f_0.$$

Proof. (i) Using $e^{-2\pi i \alpha w} = 1 - \sigma/z$ $(z = \tau_f(w))$ and $f(z) = z + z(z - \sigma)u$, we have:

$$\exp(-2\pi i \alpha F_f(w)) = e^{-2\pi i \alpha w} \left(1 - \sigma u/(1 + zu)\right)^{-1}$$

$$= (1 - \frac{\sigma}{z}) \frac{1 + zu}{1 + (z - \sigma)u} = \frac{(z - \sigma)(1 + zu)}{z(1 + (z - \sigma)u)},$$

$$\tau_f \circ F_f(w) = \frac{\sigma f(z)}{z + z(z - \sigma)u - (z - \sigma)(1 + zu)} = f(z) = f \circ \tau_f(w).$$

The other relation is obvious.

(ii) and (iv) Since $|\sigma u(z)/(1 + zu(z))| = O(\alpha)$, we can estimate the logarithm by Taylor expansion, and obtain:

$$\sup_{w \in \mathcal{D}(F_f)} \left| F_f(w) - \left(w - \frac{\sigma u(z)}{2\pi i \alpha (1 + zu(z))}\right) \right| = O(\alpha).$$

As F_f is periodic with period $\frac{1}{\alpha}$, it is enough to give the estimates in $\{w : |\operatorname{Re} \alpha w| \le 1/2\} \smallsetminus D_{R_0}$. From the facts that $\sigma(f) \to 0$, $u_f(z) \to u_{f_0}(z) = (f_0(z) - z)/z^2$ uniformly on $\mathcal{V}$ and $\tau_f \to \tau_0$ uniformly on $\{w : |\operatorname{Re} \alpha w| \le 1/2\}$, as $f \to f_0$, we conclude that when $f \to f_0$

$$w - \frac{\sigma(f) u_f(\tau_f(w))}{2\pi i \alpha(f)(1 + \tau_f(w) u_f(\tau_f(w)))} \to w + \frac{u_{f_0}(\tau_0(w))}{1 + \tau_0(w) u_{f_0}(\tau_0(w))} =$$

$$= -\frac{1}{\tau_0(w)} + \frac{f_0(\tau_0(w)) - \tau_0(w)}{\tau_0(w)^2 + \tau_0(w)(f_0(\tau_0(w)) - \tau_0(w))} = -\frac{1}{f_0(\tau_0(w))} = F_0(w)$$

and the convergence is uniform on $\{w : |\operatorname{Re} \alpha w| \le 1/2\} \smallsetminus D_{R_0}$. So (iv) is proved. This also proves (ii), since $F_0(w) = w + 1 + O(1/w)$ and one can take a larger R_0 if necessary.

(iii) Since $e^{2\pi i \alpha(f)} = f'(0) = 1 - \sigma(f) u_f(0)$, it is easy to see that $F_f(w) - (w + 1) \to 0$ as $\operatorname{Im} \alpha w \to +\infty$. Then by the periodicity of F_f, the decay must be exponential. (Use the Fourier expansion and estimate the coefficient.) $\square$

§3.4. Proof of Proposition 3.2.2.

3.4.1. Let $\mathcal{N}$ be a neighborhood of f_0 in $\mathcal{F}$ such that all the statements in §3.3 hold for $f \in \mathcal{F}$ with $\alpha(f) \neq 0$. From now on, we assume that $f \in \mathcal{N} \cap \mathcal{F}_1$, therefore $\alpha(f) \neq 0$ and $|\arg(\alpha(f))| < \pi/4$. For $\xi_1 > 0$, let

$$\Omega_f^- = \mathcal{Q}(\xi_1, -\xi_1 + \frac{1}{\alpha(f)}) \quad \text{and} \quad \Omega_f^+ = \mathcal{Q}(\xi_1 - \frac{1}{\alpha(f)}, -\xi_1) = T_f(\Omega_f^-),$$

where $\mathcal{Q}(\cdot, \cdot)$ is the region defined in 2.5.1. If ξ_1 is large enough then $\Omega_f^\pm \subset \mathcal{D}(F_f)$ for all $f \in \mathcal{N} \cap \mathcal{F}_1$. Moreover we may assume that for the same $\xi_1 > 0$,

$$\Omega_0^- = \mathcal{Q}(\xi_1, +\infty) \quad \text{and} \quad \Omega_0^+ = \mathcal{Q}(-\infty, -\xi_1),$$

where $\Omega_0^\pm$ stand for $\Omega^\pm$ defined for f_0 in §2.1.3.

Let us fix two points $w^\pm \in \Omega_0^\pm$ such that $\operatorname{Re} w^- > \xi_1$ and $\operatorname{Re} w^+ < -\xi_1 - 5/4$. Then we may assume $w^\pm \in \Omega_f^\pm$ for all $f \in \mathcal{N} \cap \mathcal{F}_1$ and $\operatorname{Re} \frac{1}{\alpha} > 2\xi_1 + 2$, taking $\mathcal{N}$ smaller if necessary.

By our assumption, $|F_f(w) - (w+1)| < \frac{1}{4}$ and $|F_f'(w) - 1| < \frac{1}{4}$ on $\Omega_f^\pm$, hence Proposition 2.5.2 can be applied. By (i), there exist fundamental regions $S_f^\pm$ ($\subset \Omega_f^\pm$) bounded by $\ell^\pm = \{w^\pm + iy | y \in \mathbb{R}\}$ and $F_f(\ell^\pm)$. By (ii), there exist univalent analytic maps $\Phi_f^- : \Omega_f^- \to \mathbb{C}$ and $\Phi_f^+ : \Omega_f^+ \to \mathbb{C}$ satisfying

$$\Phi_f^-(F_f(w)) = \Phi_f^-(w) + 1, \quad \Phi_f^+(F_f(w)) = \Phi_f^+(w) + 1,$$

and $\Phi_f^\pm(w^\pm) = \Phi_0^\pm(w^\pm)$. By (iii), $\Phi_f^\pm \to \Phi_0^\pm$ as $f \to f_0$.

Let $S_{\pm,f}$, $\ell_\pm$ be the images of $S_f^\pm$, $\ell^\pm$ by τ_f, with two points 0 and $\sigma(f)$ added. Then it is easy to check the assertion (i) in Proposition 3.2.2. For example, to prove $S_{-,f} \cap S_{+,f} = \{0, \sigma(f)\}$, it is enough to show that

$$S_f^- \cap \bigcup_{n \in \mathbb{Z}} T_f^n S_f^+ = \phi,$$

and this will be the case if $|\operatorname{Re} \frac{1}{\alpha(f)}|$ is large.

Since τ_f is injective on a neighborhood of $S_f^\pm$, $(\tau_f|_{S_f^\pm})^{-1}$ extends to a neighborhood of $S_{\pm,f}'$. Then define

$$\Phi_{\pm,f} = \Phi_f^\pm \circ (\tau_f|_{S_f^\pm})^{-1}.$$

It is easy to check (ii) and (iv) in Proposition 3.2.2. So we only need to check (iii). Before doing this, we prove the following Key lemma.

Lemma 3.4.2.

$$\Phi_f^+ \circ T_f(w) - \Phi_f^-(w) + \frac{1}{\alpha(f)} \to 0, \quad \text{when } f \to f_0.$$

Proof. The function $\Phi_f^+ \circ T_f$ is defined on $\Omega_f^- = T_f^{-1}(\Omega_f^+)$ and satisfies the same functional equation as Φ_f^-. It follows from the uniqueness in Proposition 2.5.2 (ii) that there exists a constant $d_1 = d_1(f)$ such that

$$\Phi_f^+ \circ T_f = \Phi_f^- + d_1.$$

By 3.3.3 (iii), $F_f(w) = w + 1 + O(1/w^2)$ as $\operatorname{Im}\alpha w \to \infty$, hence

$$\int_{w\pm}^w \frac{d\zeta}{F_f(\zeta) - \zeta} - w$$

converges to a constant, when w tend to ∞ satisfying $\pi/4 < \arg(w - \omega_0) < 3\pi/4$. It follows from Proposition 2.6.2 (ii) that there exist constants $c_\pm(f)$ such that

$$\Phi_f^\pm(w) = w + c_\pm(f) + o(1)$$

when w tend to ∞ within a sector of the form $\{w \mid \theta_1' < \arg(w - \omega_0) < \theta_2'\} \subset \Omega_f^\pm$, where $\pi/4 < \theta_1' < \theta_2' < 3\pi/4$ and $\omega_0 \in \mathbb{C}$.

Thus we can express the constant $d_1(f)$ as

$$d_1(f) = \lim_{w\to\infty} (\Phi_f^+ \circ T_f(w) - \Phi_f^-(w))$$
$$= \lim_{w\to\infty} [(w - \frac{1}{\alpha(f)} + c_+(f)) - (w + c_-(f))] = c_+(f) - c_-(f) - \frac{1}{\alpha(f)}.$$

Now define $\tilde{\mathcal{E}}_f(w) = \Phi_f^- \circ (\Phi_f^+)^{-1}$ on $\Phi_f^+(\Omega_f^- \cap \Omega_f^+)$, which contains at least the vertical strip $\{w \mid |\operatorname{Re} w| \le 1, |\operatorname{Im} w| \ge \eta_0\}$ for a large $\eta_0 > 0$. This η_0 can be chosen uniformly for $f \in \mathcal{N} \cap \mathcal{F}_1$. It satisfies the functional equation $\tilde{\mathcal{E}}_f(w+1) = \tilde{\mathcal{E}}_f(w)+1$ whenever the both sides are defined. By this relation, $\tilde{\mathcal{E}}_f$ can be extended to $\{w \mid |\operatorname{Im} w| > \eta_0\}$. From the asymptotic behavior of $\Phi_f^\pm$, we have

$$\tilde{\mathcal{E}}_f(w) = w + c_-(f) - c_+(f) + o(1) \quad \text{when } \operatorname{Im} w \to \infty.$$

Similarly, $\tilde{\mathcal{E}}_f(w) - w$ tends to a constant as $\operatorname{Im} w \to -\infty$.

Let $\eta_0 < \eta_1 < \eta_2$. Consider the integral of $\tilde{\mathcal{E}}_f(w) - w$ along the rectangular contour with vertices $i\eta_1, 1 + i\eta_1, 1 + i\eta_2, i\eta_2$. By Cauchy's theorem, the total integral is zero, and the integrals on the vertical sides cancel by the periodicity of $\tilde{\mathcal{E}}_f(w) - w$. Hence, we have

$$0 = \int_{i\eta_1}^{1+i\eta_1} (\tilde{\mathcal{E}}_f(w) - w)dw - \int_{i\eta_2}^{1+i\eta_2} (\tilde{\mathcal{E}}_f(w) - w)dw,$$

where the integrals are along segments. Letting $\eta_2 \to \infty$, we obtain

$$c_-(f) - c_+(f) = \int_{i\eta_1}^{1+i\eta_1} (\tilde{\mathcal{E}}_f(w) - w)dw.$$

By Proposition 2.5.2 (iii), we have $\Phi_f^- \to \Phi_0^-$ and $\Phi_f^+ \to \Phi_0^+$ as $f \to f_0$, therefore $\tilde{\mathcal{E}}_f \to \tilde{\mathcal{E}}_{f_0}$ uniformly on $[i\eta_1, 1+i\eta_1]$. Then

$$c_-(f) - c_+(f) = \int_{i\eta_1}^{1+i\eta_1} (\tilde{\mathcal{E}}_f(w) - w)dw \to \int_{i\eta_1}^{1+i\eta_1} (\tilde{\mathcal{E}}_{f_0}(w) - w)dw = 0,$$

because of the normalization for $\tilde{\mathcal{E}}_{f_0}$ (§2.4). So the assertion follows, since $\Phi_f^+ \circ T_f - \Phi_f^- + \frac{1}{\alpha(f)} = c_+(f) - c_-(f)$. $\square$

3.4.3. Normalization. Now we change the normalization for Φ_f^- and replace it by $\Phi_f^- + c_+(f) - c_-(f)$. According to Lemma 3.4.2, this replacement depends continuously on f. So it does not affect other statements of Proposition 3.2.2. After this normalization, we have:

$$\Phi_f^+ \circ T_f(w) = \Phi_f^-(w) - \frac{1}{\alpha(f)} = T_f(w) \circ \Phi_f^-(w) \; ;$$

$$\tilde{\mathcal{E}}_f(w) = w + o(1) \quad \text{as } \operatorname{Im} w \to \infty.$$

Let us prove 3.2.2 (iii). Note that $T_f(S_f^-)$ and S_f^+ are both vertical strips contained in Ω_f^+, and $T_f(S_f^-)$ is on the left side of S_f^+. So it is easy to see that for any point $w \in T_f(S_f^-)$, its forward orbit under F_f passes through S_f^+ before it goes out of Ω_f^+. Let n be the smallest integer such that $F_f^n(w) \in S_f^+$. Then

$$\Phi_f^+(F_f^n(w)) = \Phi_f^+(w) + n = \Phi_f^-(T_f^{-1}(w)) - \frac{1}{\alpha(f)} + n.$$

This would imply (iii), since $\Phi_{-,f}(z) = \Phi_f^-(T_f^{-1}(w))$ and $\Phi_{+,f}(f^n(z)) = \Phi_f^+(F_f^n(w))$ by the definition of $\Phi_{\pm,f}$. $\square$

§3.5. Proof of Proposition 3.2.3.

Lemma 3.5.1. *There exist a neighborhood $\mathcal{N}_0$ ($\subset \mathcal{N}$) of f_0 and a large $\xi_0 > 0$ (independent of f) such that if $f \in \mathcal{N}_0 \cap \mathcal{F}_1$, then*

$$\mathcal{Q}_f = \{ w \in \mathbb{C} : \, |\arg(-w - \xi_0)| < 2\pi/3 \text{ and } |\arg(w + \frac{1}{\alpha(f)} - \xi_0)| < 2\pi/3 \, \},$$

(defined in 3.2.3) is contained in $\Phi_f^+(\Omega_f^+)$.

Proof. Let us prove that Φ_f^+ maps the left boundary of Ω_f^+ to the left of $\mathcal{Q}_f$ and the right boundary of Ω_f^+ to the right of $\mathcal{Q}_f$.

By the proof of 3.3.3, we may assume that $|F_f(w) - (w + 1)| < \varepsilon < 1/4$ on Ω_f^+ for small $\varepsilon > 0$, taking $\mathcal{N}_0$ smaller and ξ_0 larger. Furthermore, Proposition 2.6.2 (i) implies that if $w \in \Omega_f^+$ and $dist(z, \partial\Omega_f^+) \geq R_1$, then

$$|(\Phi_f^+)'(w) - 1/(F_f(w) - w)| \leq C_2/dist(z, \partial\Omega_f^+).$$

Hence

$$|(\Phi_f^+)'(w) - 1| \leq |(\Phi_f^+)'(w) - 1/(F_f(w) - w)| + |1/(F_f(w) - w) - 1|$$
$$\leq C_2/dist(z, \partial\Omega_f^+) + \varepsilon/(1 - \varepsilon).$$

So taking ε small and $\xi_2 > \xi_1 > 0$ large, we can guarantee that

$$|\arg(\Phi_f^+)'(w)| < \pi/12 \quad \text{on } \partial\mathcal{Q}(\xi_2 - 1/\alpha, -\xi_2)$$

Integrating $(\Phi_f^+)'(w)$ along $\partial\mathcal{Q}(\xi_2 - 1/\alpha, -\xi_2)$, we conclude that

$$|\arg(\Phi_f^+(w) - \Phi_f^+(-\xi_2))| < \pi/3$$

on the right boundary of $\mathcal{Q}(\xi_2 - 1/\alpha, -\xi_2)$, and

$$2\pi/3 < \arg(\Phi_f^+(w) - \Phi_f^+(\xi_2 - 1/\alpha)) < 4\pi/3 \, .$$

Shrinking $\mathcal{N}_0$, we may assume that $\Phi_f^+(-\xi_2)$ is uniformly bounded for $f \in \mathcal{N}_0 \cap \mathcal{F}_1$. Therefore if we take ξ_0 (for $\mathcal{Q}_f$) large enough, $\Phi_f^+(-\xi_2)$ is on the right of $\mathcal{Q}_f$ for all $f \in \mathcal{N}_0 \cap \mathcal{F}_1$. Similarly $\Phi_f^-(\xi_2)$ is on the left of $T_f^{-1}(\mathcal{Q}_f)$ for all $f \in \mathcal{N}_0 \cap \mathcal{F}_1$. Applying T_f and using the relation $\Phi_f^+ \circ T_f = T_f \circ \Phi_f^-$ (see the normalization in 3.4.3), we conclude that $\Phi_f^+(\xi_2 - \alpha) = T_f \circ \Phi_f^-(\xi_2)$ is on the left of $\mathcal{Q}_f$.

Together with the result on the angle, this implies that Φ_f^+ maps the left (resp. right) boundary of $\mathcal{Q}(\xi_2 - 1/\alpha, -\xi_2)$ to the left (resp. right) of $\mathcal{Q}_f$. Therefore $\Phi_f^+(\mathcal{Q}(\xi_2 - 1/\alpha, -\xi_2))$ covers $\mathcal{Q}_f$, and so does $\Phi_f^+(\Omega_f^+)$. $\square$

Thus the inverse map $(\Phi_f^+)^{-1}$ is defined on $\mathcal{Q}_f$. So define

$$\varphi_f = \tau_f \circ (\Phi_f^+)^{-1} \quad \text{on} \quad \mathcal{Q}_f.$$

It is easy to check that it satisfies Proposition 3.2.3 (i).

Let $\tilde{\mathcal{R}}_f = \tilde{\mathcal{E}}_f - \frac{1}{\alpha(f)}$. Since $\tilde{\mathcal{E}}_f = \Phi_f^- \circ (\Phi_f^+)^{-1}$ and $\Phi_f^+ \circ T_f = T_f \circ \Phi_f^-$, there are other expressions of $\tilde{\mathcal{R}}_f$:

$$\tilde{\mathcal{R}}_f = T_f \circ \Phi_f^- \circ (\Phi_f^+)^{-1} = \Phi_f^+ \circ T_f \circ (\Phi_f^+)^{-1}.$$

As we have seen in the proof of Lemma 3.4.2, $\tilde{\mathcal{E}}_f$ hence $\tilde{\mathcal{R}}_f$ are defined in $\{w : |\operatorname{Im} w| > \eta_0\}$ for some $\eta_0 > 0$. It is easy to see that

$$\tilde{\mathcal{R}}_f(z+1) = \tilde{\mathcal{R}}_f(z) + 1;$$

$\tilde{\mathcal{R}}_f(z) - z$ tends to $-1/\alpha(f)$ when $\operatorname{Im} z \to \infty$, to a constant when $\operatorname{Im} z \to -\infty$.

To prove (iii), suppose that $w, w' = \tilde{\mathcal{R}}_f(w) + n \in \mathcal{Q}_f$ for some $n \in \mathbf{Z}$. If $n \geq 0$,

$$\begin{aligned}
\varphi_f(w') &= \tau_f \circ (\Phi_f^+)^{-1}(\Phi_f^+ \circ T_f \circ (\Phi_f^+)^{-1}(w) + n) \\
&= f^n \circ \tau_f \circ T_f \circ (\Phi_f^+)^{-1}(w) \\
&= f^n \circ \tau_f \circ (\Phi_f^+)^{-1}(w) = f^n(\varphi_f(w)).
\end{aligned}$$

Similarly we have $f^{-n}(\varphi_f(w')) = \varphi_f(w)$ for $n < 0$.

Finally (iv) follows from the continuity of $\Phi_f^\pm$ etc.

Thus Proposition 3.2.3 is proved. $\square$

§4. GLOBAL BIFURCATION

§4.1. Global dynamics and extended Fatou coordinates.

When the dynamics is defined globally, i.e., in the case of rational maps or entire functions, the bifurcation of parabolic fixed points can have global effects, which are more interesting and important. To study such a global phenomenon, we need to know the connection between the global orbits and the orbits near the fixed point. One way to treat this is to extend the Fatou coordinates and the transition map to a larger region, or if possible to the maximal domain of extension.

Such an extension can be determined from the functional equations which the Fatou coordinates etc. satisfy, for example, $\Phi_- \circ f_0^* = \Phi_- + 1 = T \circ \Phi_-$. The only problem is to control the domain of definition in order to obtain a single-valued function. For example, for the unperturbed map $f_0(z) = z + z^2 + \cdots$, it is easier to extend Φ_- than Φ_+, and in fact we will extend $\varphi_0 = (\Phi_+)^{-1}$ instead of Φ_+.

Another importance of the extension is that the maximal extension often has good properties– such as being a branched covering, belonging to a compact class of maps. See §5, for example. Then we can extract more information on the Fatou coordinates and the transition map.

For a perturbed map $f(z) = z + O(e^{2\pi i\alpha})z + O(z^2)$, it is more difficult to choose the domain to extend. And we cannot decide the maximal extension. In this case, if we try to extend them just by analytic continuation, we will soon encounter multi-valuedness of the function. However it is still possible to choose an appropriate domain so that the continuity of Fatou coordinates etc with respect to the map f is guaranteed. Since the domain of definition of the Fatou coordinate varies depending on the map f, we need to use the "compact open topology together with the domain of definition" (defined 2.5.1), in order to express the continuity.

After extending the domains of definition, we can formulate a global version of Proposition 3.2.3. We will use this form in the application.

§4.2. The extension for the unperturbed map $f_0 \in \mathcal{F}_0$.

4.2.1. Let $f_0 \in \mathcal{F}_0$, i.e. f_0 is defined and analytic in a neighborhood of 0 and satisfies
$$f_0(0) = 0, \quad f_0'(0) = 1 \quad \text{and} \quad f_0''(0) = 1.$$

We denote the domain of definition of f_0 by $\mathcal{D}(f_0)$, which we need to take into the consideration. We will also use other notations defined in §§2 and 3, such as $\Phi_{\pm,0}$, $\tilde{\mathcal{E}}_{f_0}$, $\Omega_{\pm,0}$, etc.

4.2.2. Extension of φ_0.
The map $\varphi_0 = (\Phi_{+,0})^{-1}$ can be extended analytically to a region $\mathcal{D}(\varphi_0)$ which satisfies the following:
(i) $\mathcal{D}(\varphi_0)$ contains $\mathcal{Q}_0 = \{w \in \mathbb{C}| \ \pi/3 < \arg(w + \xi_0) < 5\pi/3 \ \}$ and $\{w| \ |\operatorname{Im} w| > \eta_0\}$ for large $\xi_0, \eta_0 > 0$.
(ii) If $w, w + 1 \in \mathcal{D}(\varphi_0)$, then
$$\varphi_0(w + 1) = f_0 \circ \varphi_0(w).$$

Moreover in the above formula, one side is defined if the other side is defined. More precisely , $w+1 \in \mathcal{D}(\varphi_0)$ if and only if $w \in \mathcal{D}(\varphi_0)$ and $\varphi_0(w) \in \mathcal{D}(f_0)$.
(iii) If f_0 is a rational map or an entire function, then $\mathcal{D}(\varphi_0) = \mathbb{C}$.

Proof. (i) If follows from the definition and the estimate in Proposition 2.2.1 that Q_0 is contained in $\Phi_{0,-}(\Omega_+)$ on which φ_0 was originally defined. Then we take (ii) as the definition of $\mathcal{D}(\varphi_0)$, in other words, $w \in \mathcal{D}(\varphi_0)$ if and only if there is an integer $n \geq 0$ such that $w - n \in Q_0$ and $f_0^j(\varphi_0(w-n)) \in \mathcal{D}(f_0)$ for $j = 0, \ldots, n-1$. In this case, we define $\varphi_0(w) = f_0^n(\varphi_0(w-n))$. This is well-defined, i.e., the definition does not depend on the choice of n because of the relation on φ_0 in Q_0.

It follows also from the estimate in Proposition 2.2.1 that

$$\{w \mid |\operatorname{Im} w| > \eta + 2|\operatorname{Re} w|\} \subset \Phi_{+,0}(\Omega_- \cap \Omega_+) = \Phi_0^+(\Omega^- \cap \Omega^+)$$

for large $\eta > 0$. Hence if $|w| > \eta + 1$ then there is an integer m such that $w + m \in \Phi_{+,0}(\Omega_-)$. Therefore by 2.1.3, $f_0^n(\varphi_0(w+m))$ is defined for all $n \geq 0$. It follows from the definition that $w \in \mathcal{D}(\varphi_0)$. Thus (i) is proved.

(iii) is a consequence of (ii), since we can iterate rational maps and entire functions anyway. $\square$

4.2.3. Extension of $\Phi_{-,0}$. Let $\mathcal{B}$ be the basin for the fixed point 0, i.e., $z \in \mathcal{B}$ if and only if there is a neighborhood of z on which f^n ($n = 1, 2, \ldots$) are defined and converge to 0 uniformly.

Then $\Phi_{-,0}$ can be extended to $\mathcal{B}$, which contains Ω_-. If $z \in \mathcal{B}$, then $f_0(z) \in \mathcal{B}$ and

$$\Phi_{-,0} \circ f_0(z) = \Phi_{-,0}(z) + 1.$$

Proof. If $z \in \mathcal{B}$, then $f_0^n(z) \in \Omega_-$ for some $n \geq 0$, by Lemma 2.1.6. So let $\Phi_{-,0}(z) = \Phi_{-,0}(f_0^n(z)) - n$. Then it is easy to see that this is well-defined, and satisfies the functional equation. $\square$

For simplicity of notation, let us write $\Phi_0 = \Phi_{-,0}$.

4.2.4. Extension of $\tilde{\mathcal{E}}_{f_0}$. Let $\tilde{\mathcal{B}} = \varphi_0^{-1}(\mathcal{B})$, then $T(\tilde{\mathcal{B}}) = \tilde{\mathcal{B}}$ and $\tilde{\mathcal{B}}$ contains $\{w \mid |\operatorname{Im} w| > \eta_0\}$ for some $\eta_0 > 0$.

Now define $\tilde{\mathcal{E}}_{f_0} : \tilde{\mathcal{B}} \to \mathbb{C}$ by

$$\tilde{\mathcal{E}}_{f_0} = \Phi_0 \circ \varphi_0.$$

This definition coincides with the original definition in 2.4. It satisfies

$$\tilde{\mathcal{E}}_{f_0}(w+1) = \tilde{\mathcal{E}}_{f_0}(w) + 1 \quad \text{for } w \in \tilde{\mathcal{B}}.$$

We always assume that $\tilde{\mathcal{E}}_{f_0}$ is normalized, i.e.,

$$\tilde{\mathcal{E}}_{f_0}(w) = w + o(1) \quad \text{as } \operatorname{Im} w \to \infty.$$

The induced map $\mathcal{E}_{f_0} = \pi \circ \tilde{\mathcal{E}}_{f_0} \circ \pi^{-1} : \pi(\tilde{\mathcal{B}}) \to \mathbb{C}^*$ is well-defined. Moreover it extends to 0 and ∞ analytically by $\mathcal{E}_{f_0}(0) = 0$ and $\mathcal{E}_{f_0}(\infty) = \infty$, and $\mathcal{E}'_{f_0}(0) \neq 0$, $\mathcal{E}'_{f_0}(\infty) \neq 0$. So $\mathcal{D}(\mathcal{E}_{f_0}) = \pi(\tilde{\mathcal{B}}) \cup \{0, \infty\}$.

4.2.5 Inverse orbits and φ_0.
There is a natural interpretation of φ_0 in terms of inverse orbits.

Definition. For a mapping f, a sequence of points $\{z_j\}_{j=0}^{\infty}$ is called *an inverse orbit* (of z_0) for f, if $z_j \in \mathcal{D}(f)$ and $f(z_j) = z_{j-1}$ for $j \geq 1$.

Lemma 4.2.6. For $w \in \mathcal{D}(\varphi_0)$, let $z_j = \varphi_0(w - j)$ $(j = 0, 1, \ldots)$. Then $\{z_j\}_{j=0}^{\infty}$ is an inverse orbit for f_0 converging to 0. This gives a one to one correspondence between $\mathcal{D}(\varphi_0)$ and the set of inverse orbits converging to 0, except the orbit $z_j = 0$ $(j = 0, 1, \ldots)$.
Moreover if z_j $(j \geq 1)$ are not critical points, then $\varphi_0'(w) \neq 0$.

Proof. If $w \in \mathcal{D}(\varphi_0)$, then $w - j \in \mathcal{D}(\varphi_0)$ $(j = 0, 1, \ldots)$ and $\varphi_0(w - j) \to 0$ by (4.1.2), so the first statement is obvious. Let $\{z_j\}$ be an inverse orbit converging to 0 and suppose $z_j \neq 0$ for some j. For large j, say for $j \geq j_0$, z_j belongs to $\Omega_- \cup \Omega_+$ (see (4.1.1)). But f_0^n tends to 0 uniformly on Ω_-, so there exists j_0 such that $z_j \in \Omega_+$ for $j \geq j_0$. Let $w = (\varphi_0|_{\mathcal{Q}_0})^{-1}(z_j) + j$ $(j \geq j_0)$. It is easy to see that w does not depend on $j \geq j_0$ and corresponds to the inverse sequence $\{z_j\}$. If w and w' give the same inverse sequence, take $j \geq 0$ such that $w - j, w' - j \in \mathcal{Q}_0$, then we have $w = w'$ by the injectivity of φ_0 on $\mathcal{Q}_0$.
The last statement follows from the facts that $\varphi_0(w) = f_0^n \circ \varphi_0(w - n)$ for $w \in \mathcal{D}(\varphi_0)$ and that $\varphi_0' \neq 0$ in $\mathcal{Q}_0$. $\square$

4.2.7. Grand orbits and Φ_0. There is a similar interpretation for Φ_0:
$\Phi_0(z) = \Phi_0(z')$ if and only if there is an $n \geq 0$ such that $f_0^n(z) = f_0^n(z')$;
$\Phi_0(z) \equiv \Phi_0(z')$ *(mod $\mathbb{Z}$)* if and only if there are $n, m \geq 0$ such that $f_0^n(z) = f_0^m(z')$.
The equivalence class for the latter one is called a *grand orbit*.

§4.3. Extension of φ_f and $\tilde{\mathcal{R}}_f$ for perturbed maps.

4.3.1. Extension of φ_f. Let $f \in \mathcal{N}_0 \cap \mathcal{F}_1$ as in §3. Then $\varphi_f : \mathcal{Q}_f \to \mathbb{C}$ was defined. We extend it to a region $\mathcal{D}(\varphi_f)$ defined by:

$$\mathcal{D}(\varphi_f) = \left\{ w \in \mathbb{C} \;\middle|\; \begin{array}{l} w - n \in \mathcal{Q}_f \text{ for an integer } n \geq 0 \text{ and} \\ f^j(\varphi_f(w - n)) \in \mathcal{D}(f) \text{ for } j = 0, \ldots, n - 1 \end{array} \right\}.$$

The extension of φ_1 is automatically determined by $\varphi_f(w) = f^n(\varphi_f(w-n))$ for $w - n \in \mathcal{D}(\varphi_f)$ with n in the definition. Obviously $\mathcal{Q}_f \subset \mathcal{D}(\varphi_f)$ and if $w, w + 1 \in \mathcal{D}(\varphi_f)$, then $\varphi_f(w) \in \mathcal{D}(f)$ and

$$\varphi_f(w + 1) = f \circ \varphi_f(w).$$

Continuity. *When $f \in \mathcal{N}_0 \cap \mathcal{F}_0$ and $f \to f_0$, we have $\varphi_f \to \varphi_0$, with respect to the compact open topology together with the domain of definition.*

Proof. Let K be any compact set contained in $\mathcal{D}(\varphi_0)$. There exists an $n \geq 0$ such that $T^{-n}(K) = \{w - n | w \in K\} \subset \mathcal{Q}_0$. By definition of $\mathcal{D}(\varphi_0)$, $f_0^j(K) \subset \mathcal{D}(f_0)$ for $j = 0, 1, \ldots, n$. If f is sufficiently close to f_0, then $T^{-n}(K)$ is contained in $\mathcal{Q}_f$ and $f^j(K) \subset \mathcal{D}(f)$ for $j = 0, 1, \ldots, n$. This implies that $K \subset \mathcal{D}(\varphi_f)$. Moreover $f^n(\varphi_f(w - n)) \to f_0^n(\varphi_0(w - n))$, since Proposition 3.2.3 already guarantees the continuity on $\mathcal{Q}_f$: $\varphi_f|_{\mathcal{Q}_f} \to \varphi_0|_{\mathcal{Q}_0}$. $\square$

Remark. We extend φ_f only to the right of $\mathcal{Q}_f$. It is possible to extend to the left to some extent, but there is a danger of the multivaluedness.

4.3.2. Extension of $\Phi_f = \Phi_f^- \circ \tau_f^{-1}$. For f_0, we had $\Phi_0 = \Phi_0^- : \mathcal{B} \to \mathbb{C}$. It is not easy to define its counterpart for f. Let

$$ S_f^* = \{ w \in \mathbb{C} : \xi_0 < \operatorname{Re} w < \operatorname{Re}(\frac{1}{3\alpha(f)}) \} $$

where ξ_0 is given in Proposition 3.2.3. Then define a region

$$ \mathcal{D}(\Phi_f) = \left\{ z \in \mathcal{D}(f) \,\middle|\, \begin{array}{l} \text{there exists an integer } n \text{ such that} \\ f^n(z) \in \tau_f(S_f^*) \text{ and } 0 \leq n \leq \operatorname{Re}(\frac{1}{3\alpha(f)}) \end{array} \right\} $$

Note that τ_f is injective on S_f^*. We define $\Phi_f : \mathcal{D}(\Phi_f) \to \mathbb{C}$ by:

$$ \Phi_f(z) = \Phi_f^- \circ (\tau_f|_{S_f^*})^{-1}(f^n(z)) - n, $$

where n is the integer appeared in the definition of $\mathcal{D}(\Phi_f)$.

Claim. *Φ_f is well defined and satisfies*

$$ \Phi_f(f(z)) = \Phi_f(z) + 1 \quad \text{if } z, f(z) \in \mathcal{D}(\Phi_f). $$

Moreover $\varphi_f \to \varphi_0$, as $f \in \mathcal{N}_0 \cap \mathcal{F}_0$ and $f \to f_0$.

Proof. First note that if $0 \leq n < m \leq \operatorname{Re}(\frac{1}{3\alpha(f)})\}$ and $f^n(z), f^m(z) \in \tau_f(S_f^*)$, then for all j with $n \leq j \leq m$, $f^j(z) \in \tau_f(S_f^*)$. This can be proved by looking at $F_f^{j-n}(\tau_f|_{S_f^*})^{-1}(f^n(z))$.

This implies the well-definedness and the functional equation. The continuity can be proved as in 4.3.1. $\square$

4.3.3. The return map $\tilde{\mathcal{R}}_f$. We redefine the return map $\tilde{\mathcal{R}}_f$ as follows. First define:

$$\mathcal{D}_1 = (\varphi_f)^{-1}(\mathcal{D}(\Phi_f))$$

$$\tilde{\mathcal{R}}_f(w) = \Phi_f(\varphi_f(w)) - \frac{1}{\alpha(f)} \quad \text{on } \mathcal{D}_1.$$

If $w, w+1 \in \mathcal{D}_1$, then

$$\tilde{\mathcal{R}}_f(w+1) = \tilde{\mathcal{R}}_f(w) + 1.$$

Finally define

$$\mathcal{D}(\tilde{\mathcal{R}}_f) = \{w+n | w \in \mathcal{D}_1, |\Phi_f(\varphi_f(w)) - w| < \text{Re}(\frac{1}{2\alpha(f)}) \text{ and } n \in \mathbb{Z}\};$$

and

$$\tilde{\mathcal{R}}_f(w+n) = \tilde{\mathcal{R}}_f(w) + n \quad \text{if } w \in \mathcal{D}_1.$$

Then we have:

Claim. $\tilde{\mathcal{R}}_f$ *is well-defined and satisfies*

$$\tilde{\mathcal{R}}_f(w+1) = \tilde{\mathcal{R}}_f(w) + 1 \quad \text{for } w \in \mathcal{D}(\tilde{\mathcal{R}}_f).$$

It coincides with the definition in §3. Moreover $\tilde{\mathcal{R}}_0 + 1/\alpha(f) \to \tilde{\mathcal{E}}_{f_0}$, as $f \in \mathcal{N}_0 \cap \mathcal{F}_0$ and $f \to f_0$.

Proof. This follows from 4.3.1 and 4.3.2. $\square$

§4.4. A global version of Proposition 3.2.3.

Now we can state a global version of Proposition 3.2.3 with extended domain of definitions.

Proposition 4.4.1. *Let $f_0 \in \mathcal{F}_0$. There exists a neighborhood $\mathcal{N}_0$ of f_0 such that if $f \in \mathcal{N}_0 \cap \mathcal{F}_1$, then there exist large $\xi_0, \eta_0 > 0$, and analytic maps $\varphi_f : \mathcal{D}(\varphi_f) \to \overline{\mathbb{C}}$ and $\tilde{\mathcal{R}}_f : \mathcal{D}(\tilde{\mathcal{R}}_f) \to \mathbb{C}$ satisfying the following:*
(i) $\mathcal{D}(\varphi_f)$ contains the set

$$\mathcal{Q}_f = \{w \in \mathbb{C} : |\arg(-w - \xi_0)| < 2\pi/3 \text{ and } |\arg(w + \frac{1}{\alpha(f)} - \xi_0)| < 2\pi/3 \},$$

and $\varphi_f(\mathcal{D}(\varphi_f)) \subset \mathcal{D}(f)$;
If $w, w+1 \in \mathcal{D}(\varphi_f)$, then

$$\varphi_f(w+1) = f \circ \varphi_f(w);$$

$\varphi_f(w) \to 0$ when $w \in \mathcal{Q}_f$ and $\text{Im}\, w \to +\infty$; $\varphi_f(w) \to \sigma(f)$ when $w \in \mathcal{Q}_f$ and $\text{Im}\, w \to -\infty$.

(ii) $\mathcal{D}(\tilde{\mathcal{R}}_f)$ contains $\{w \in \mathbb{C} : |\text{Im}\, w| > \eta_0\}$ and is invariant under $T(w) = w + 1$; If $w \in \mathcal{D}(\tilde{\mathcal{R}}_f)$, then

$$\tilde{\mathcal{R}}_f(w+1) = \tilde{\mathcal{R}}_f(w) + 1;$$

$\tilde{\mathcal{R}}_f(w) - w$ tends to $-1/\alpha(f)$ when $\text{Im}\, w \to \infty$, to a constant when $\text{Im}\, w \to -\infty$.

(iii) If $w \in \mathcal{D}(\tilde{\mathcal{R}}_f) \cap \mathcal{D}(\varphi_f)$ and $w' = \tilde{\mathcal{R}}_f(w) + n \in \mathcal{D}(\varphi_f)$ for some $n \in \mathbb{Z}$, then

$$f^n(\varphi_f(w)) = \varphi_f(w') \text{ if } n \geq 0, \quad f^{-n}(\varphi_f(w')) = \varphi_f(w) \text{ if } n < 0.$$

Moreover if $|\arg(w' + \frac{1}{2\alpha(f)} - w)| < 2\pi/3$, then $n > 0$.

(iv) when $f \in \mathcal{F}_1$ and $f \to f_0$,

$$\varphi_f \to \varphi_0 \quad \text{and} \quad \tilde{\mathcal{R}}_f + \frac{1}{\alpha(f)} \to \tilde{\mathcal{E}}_{f_0}.$$

Proof. This follows from Proposition 3.2.3, 3.2.4, 4.3.1-4.3.3. Note that the condition $|\Phi_f(\varphi_f(w)) - w| < \text{Re}(\frac{1}{2\alpha(f)})$ was added in the definition of $\mathcal{D}(\tilde{\mathcal{R}}_f)$ in order to guarantee the last statement of (iii). $\square$

§4.5. Branched covering and transition map.

4.5.1. In this section, we study the property of the maximal extension of the transition map for analytic maps which are branched coverings on some domain.

Definition. An analytic mapping $f : U \to V$, where U and V are domains of $\overline{\mathbb{C}}$, is called a branched covering, if for any $z \in V$, there is a neighborhood W of z such that for any connected component D of $f^{-1}(W)$, $f|_D$ is proper and $f : D - f^{-1}(z) \to W \smallsetminus \{z\}$ is a covering in ordinary sense.

Definition. (Immediate parabolic basin) Let ζ be a parabolic fixed point of an analytic function f. A connected open set $B \subset \mathcal{D}(f)$ is called *an immediate parabolic basin* of ζ for f, if B is a connected component of the parabolic basin of ζ for f (see (4.1.2)) such that $f(B) = B$ and $f : B \to B$ is proper (hence a branched covering of finite degree).

For a rational map, a parabolic periodic point always has an immediate parabolic basin and it coincides with the ordinary definition. However a parabolic fixed point for an analytic map in general may not have any immediate parabolic basin.

Lemma 4.5.2. *Suppose that f is defined and analytic in a neighborhood of 0 and $f'(0) = 1$, $f^{(k)}(0) = 0$ $(1 < k \leq q)$, $f^{(q+1)}(0) \neq 0$ $(q \geq 1)$, i.e. $q+1$ is the order of $f(z) - z$ at 0. Then f has at most q immediate parabolic basins of 0, each of which contains at least one critical value of f, except for a parabolic Möbius transformation. Moreover if an immediate parabolic basin contains only one critical point (or critical value), then it is simply connected.*

Proof. This is a well-known argument for rational maps. Let us see it briefly. By a local analysis at 0 (see for example or [Mi]), it can be shown that there exist q disjoint simply connected open sets $V^{(1)}, \ldots, V^{(q)}$ $(\subset \mathcal{D}(f))$ ("attracting petals") such that $0 \in \partial V^{(i)}$; $f(\overline{V}^{(i)}) \subset V^{(i)} \cup \{0\}$; f is injective on $V^{(i)}$; a point $z \in \mathcal{D}(f)$ belongs to the parabolic basin of 0 if and only if $f^n(z) \in \cup_i V^{(i)}$ for some $n \geq 0$; and the orbit spaces $V^{(i)}/\sim$ (where $z \sim f(z)$ if $z, f(z) \in V^{(i)}$), are isomorphic to $\mathbb{C}$.

Let B be an immediate basin of 0, and $B^\sharp$ the parabolic basin. By the above, $B \cap \cup_i V^{(i)} \neq \phi$ and $\cup_i V^{(i)} \subset B^\sharp$. Since B is a component of $B^\sharp$, B contains one of $V^{(i)}$'s. So there are at most q immediate basins.

Define $V_n^{(i)}$ $(n = 0, 1, \ldots)$ inductively, as follows: set $V_0^{(i)} = V^{(i)}$; if $f(V_n^{(i)}) \subset V_n^{(i)}$, $V_{n+1}^{(i)}$ is the component of $f^{-1}(V_n^{(i)})$ containing $V_n^{(i)}$, which exists and $f(V_{n+1}^{(i)}) = V_n^{(i)} \subset V_{n+1}^{(i)}$. It is easy to see that for each i, $\cup_{n \geq 0} V_n^{(i)}$ is a component of $B^\sharp$. Now suppose $B = \cup_{n \geq 0} V_n^{(i)}$ is an immediate parabolic basin, hence by definition, $f : B \to B$ is a branched covering. If B contains no critical value or no critical point, then $f : V_{n+1}^{(i)} \to V_n^{(i)}$, hence $f^n : V_n^{(i)} \to V^{(i)}$ are covering maps, therefore they induces a covering map $B \to V^{(i)}/\sim$. Moreover $V_n^{(i)}$ hence B are simply connected. Therefore B must be isomorphic to $\mathbb{C}$ and f is a Möbius transformation.

Similarly, if B contains only one critical point or critical value, then one can show inductively that $V_n^{(i)}$ are simply connected, hence so is B. $\square$

4.5.3. We suppose now that $f_0 \in \mathcal{F}$ is a function satisfying $f_0'(0) = 1$ and $f_0''(0) = 1$. So we have φ_0 and $\mathcal{E}_{f_0}$ as in §§2 and 4. Let us denote $g_0 = \mathcal{E}_{f_0}$.

Suppose that f_0 has an immediate parabolic basin B of 0. The above argument applies to $V^{(1)} = \Omega_-$ defined in §2.1, hence B is unique and contains Ω_-. Let $B^\sharp$ be the (whole) parabolic basin of 0, and define $\tilde{B} = \varphi_0^{-1}(B)$ and $\tilde{B}^\sharp = \varphi_0^{-1}(B^\sharp)$. By 4.2.4, $\{w| \, |\operatorname{Im} w| > \eta_0\} \subset \tilde{B}^\sharp$. Since for any $w \in \tilde{B}^\sharp$ there is $n \geq 0$ such that $T^n(w) \in \tilde{B}$, we have

$$\{w| \, |\operatorname{Im} w| > \eta_0\} \subset \tilde{B}.$$

Denote by $\tilde{B}^u$ (resp. by $\tilde{B}^\ell$) the component of $\tilde{B}$ containing $\{w| \operatorname{Im} w > \eta\}$ (resp. $\{w| \operatorname{Im} w < -\eta\}$). (The superscript "$u$" stands for upper, and

"ℓ" for lower.) In general, $\tilde{B}^u$ and $\tilde{B}^\ell$ may coincide. Obviously $T\tilde{B}^u = \tilde{B}^u$, $T\tilde{B}^\ell = \tilde{B}^\ell$. Then define $B^u = \pi(\tilde{B}^u)$, $B^\ell = \pi(\tilde{B}^\ell)$.

Definition. Let $\mathcal{F}_0^*$ be the set of functions $f \in \mathcal{F}$ such that $f'(0) = 1$, $f''(0) = 1$ and f has an immediate parabolic basin which contains only one critical point of f. Then this basin is automatically simply connected by Lemma 4.5.2.

Proposition 4.5.4. *Let $f_0 \in \mathcal{F}_0$ and $B, \tilde{B}^u, \tilde{B}^\ell$ as above. Then $g_0 : B^u \cup B^\ell \to \mathbb{C}^*$ is a branched covering of infinite degree, ramified only over one point $v \in \mathbb{C}^*$.*

The sets $\tilde{B}^u$, $\tilde{B}^\ell$, $B^u \cup \{0\}$, $B^\ell \cup \{\infty\}$ are simply connected, and $\dot{B}^u \cap B^\ell = \phi$.

Proof. Let us first show that $\Phi_0 : B \to \mathbb{C}$ is a branched covering. In fact, $\Phi_0|_{\Omega_-}$ is injective, hence $\Phi_0 : f_0^{-n}(\Omega_-) \cap B \to T^{-n}\Phi_0(\Omega_-)$ is a branched covering ($n \geq 0$). So it follows that Φ_0 on B is a branched covering.

Now let us show that $\varphi_0 : \tilde{B}^u \cup \tilde{B}^\ell \to B$ is a branched covering. Let z be a point in B. Take simply connected neighborhoods U, U' of z such that $\overline{U} \subset U'$ and U' contains at most one forward orbit of the critical point. Let z' be a point in $\varphi_0^{-1}(U) \cap (\tilde{B}^u \cup \tilde{B}^\ell)$. Let U_n' be the component of $f_0^{-n}(U')$ containing $\varphi_0(z' - n)$ ($n = 0, 1, \ldots$). Then for some $m \geq 0$, U_n' ($n \geq m$) do not contain the critical point. Hence there exist inverse branches $f_{0,U_m'}^{(-k)} : U_m' \to U_{m+k}'$ of f_0^k. The family $\{f_{0,U_m'}^{(-k)}\}$ is normal, since it avoids at least three values (0 and the orbit of the critical point). Moreover it converges to 0 uniformly on compact sets, since it does so near $\varphi_0(z' - m)$. Hence there exists an $n \geq m$ such that $U_n = f_{0,U_m'}^{(-n+m)}(U_m) \subset \Omega_+ = \varphi_0(\mathcal{Q}_0)$. Let $V = T^n \circ (\varphi_0|_{\mathcal{Q}_0})^{-1}(U_n)$. Then $z' \in V$ and $\varphi_0|_V = (f_0^n|_{U_n}) \circ (\varphi_0|_{\mathcal{Q}_0}) \circ T^{-n}$. So $\varphi_0 : V \to U$ is a branched covering with at most one critical point, since U contains at most one critical orbit. This shows that each component of $\varphi_0^{-1}(U)$ is either unramified or ramified over a common point in U, hence $\varphi_0 : \tilde{B}^u \cup \tilde{B}^\ell \to B$ is a branched covering.

Therefore $\tilde{\mathcal{E}}_{f_0} = \Phi_0 \circ \varphi_0 : \tilde{B}^u \cup \tilde{B}^\ell \to \mathbb{C}$ and $\mathcal{E}_{f_0} : B^u \cup B^\ell \to \mathbb{C}^*$ are branched coverings. Moreover it is easy to see from the above that $\mathcal{E}_{f_0}$ is ramified only over $v = \pi \circ \Phi_0(c)$, where c is the unique critical point.

Now let U be a component of $f_0^{-1}(f_0(\Omega_-))$ contained in B, different from Ω_-. Then we have $f_0^n(U) \cap U = \phi$ ($n \geq 1$). Hence for any component V of $\varphi_0^{-1}(U)$, $\pi|_V$ is injective (by the functional equation for φ_0). On the other hand, $\pi \circ \Phi_0 : U \to \mathbb{C}^*$ is infinite to one, since $\pi \circ \Phi_0|_U = \pi \circ \Phi_0|_{\Omega_-} \circ f_0|_U$ and $\pi \circ \Phi_0|_{\Omega_-}$ is infinite to one. Hence g_0 is of infinite degree.

Let us show the simple connectivity of $\tilde{B}^u$ (or $\tilde{B}^\ell$). If γ is a closed curve in $\tilde{B}^u$, $\gamma' = T^{-n}\gamma \subset \mathcal{Q}_0$ for some $n \geq 0$. Let W be a region bounded by $\varphi_0(\gamma')$, not containing 0. Since both the immediate basin B and $\Omega_+ = \varphi_0(\mathcal{Q}_0)$ are simply connected and do not contain 0, we have $W \subset B \cap \Omega_+$. It follows that γ' is trivial in $\tilde{B}^u$ and so is γ. Therefore $\tilde{B}^u$ is simply connected.

Then $B^u \cup \{0\}$ (or $B^\ell \cup \{0\}$) is also simply connected, since the fundamental group of B^u is generated by a curve around 0, which is trivial in $B^u \cup \{0\}$.

Finally, let us show that $\tilde{B}^u$ and $\tilde{B}^\ell$ are different components of $\tilde{B}$. Suppose $\tilde{B}^u = \tilde{B}^\ell$. Then $\tilde{B}^u = \tilde{B}^\ell = \mathbb{C}$, since $\tilde{B}^u$ and $\tilde{B}^\ell$ are invariant under T, simply connected, and contain half planes. Therefore B contains Ω_- and $\Omega_+ = \varphi_0(\mathcal{Q}_0)$, and the union is a punctured neighborhood of 0. But B is simply connected punctured neighborhood of 0, so $B = \overline{\mathbb{C}} \smallsetminus \{0\}$. Hence f_0 is a parabolic Möbius transformation, since it is analytic on $\overline{\mathbb{C}}$ and has no periodic point in B. Then f_0 has no critical point and this contradicts the assumption. Thus we have $\tilde{B}^u \cap \tilde{B}^\ell = \phi$, hence $B^u \cap B^\ell = \phi$. $\square$

By the normalization in 2.4, we have $g_0'(0) = 1$. So 0 is again a parabolic fixed point of g_0.

Lemma 4.5.5. *Let $f_0 \in \mathcal{F}_0$ and $g_0 = \mathcal{E}_{f_0}$. Then $g_0''(0) \neq 0$ and g_0 has a simply connected immediate parabolic basin which contains only one critical point. In other words, g_0 belongs to $\mathcal{F}_0$ after a linear scaling of the coordinate. Moreover $g_0^n(v)$ $(n = 0, 1, \ldots)$ are defined and $g_0^n(v) \to 0$ $(n \to \infty)$, where v is the unique critical value of g_0.*

Proof. Obviously $g_0 \not\equiv 0$. So there exist attracting petals $V^{(i)}$ $(i = 1, \ldots, q)$ for g_0 as in the proof of Lemma 4.5.2, where $q + 1 = ord(g_0(z) - z)$. Since $g_0 : B^u \cup B^\ell \to \mathbb{C}^*$ is a branched covering, we can also construct $V_n^{(i)}$ and $g_0 : V_{n+1}^{(i)} \to V_n^{(i)}$ is a branched covering with at most one critical point. Then $V_n^{(i)}$ are simply connected and $\deg g_0|_{V_n^{(i)}}$ $(n = 0, 1, \ldots)$ are eventually constant. It follows that $B_i = \cup_i V_n^{(i)}$ are simply connected and $g_0 : B_i \to B_i$ is a branched covering with at most one critical point. Hence B_i are immediate parabolic basins. By Lemma 4.5.2, we have $q = 1$, i.e., $g_0''(0) \neq 0$. The rest follows easily. $\square$

APPENDIX : HOLOMORPHIC DEPENDENCE
OF THE FATOU COORDINATES (BY TAN LEI)

We show here that the Fatou coordinates depend analytically on the parameter, with a proof given by A. Douady.

Proposition A.1. *In the setting of Proposition 3.2.2 (resp. 2.5.2), assume in addition that $f_s(z)$ (resp. $F_s(z)$) is a family of holomorphic maps for $s \in \Delta \cup \{0\} \subset \mathbb{C}^k$, with $\Delta \subset \mathbb{C}^k$ an open connected set containing 0 in the boundary, such that f_s (resp. $F_s(z)$) is continuous for $s \in \Delta \cup \{0\}$*

and analytic in $s \in \Delta$, then the Fatou coordinates $\Phi_{\pm, f_s} := \Phi_{\pm, s}$ (resp. $\Phi_s := \Phi_{F_s}$) depend analytically on s for $s \in \Delta$.

Proof. We will at first deal with the situation in Proposition 2.5.2. In its proof we replace F by F_s and $h_1(w)$ by $h_{1,s}(w)$. The definition of $h_{1,s}$ indicates that it is continuous in (s, w), quasi-conformal in w and holomorphic in s. Therefore the complex structure $\sigma_s = h_{1,s}^* \sigma_0$ is holomorphic in s. By Ahlfors-Bers theorem, $h_{2,s}(w)$, the solution of σ_s (normalized as in the proof of Proposition 2.5.2), is again continuous in (s, w), quasi-conformal in w and holomorphic in s. Now define $\Phi(s, z) = \Phi_s(z) = h_{2,s} \circ h_{1,s}^{-1}(z)$, as in the proof of Proposition 2.5.2. It is continuous in (s, z) and holomorphic in z and satisfies the other properties of Proposition 2.5.2. It remains to show that $\Phi(s, z)$ is holomorphic in (s, z).

Fix $\widehat{p} = (\widehat{s}, \widehat{z})$ a base point. Relative to it we define $\widehat{w} = h_{1,\widehat{s}}^{-1}(\widehat{z})$ and three holomorphic functions $P(z) = \Phi(\widehat{s}, z)$, $Q(s) = h_{1,s}(\widehat{w})$ and $H(s) = h_{2,s}(\widehat{w})$. Note that $(\widehat{s}, z)$ for z close to $\widehat{z}$ and $(s, Q(s))$ for s close to $\widehat{s}$ are two local complex varieties in the space of (s, z), both passing through $\widehat{p}$, and being transversal there. Furthermore $P(z)$ is the value of Φ along the leaf $(\widehat{s}, z)$, and, as $\Phi(s, Q(s)) = H(s)$, the function $H(s)$ is the value of Φ along the leaf $(s, Q(s))$.

(The following argument applies to maps in a more general setting. It says roughly that if a continuous map from (an open set of) $\mathbb{C}^{k+1}$ to $\mathbb{C}$ is complex differentiable along two transversal continuous foliations of complex varieties, then the map is in fact holomorphic. It mimics strongly the proof for the case where the map is differentiable in each variable.)

Assume at first s varies in a open set of $\mathbb{C}$ (i.e. $k = 1$).

Formal differential. Define at first a formal differential $\delta\Phi_{\widehat{p}}$ as follows: it is the $\mathbb{C}$-linear map from $\mathbb{C}^2$ to $\mathbb{C}$ (having the form $(s, z) \mapsto as + bz$) which maps the vector $(0, 1)$ to $P'(\widehat{z})$, and the vector $(1, Q'(\widehat{s}))$ to $H'(\widehat{s})$. Indeed, this defines the linear map (i.e. the values of a and b) uniquely, as the vectors $(0, 1)$ and $(1, Q'(\widehat{s}))$ are transversal.

Continuity. Now we let the base point $\widehat{p}$ vary. By continuity of $h_{1,s}(w)$ and $h_{2,s}(w)$ we conclude that $\widehat{w}$ is continuous in $\widehat{p}$, and the three functions $P(z)$, $Q(s)$ and $H(s)$ all depend continuously in $\widehat{p}$.

By Cauchy formula for holomorphic functions, the derivatives of these three holomorphic functions also depend continuously on $\widehat{p}$. Therefore $P'(\widehat{z})$, $Q'(\widehat{s})$ and $H'(\widehat{s})$ are all continuous in $\widehat{p}$. It follows that the formal differential $\delta\Phi_{\widehat{p}}$ (i.e. the values a and b) depends also continuously on $\widehat{p}$.

Integration along the leaves. Let p and p_0 be in $\mathbb{C}^2$ and ε apart (with ε small). We may connect p to p_0 by an arc $\gamma(t)$, $t \in [0, 1]$ which goes parallel to the z-axis for $t \in [0, 1/2]$, and then follows the curve $(s, h_{1,s}(w_1))$ for some w_1. We use dot-derivative to denote derivative relative to t. We claim that

$$\Phi(p) - \Phi(p_0) = \int_0^1 \delta\Phi_{\gamma(t)}\, \dot\gamma(t)\, dt\ .$$

Proof. Transversality implies that $\int_0^1 \|\dot\gamma(t)\|\, dt \le C\epsilon$ where $C < \infty$. Let $p_1 = \gamma(1/2)$.

Assume $t \in [0, 1/2]$. We have $\gamma(t) = (s_0, z(t))$ and $\Phi(\gamma(t)) = P_{p_0}(z(t))$, with dot-derivative equal to $P'_{p_0}(z(t)) \cdot \dot z(t) = \delta\Phi_{\gamma(t)}\dot\gamma(t)$. Here we use the trivial fact that $P_{p_0}(z) = P_{\gamma(t)}(z)$. So $\Phi(p_1) - \Phi(p_0) = \int_0^{1/2} \delta\Phi_{\gamma(t)}\dot\gamma(t)\, dt$.

Now let t vary in $[1/2, 1]$. As $\gamma(t)$ follows the curve $(s, h_{1,s}(w_1))$ with w_1 unchanged, we have the same functions $Q(s)$ and $H(s)$ along the curve. So $\gamma(t) = (s(t), Q(s(t)))$, and $\Phi(\gamma(t)) = H(s(t))$ with dot-derivative equal to $H'(s(t)) \cdot \dot s(t) = \delta\Phi_{\gamma(t)}\dot\gamma(t)$, here we use the fact that $\dot\gamma(t) = (1, Q'(s(t))) \cdot \dot s(t)$, and $\delta\Phi_{\gamma(t)}$ maps $(1, Q'(s(t))$ to $H'(s(t))$. So $\Phi(p) - \Phi(p_1) = \int_{1/2}^1 \delta\Phi_{\gamma(t)}\dot\gamma(t)\, dt$.

The formal differential is equal to the differential. For p, p_0 as above,

$$|\Phi(p) - \Phi(p_0) - \delta\Phi_{p_0}(p - p_0)| = \left| \int_0^1 (\delta\Phi_{\gamma(t)} - \delta\Phi_{p_0})\, \dot\gamma(t)\, dt \right|$$

$$\le \int_0^1 \|\delta\Phi_{\gamma(t)} - \delta\Phi_{p_0}\| \cdot \|\dot\gamma(t)\|\, dt \le o(\epsilon)\ .$$

Letting $\epsilon \to 0$ we see that Φ is complex differentiable at p_0 with differential $d\Phi_{p_0} = \delta\Phi_{p_0}$, and with $d\Phi_{p_0}$ depending continuously on p_0.

Assume now that $s = (s_1, \cdots, s_k)$ varies in an open set of $\mathbb{C}^k$. The same argument as above shows that Φ is holomorphic in each variable s_j. As Φ is continuous in (s, z), it is in fact holomorphic.

Finally we come to the setting of Proposition 3.2.2. For f_s a family of maps the approximate Fatou coordinates for f_s will provide a holomorphic family $F_s(z)$ satisfying the conditions of Propositions 2.5.2 and A.1. Therefore the Fatou coordinates $\Phi_{\pm, s}(z)$ are holomorphic for $s \in \Delta$.

REFERENCES

[A] Ahlfors, L., Lectures on quasiconformal mappings, Van Nostrand, 1966.

[Ca] Camacho, C., On the local structure of conformal mappings and holomorphic vector fields, Astérisque 59-60 (1978) 83-94

[Do] Douady, A., Does a Julia set depend continuously on the polynomial? Proceedings of Symposia in Applied Mathematics, Vol. 49, 1994.

[DH] Douady, A. and Hubbard, J. H., Étude dynamique des polynômes complexes, Publ. Math. d'Orsay, 1er partie, 84-02; 2me partie, 85-04.

[Fa] Fatou P., Sur les équations fonctionnelles, Bull. Soc. Math. France 47 (1919) 161-271 and 48 (1920) 33-94, 208-314.

[La] Lavaurs, P., Systèmes dynamiques holomorphes: explosion de points périodiques paraboliques, Thèse de doctorat de l'Université de Paris-Sud, Orsay, France, 1989.

[Le] Leau, L., Etude sur les equations fonctionnelles à une ou plusieures variables, Ann. Fac. Sci. Toulouse 11 (1897).

[Ma] Mandelbrot, B., On the dynamics of Iterated maps V: Conjecture that the boundary of the M-set has a fractal dimension equal to 2, p. 235-238, Chaos, Fractals and Dynamics, Eds. Fischer and Smith, Marcel Dekker, 1985.

[Mi] Milnor, J., Dynamics in one complex variables: Introductory lectures, Preprint SUNY Stony Brook, Institute for Mathematical Sciences, 1990.

[P] Pommerenke, Ch., Univalent functions, Vandenhoeck & Ruprecht, Göttingen, 1975.

[Sh] Shishikura, M., The Hausdorff dimension of the boundary of the Mandelbrot set and Julia sets, *Annals of Math.* **147**(1998), 225–267.

Hiroshima University, Department of Mathematics, Kagamiyama, Higashi-Hiroshima 739-8526, JAPAN
email : mitsu@math.sci.hiroshima-u.ac.jp

References

[Intro] Tan L., Introduction.

[Hubbard] J. Hubbard, Preface.

[McMullen] C. McMullen, *The Mandelbrot set is universal.*

[Douady-Buff-Devaney-Sentenac] A. Douady, with the participation of X. Buff, R. Devaney and P. Sentenac, *Baby Mandelbrot sets are born in the cauliflowers.*

[Haïssinsky] P. Haïssinsky, *Modulation dans l'ensemble de Mandelbrot.*

[Milnor] J. Milnor, *Local connectivity of Julia sets: Expository lectures.*

[Roesch] P. Roesch, *Holomorphic motions and puzzles (following M. Shishikura)*

[Tan] Tan L., *Local properties of the Mandelbrot set at parabolic points.*

[Petersen-Ryd] C. Petersen and G. Ryd, *Convergence of rational rays in parameter spaces.*

[Luzzatto] S. Luzzatto, *Bounded recurrence of critical points and Jakobson's Theorem.*

[Petersen] C. Petersen, *The Herman-Światek theorems with applications.*

[Jellouli1] H. Jellouli, *Perturbations d'une fonction linéarisable.*

[Jellouli2] H. Jellouli, *Indice holomorphe et multiplicateur.*

[Shishikura-Tan] M. Shishikura and Tan L., *An alternative proof of Mañé's theorem on non-expanding Julia sets.*

[Yin] Yin Y.-C., *Geometry and dimension of Julia sets.*

[Shishikura1] M. Shishikura, *On a theorem of Mary Rees for the matings of polynomials.*

[Douady] A. Douady, d'après des notes de X. Buff, *Le théorème d'intégrabilité des structures presque complexes.*

[Shishikura2] M. Shishikura, *Bifurcation of parabolic fixed points.*

365

Printed in the United States
By Bookmasters